Introduction to Proof Through Number Theory

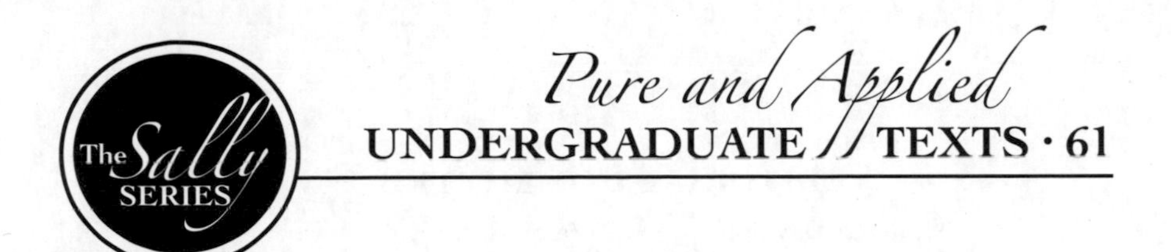

Introduction to Proof Through Number Theory

Bennett Chow

Providence, Rhode Island USA

2020 *Mathematics Subject Classification.* Primary 00-XX, 03-XX, 05-XX, 11-XX, 97-XX; Secondary 68-XX.

For additional information and updates on this book, visit
www.ams.org/bookpages/amstext-61

Library of Congress Cataloging-in-Publication Data

Names: Chow, Bennett, author.
Title: Introduction to proof through number theory / Bennett Chow.
Description: Providence, Rhode Island : American Mathematical Society, [2023] | Series: Pure and applied undergraduate texts, 1943-9334 ; volume 61 | Includes bibliographical references and index.
Identifiers: LCCN 2022041609 | ISBN 9781470470272 (paperback) | ISBN 9781470472580 (ebook)
Subjects: LCSH: Proof theory. | Number theory. | AMS: General. | Mathematical logic and foundations. | Combinatorics. | Number theory. | Mathematics education. | Computer science.
Classification: LCC QA9.54 .C486 2022 | DDC 511.3/6–dc23/eng20221121
LC record available at https://lccn.loc.gov/2022041609

∞ The paper used in this book is acid-free and falls within the guidelines established to ensure permanence and durability.
Visit the AMS home page at https://www.ams.org/

10 9 8 7 6 5 4 3 2 1   28 27 26 25 24 23

Contents

Preface

Figure 0.0.1. The impossible cube. In mathematics and in life, although it sounds contradictory, the impossible is possible. Credit: Wikimedia Commons, authors: Original—Maksim, vector—Boivie. Licensed under Creative Commons Attribution Share Alike 3.0 Unported (https://creativecommons.org/licenses/by-sa/3.0/deed.en) license.

Philosophy about learning and teaching

Lighten up about mathematics!

Have fun. What activities do people get good at without the process of improving being unpleasant? At the top of the list of such activities are perhaps *games.*[1] Why is that? Because they are fun! This leads us to the conclusion (inductive reasoning) that one of the main difficulties in exposition and teaching is to make the subject matter and presentation fun for the reader and student.

Walk the path. How do we teach infants how to speak? Well, we don't say: "This is the first rule of grammar: ...." Likewise, it doesn't make sense to start teaching mathematical proofs by saying: "This the first rule of proving theorems: ...."

[1] Disclaimer: Most of the author's childhood was spent playing sports, pinball, video games, and chess.

There's a difference between knowing the path and walking the path.
– Morpheus

If we have to choose, we would rather have you walk the path of proving theorems than know the path of proving theorems. But, fortunately, you do not have to choose and you can do both.

Make it natural and balanced. The best way to learn something is to make the process natural. To balance conscious thinking with subconscious thinking. Not to overthink things. To focus, but not to rush. To wander, but to come back. On one hand, ideas arise naturally, whose origins wherewith we do not know. On the other hand, ideas come from ideas. After all, imitation is the sincerest form of flattery!

Don't be formulaic. Cookbooks are comfortable to read. And by following them, we can make delicious meals! However, if we make this a cookbook on how to prove theorems, then aren't we doing the same thing as making a cookbook on how to integrate functions, except perhaps on a higher level?

Practical advice. These are our ideas, but we suggest that you take them with a grain of salt and follow your own path!

(1) You don't have to understand everything at once. Being confused and struggling are natural parts of learning.
(2) We learn by watching and we learn by doing. With respect to watching, this book may give more detailed proofs of elementary results than other books on mathematical reasoning. We compensate this by including lots of solved problems as well as exercises with hints given at the end of each chapter.
(3) Success is mostly based on passion, patience, and perseverance. In this sense, we support the Mamba Mentality or John Wick level of dedication, tempered with a relaxed attitude. If it is not for this book, then whatever you are interested in!
(4) No matter what our level is, we have the ability to greatly improve in learning a subject. Useful are:
    (a) Interest and curiosity in the subject. Ask questions while you are learning and reading!
    (b) Repetition. Imagine you are getting good at a video game. What do you do? Keep playing!
    (c) Watch and imitate (see the corresponding Yogi-isms). In the video game analogy, watch good players play and copy their techniques.

Content of this book

Je suis désolé. ... I've heard it all before.
– From "Sorry" by Madonna

A word to the wise: If you read this book, you will have to endure bad math puns and jokes and out-of-date pop culture references (we old-timers prefer to think of these as the classics).

Now that we got that out of the way, foremost, the material in this book is not original. Anyone potentially teaching from this book will have heard it all before. What distinguishes this book is more about what material we've selected, how we've decided to pedagogically present the material, and the informal style of writing while trying not to compromise the mathematical content of the book.

In this book we learn how to read and write mathematical proofs. One of the most beautiful subjects in mathematics is number theory. The subject of number theory ranges from the most elementary mathematics to the deepest mathematics. Number theory also provides an arena in which students already have a strong intuition for the objects of study, so this choice makes it possible to build on that existing knowledge. These properties make it an ideal subject for us to study. One reason why mathematics can be deep and advanced is that it can be abstract and complicated. We will see how abstraction can be introduced to simplify proofs and how, in order to understand concrete problems, we are forced to understand abstract notions. Complicated proofs can be broken down into smaller and far less complicated proofs. Deconstructing complicated proofs can be helpful in their understanding.

Some of the specific topics we cover are logic and implications, set theory, the arithmetic of integers, prime numbers, and algebraic structures.

Some of the themes in understanding mathematics we emphasize are conceptualization and visualization.

This book is suitable to be used as a quarter or semester college course on mathematical reasoning provided that one skips sections (and chapters!). The teacher should decide what sections to skip according to personal taste. We have put an asterisk * after certain sections which we feel may be of less priority.

Style of this book

To keep the discussion in the book lively, we do not always proceed in a linear manner. Indeed, you are likely familiar with many of the elementary rules and assumptions we use in mathematical reasoning. If we start from the beginning, you will likely be bored, at least temporarily. On the other hand, when we discuss concepts and methods that have not been introduced before, there is the risk you may not (at least initially) know what we are talking about. To remedy this, we will, when necessary or helpful, give forward references (i.e., references to material later in the book) or Wikipedia references (on the World Wide Web), which are also hyperlinked (such as the aforementioned World Wide Web and the word "hyperlinked"!) to missing items. This will facilitate looking up information in the e-version of this book.

We also include some awful jokes, bad puns, and esoteric popular culture references. Their intent is to make the book more lively, interesting, and broad. Mainly, we would like to encourage you to think about mathematics in your own way, especially writing your own proofs!

Problem solving

One of the best ways to learn mathematics (or just about anything for that matter!) is to do problems. Problems test our understanding. Problems force us to learn the material better. Problems challenge us. Problems are fun (or not, depending on your viewpoint!).

Polya [**Pol14**] has the following suggestions for "How to solve it":

(1) "You have to understand the problem."
(2) "Find the connection between the data and the unknown."
(3) "Carry out your plan."
(4) "Examine the solution obtained."

LaTeX

This book is written in LaTeX, which is a mathematics word processing software. You may consider learning LaTeX to be able to type up solutions to problems! Here is a sample of how it works. Consider the following statements:

I love the equation $e^{\pi i} + 1 = 0$. Beyond amazing is

$$\int_a^b f'(x)dx = f(b) - f(a). \tag{0.1}$$

Here is the LaTeX code we used to write this:

```
I love the equation $e^{\pi i} + 1 = 0$.  Beyond amazing is
\begin{equation}
    \int_a^b f'(x) dx = f(b) - f(a) .
\end{equation}
```

For the reader who is interested in learning LaTeX, an internet search will yield many sources. One source is https://www.overleaf.com/learn.

Origins

This book started as notes for a mathematical proofs class we taught for many years at the University of California San Diego using the book by Eccles [**Ecc97**]. The first few years, we used the book by Fletcher and Patty [**FP96**]. As such, this book is largely influenced by Eccles's book.

Further reading

There are many directions the student may pursue after, or even before(!), reading this book. Below are a handful of classics, out of the many wonderful mathematics books in the literature:

- Munkres, James R., *Topology. Second edition.* Prentice Hall, Inc., Upper Saddle River, NJ, 2000. xvi + 537 pp.
- Massey, William S., *Algebraic topology: An introduction.* Harcourt, Brace & World, Inc., New York, 1967. xix + 261 pp.
- Rudin, Walter, *Principles of mathematical analysis. Third edition.* International Series in Pure and Applied Mathematics. McGraw-Hill Book Co., New York-Auckland-Düsseldorf, 1976. x + 342 pp.

- Spivak, Michael, *Calculus on manifolds. A modern approach to classical theorems of advanced calculus.* W. A. Benjamin, Inc., New York-Amsterdam, 1965. xii + 144 pp.
- Ahlfors, Lars V., *Complex analysis. An introduction to the theory of analytic functions of one complex variable. Third edition.* International Series in Pure and Applied Mathematics. McGraw-Hill Book Co., New York, 1978. xi + 331 pp.
- Serre, Jean-Pierre, *A course in arithmetic.* Translated from the French. Graduate Texts in Mathematics, No. 7. Springer-Verlag, New York-Heidelberg, 1973. viii + 115 pp.

Acknowledgments

We would like to thank the countless undergraduate students, graduate teaching assistants, colleagues, teachers, and friends that we have greatly benefited from over our career. Special thanks to AMS publisher Sergei Gelfand, AMS book acquisitions editors Eriko Hironaka and Ina Mette. We are grateful to Ina for all of her splendid work and help as our book editor. We are deeply indebted to MAA Press acquisitions editor Stephen Kennedy for reading through the whole book and for his countless suggestions and fantastic help in greatly improving the book. We would like to thank the editors of the AMS Pure and Applied Undergraduate Texts series and the anonymous reviewers for all of their help and suggestions, which have greatly improved the book. We would like to thank Marcia Almeida for her assistance. Special thanks to Arlene O'Sean for her magnificent copy editing.

Special thanks to Ed Dunne, John Eggers, Brett Kotschwar, Mat Langford, Peng Lu, Zilu Ma, Fadi Twainy, Deane Yang, and Yu Yuan for their helpful suggestions and encouragement.

Ben would like to thank his wife, Jingwei, his daughters, Michelle, Isabelle, and Gloriana, his brother, Peter, and his parents for their encouragement. He is especially grateful to his wife for her support and patience through the writing process. Ben dedicates this book to his parents, Yutze Chow and Wanlin Wu Chow.

Bennett Chow
University of California San Diego

Notations and Symbols

$\square$	end of proof symbol
$=$	equals
$:=$	defined to be equal to
$\stackrel{☹}{=}$	a wishful thinking equality; usually wrong
☺	smiley face, dad joke
$\equiv$	congruent to
$\sim$	related to (for a relation)
$<$	less than
$\leq$	less than or equal to; if H and G are groups, $H \leq G$ denotes H is a subgroup of G
$>$	greater than
$\geq$	greater than or equal to
$\in$	an element of
$\notin$	not an element of
$\subset$	a subset of
$\supset$	subset, in the reverse direction
$\cap$	intersection
$\cup$	union
$\bigcap_{i \in I} X_i$	intersection of a family of sets
$\bigcup_{i \in I} X_i$	union of a family of sets
$+$	addition
$\cdot$	multiplication
$\cdot, +, \times, \odot, \oplus$	group multiplication
$\sum$	sum

$\prod$	product
$\wedge$	the logical connective "and"
$\vee$	the logical connective "or"
$\neg$	the logical connective "not"
$\Rightarrow$	the logical connective "implies"
$\Leftrightarrow$	the logical connective "if and only if" (biconditional)
$\Leftarrow$	implies, in the reverse direction
$\circ$	composition of functions
$\exists$	exists
$\forall$	for all
$\nexists$	does not exist
$\emptyset$	the empty set
$\binom{n}{k}$	binomial coefficient; read as "n choose k"
!	factorial if after a number; if after a word, wow! ☺
$\sqrt{x}$	the (non-negative) square root of a non-negative real number x
χ_A	the characteristic function of a subset A
$\aleph_0$	the cardinality of the set of integers
$\aleph_1$	the cardinality of the set of real numbers
$\varphi(m)$	Euler's totient function
$a\|b$	a divides b
$[a, b]$	the closed interval from a to b
(a, b)	the open interval from a to b
(a, ∞)	the set of real numbers $x > a$
$(-\infty, b)$	the set of real numbers $x < b$
$\{a_n\}_{n=1}^{\infty}$	a sequence
$[a]_m$	the congruence class of a modulo m
a^{-1}	if a is a group element, the inverse of a
$\langle a \rangle$	if a is a group element, the subgroup generated by a (see (8.110))
$A - B$	the set difference A minus B
arg	argument of a complex number
Arg	principal value of the argument
$B_r(x)$	the open ball of radius r centered at x
$\bar{B}_r(x)$	the closed ball of radius r centered at x
$\partial B_r(x)$	the boundary of the ball, a.k.a. sphere of radius r centered at x
^{c}	the complement of a subset
$\mathbb{C}$	the set of complex numbers
$\mathrm{ComDiv}(a)$	set of common positive divisors of a and b
$\mathrm{Div}(a)$	set of positive divisors of a
e	if the base of an exponential, Euler's number; approximately 2.71828; e is also used to denote the identity element of a group

E	the set of even integers
$f : X \to Y$	a function from X to Y
f_n	(usually) the n-th Fibonacci number
$\mathrm{Fun}(X, Y)$	the set of functions from X to Y, a.k.a. Y^X
$f\vert_A$	the restriction of a function to a subset A of the domain
$f^{-1}(B)$	the pre-image of a subset B of the codomain
$f^{-1}(y)$	$f^{-1}(\{y\})$
G_f	the graph of a function f
gcd	greatest common divisor
$\mathbb{H}$	the set of quaternions
i	if referring to a complex number, the imaginary unit
I_X	the identity function of a set X
$\mathrm{Im}(f)$	the image of a function f
inf	infimum
$\mathrm{Inj}(X, Y)$	set of injections from a set X to a set Y
lcm	least common multiple
lim	limit
ln	natural logarithm
max	the maximum
min	the minimum
mod	modulo
M_p	the p-th Mersenne prime
$m\mathbb{Z}$	the set of integers that are multiples of m
$\mathbb{N}_n$	the set $\{1, 2, 3, \ldots, n\}$
O	the set of odd integers
p	typically a prime number
$\mathcal{P}(X)$	the power set of X
$\mathcal{P}_k(X)$	the set of k-element subsets of X
$\mathbb{Q}$	the set of rational numbers
$\mathbb{Q}^+$	the set of positive rational numbers
$\left(\frac{q}{m}\right)$	Legendre symbol
R_m	$\{0, 1, 2, \ldots, m-1\}$, the set of remainders modulo m
R_m^*	set of elements of R_m coprime to m
$\mathbb{R}$	the set of real numbers
$\mathbb{R}^+$	the set of positive real numbers
$\mathbb{R}^{\geq}$	the set of non-negative real numbers
$\mathbb{R}^n$	n-dimensional (real) Euclidean space
$\mathbf{r}$	the remainder function $\mathbf{r} : \mathbb{Z} \to R_m$, also denoted by $\mathbf{r}_m$
S^n	the unit n-dimensional sphere
$S^n(r)$	the n-dimensional sphere of radius r

sup	supremum
$X \times Y$	the cartesian product of X and Y
$[x]$	the equivalence class of $x \in X$, where X has an equivalence relation $\sim$
X^n	the n-fold cartesian product of X
$\mathbf{x} \cdot \mathbf{y}$	the Euclidean dot product
$\lvert\mathbf{x}\rvert$	the Euclidean norm
$\lvert X\rvert$	if X is a finite set, its cardinality
(x, y)	an ordered pair
$(x_1, x_2, \ldots, x_n)$	an (ordered) n-tuple
X_I	$\bigcap_{i \in I} X_i$ if I is a set of indices of a family $\{x_i\}$
$x \mapsto \lfloor x \rfloor$	the floor function
$x \mapsto \lceil x \rceil$	the ceiling function
$\mathbb{Z}$	the set of integers
$\mathbb{Z}^+$	the set of positive integers
$\mathbb{Z}^{\geq}$	the set of non-negative integers
$\mathbb{Z}_m$	$\{[0]_m, [1]_m, [2]_m, \ldots, [m-1]_m\}$, the set of congruence classes modulo m
$\mathbb{Z}_m^*$	set of elements of $\mathbb{Z}_m$ with inverses (see (8.28))
$\overline{z}$	the complex conjugate of z (see (8.171))

Down the rabbit-hole.[1]

[1] From "Alice's Adventures in Wonderland" by Lewis Carroll. Wikimedia Commons, Public Domain.

Chapter 1

Evens, Odds, and Primes: A Taste of Number Theory

Figure 1.0.1. A double rainbow at Yellowstone National Park. Photo by Jingwei Xia.

Trinity: It's so beautiful. – From "The Matrix Resurrections"

Goals of this chapter: To demonstrate that we can, without much background, understand some elementary aspects of a deep subject: prime numbers. To see some of the pitfalls of inductive reasoning. To give examples of deductive reasoning leading to the proofs of results. After reading this chapter, the reader should be able to write some simple proofs such as in the exercises.

Figure 1.0.2. Euclid (fl. 300 BC). Detail from Raphael's *The School of Athens*. Wikimedia Commons, Public Domain.

In this introductory chapter, we take a closer look at something we are all familiar with: *even* and *odd* integers. Perhaps a bit less familiar, but still something we know about, are *prime* numbers. Do you know Euclid's Theorem that there are infinitely many prime numbers? If you know it, congratulations; you are truly knowledgeable.[1] If not, you will first learn the proof of this wonderful theorem in this book.[2]

On the other hand, do you know the proof that there are infinitely many *twin primes*? As of this writing, *nobody* knows whether this statement is true or false although remarkable progress has been made on this *conjecture* (see §1.9* below for the definition of "twin prime" and for the statement of the conjecture). The subject of prime numbers, while fascinating, at advanced levels can be not at all easy. So in this chapter we consider results that are either easy to state, easy to prove, or both.

Math, perhaps like life, is a game. Imagine beginning to play a video game with 40 levels. At the beginning, level 38 looks very, maybe impossibly, difficult. Math, like some video games, can be thought of as being built on levels. Our goal will be to get good at the *game of math*.

If we approach the game of math very formally, then proceeding through the levels of our game will necessarily take a lot of time and a lot of pages. So, typically, in a book we jump over some of the elementary levels. In this book, to make the discussion more lively, we give you, the reader, a sneak peak of the higher levels, including some famous unsolved conjectures!

Instead of beginning with a more formal treatment of the basics of logic, we will first learn how to prove theorems by example. Our philosophy is that the easiest way to learn how to prove statements is by seeing how it is done. We encourage you to think independently and to work out proofs with the minimum amount of help that you need to work it out in a reasonable amount of time. For example, while

[1]This type of sentence is patterned after the pattern of speech of chess Youtuber agadmator.

[2]That is, if you do not first click on the Wikipedia page link for this theorem! Euclid's Theorem is also Theorem 4.12 in this book.

you are reading a proof, you may have an aha moment, where you get the idea of the proof and feel that you can finish off the proof by yourself. When this happens, we hope that you try to do this. At least at the beginning, it may be helpful to check your proof with the proof in the book or any other source. Typically, the proofs will be rather similar. If they are not, then you have an independent proof!

In this chapter, as in every chapter, *exercises* are sprinkled throughout the text. We strongly suggest you work as many problems as you can. After straining for at least a few minutes, you may look at the hints at the end of the chapter.

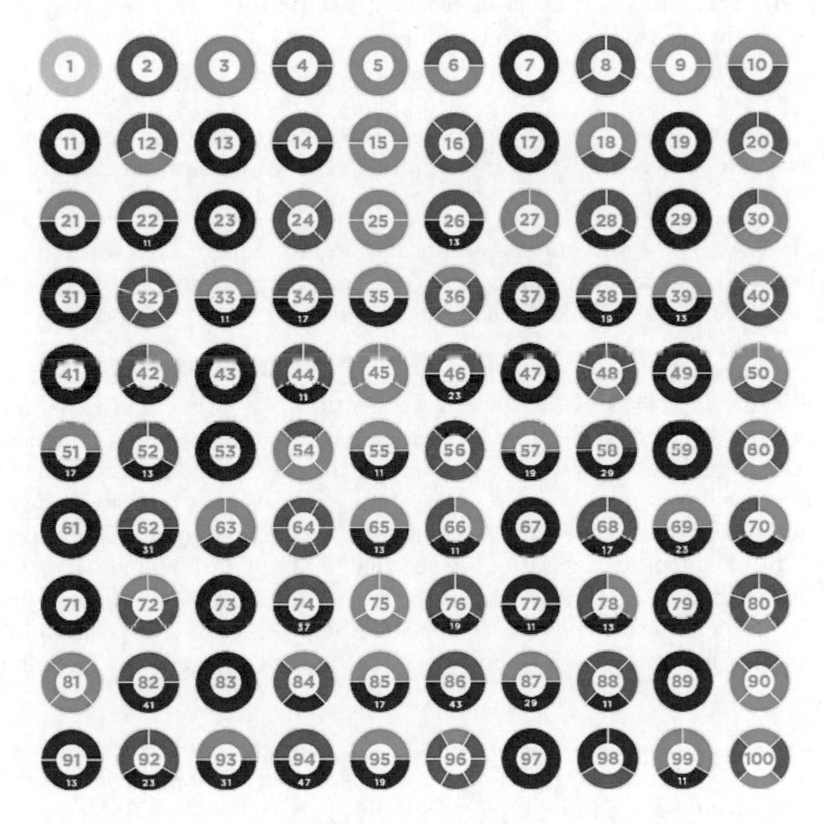

Figure 1.0.3. Can you decode the color coding of the integers from 1 to 100? This image is from Dan Finkel's TEDx talk "Five Principles of Extraordinary Math Teaching".

1.1. A first excursion into prime numbers

Not all math puns are awful, just sum.

In this section we learn what a prime number is, we see all of the primes less than 100, and we formulate some naive conjectures on primes, which we prove are false.

1.1.1. Prime numbers. Let $\mathbb{Z}$ denote the set of integers, including the positive ones, the negative ones, and zero, denoted by 0. For example, -17 and 8 are integers, but π, read as "pi", is not. On this set $\mathbb{Z}$ we have the operations of addition $+$ and multiplication $\cdot$ or $\times$. For example $2+2=2\cdot 2=4$ and $2\cdot 3=6$.

1.1.1.1. *Definition of prime number.* Let us consider positive integers. Primes are their multiplicative building blocks. We see that some numbers factor, such as $6=2\cdot 3$, $12=4\cdot 3$, and $35=5\cdot 7$. Other numbers don't factor, such as 17: we can write $17=1\cdot 17=17\cdot 1$, but there is no other way to write 17 as a product of two positive integers. With this in mind, we make the following:

Definition 1.1. We say that an integer $p\geq 2$ is **prime** if the only way to factor p as the product of two positive integers is

$$p=1\cdot p=p\cdot 1.$$

That is, the only way of "breaking up" p as the product of positive integers is the trivial way. In other words, an integer at least 2 is prime **if and only if** it is "indivisible" multiplicatively.

Remark 1.2. By "if and only if" we mean "exactly when". We will see a formal definition of "if and only if", which we do not need here, in Chapter 3 below. When the word "if" is used to make a *definition*, we actually mean "if and only if".

For example, 2 is prime since the only way to factor it as the product of positive integers is $2=1\cdot 2=2\cdot 1$. Similarly, we see that 3 is prime since the only way to factor it as the product of positive integers is $3=1\cdot 3=3\cdot 1$. One the other hand, even though the only way to factor 1 as the product of two positive integers is $1=1\cdot 1$, we have that 1 is not a prime simply because $1<2$. As will be evident later, one makes this choice that 1 is not a prime so that many results involving primes will be smoother to state. We see that 4 is not prime since $4=2\cdot 2$ and since the positive integers $2,2$ are not $1,4$, in either order. Next, 5 is prime as $5=1\cdot 5=5\cdot 1$ is the only way to factor 5 as the product of two positive integers. But 6 is not prime since $6=2\cdot 3$ and since the positive integers $2,3$ are not $1,6$, in either order.

Exercise 1.1. *Explain the color coding of the first one hundred positive integers in Figure* 1.0.3. *The topic of* §1.6 *is a hint! If you are having difficulty with this exercise, continue reading and come back to it.*

Certain elementary facts about integers will be useful.

Solved Problem 1.3. *Prove that if a is a positive integer satisfying $a\neq 1$, then $a>1$, and in fact $a\geq 2$.*

Solution. Firstly, for this proof, we assume the basic facts that 1 is the smallest positive integer and 2 is the next smallest positive integer. In particular, there are no integers strictly between 0 and 1, and there are no integers strictly between 1 and 2.

Let a be a positive integer satisfying $a\neq 1$. Since $a>0$ and a is an integer, we have $a\geq 1$. Indeed, there is no integer a satisfying $0<a<1$. Now, since $a\geq 1$ and $a\neq 1$, we obtain $a>1$. Now, since there is no integer a satisfying $1<a<2$, we conclude that $a\geq 2$. □

The $\square$ symbol indicates the end of the solution or proof. That was a relatively easy proof. And it was indeed a proof. That is, it was

a logical argument which demonstrated the truth of a statement.

Part of the key to our success was that we clearly understood beforehand what it means to be an integer. As Polya said, "You have to understand the problem."

Exercise 1.2. *Let n be a positive integer, and suppose that a and b are integers such that $n = ab$. Prove:* ***If*** *a is equal to 1 or n,* ***then*** *b is equal to n* ***or*** *1, respectively. Hint: Show that if $a = 1$, then $b = n$. Similarly for 1 and n switched.*

Remark 1.4. In the above, we boldfaced the "if-then" and "or" natures of the statement. **If-then statements** are called *implications* (a.k.a. conditional statements), and **or** is an example of a logical connective, all of which is studied in more detail in Chapter 3 below. But we assume that since such statements are common, even in elementary mathematics, you are comfortable with their meanings.

The exercise above is a short proof, but there is logical reasoning going on here! Usually we give hints to the exercises at the end of the chapter. However, since this is our first "proof" exercise, we solve part of Exercise 1.2 here:

Suppose $a = 1$. Then $n = a \cdot b = 1 \cdot b = b$. This proves: If $a = 1$, then $b = n$.

Characterization of non-primes: So what does it mean for an integer n *not* to be prime? Firstly, if $n \leq 1$, then n is not prime. So let us assume that $n \geq 2$. The following is a tweak to the characterization of not being prime.

Lemma 1.5. *An integer $n \geq 2$ is not a prime if and only if there exist integers $a > 1$ and $b > 1$ such that $n = ab$.*

Proof. By Definition 1.1, n is not prime precisely when there is another way to factor n by positive integers besides $1 \cdot n$ and $n \cdot 1$; that is, there exist positive integers a and b such that

$$n = ab, \tag{1.1}$$

where a, b are not equal to $1, n$ in any order.

Claim. The condition above that a, b are not equal to $1, n$ in any order *is equivalent to* the condition

$$a \neq 1 \quad \text{and} \quad b \neq 1. \tag{1.2}$$

Proof of the claim. Firstly, if $a \neq 1$ and $b \neq 1$, then clearly a, b are not equal to $1, n$ in any order.

Secondly and conversely, suppose that a, b are not equal to $1, n$ in any order. **Then suppose for a contradiction that** $a = 1$. By Exercise 1.2, we have $b = n$, contradicting that a, b are not equal to $1, n$ in any order. Since we have arrived at a *contradiction* to our last assumption that $a = 1$, we must have that $a \neq 1$.

The proof that $b \neq 1$ is exactly analogous to the proof that $a \neq 1$ by switching the roles of a and b in the proof. So we have proved the claim.

Since a and b are positive, by Solved Problem 1.3, condition (1.2) is in turn equivalent to the condition that

$$a > 1 \quad \text{and} \quad b > 1. \tag{1.3}$$

This and the proved claim above complete the proof of the lemma. □

Remark 1.6. This was the first time we proved something by the method of *contradiction.* Namely, at one point in the proof we assumed that $a = 1$, and from this assumption we logically deduced a contradiction. This proves that the assumption $a = 1$ is false, so necessarily we have $a \neq 1$. We will discuss proof by contradiction in more detail in Chapter 3.

For Lemma 1.5 we walked the path of a proof. By going over the logic of the proof, you will know the path of the proof better. There is no single right balance of walking and knowing proofs. The way you choose is your path.

Exercise 1.3. *Given that $n = ab$, show that condition* (1.3) *on a and b is equivalent to the condition*

$$1 < a < n. \tag{1.4}$$

This is also equivalent to $1 < b < n$. Hint: Use some basic facts about inequalities.

1.1.1.2. *Composite numbers.* Non-prime integers greater than 1 are also called **composite numbers**. So, by Lemma 1.5, an integer $n > 1$ is composite if and only if there exist integers $a, b > 1$ such that $n = ab$.

If $n = ab$, where a and b are integers, then we say that a and b are **divisors** of n. Another way to say this is: An integer a is a *divisor* of an integer n if there is some integer b such that $n = ab$. We also say that a **divides** n. We can think of a dividing n as meaning that a is "contained" in n from the point of view of multiplication.

Example 1.7. (1) 7 is a divisor of 56 since $7 \cdot 8 = 56$ and 56 is a composite number.

(2) 1 is a divisor of any integer n since $1 \cdot n = n$.

Exercise 1.4. *Show that* 0 *is not a divisor of any non-zero integer.*

Lemma 1.8. *If a and n are positive integers and if a is a divisor of n, then*

$$1 \leq a \leq n. \tag{1.5}$$

Proof. Let a and n be positive integers such that a divides n. Since a is a positive integer, we have $a \geq 1$ (it is not possible that $0 < a < 1$). We also have that there exists a positive integer b such that $ab = n$. Since $b \geq 1$, this implies that $a = \frac{n}{b} \leq \frac{n}{1} = n$. □

Notice how most mathematical proofs, in contrast to computer programming, have (hopefully small) jumps in logic. For example, in the proof above, we used the "elementary" fact that since $n > 0$ and $b \geq 1 > 0$, we have $\frac{n}{b} \leq \frac{n}{1}$. For a justification of this fact, see (1.18) below.

Given a positive integer n, we call 1 and n **trivial divisors** of n. If $1 < a < n$ is a divisor of n, then we call a a **non-trivial divisor** of n.

Example 1.9. 1 and 15 are trivial divisors of 15, whereas 3 and 5 are non-trivial divisors of 15.

Exercise 1.5. *Let $n = ab$ be a positive integer. Prove that if a is a non-trivial divisor of n, then b is a non-trivial divisor of n.*

By Exercise 1.3, we have the following further tweak of the characterization of composite numbers.

Corollary 1.10. *Let $n \geq 2$ be an integer. The integer n is a composite number if and only if there exists a divisor a of n satisfying $1 < a < n$, that is, if and only if n has a non-trivial divisor.*

Proof. Let $n \geq 2$ be a composite number. By Lemma 1.5, this is equivalent to there existing integers $a > 1$ and $b > 1$ such that $n = ab$.

By Exercise 1.3, this implies that a is a non-trivial divisor of n.

Conversely, suppose that $n \geq 2$ has a non-trivial divisor a. Since a is a divisor of n, there exists an integer b such that $n = ab$. Since $1 < a < n$, by Exercise 1.3 again, this implies that $1 < b < n$. This proves that n is a composite number. □

So, a composite number is an integer at least 2 which has at least one non-trivial divisor. On the other hand, a prime is an integer at least 2 which has no non-trivial divisors. For example, to prove that 5 is prime (as we claimed earlier), we just need to check that none of $2, 3, 4$ divide 5. To reprove that 6 is a composite number, we just need to observe that 2 divides 6 and that $1 < 2 < 6$.

A visualization of prime versus composite number is given in Figure 1.1.1.

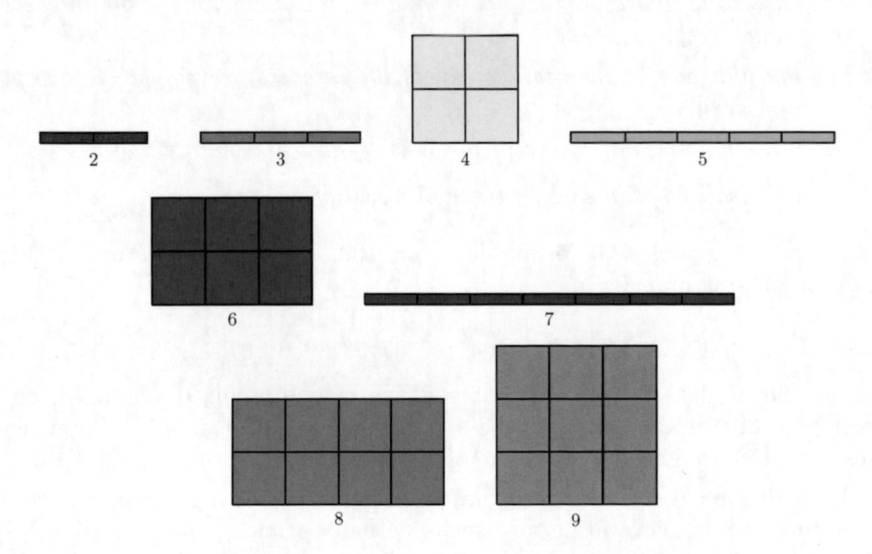

Figure 1.1.1. Visualizing the integers 2 through 9 as primes or composite numbers. The composite numbers are $4 = 2 \cdot 2$, $6 = 2 \cdot 3$, $8 = 2 \cdot 4$, $9 = 3 \cdot 3$.

1.1.1.3. *The primes under* 100. Having found which integers at most 6 are primes, if we continue in this way of checking for non-trivial divisors, then we can find the primes less than 100.

Exercise 1.6. *List all of the primes less than* 100. *Hint: We give the answer momentarily, but don't peek!*

By Corollary 1.10, an integer $p \geq 2$ is prime if and only if p does not have a divisor a satisfying $1 < a < p$.

It is cool to picture primes in red and non-primes in green as in Figure 1.1.2.

01	02	03	04	05	06	07	08	09	10
11	12	13	14	15	16	17	18	19	20
21	22	23	24	25	26	27	28	29	30
31	32	33	34	35	36	37	38	39	40
41	42	43	44	45	46	47	48	49	50
51	52	53	54	55	56	57	58	59	60
61	62	63	64	65	66	67	68	69	70
71	72	73	74	75	76	77	78	79	80
81	82	83	84	85	86	87	88	89	90
91	92	93	94	95	96	97	98	99	100

Figure 1.1.2. The first 100 positive integers: primes are in red and composite numbers are in green.

Exercise 1.7. *We make the following observations about Figure* 1.1.2.

(1) *For the second, fourth, sixth, eighth, and tenth columns, all of the integers are composite except for the number* 2.

(2) *For the fifth and tenth columns, all of the integers are composite except for the number* 5.

(3) *All of the multiples of* 3, *except for* 3 *itself, are composite numbers.*

Explain the reason for why each of these observations is true.

Now we belatedly list the primes less than 100. They are, as pictured in Figure 1.1.2, in increasing order:

(1.6)
$$2, 3, 5, 7, 11, 13, 17, 19, 23, 29, 31, 37, 41, 43, 47, 53, 59, 61, 67, 71, 73, 79, 83, 89, 97.$$

There are 25 in all, so assuming that you live to one hundred years old, on one quarter of your birthdays you will be a prime number of years old. So at age 97 you can say that you are in the prime of your life!

Although, we have listed the primes less than 100, it is a good exercise to give a *proof* that the first several in our list are actually primes.

Exercise 1.8. *Prove that each of the integers* 5, 7, 11, 13, 17 *is a prime. That is, show that each of these integers does not have any non-trivial positive divisors. You may use Theorem* 1.34 *below.*

We remark that the prime 17 is a rather interesting integer. The address of the Mathematical Sciences Research Institute (MSRI) is 17 Gauss Way in Berkeley.

Remark 1.11. To see that 5 is not a prime, we calculate that $\frac{5}{2} = 2.5$, $\frac{5}{3} = 1 + \frac{2}{3}$, and $\frac{5}{4} = 1.25$, none of which are integers. (Namely, $2 < \frac{5}{2} < 3$, $1 < \frac{5}{3} < 2$, and $1 < \frac{5}{4} < 2$. Any real number strictly between two consecutive integers is not an integer.) Therefore, none of $2, 3, 4$ divide 5. This proves that 5 is prime.

1.1.2. Conjectures on primes—to prove or not to prove. That is the question!

To understand any concept, such as primes, we should ask questions and guess and check statements pertaining and related to this concept. A guessed statement is called a conjecture.

1.1.2.1. *Naive conjectures.* Let us make a "gonzo" conjecture based on wishful thinking.

An elementary fact is that all primes greater than 2 are odd; we will prove this in Theorem 1.19 (or Corollary 1.23) below, when we discuss even and odd in more detail. Moreover, positive powers of 2 are even, such as $2^1 = 2$, $2^2 = 4$, etc. So $2^n - 1$ is odd for every positive integer n. Naively, based on these observations we can ask:

Is $2^n - 1$ always prime for every positive integer n?

Let us try the first few cases. For $n = 1$, we have $2^n = 2 - 1 = 1$, which is not prime. So we add the hypothesis that $n \geq 2$. Now, for $n = 2$, we have $2^n - 1 = 3$, which is prime. For $n = 3$, we have $2^n - 1 = 7$, which is prime again.

So we are quick (perhaps too quick!) to pose the following.

Conjecture 1.12. *If $n \geq 2$ is an integer, then $2^n - 1$ is a prime.*

False conjectures are often easy to disprove. This statement is no exception! For $n = 4$, we have $2^n - 1 = 15$, which is not prime since $15 = 3 \cdot 5$. This suffices to disprove the conjecture, for there exists an integer at least 2, namely 4, for which the conjectured statement is false. In other words, if the conjecture were true, then 15 would be prime, but it isn't. So the conjecture is false.

So, by the logical argument above, which is not much of an argument(!), we have proved:

Proposition 1.13. Not all integers of the form $2^n - 1$ are prime, where $n \geq 2$ is an integer.

Moreover (we leave it to you to check this), not all integers of the form $2^n + 1$ are prime, where $n \geq 0$ is an integer.

One of the goals of the book by Fletcher and Patty [**FP96**] is to develop the reader's ability to "... distinguish mathematical thinking from wishful thinking." This is very important, and we would like to add to this statement that we would like to encourage wishful thinking (that is, conjecture making) with the qualification that the reader should logically check the wishful thinking to see if they can determine whether their conjecture is true or false.

Exercise 1.9. *Prove or disprove the following conjecture: For every non-negative integer n, $3^n + 2$ is a prime number.*

So, you have to decide whether the conjecture above is true or false. How did your thinking about the truth or falsehood of the statement evolve as you thought about the exercise?

In general, mathematical results and conjectures come from observation. In a sense, mathematics is an experimental science. We look at mathematical objects and structures, such as prime numbers, and we search for patterns. We then make hypotheses about the existence of patterns based on our observations. Before we prove or disprove them, these hypotheses may be true or false for all we know. To prove a conjecture, we need to come up with a logical argument that establishes the statement of the conjecture. Typically, to disprove a conjecture, we just need to find a **counterexample**, that is, an object that does not fit the pattern we conjectured. More precisely, a counterexample to a statement is an example (a.k.a. special case) for which the statement is false.

The previous paragraph exhibits a rational way of approaching conjectures. A less rational approach, as we saw in Conjecture 1.12, is what we describe as "gonzo mathematics": rather arbitrarily hypothesizing patterns based on only a small amount of observation. In other words, jumping to conclusions. Here, due to mostly wishful thinking with only a little empirical evidence, we hypothesize patterns usually guided by beauty and simplicity, which are subjective characteristics.

One advantage of making gonzo conjectures is that we gain experience and feedback. By determining whether they are true or false, we often are inspired to make better and less gonzo (i.e., less naive) conjectures.

Thing One to Thing Two: *Did you know that gullible isn't a word?*

As in the imagined conversation above (between two Dr. Seuss characters), it is perhaps just as important to know what is false as it is to know what is true!

1.1.2.2. *Less naive conjectures that are still false.* Going back to Conjecture 1.12, we may observe that 2 and 3 are primes, whereas 4 is a composite number. Is it possible that Conjecture 1.12 is true for n prime? Let's see. For n equal to the primes $2, 3, 5, 7$ we obtain for $2^n - 1$ the integers

$$3, 7, 31, 127,$$

which are all primes. (We show below that 127 is a prime.) In general, the larger the number, the harder it is to check whether or not it is a prime.

Despite the optimistic calculations above, can you disprove the following conjecture?

Conjecture 1.14. *If $p \geq 2$ is a prime, then $2^p - 1$ is a prime.*

To see if this conjecture has a chance to be true, we check the primeness of $2^p - 1$ for the next prime $p = 11$. We obtain the integer 2047, which is equal to $23 \cdot 89$.[3] So we conclude that Conjecture 1.14 is false! So, it was a bit harder to

[3] Indeed, $23 \times 100 = 2300$. So $23 \times 90 = 2300 - 230 = 2070$. Thus $23 \times 89 = 2070 - 23 = 2047$. Quick maths!

disprove this conjecture as compared to Conjecture 1.12, but by checking a few more primes we were able to do it.

Summarizing, we have proved:

Theorem 1.15. *The following statement is false:*

"If $p \geq 2$ is a prime, then $2^p - 1$ is a prime."

In other words, there exists a prime number p such that $2^p - 1$ is not a prime. In particular this is the case for $p = 11$.

Here are $2^p - 1$, color coded, for the the primes $p < 20$:

$$3, 7, 31, 127, 2047, 8191, 131071, 524287. \tag{1.7}$$

The moral of the story is that sometimes one aptly chosen example is enough to disprove a theorem, so it often makes sense to look for such an example. Even if you don't find one, often the work of checking several examples may lead you to a pattern that leads to a proof. So, having made a conjecture, it makes sense to check the validity of a number of examples of the stated conjecture. Many conjectures fall by the wayside simply from checking some examples!

1.1.2.3. *A true conjecture about primes.* Now that we have had so many misfires in formulating conjectures, it is nice to state true conjectures. A wonderful and true result is:

Theorem 1.16. *If n is a positive integer such that $2^n - 1$ is a prime, then n itself is prime!*

For the proof of this, see Exercise 1.10 or Theorem 3.40 below.

To analyze the implication in Theorem 1.16, we'll need some elementary logic, which we briefly discuss here and which will be discussed in more detail in Chapter 3. Consider an implication of the general form:

If P, then Q.

This statement is equivalent to its **contrapositive** implication (see §3.7 below):

If not Q, then not P.

For example, let p be an integer greater than 2, and consider the implication:

If p is prime, then p is odd.

Then this implication is equivalent to its contrapositive:

If p is not odd (that is, p is even), then p is not prime.

(Recall that we are assuming that $p > 2$.)

The special case of the logical equivalence above that we find useful is: The implication in Theorem 1.16 is equivalent to its "contrapositive":

Let n be a positive integer. If n is a composite number, then $2^n - 1$ is a composite number.

Regarding this last statement, which we have not proved yet and leave as an exercise below, let us look at some examples. We have that $2^6 - 1 = 63$ and $63 = 3^2 \cdot 7$. We observe that the non-trivial divisors of 6 are 2 and 3 and that $2^2 - 1 = 3$ and $2^3 - 1 = 7$ are divisors of $63 = 2^6 - 1$. Next, let us consider the composite number 10 and we compute that $2^{10} - 1 = 1023$ and that $1023 = 3 \cdot 11 \cdot 31$. Again, we observe that the non-trivial divisors 2 and 5 of 10 have the property that $2^2 - 1 = 3$ and $2^5 - 1 = 31$ are divisors of $1023 = 2^{10} - 1$. We might as well try one more example to see the pattern more clearly: we compute that $2^{12} - 1 = 4095$ and that $4095 = 3^2 \cdot 5 \cdot 7 \cdot 13$. The non-trivial divisors 2, 3, 4, and 6 of 12 satisfy

$$2^2 - 1 = 3, \quad 2^3 - 1 = 7, \quad 2^4 - 1 = 15 = 3 \cdot 5, \quad 2^6 - 1 = 63 = 3^2 \cdot 7,$$

which are all divisors of $4095 = 2^{12} - 1$. All of this leads us to:

Conjecture 1.17. *If we have positive integers a, b, n satisfying $n = ab$, then $2^a - 1$ and $2^b - 1$ both divide $2^n - 1$. Thus, if n is a composite number, then $2^n - 1$ is a composite number.*

Exercise 1.10. *Prove the conjecture. Hint: Apply polynomial long division, where the polynomial variable "x" is equal to 2^a. Namely, divide the polynomial $2^n - 1 = x^b - 1$ (explain why this equality is true) by the linear polynomial $2^a - 1 = x - 1$. Then explain why this proves the conjecture.*

Since the conjecture implies Theorem 1.16, *this exercise provides a proof of the theorem.*

1.2. Even and odd integers

What is more elementary than prime numbers, but nonetheless interesting, are the notions of even and odd. We discuss elementary properties of even and odd integers, including proving properties relating to addition, multiplication, and prime numbers.

1.2.1. Properties of even and odd integers.

Question: Do you know what's odd?
Answer: Numbers that aren't divisible by two.
Rebuttal: What are the odds you are even right?

Statements involve concepts. We discuss concepts by making definitions. Even though we all know what even and odd integers are, we now give their formal definitions.

Definition 1.18. An integer n is **even** if there exists an integer a such that $n = 2a$.

For example, 98 is even since $98 = 2 \cdot 49$. That is, the integer a that exists satisfying $98 = 2a$ is $a = 49$.

Two is the only even prime:

Theorem 1.19. *If $n > 2$ is an even integer, then n is not prime.*

Proof. Since n is even, there exists an integer b such that

$$n = 2b. \tag{1.8}$$

Since $2b = n > 2$, we have $b > 1$. This and $2 > 1$ imply that n is not prime (by Lemma 1.5). □

Mini-analysis of the proof above: The proof is an example of a "direct" proof. Our aim was to show that n is the product of two integers $a, b > 1$. Using our hypothesis, we were able to do this with $a = 2$.

Two mice, Mickey and Minnie, not quite satisfied with the mini-analysis (perhaps Mickey wasn't satisfied with the earlier proofs either!), have a long and sometimes silly Socratic dialogue about the proof: [4]

Mickey: How did you do that? It looked like magic. I'm not quite convinced.

Minnie: Well, I looked at the statement and saw that we have to prove that each even integer greater than 2 is not prime. So firstly, I took such an integer.

Mickey: How do you take such an integer? Is there a store where I can buy them?

Minnie: Actually, I conjured them out of thin air by using the word "let". So we started with:

Let $n > 2$ be an even integer.

This is our hypothesis.

Mickey: Seems impenetrable to me. How can you conjure something out of nothing?

Minnie: It is like conjuring anything. For example, assume that we have a unicorn. We can always assume this. However, the key is to *prove* something about the unicorn. Since unicorns are mythical objects, we won't get very far. So let's stick to math.

Mickey: I can agree to that!

Minnie: Now, please give me an even integer.

Mickey: Okay, how about the number 2?

Minnie: Sorry! We want an integer that is greater than 2.

Mickey: Okay, then, let's take 46.

Minnie: 46 is good. We have $46 = 2 \cdot 23$, so it is not prime since $2 > 1$ and $23 > 1$. Please give me another.

Mickey: Let's try 1363.

Minnie: That's not even.

Mickey: How do you know?

Minnie: Well, $1363 = 2 \cdot 681.5$, but 681.5 is not an integer, so 1363 is not even.

Mickey: Alright, I'm feeling bold, so now let's take n, where n is even and greater than 2.

Minnie: Great! Since n is even, by definition:

There is some integer b such that $n = 2b$.

[4] Silliness is not restricted to Socratic dialogues. For example, there is the Ministry of Silly Walks.

And not only that, since $n > 2$, we also get

$$b = \frac{n}{2} > 1.$$

Mickey: That is all fine and dandy. Now what?

Minnie: Now we need to observe that we are actually done with the proof! Why is that? Because we have proved that our hypothesized integer n is the product of two integers: $a = 2 > 1$ and $b > 1$. By the definition of prime number (more precisely, Lemma 1.5), we see that this implies that n is not a prime number!

Mickey: So I give you an integer that is greater than 2 and equal to 2 times *something*. You tell me that this *something* is greater than 1. Since both 2 and this *something* are greater than 1, my integer, which is their product, is not prime.

Minnie: You've got it, way to go!

Mickey: You are awesome! ♡

Minnie: Thanks, so are you! ♡

As in the joke quoted at the beginning of this subsection, we have:

Definition 1.20. An integer n is **odd** if n is not even. The **parity** of an integer refers to whether it is even or odd.

For example:

Lemma 1.21. *The integer* 3 *is odd.*

Proof. Suppose for a contradiction that 3 is even. Then $3 = 2a$ for some integer a. Then $a = 1.5$, which is not an integer.[5] So we have a contradiction to the supposition that 3 is even. Now, a contradiction cannot follow from a true statement. So we conclude that 3 is not even. □

The argument above can generalized as follows.

Lemma 1.22. *Any even integer plus* 1 *is odd. In other words, if n is even, then $n+1$ is odd.*

Proof. Let n be an even integer. Then $n = 2k$ for some integer k. Suppose for a contradiction that $n+1$ is not odd; that is, suppose $n+1$ is even. Then there exists an integer ℓ such that $n+1 = 2\ell$. We conclude that $2k+1 = n+1 = 2\ell$, which implies that

$$1 = 2\ell - 2k = 2(\ell - k).$$

Hence, since $\ell - k$ is an integer, we have that 1 is even, a contradiction. (Alternatively, $\ell - k = \frac{1}{2}$ is not an integer, a contradiction.) Therefore $n+1$ is odd. □

Finally, we remark that we should be careful with how we state results. By Theorem 1.19, each of the statements in the following corollary, when interpreted correctly, is true!

Corollary 1.23. *All prime numbers are odd except one. Actually, all prime numbers are odd except two.*

[5] Alternatively, since $2 < 3 = 2a < 4$, we have $1 < a < 2$, so that a is not an integer.

Indeed, in the first sentence we mean "... except *one* of the prime numbers", while in the second sentence we mean "... except the prime number *two*"! So, in the statements of the corollary we were ambiguous. Having learned our lesson, from now on we will be more careful to write clear mathematical statements.

If you are not sure about the proof of this corollary, for a hint see Exercise 3.28 below in the chapter on implications and all that.

1.2.2. Even and odd and addition.

Question: How do you make seven an even number?

Answer: Remove the "s".

In this subsection and the next, we discuss the properties of even and odd with respect to the arithmetic operations of addition and multiplication. To wit, we answer the question: How does parity *interact* with addition and multiplication? Here are the answers.

Theorem 1.24. *The sum of two even integers is even. In other words, if a and b are even integers, then the integer $a+b$ is even.*

Proof. Suppose that a and b are even integers. Then there exist integers k and ℓ such that $a = 2k$ and $b = 2\ell$. We *calculate* that

$$a + b = 2k + 2\ell = 2(k + \ell). \tag{1.9}$$

Since $k+\ell$ is an integer (the sum of two integers is an integer), by (1.9) we conclude that $a + b$ is even. □

Regarding the proof of Theorem 1.24, if you prefer to think visually, we offer Figure 1.2.1.

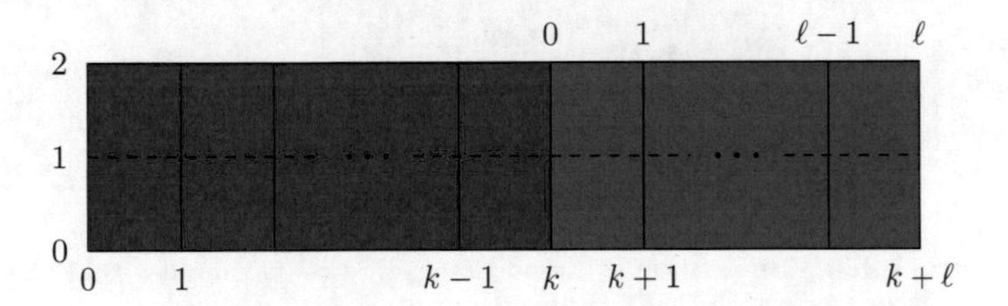

Figure 1.2.1. A visual proof that an even number plus an even number is even: $a = 2k$ and $b = 2\ell$ implies $a + b = 2(k + \ell)$.

Theorem 1.25. *The sum of an even integer and an odd integer is odd.*

Proof 1 of Theorem 1.25. This proof uses a fact which we will not prove until a fair bit later in the book. Thankfully, this fact is something you are most likely familiar with and take for granted as being true.

Fact 1.26. *If a is an odd integer, then there exists an integer k such that*

$$a = 2k + 1.$$

That is, any odd integer can be written as an even number plus 1.

As an example, 17 is odd, and $17 = 2 \cdot 8 + 1$. See Corollary 1.41 below for how this fact follows from the "Division Theorem" (Theorem 4.1 below). We now *boldly* proceed to prove the theorem.

To boldly go where no person has gone before! – Star Trek

Let a be an odd integer and let b be an even integer. Since a is odd, by the fact above, there exists an integer k such that $a = 2k + 1$. Since b is even, there exists an integer ℓ such that $b = 2\ell$. We calculate that

$$a + b = (2k + 1) + 2\ell = 2(k + \ell) + 1. \tag{1.10}$$

Since $k + \ell$ is an integer, by (1.10) and Lemma 1.22, we have that $a + b$ is odd. □

Proof 2, by contradiction, of Theorem 1.25. By the commutativity of addition, we may assume that the first integer is even and the second integer is odd. So let a be an even integer and let b be an odd integer. Then there exists an integer k such that $a = 2k$.

Suppose for a contradiction that $a+b$ is not odd; that is, it is even. Then there exists an integer m such that $a + b = 2m$. We calculate that

$$b = (a + b) - a = 2m - 2k = 2(m - k). \tag{1.11}$$

Since $m - k$ is an integer (the difference of two integers is an integer), we conclude that b is even. This is a contradiction to our assumption. Therefore we conclude that $a + b$ is odd! □

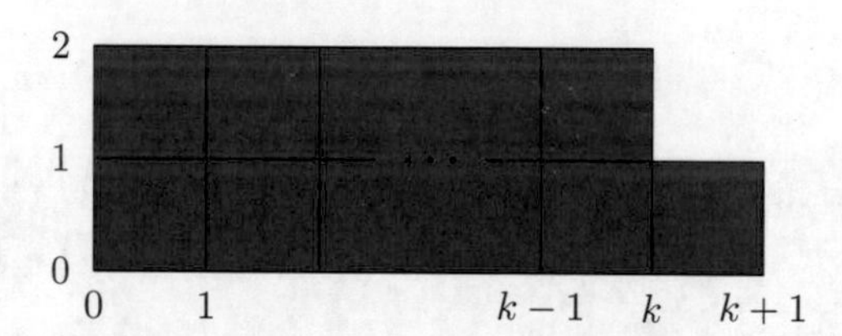

Figure 1.2.2. Visualization of an odd integer: a is odd implies that there exists an integer k such that $a = 2k + 1$.

Theorem 1.27. *The sum of two odd integers is even.*

Proof. Let a and b be odd integers. By this hypothesis and Fact 1.26, there exist integers k and ℓ such that $a = 2k + 1$ and $b = 2\ell + 1$. We calculate that

$$a + b = (2k + 1) + (2\ell + 1) = 2(k + \ell + 1). \tag{1.12}$$

Since $k + \ell + 1$ is an integer, we conclude from (1.12) that $a + b$ is even. □

Exercise 1.11. *Give visual proofs of Theorems* 1.25 *and* 1.27 *where the drawing is analogous to Figure* 1.2.1. *See also Figure* 1.2.2.

1.2.3. Even and odd multiplication. Observe that 6 is an even integer, and for every integer b, we have that $6b = 2(3b)$. Since $3b$ is an integer, we conclude that $6b$ is an even integer for every integer b. In general, we have the following.

Theorem 1.28. *The product of any integer and an even integer is even.*

Proof. Without loss of generality, we assume that the first integer is even. Let a be an even integer and let b be an integer. Since a is even, there exists an integer k such that $a = 2k$. We calculate that

$$ab = (2k)b = 2(kb). \tag{1.13}$$

Of course, kb is an integer. So (1.13) proves that ab is even. □

Exercise 1.12. *Show that Theorem* 1.28 *may be restated as:*

If 2 *divides* a *and if* b *is an integer, then* 2 *divides* ab.

A visual proof of Theorem 1.28 is given by Figure 1.2.3.

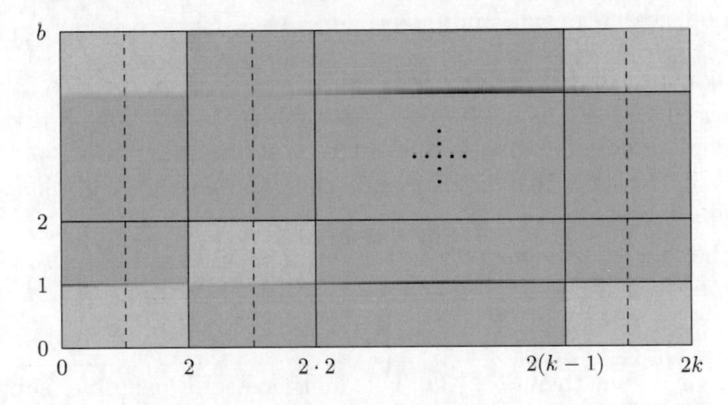

Figure 1.2.3. A visual proof of Theorem 1.28: Suppose $a = 2k$ for some integer k. We see that there are an even number of total squares in the rectangle representing the product. Can you think of an easier visual proof by dividing the $a \times b$ rectangle into only two subrectangles?

Theorem 1.29 (An odd fact). *The product of two odd integers is odd.*

Proof. Let a and b be odd integers, by Fact 1.26 there exist integers k and ℓ such that $a = 2k + 1$ and $b = 2\ell + 1$. We calculate that

$$ab = (2k + 1)(2\ell + 1) = 2(2k\ell + k + \ell) + 1. \tag{1.14}$$

Since $2k\ell + k + \ell$ is an integer, from this we conclude that ab is odd. □

Exercise 1.13. *Give a visual proof of Theorem* 1.29.

By taking the two integers to be equal in each of the previous two theorems, we have:

Corollary 1.30. (1) *The square of an even integer is even.*

(2) *The square of an odd integer is odd.*

Direct proof of part (1) of Corollary 1.30. Let n be an even integer. By definition, $n = 2k$ for some $k \in \mathbb{Z}$. We compute that

$$n^2 = (2k)^2 = 2 \cdot 2k^2. \tag{1.15}$$

Since $2k^2 \in \mathbb{Z}$ (the product of integers is an integer), we conclude that n^2 is even. □

Exercise 1.14. *Prove: The cube of an even integer is even. The cube of an odd integer is odd.*

Exercise 1.15. *Prove that if the product of two integers is even, then at least one of the integers is even.*

1.3. Calculating primes and the sieve of Eratosthenes

Now we return to the subject of primes, our first love. Wouldn't it be nice to be able to rather easily figure out which numbers are primes? Fortunately, for numbers that are not too large, this is possible even if you are cast away on a deserted island with only food, water, pencil, and paper!

1.3.1. A shortcut for proving that an integer is a prime. To solve Exercise 1.8, for the prime 11 for example, you presumably checked for all divisors a with $1 < a < 11$ and came up empty. Interestingly, you only need to know that 2 and 3 do not divide 11 to show that 11 is not a prime. You will see why this is true from our discussion below.

Recall the following elementary fact about inequalities: Let a, b, c be real numbers and suppose that $c \geq 0$.

$$\text{If } b \geq a, \text{ then } cb \geq ca. \tag{1.16}$$

For example, if $b \geq a$, then $3b \geq 3a$. For another useful example, suppose that a and b are positive real numbers. By taking $c = \frac{1}{ab}$, we obtain:

$$\text{If } b \geq a > 0, \text{ then } \frac{1}{a} \geq \frac{1}{b} > 0. \tag{1.17}$$

By combining (1.16) and (1.17), we obtain:

$$\text{If } b \geq a > 0 \text{ and } m \geq 0, \text{ then } \frac{m}{a} \geq \frac{m}{b}. \tag{1.18}$$

Another result that we will find useful is:

Fact 1.31. *For every integer $n \geq 2$, there exists a prime p dividing n.*

This is Corollary 1.55 below. In particular, if n is a prime, then we simply take $p = n$. On the other hand, if n is a composite number, then this says that n has a prime divisor. For example, for $n = 24$ we may take $p = 2$ or $p = 3$.

We now return to the unfinished business from §1.1.2.2 of showing that 127 is a prime. Suppose that $127 = ab$, where a and b are positive integers. **Without loss of generality**, we may assume that $b \geq a$. We then get that (simply take $c = a$ in (1.16))

$$127 = ab \geq aa = a^2. \tag{1.19}$$

Since $a^2 \leq 127$, by taking square roots, we obtain

$$a \leq \sqrt{127}. \tag{1.20}$$

More generally, if $a^2 \leq b^2$ and $b \geq 0$, then $a \leq b$. In this paragraph we proved that **if** 127 is the product of two (non-trivial) factors, **then** the smaller of the two is at most the square root of 127.

From (1.20) we deduce that

$$a \leq \sqrt{127} < \sqrt{144} = 12. \tag{1.21}$$

Since $a < 12$ is an integer, we have that $a \leq 11$. To summarize, we have proved that if 127 is a composite number, then 127 has a divisor a satisfying $1 < a \leq 11$. In other words, if 127 does not have a divisor a satisfying $1 < a \leq 11$, then 127 is prime. Indeed this is the case and we leave it to you, the reader, to check this.

We can make the job of proving that 127 is a prime even easier. We have the following:

Fact 1.32. *If* 127 *is not prime, then* 127 *has a* **prime** *divisor* $1 < p \leq 11$. *Equivalently: If* 127 *does not have a* **prime** *divisor* $1 < p \leq 11$, *then* 127 *is prime.*

The reason this is true is as follows. As we have seen above, if 127 is not prime, then it has a divisor a satisfying $1 < a \leq 11$. Since $a > 1$, there exists a prime p dividing a (see Corollary 1.55 below). Since p divides a, we have $p \leq a \leq 11$. Also because p divides a and since a divides 127, we have that p divides 127 (by the "transitivity of division"). This proves Fact 1.32. □

For a more general and slightly more detailed version of this argument, see the proof of Theorem 1.34 below.

Exercise 1.16. *Prove that* 11 *is prime by showing that it is not divisible by* 2 *or by* 3. *Include a justification of why this suffices.*

Does this exact same method work for every integer strictly between 1 *and* 16*? Explain why or why not.*

As we have seen in the discussion above, taking square roots is useful.

> I poured root beer into a square cup. Now I have beer.

A mathematician translates this quote to the equation

$$(\sqrt{\text{beer}})^2 = \text{beer}. \ ☺$$

Generalizing what we did for the number 127 above, we observe the following.

Lemma 1.33. *An integer* $n > 1$ *is composite if and only if there exists a divisor* a *of* n *satisfying* $1 < a \leq \sqrt{n}$.

Proof. An integer $n > 1$ is composite if and only if there exist integers $a > 1$ and $b > 1$ such that $ab = n$. Without loss of generality, we may assume that $a \leq b$. This implies that $a^2 \leq ab = n$, so that $a \leq \sqrt{n}$.

Conversely, suppose that $n > 1$ is an integer with a divisor a of n satisfying $1 < a \leq \sqrt{n}$. Since $n > 1$, we have $1 < a < n$. By Corollary 1.10, this proves that n is a composite number. □

In fact, we have an even nicer statement:

Theorem 1.34. *An integer $n > 1$ is composite if and only if there exists a prime divisor p of n satisfying $p \leq \sqrt{n}$.*

Equivalently, an integer $n > 1$ is prime if and only if for every prime $p \leq \sqrt{n}$, p does not divide n.

Proof. Let $a > 1$ be an integer. By Corollary 1.55, a has a prime divisor p; that is, there exists a prime p such that $a = pk$ for some integer k.

Now assume in addition that $a \leq \sqrt{n}$ and a is a divisor of n. Then there exists a positive integer b such that $n = ab$. Thus,

$$n = ab = pk \cdot b = p \cdot kb,$$

where p is prime and $p \leq kb$. Therefore, p is a prime divisor of n satisfying $p \leq \sqrt{n}$.

To summarize, we have proved: If $1 < a \leq \sqrt{n}$ is a divisor of n, then there exists a prime divisor p of n satisfying $p \leq \sqrt{n}$.

Conversely, clearly if there exists a prime divisor p of n satisfying $p \leq \sqrt{n}$, then there exists a divisor a of n satisfying $1 < a \leq \sqrt{n}$, namely $a = p$.

Finally, by Lemma 1.33, an integer $n > 1$ being composite is equivalent to there existing a divisor a of n satisfying $1 < a \leq \sqrt{n}$. So we are done. □

Remark 1.35. Regarding the equivalence of the two statements in the theorem above, we observe it is true by the following general fact, which will be proved in Chapter 3. A statement of the form "P if and only if Q" is equivalent to the statement "(not P) if and only if (not Q)".

Theorem 1.34 allows us to efficiently decide for example if integers less than $1024 = 2^{10}$ are prime, as we only need to see if they are divisible by the primes less than $32 = 2^5$.

We may also use Theorem 1.34 to help find the prime factorizations of numbers. For example, to find the prime factorization of 2074, by Theorem 1.34 we know that if 2074 is not prime, then it has a prime divisor at most $\sqrt{2074} \approx 45.5$. Indeed, since 2074 is even, 2 is a divisor of 2074. Dividing 2074 by 2 yields 1037. Now, if 1037 is not prime, then it has a prime divisor at most $\sqrt{1037} \approx 32.2$. It turns out that 17, which is less than $\sqrt{1037}$, is a prime divisor of 1037. Dividing 1037 by 17 yields 61. Finally, 61 is a prime, which can be verified by showing that it has no prime divisors at most $\sqrt{61}$. Since $\sqrt{61} < 8$, to show that 61 is prime, we just need to check that none of the primes $2, 3, 5, 7$ divide 61, which is true. By combining all of the above, we obtain

$$2074 = 2 \cdot 1037 = 2 \cdot 17 \cdot 61, \tag{1.22}$$

where the last equality gives the prime factorization of 2074.

Exercise 1.17. *Let a and b be positive real numbers satisfying $ab = 10000$. Prove that $a \leq 100$ or $b \leq 100$.*

Hint: Can you derive a contradiction if $a > 100$ and $b > 100$? Recall that we used proof by contradiction in the proof of Lemma 1.21.

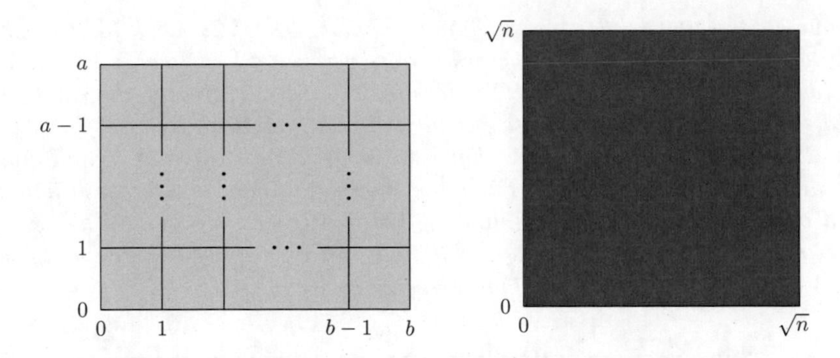

Figure 1.3.1. If $n = ab$ and $a \leq b$, then $a \leq \sqrt{n}$. The rectangle and the square have the same area n.

Commentary: You may have to use some (elementary) rules of logic, which you are likely familiar with. These logic rules are also discussed more formally later in the book. See, e.g., Chapter 3.

Exercise 1.18. *By the proof of Lemma 1.33, we have: If $n = ab$, where a, b are positive integers, then at least one of a, b is less than or equal to $\sqrt{n}$.*

Now, suppose that $n = abc$, where a, b, c are positive integers. Prove that at least one of a, b, c is less than or equal to $\sqrt[3]{n}$.

Example 1.36. If n is a positive integer less than or equal to 100, then by Theorem 1.34 we have that n is prime if and only if it is not divisible by any of $2, 3, 5, 7$ since these are the primes at most $10 = \sqrt{100}$.

If n is a positive integer less than 49, then n is prime if and only if it is not divisible by any of $2, 3, 5$.

If n is a positive integer less than 25, then n is prime if and only if it is not divisible by any of $2, 3$.

In this way, it is quite easy to come up with the list of primes in (1.6).

Figure 1.3.2. Eratosthenes (276 BC–194 BC). Wikimedia Commons, Public Domain.

Exercise 1.19. *Verify that the primes less than 100 are given by the list in* (1.6). *A convenient way to carry this out is to use the sieve of Eratosthenes. Namely,* *cross out all the non-trivial (greater than* 1*) multiples of* 2*. Look for the smallest integer greater than* 2 *that isn't crossed out, which is* 3*. Now cross out all the non-trivial multiples of* 3*. Look for the smallest integer greater than* 3 *that isn't crossed out, which is* 5*, and cross out all the non-trivial multiples of* 5*. Look for the smallest integer greater than* 5 *that isn't crossed out, which is* 7*, and cross out all the non-trivial multiples of* 7*. Since the smallest integer greater than* 7 *that isn't crossed out is* 11*, which is greater than* $\sqrt{100} = 10$*, we stop at* 7*. The remaining integers are primes (don't count* 1*). (Note that in this way we have actually verified that* $2, 3, 5, 7$ *are the primes less than or equal to* 10*.)*

01	02	03	04	05	06	07	08	09	10
11	12	13	14	15	16	17	18	19	20
21	22	23	24	25	26	27	28	29	30
31	32	33	34	35	36	37	38	39	40
41	42	43	44	45	46	47	48	49	50
51	52	53	54	55	56	57	58	59	60
61	62	63	64	65	66	67	68	69	70
71	72	73	74	75	76	77	78	79	80
81	82	83	84	85	86	87	88	89	90
91	92	93	94	95	96	97	98	99	100

Figure 1.3.3. Cross out all of the multiples of 2, 3, 5, 7 between 1 and 100. Compare the integers remaining with a list of primes between 1 and 100. Are these two sets of integers the same?

Exercise 1.20. *Find the primes between* 101 *and* 200 *using the sieve of Eratosthenes (you can read about this at the "Sieve of Eratosthenes" Wikipedia link, but we give a brief description below). See also Figure* 1.6.1 *for the primes between* 1 *and* 100*. In short, the sieve works by crossing out all multiples of at least* 2 *or more of the primes starting with the lowest prime first. For example, for the prime* 2 *we cross out*

$$100, 102, 104, 106, 108, \ldots, 200.$$

For the prime 3*, we cross out*

$$102, 105, 108, 111, 114, \ldots, 198.$$

(Note that there are redundancies.) Did considering primes greater than 13 *help?*

Hints: See Figure 1.3.4 *and use Theorem* 1.34.

101	102	103	104	105	106	107	108	109	110
111	112	113	114	115	116	117	118	119	120
121	122	123	124	125	126	127	128	129	130
131	132	133	134	135	136	137	138	139	140
141	142	143	144	145	146	147	148	149	150
151	152	153	154	155	156	157	158	159	160
161	162	163	164	165	166	167	168	169	170
171	172	173	174	175	176	177	178	179	180
181	182	183	184	185	186	187	188	189	190
191	192	193	194	195	196	197	198	199	200

Figure 1.3.4. Cross out all of the multiples of 2, 3, 5, 7, 11, and 13 between 101 and 200. Compare the integers remaining with a list of primes between 101 and 200. Are these two sets of integers the same?

Exercise 1.21. *Disprove the following conjecture (i.e., prove that the statement is false): For every non-negative integer n, $F_n := 2^{2^n} + 1 := 2^{(2^n)} + 1$ is a prime. You may wish to use some sort of computer aid (instead of working it out with paper and pencil).*

Remarkably, F_n is known to be composite for all $5 \le n \le 32$. (Do not try to prove this full statement!) Why can we characterize this as extrapolation gone wrong?

1.4. Division

Primes are intimately related to multiplication and division. In this section we further consider division and some more of its elementary properties, and we also learn about some basic properties of the greatest common divisor of two integers.

1.4.1. Factoring 1.

"One and one make one." – From "Bargain" by The Who

The rock group, The Who, were evidently thinking about multiplication. And what can be simpler than the multiplicative identity? The following, which we could have proved earlier(!), says that there is no other way to make *one* via multiplication of positive integers. So a mathematician would amend The Who's lyrics by adding "uniquely"!

Theorem 1.37. *Suppose that a and b are positive integers satisfying $ab = 1$. Then $a = b = 1$.*

Proof. Let a and b be positive integers satisfying $ab = 1$. Suppose for a contradiction that $a > 1$. Then, since $b \geq 1$ and by (1.16), we have

$$1 = a \cdot b \geq a \cdot 1 = a > 1, \tag{1.23}$$

which is a contradiction. Thus $a \leq 1$, and hence $a = 1$ since $a > 0$. This implies that

$$b = \frac{1}{a} = 1. \tag{1.24}$$

□

Have you thought of any alternate proofs of Theorem 1.37? Figure 1.4.1 gives us an idea for one.

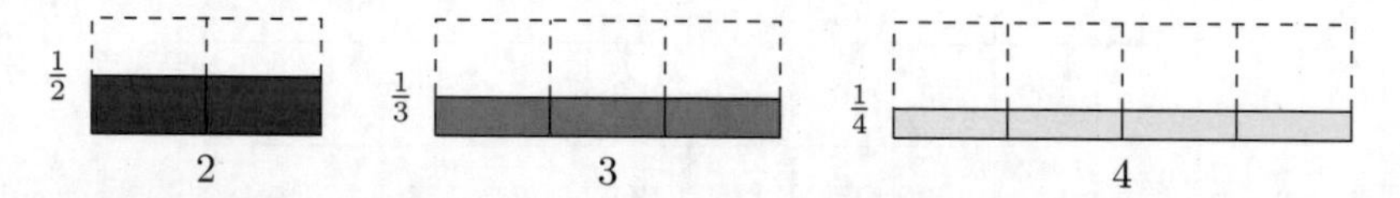

Figure 1.4.1. Visual idea for a proof of Theorem 1.37. The area of each colored region is equal to 1.

Alternate proof of Theorem 1.37. Suppose for a contradiction that $a > 1$. Since $\frac{1}{a}$ is a positive real number, we then have

$$1 = \frac{1}{a} \cdot a > \frac{1}{a} \cdot 1 = \frac{1}{a} = b > 0. \tag{1.25}$$

This is a contradiction since there is no integer strictly between 0 and 1. □

Given a result, it is usually interesting to consider special cases. In this case, for variety and to add a twist, we consider *negative* numbers instead of positive numbers in our special case:

Corollary 1.38. *There exists a unique negative integer n such that $n^2 = 1$. Namely, $n = -1$.*

Proof. Suppose that n is an integer satisfying $n^2 = 1$. Then $|n|$ is a non-negative integer satisfying

$$|n|^2 = n^2 = 1. \tag{1.26}$$

Suppose that $|n| = 0$. Then $0 = 0^2 = |n|^2 = 1$, which is a contradiction. Thus $|n|$ is a positive integer. So, by the preceding theorem, since $|n|\,|n| = 1$, we conclude that $|n| = |n| = 1$. Finally, this implies that $n = -1$ or $n = 1$. But since n is negative, we must have that $n = -1$. Since $(-1)^2 = 1$, we have proved that $n = -1$ is the unique negative integer with $n^2 = 1$. □

Here is an **Alternate proof of Corollary 1.38**, which implicitly assumes a couple of more facts but is shorter: By hypothesis, we have

$$1 = \sqrt{1} = \sqrt{n^2} = |n|. \tag{1.27}$$

This implies that $n = 1$ or $n = -1$. Since n is negative, $n = -1$. So we are done since $(-1)^2 = 1$.

Exercise 1.22. *Prove that there exists a unique positive integer n such that $n^2 = 1$.*

1.4.2. Division and its elementary properties. Recall that we say that an integer a **divides** an integer b if there exists an integer k such that

$$ak = b. \tag{1.28}$$

We also say that b is a **multiple** of a. Recall that we say that a is a **divisor** of b.

For example, 15 is a multiple of 3 since $3 \cdot 5 = 15$. And 3 is a divisor of 15.

By definition, an integer n is even if and only if n is a multiple of 2. An integer m is odd if and only if m is not a multiple of 2.

Figure 1.4.3 visualizes that 7 does not divide 139.

Theorem 1.37 says that the only positive divisor of 1 is 1 itself.

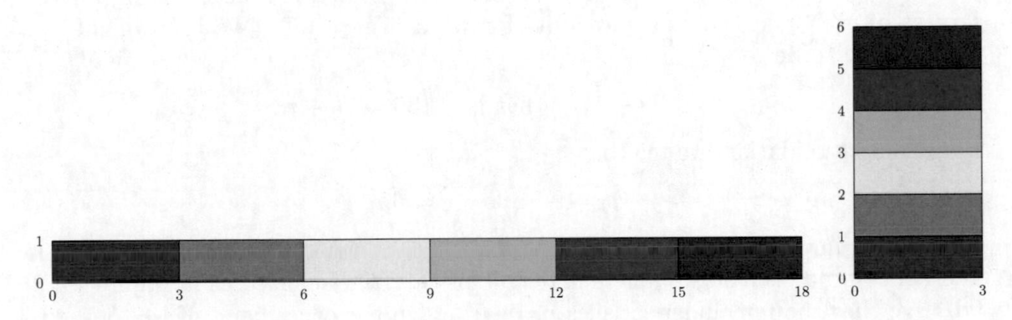

Figure 1.4.2. From these rectangles, we see that 18 is divisible by 3 and 6. The only other non-trivial way to factor 18 is as 2 times 9. So the positive divisors of 18 are $\mathrm{Div}(18) = \{1, 2, 3, 6, 9, 18\}$.

Observe that 7 divides 14 and that 5 divides 15, so the product $7 \cdot 5$ divides the product $14 \cdot 15$. More generally, we have the following.

Solved Problem 1.39 (Division and products). *All quantities are integers. If a divides b and if c divides d, then ac divides bd.*

Solution. Since a divides b and c divides d, there exist integers k and ℓ such that

$$b = ka, \qquad d = \ell c. \tag{1.29}$$

Thus, multiplying these two integers, we obtain

$$bd = ka\ell c = k\ell ac \tag{1.30}$$

(where we used the commutativity of multiplication for the second equality). Since $k\ell$ is an integer, we conclude that ac divides bd by the definition of divides.

Exercise 1.23. *Show that if an integer a divides an integer b and if c is an integer, then a divides bc. Hint: You may pattern your proof after the proof of Theorem* 1.28.

Another way to say this is: If a divides b and if d is a multiple of b, then a divides d.

Show that, as a special case of this statement, we have: If a divides b, then a divides b^2.

Exercise 1.24. *Observe that* 3 *divides* 6*, and related to this,* 9 *(the square of* 3*) divides* 36 *(the square of* 6*).*

Generalizing this, prove: If 3 *divides* b*, then* 9 *divides* b^2*.*

Generalizing again, prove: If a *divides* b*, then* a^2 *divides* b^2*.*

Although the following has been observed earlier in this chapter, it is good for you to work it out again as an exercise.

Exercise 1.25. *Prove: If* a *divides* b *and if* b *divides* c*, then* a *divides* c*.*

1.4.3. The statement of the Division Theorem. Consider for example the problem of dividing 139 by 7 to get a remainder. That is, we are looking for the *largest* integer q, called the **quotient**, for which $7q \le 139$. We then define the **remainder** to be

$$r = 139 - 7q; \quad \text{that is,} \quad 139 = 7q + r. \tag{1.31}$$

By a short calculation, we see that

$$q = 19 \quad \text{and} \quad r = 6.$$

Indeed, we have $139 = 7 \cdot 19 + 6$, whereas $q = 20$ would yield $7q = 140 > 139$ and hence produce a negative remainder. This can be visualized as in Figure 1.4.3. Observe that the remainder satisfies $r = 6 < 7$. On the other hand, if we took q to be smaller, e.g., $q = 18$, then we would obtain a remainder $r \ge 7$, e.g., in this case $r = 13$. To summarize, $q = 19$ and $r = 6$ are the unique integers satisfying

$$139 = 7q + r \quad \text{and} \quad 0 \le r < 7.$$

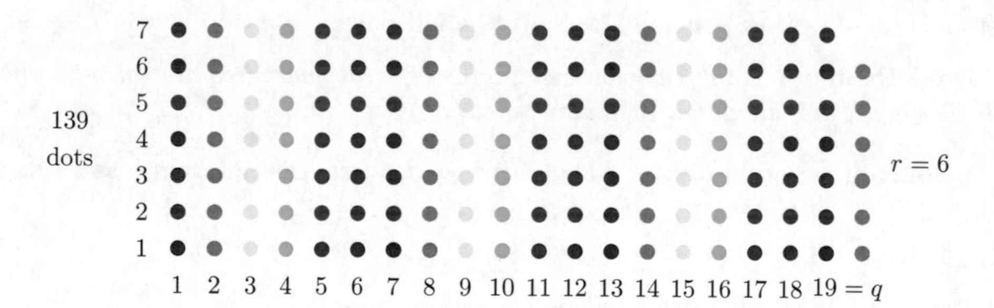

Figure 1.4.3. A visual rendering of the Division Theorem example that $139 = 7q + r$ has quotient $q = 19$ and remainder $r = 6$.

The general statement of the **Division Theorem**, which we prove in Chapter 4, is the following.

Theorem 1.40. *Let* a *be an integer and let* m *be a positive integer. Then there are unique integers* q *and* r *such that*

$$a = m\,q + r \quad \text{and} \quad 0 \le r < m. \tag{1.32}$$

That is, if we divide an integer a *by a positive integer* m*, then we get a unique quotient* q *and remainder* r*.*

By considering the special case $m = 2$ of the Division Theorem and since $0 \leq r < 2$ means the same thing as $r = 0$ or $r = 1$, we have:

Corollary 1.41. *Let a be an integer. Then there are unique integers q and r such that*

$$a = 2q + r \quad \text{and} \quad r = 0 \text{ or } r = 1. \tag{1.33}$$

Furthermore, $r = 0$ if and only if a is even, and $r = 1$ if and only if a is odd.

Example 1.42. We can use the Division Theorem to show that the integer 139 is a prime. To see this, we just need to check that none of the primes $2, 3, 5, 7, 11$ divide 139, since these are the primes less than 12, and $12^2 = 144 > 139$. We can see this as follows. By the Division Theorem, we may divide 139 by a prime p to get

$$139 = p \cdot q + r, \tag{1.34}$$

where the remainder r satisfies

$$0 \leq r < p.$$

Then the table in Figure 1.4.4 shows that none of the remainders r is equal to zero. Hence 139 is prime.

p	q	r
2	69	1
3	46	1
5	27	4
7	19	6
11	12	7

Figure 1.4.4. Table of values of primes $p \leq \sqrt{139}$, factors q, and remainders r, where $139 = p \cdot q + r$. Since the last column has no zeroes, 139 is a prime.

For example, the first row of Figure 1.4.4 is because $139 = 2 \cdot 69 + 1$ and $0 \leq 1 < 2$.

Remark 1.43. Without using the Division Theorem, one can show that 139 is prime by verifying that $\frac{139}{p}$ is not an integer for each of $p = 2, 3, 5, 7, 11$.

1.5. Greatest common divisor

Given two integers, what do they have in common from the point of view of *multiplication*? The answer is given by the greatest common divisor. We begin by discussing examples and then we give the formal definition of the greatest common divisor.

1.5.1. Definition of the greatest common divisor.

What do the integers 51 and 68 have in common from the point of view of multiplication? Firstly, we observe that 17 divides 51 and 51 is a multiple of 17. This is because $17 \cdot 3 = 51$. By inspection, we see that the set of positive divisors of 51 is

$$\text{Div}(51) := \{1, 3, 17, 51\}. \tag{1.35}$$

Secondly, we also observe that 17 divides 68. The set of positive divisors of 68 is

$$\mathrm{Div}(68) := \{1, 2, 4, 17, 34, 68\}. \tag{1.36}$$

If we compare the lists of divisors for both 51 and 68, we see that the common divisors of 51 and 68 are 1 and 17. Thus, the *greatest common divisor* of 51 and 68 is the number 17. We write this as

$$\gcd(51, 68) = 17. \tag{1.37}$$

That is, the largest integer that divides both 51 and 68 is the number 17.

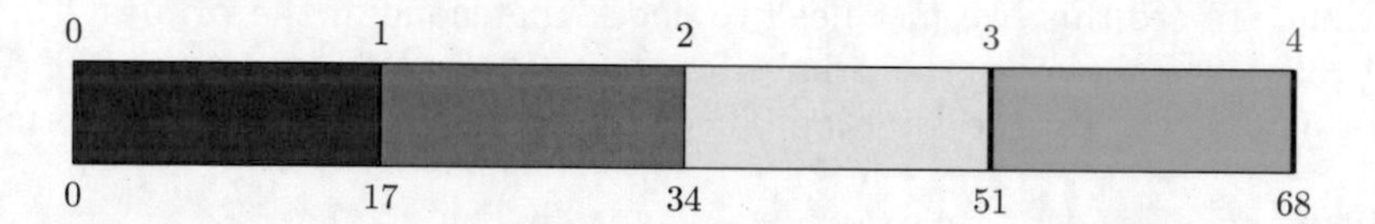

Figure 1.5.1. The gcd of 51 and 68 is 17: after factoring 51 and 68 by 17 we get 3 and 4, which are coprime.

From the examples of 51 and 68, you may have anticipated the following.

Exercise 1.26. *Prove that a divisor of a positive integer b that is not equal to b cannot be more than one-half of b.*

We will find the notion of the maximum of two (or sometimes more) numbers useful. We embark on a brief excursion regarding the maximum. The **maximum** of two real numbers a and b is defined by

$$\max\{a, b\} := \begin{cases} a & \text{if } a \geq b, \\ b & \text{if } a < b. \end{cases} \tag{1.38}$$

Similarly, we define the **minimum** of two real numbers a and b by

$$\min\{a, b\} := \begin{cases} b & \text{if } a \geq b, \\ a & \text{if } a < b. \end{cases} \tag{1.39}$$

Solved Problem 1.44 (Being bigger than the bigger of the two is the same as being bigger than both!). *Let a and b be real numbers.*

(1) *Prove that* $\max\{a, b\} \geq a$ *and* $\max\{a, b\} \geq b$.

(2) *Prove that for every $x \in \mathbb{R}$,*

$$x \geq \max\{a, b\} \quad \textit{if and only if} \quad (x \geq a \textit{ and } x \geq b). \tag{1.40}$$

Solution. Let $m := \max\{a, b\}$.

(1) *Case i*: $a \geq b$. In this case, $m = a \geq b$, so $m \geq a$ and $m \geq b$.

Case ii: $a < b$. In this case, $m = b > a$, so again $m \geq a$ and $m \geq b$.

Since for every real number a and b, we have $a \geq b$ or $a < b$, we have proved (1). Aside: Two elementary facts which we have used are: $x = y$ implies $x \geq y$ and $x > y$ implies $x \geq y$.

(2) (**Only if**) Suppose $x \geq m$. Then, by part (1) we have $x \geq m \geq a$ and $x \geq m \geq b$. Thus, by the transitivity of the relation $\geq$, we conclude that $x \geq a$ and $x \geq b$.

(If) Suppose $x \geq a$ and $x \geq b$.

Case i: $a \geq b$. In this case, $m = a \leq x$, so $x \geq m$.

Case ii: $a < b$. In this case, $m = b \leq x$, so again $x \geq m$.

This proves (2). □

Exercise 1.27. *Let a and b be real numbers.*

(1) *Prove that* $\min\{a, b\} \leq a$ *and* $\min\{a, b\} \leq b$.

(2) *Prove that for every* $x \in \mathbb{R}$,

(1.41) $$x \leq \min\{a, b\} \quad \text{if and only if} \quad (x \leq a \text{ and } x \leq b).$$

Solved Problem 1.45. *Let a and b be positive odd integers greater than* 1. *Prove that*

(1.42) $$ab \geq \max\{3a, 3b\}.$$

Solution. Since a is an odd integer and $a > 1$, we have $a \geq 3$. Indeed, 2 is even, and hence not odd. For the same reasons, $b \geq 3$. Thus by (1.16),

(1.43) $$ab \geq 3b \quad \text{and} \quad ab \geq a3.$$

Since we have both $ab \geq 3b$ and $ab \geq 3a$, we conclude by Solved Problem 1.44(2) that $ab \geq \max\{3a, 3b\}$. □

Exercise 1.28. *Let a and b be positive integers. Prove that*

(1.44) $$ab \geq \max\{a, b\}.$$

Hint: Any positive integer is at least 1.

Example 1.46 (Positive divisors of an integer)**.** If a positive integer a divides 1575, then $a \leq 1575$ by Lemma 1.8. One can also calculate (or at least verify!) that

(1.45) $$1575 = 3^2 \cdot 5^2 \cdot 7.$$

In particular, the prime divisors of 1575 are 3, 5, and 7. One can check that the positive divisors of 1575 are

(1.46) $$1,\ 3,\ 3^2,\ 5,\ 3 \cdot 5,\ 3^2 \cdot 5,\ 5^2,\ 3 \cdot 5^2,\ 3^2 \cdot 5^2,$$
$$7,\ 3 \cdot 7,\ 3^2 \cdot 7,\ 5 \cdot 7,\ 3 \cdot 5 \cdot 7,\ 3^2 \cdot 5 \cdot 7,\ 5^2 \cdot 7,\ 3 \cdot 5^2 \cdot 7,\ 3^2 \cdot 5^2 \cdot 7.$$

Calculating the numerical values of this list of positive divisors yields

(1.47) $$1,\ 3,\ 9,\ 5,\ 15,\ 45,\ 25,\ 75,\ 225,$$
$$7,\ 21,\ 63,\ 35,\ 105,\ 315,\ 175,\ 525,\ 1575.$$

Observe that the integers in (1.46) are exactly of the form

(1.48) $$3^a \cdot 5^b \cdot 7^c,$$

where $0 \leq a \leq 2$, $0 \leq b \leq 2$, and $0 \leq c \leq 1$. That is, the list of integers in (1.46) consists of $3^a \cdot 5^b \cdot 7^c$, where the triples (a, b, c) are given by

(1.49)
$$(0,0,0),\ (1,0,0),\ (2,0,0),\ (0,1,0),\ (1,1,0),\ (2,1,0),\ (0,2,0),\ (1,2,0),\ (2,2,0),$$

(1.50)
$$(0,0,1),\ (1,0,1),\ (2,0,1),\ (0,1,1),\ (1,1,1),\ (2,1,1),\ (0,2,1),\ (1,2,1),\ (2,2,1).$$

Exercise 1.29. *Can you come up with a conjecture about the relationship between factoring a number as a product of primes and the list of positive divisors of that number?*

Given a positive integer a, let $\mathrm{Div}(a)$ denote the set of positive divisors of a.

Given $m \in \mathbb{Z}^+$, let

$$\mathbb{N}_m := \{1, 2, 3, \ldots, m\} \tag{1.51}$$

be the set of the first m positive integers.

By Lemma 1.8, any positive divisor of a is an integer between 1 and a, inclusive, so that

$$\mathrm{Div}(a) \subset \mathbb{N}_a. \tag{1.52}$$

Given two non-zero integers a and b, the set of **common divisors** of a and b is

$$\mathrm{ComDiv}(a, b) = \mathrm{Div}(a) \cap \mathrm{Div}(b). \tag{1.53}$$

Here, the symbol $\cap$ is denotes "intersection". The **intersection** of two sets A and B is defined to be the set of elements that are in both sets.[6] So $\mathrm{ComDiv}(a, b)$ is the set of integers c such that c is a divisor of both a and b. For example, by (1.35) and (1.36), we see that

$$\mathrm{ComDiv}(51, 68) = \{1, 17\}$$

since 1 and 17 are the only integers in both $\mathrm{Div}(51)$ and $\mathrm{Div}(68)$.

We leave it to you to check that

$$\mathrm{ComDiv}(18, 24) = \{1, 2, 3, 6\}, \tag{1.54}$$

and so $\gcd(18, 24) = 6$.

Observe that for all positive integers a and b,

$$\mathrm{ComDiv}(a, b) \subset \mathbb{N}_a \cap \mathbb{N}_b = \mathbb{N}_{\min\{a,b\}}. \tag{1.55}$$

For example, $\mathrm{ComDiv}(51, 68) \subset \mathbb{N}_{51}$.

If one of the integers, say a, is equal to zero, then we have the following:

$$\mathrm{Div}(0) = \mathbb{Z}, \qquad \mathrm{ComDiv}(0, b) = \mathrm{Div}(b). \tag{1.56}$$

In particular, for every two non-negative integers a and b, where at most one of them is zero, there always exists a **greatest common divisor**, which we denote by

$$\gcd(a, b). \tag{1.57}$$

Exercise 1.30. *Show that* $\gcd(5 \cdot 17, 23 \cdot 17) = 17$.

A potentially divisive conversation that ends up finding common ground:

> Alpha: What do we have in common?
> Beta: I don't know. What are your divisors?

And the conversation continues. Can you fill in how it might go?

[6] We will discuss set theory in more detail, including the notation of intersection, in Chapter 5.

Solved Problem 1.47 (Odd integers have nothing in common with 2 multiplicatively). *Let n be an odd integer. Prove that* $\gcd(2, n) = 1$.

Solution. We have that $g := \gcd(2, n)$ is a positive integer dividing 2. Thus $g = 1$ or $g = 2$. Suppose for a contradiction that $g = 2$. Then 2 divides n, which implies that n is even, which contradicts the assumption that n is odd. Therefore $g \neq 2$, so that $g = 1$. □

Solved Problem 1.48 (The gcd and multiples). *Prove that if a is a positive integer, then*

$$\gcd(a, 6) \leq \gcd(a, 30). \tag{1.58}$$

Solution. Let $g := \gcd(a, 6)$. Then g divides a, and g divides 6. Since 6 divides 30, this implies that g divides 30. Therefore g divides both a and 30, which implies that $g \leq \gcd(a, 30)$ by the definition of the greatest common divisor of a and 30. □

The property above generalizes as follows.

Exercise 1.31. *Let $\mathbb{Z}^+$ be the domain of discourse; that is, the variables a, b, d we discuss below are all assumed to be positive integers. Prove that if d divides a, then*

$$\gcd(d, b) \leq \gcd(a, b).$$

1.5.2. Rational numbers. So far, we have stayed in the universe of integers. It is time to expand this universe!

1.5.2.1. *The definition of a rational number.* We say that a real number r is a **rational number** if it can be expressed in the form

$$r = \frac{a}{b}, \tag{1.59}$$

where a and b are integers. For this to make sense, we assume that $b \neq 0$. For example, $-\frac{17}{3} = \frac{17}{-3}$ is a rational number.

Example 1.49. (1) Integers are rational numbers. Indeed, if a is an integer, then we may write it as $a = \frac{a}{1}$.

(2) Suppose that $a = mq + r$, where $0 < r < m$. Then $\frac{a}{m}$ is a rational number which is not an integer. Indeed, we have $q < \frac{a}{m} < q + 1$.

1.5.2.2. *Fractions in lowest terms.* We can understand rational numbers by using division. Let $g = \gcd(a, b)$. Given any rational number $\frac{a}{b}$, we define its **lowest terms** to be the fraction

$$\frac{a/g}{b/g}. \tag{1.60}$$

Here, we are considering a/g and b/g as integers. For example, the lowest terms of $\frac{51}{68}$ is $\frac{3}{4}$ since $51/17 = 3$ and $68/17 = 4$. Observe that $\gcd(3, 4) = 1$.

The following result explains why the lowest terms of a fraction has the gcd of the numerator and denominator equal to 1.

Theorem 1.50. *For all positive integers a and b, we have that*

$$\gcd\left(\frac{a}{g}, \frac{b}{g}\right) = 1, \tag{1.61}$$

where $g = \gcd(a, b)$.

Proof. Suppose that c is a common divisor of $\frac{a}{g}$ and $\frac{b}{g}$. By definition, this means that there exist integers k and ℓ such that

$$ck = \frac{a}{g}, \qquad c\ell = \frac{b}{g}. \tag{1.62}$$

By multiplying these equations by g, we may rewrite this as

$$(gc)k = a, \qquad (gc)\ell = b. \tag{1.63}$$

This tells us that gc is a common divisor of a and b. By the definition of g as the greatest common divisor of a and b, we obtain that

$$gc \leq g. \tag{1.64}$$

Since $g > 0$, **this implies that $c \leq 1$.** This proves that the greatest common divisor of $\frac{a}{g}$ and $\frac{b}{g}$ is 1. □

Figure 1.5.1 exhibits the fact that

$$\frac{51}{68} = \frac{3}{4}.$$

Exercise 1.32. *If $\frac{a}{b}$ is in lowest terms, can both a and b be even? Explain your answer.*

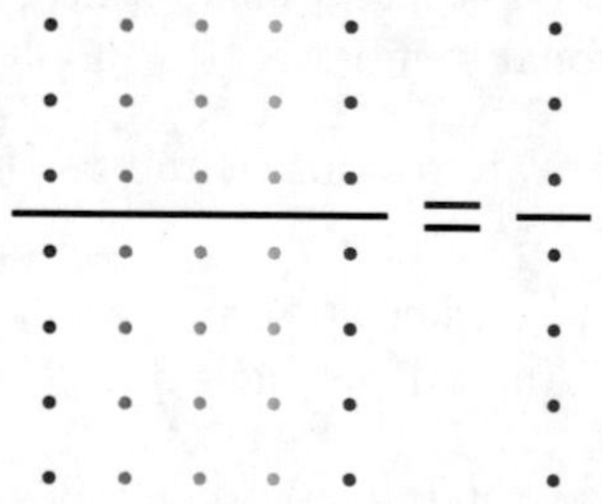

Figure 1.5.2. Visualizing Theorem 1.50: Imagining the rational number $\frac{15}{20}$ in lowest terms by dividing both the numerator and denominator by $\gcd(15, 20) = 5$ to obtain $\frac{3}{4}$, where $\gcd(3, 4) = 1$.

Exercise 1.33. *Find the lowest terms expressions for the following rational numbers:*

$$\frac{68}{51}, \frac{936}{324}, \frac{1748}{4199}, \frac{1139163}{987654321}. \tag{1.65}$$

1.5.3. Coprime integers. Just as it is interesting when two integers have something in common, it is interesting when two integers have nothing to do with each other, multiplicatively that is.

Definition 1.51. We say that integers a and b, not both zero, are **coprime** if $\gcd(a, b) = 1$.

In particular, two distinct primes are necessarily coprime. (For example, we have $\gcd(5, 17) = 1$.) Indeed, suppose that p and q are distinct (not equal) primes. Let $g = \gcd(p, q)$ be the greatest common divisor of p and q. Since g is a positive divisor of the prime p, we have $g = 1$ or $g = p$. Suppose for a contradiction that $g \neq 1$. Then $g = p$. Since g divides q, this implies that p divides q. Since q is a prime and since $p > 1$, we conclude that $p = q$, a contradiction. Therefore $g = 1$.

Exercise 1.34. *Let p and q be non-equal prime numbers. Prove that* $\gcd(p^2, q^2) = 1$.

More generally, it turns out that two integers are coprime if (and only if) they do not have any prime divisors in common. For example, the two integers

$$2^7 \cdot 3^4 \cdot 11 \quad \text{and} \quad 5^2 \cdot 7^8 \cdot 19$$

are coprime, for the prime divisors of the first integer are $2, 3, 11$ and the prime divisors of the second integer are $5, 7, 19$, and these two sets of primes contain no common primes. In other words, positive integers m and n are coprime means that if all the primes making up m disappeared from the list of all primes, then n would not be affected.

Flippantly, we may say that for molecules, water H_2O and salt NaCl are coprime, but carbon dioxide CO_2 and methane CH_4 are not coprime. ☺

Theorem 1.50 says that for every two integers a and b, where at most one of them is zero:

The integers a/g and b/g are coprime, where g is the gcd of a and b.

See Chapter 4 for more discussion about coprime integers.

1.5.4. Division and integral linear combinations. So far, we have studied some elementary properties of division restricted to the realm of the binary operation of multiplication. It is time to add the binary operation of addition into the mix.

Regarding division, observe that 7 divides 21 and 7 divides 35. From this we can show 7 divides $21m + 35n$ for every pair of integers m, n. Indeed, let m and n be integers. We calculate that

$$21m + 35n = 7(3m + 5n).$$

Since $3m + 5n$ is an integer, we conclude that 7 divides $21m + 35n$.

We will find the following generalization of this rather useful. Why this is useful should not be immediately apparent to you!

Theorem 1.52. *Suppose that c is a common divisor of integers a and b. Then for all integers m and n, we have that c divides the* ***integral linear combination***

$$ma + nb$$

of a and b.

Proof. By hypothesis, there exist integers k and ℓ such that $ck = a$ and $c\ell = b$. Hence

$$ma + nb = m(ck) + n(c\ell) = c(mk + n\ell). \tag{1.66}$$

Since $mk + n\ell$ is an integer, by the definition of division this proves the desired conclusion. □

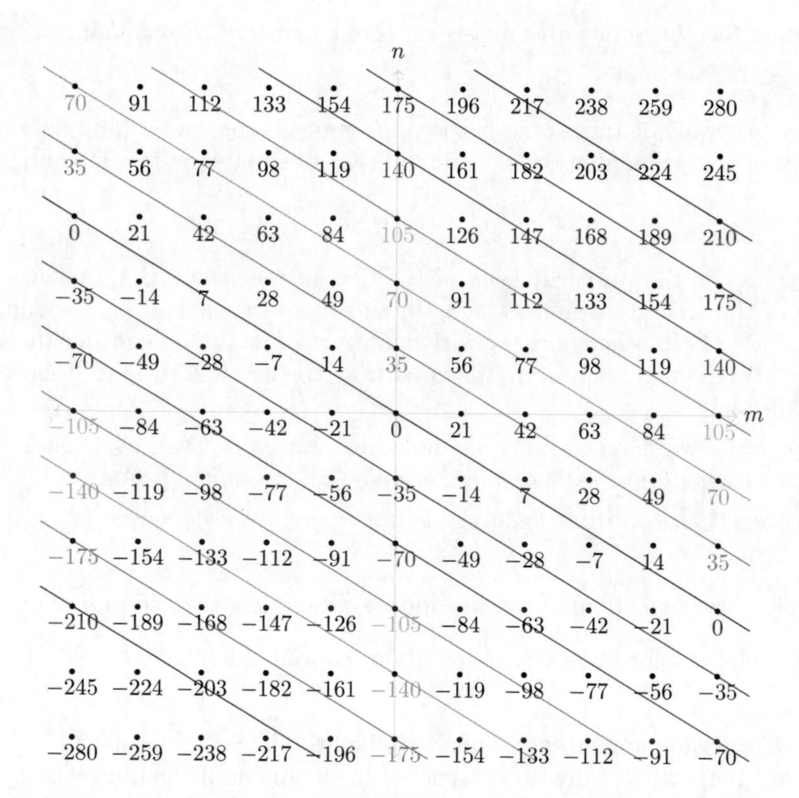

Figure 1.5.3. Visualizing Theorem 1.52: At each point (m, n) the integral linear combination $21m + 35n$ is displayed right below it. For example, $-133 = (-3)21 + (-2)35$ is displayed right below the point $(-3, -2)$. Since $\gcd(21, 35) = 7$, each integral linear combination is a multiple of 7. Horizontally we count by 21's and vertically we count by 35's.

1.5.5. Example of a linear Diophantine equation.

It is easy to solve linear equations for real numbers. What about for integers? We consider this question in detail in Chapter 4. For now, we just consider an example.

Solved Problem 1.53 (Solving a linear Diophantine equation by hand). *Find integers m and n such that*

$$9m + 11n = 1. \tag{1.67}$$

That is, find an integral linear combination of 9 *and* 11 *that is equal to* 1.

Solution. Here is a simple approach. Counting by 9's yields

$$9, 18, 27, 36, 45, \ldots, \tag{1.68}$$

while counting by 11's yields

$$11, 22, 33, 44, \ldots. \tag{1.69}$$

Noticing that 45 is one more than 44, we see that

$$9(5) + 11(-4) = 1, \tag{1.70}$$

so that $m = 5$, $n = -4$ gives a solution to $9m + 11n = 1$.

Since we were not asked to find *all* solutions to the equation, we just found a *particular* solution. See the following two exercises for examples of the more general question.

Exercise 1.35. *Verify that for every integer k,*

$$m = 5 + 11k, \qquad n = -4 - 9k \tag{1.71}$$

is a solution to (1.67). *We will see in Chapter* 4 *below that this is the general solution (that is, there are no other solutions)!*

Exercise 1.36. *By trial and error, find integers m and n such that*

$$13m + 19n = 1. \tag{1.72}$$

Using the above, find an integer solution to

$$13m + 19n = 5. \tag{1.73}$$

Call your solution (m_0, n_0). Is (m_0+19, n_0-13) also an integer solution? Can you find infinitely many integer solutions?

1.6. Statement of prime factorization

Have you answered the question in the caption for Figure 1.0.3 at the beginning of this chapter? If not, this section will help!

1.6.1. Atomic theory.

Primes are the building blocks (dare we say atoms?) for the set of positive integers from the point of view of multiplication.

It turns out that (we prove this in later chapters):

Theorem 1.54. *Any integer $n \geq 2$ may be written (uniquely) as the product of primes.*

This is called the **prime factorization** of n, and this result is also known as the Fundamental Theorem of Arithmetic. Here are the prime factorizations of the integers from 2 to 19:

$$\begin{array}{cccccc} 2 = 2^1, & 3 = 3^1, & 4 = 2^2, & 5 = 5^1, & 6 = 2 \cdot 3, & 7 = 7^1, \\ 8 = 2^3, & 9 = 3^2, & 10 = 2 \cdot 5, & 11 = 11^1, & 12 = 2^2 \cdot 3, & 13 = 13^1, \\ 14 = 2 \cdot 7, & 15 = 3 \cdot 5, & 16 = 2^4, & 17 = 17^1, & 18 = 2 \cdot 3^2, & 19 = 19^1. \end{array}$$

As a reminder (we've referred to this result earlier), we have:

Corollary 1.55. *For every integer $n \geq 2$, there exists a prime p divisor of n.*

Proof. Let $n \geq 2$ be an integer. By Theorem 1.54, there exist primes $p_1, p_2, \ldots, p_k$ (where the primes may not be distinct) such that

$$n = p_1 \cdot p_2 \cdots p_k. \tag{1.74}$$

So we have that p_1 is a prime divisor of n. □

Exercise 1.37. *Find the prime factorizations of all non-prime integers $2 \leq n \leq 100$. Hint: In §1.6 we have considered $n < 20$. So you may start with $n = 20$. Consider the green integers in Figure 1.1.2. Hint: We give the answer below, but again no peeking!*

1	2	3	4	5	6	7	8	9	10
11	12	13	14	15	16	17	18	19	20
21	22	23	24	25	26	27	28	29	30
31	32	33	34	35	36	37	38	39	40
41	42	43	44	45	46	47	48	49	50
51	52	53	54	55	56	57	58	59	60
61	62	63	64	65	66	67	68	69	70
71	72	73	74	75	76	77	78	79	80
81	82	83	84	85	86	87	88	89	90
91	92	93	94	95	96	97	98	99	100

Figure 1.6.1. A partially drawn visualization of the prime factorization for the integers 1 to 100. (The filled squares are the integers 2 through 20, as well as the primes and squares less than 50.) The primes are framed in thick lines. Note that no prime greater than 50 is a divisor of a composite number less than or equal to 100.

Exercise 1.38. *Find the prime factorization of 3233. You can use an online applet.*

With the help of Figure 1.1.2, we give a table of prime factorizations of the integers from 2 to 100, inclusive:

(1.75)

$$\begin{array}{cccccccccc}
 & 2 & 3 & 2^2 & 5 & 2\cdot 3 & 7 & 2^3 & 3^2 & 2\cdot 5 \\
11 & 2^2 3 & 13 & 2\cdot 7 & 3\cdot 5 & 2^4 & 17 & 2\cdot 3^2 & 19 & 2^2 5 \\
3\cdot 7 & 2\cdot 11 & 23 & 2^3 3 & 5^2 & 2\cdot 13 & 3^3 & 2^2 7 & 29 & 2\cdot 3\cdot 5 \\
31 & 2^5 & 3\cdot 11 & 2\cdot 17 & 5\cdot 7 & 2^2 3^2 & 37 & 2\cdot 19 & 3\cdot 13 & 2^3 5 \\
41 & 2\cdot 3\cdot 7 & 43 & 2^2 11 & 3^2 5 & 2\cdot 23 & 47 & 2^4 3 & 7^2 & 2\cdot 5^2 \\
3\cdot 17 & 2^2 13 & 53 & 2\cdot 3^3 & 5\cdot 11 & 2^3 7 & 3\cdot 19 & 2\cdot 29 & 59 & 2^2 3\cdot 5 \\
61 & 2\cdot 31 & 3^2 7 & 2^6 & 5\cdot 13 & 2\cdot 3\cdot 11 & 67 & 2^2 17 & 3\cdot 23 & 2\cdot 5\cdot 7 \\
71 & 2^3 3^2 & 73 & 2\cdot 37 & 3\cdot 5^2 & 2^2 19 & 7\cdot 11 & 2\cdot 3\cdot 13 & 79 & 2^4 5 \\
3^4 & 2\cdot 41 & 83 & 2^2 3\cdot 7 & 5\cdot 17 & 2\cdot 43 & 3\cdot 29 & 2^3 11 & 89 & 2\cdot 3^2 5 \\
7\cdot 13 & 2^2 23 & 3\cdot 31 & 2\cdot 47 & 5\cdot 19 & 2^5 3 & 97 & 2\cdot 7^2 & 3^2 11 & 2^2 5^2
\end{array}$$

We prove the "existence" part of the Prime Factorization Theorem 1.54 (a.k.a. Fundamental Theorem of Arithmetic) in §2.4.3 below. The "uniqueness" part is proved in §4.8.

1.6.2. Euclid's theorem that there are infinitely many primes. Bountiful primes! We can use division and a clever idea (using the power of proof by contradiction!) to obtain the following result stated at the beginning of this chapter.

Theorem 1.56 (Euclid's Theorem). *There are infinitely many primes.*

Proof. Suppose for a contradiction that there are only a finite number of primes in totality. Let n be the total number of primes, and denote the distinct primes by

$$p_1, p_2, \ldots, p_n. \tag{1.76}$$

Let

$$m = p_1 p_2 \cdots p_n + 1. \tag{1.77}$$

Since $m \geq 2$, by the Prime Factorization Theorem 1.54, there is some prime p_i, where $1 \leq i \leq n$, which divides m.

Since p_i also divides $p_1 p_2 \cdots p_n$, by Theorem 1.52 we obtain that p_i divides the difference

$$m - p_1 p_2 \cdots p_n = 1. \tag{1.78}$$

Since $p_i > 1$, this is a contradiction to Theorem 1.37. Hence there are infinitely many primes. □

1.6.3. How to make a quick buck. Primes have a number of important real-world applications. And they are fun to boot! Here is a game you can play. Choose two really large primes p and q, and multiply them together to get pq. Tell your friend the integer pq and also tell them that this integer is the product of two primes and ask them to find the two primes. If you can find two primes large enough so that they can never win the game, then you have a real world application: The RSA algorithm in cryptography is based on the difficulty of finding the prime

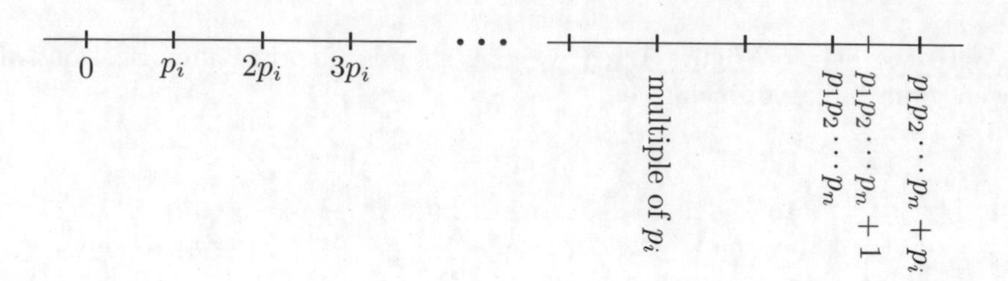

Figure 1.6.2. If there are only a finite number of primes, there exists an i such that $p_1p_2 \cdots p_n + 1$ divided by p_i has remainder both 0 and 1, which is a contradiction.

factorization for a large number that is the product of two large primes (called a semiprime). The basics of the RSA algorithm are discussed in §6.11* below.

If you are good at factoring semiprimes, you can make money: RSA Factoring Challenge. For example, the first RSA factoring challenge number is

$$\begin{gathered}1522605027922533360535618378132637429718068114961 3\\ 80688657908494580122963258952897654000350692006139.\end{gathered}$$

Sorry for writing a single number on two lines; it was too long to write on one line! This number was factored as

$$\begin{gathered}37975227936943673922808872755445627854565536638199\\ \times 40094690950920881030683735292761468389214899724061\end{gathered}$$

on April Fool's Day (of all days!) of 1991 by Arjen K. Lenstra.

You will make 200,000 USD if you can find the factorization of the following number (if it has not already been done by the time you read this):

2519590847565789349402718324004839857142928212620403202777713783604366 20
2070759555626401852588078440691829064124951508218929855914917618450280 84
8912007284499268739280728777673597141834727026189637501497182469116507 76
1337985909570009733045974880842840179742910064245869181719511874612151 51
7265463228221686998754918242243363725908514186546204357679842338718477 44
4792073993423658482382428119816381501067481045166037730605620161967625 61
3384414360383390441495263443219011465754445417842402092461651572335077 87
0774981712577246796292638635637328991215483143816789988504044536402352 73
81951378636564391212010397122822120720357.

This number is indeed the product of two primes p and q. Presumably the presenters of the prize are the only ones who know the answer and they are bound to secrecy. Your challenge is to find p and q.

One of the difficulties is current computational power. This is related to Moore's Law.

1.7*. Perfect numbers

Question: What is the perfect number of hours to work per day?
Answer: Six! We will see why below.

Of course, primes are not the only natural numbers that have special properties. Observe that some numbers, like 36, have "lots" of divisors, whereas other numbers, like 39, have "few" divisors. Besides simply counting the number of positive divisors, we can sum the positive divisors of a number. For example, since the divisors of 36 are $1, 2, 3, 4, 6, 9, 12, 18, 36$, the sum of the positive divisors of 36 not counting 36 itself is "too much" in the sense that

$$1 + 2 + 3 + 4 + 6 + 9 + 12 + 18 = 55 > 36.$$

On the other hand, the divisors of 39 are $1, 3, 13, 39$, and their sum not counting 39 is "too little" in the sense that

$$1 + 3 + 13 = 17 < 39.$$

When something is neither too much nor too little, we say that is perfect, so:

Definition 1.57. We say that a positive integer n is a **perfect number** if n is equal to the sum of its positive divisors less than n.

For example, 6 is a perfect number since its positive divisors are $1, 2, 3, 6$ and

$$1 + 2 + 3 = 6. \tag{1.79}$$

Perfect! We also have that 28 is a perfect number since its positive divisors are $1, 2, 4, 7, 14, 28$ and

$$1 + 2 + 4 + 7 + 14 = 28. \tag{1.80}$$

Perfect again!

Here are the first few perfect numbers:

$$6, 28, 496, 8128, 33550336, 8589869056, 137438691328. \tag{1.81}$$

For example, the positive divisors of 496 are

$$1, 2, 4, 8, 16, 31, 62, 124, 248, 496, \tag{1.82}$$

and

$$496 = 1 + 2 + 4 + 8 + 16 + 31 + 62 + 124 + 248. \tag{1.83}$$

Remarkably, it is unknown if there exist *odd* perfect numbers.

Conjecture 1.58 (Nicomachus). *All perfect numbers are even.*

Exercise 1.39. (1) *Prove that if p is a prime, then p is not a perfect number.*

(2) *Prove that if p and q are primes, then pq is a perfect number if and only if $pq = 6$, i.e., one of p, q is 2 and one of p, q is 3.*

Exercise 1.40. *We know that 6 and 28 are perfect numbers. Prove that there are no perfect numbers n satisfying $6 < n < 28$.*

Figure 1.7*.1. Nichomachus. Line engraving by P. Ghigi after L. Agricola after Raphael. Wellcome Collection, https://wellcomecollection.org/works/v8swpwtt. Licensed under Creative Commons, https://creativecommons.org/publicdomain/mark/1.0. Public Domain.

1.8*. One of the Mersenne conjectures

It is cool to look at primes of various special forms. Remarkably, this topic is related to perfect numbers. A **Mersenne prime** is a prime p that is equal to one less than a power of two. That is, a prime of the form

$$p = 2^n - 1,$$

where n is some positive integer. Recall, by Theorem 1.16, n must be a prime for p to be a prime. For example, since 100 is not a prime, we know that $2^{100} - 1$ is not a prime.

The Mersenne primes corresponding to the integers

$$2, 3, 5, 7, 13, 17, 19, 31 \tag{1.84}$$

Figure 1.8*.1. Marin Mersenne (1588–1648). Wikimedia Commons, Public Domain.

are

$$3, 7, 31, 127, 8191, 131071, 524287, 2147483647, \tag{1.85}$$

respectively.

Exercise 1.41. *For n equal to the primes 11, 23, and 29, can you find the (non-trivial) prime factorizations of* $p = 2^n - 1$*? You may wish to use an online applet for this!*

Let q be a prime with the property that $M_q := 2^q - 1$ is a Mersenne prime. After the first 8 Mersenne primes in (1.85), the next 8 Mersenne primes are

$$M_{61}, M_{89}, M_{107}, M_{127}, M_{521}, M_{607}, M_{1279}, M_{2203}. \tag{1.86}$$

These integers get large quite quickly. For example, the sixteenth Mersenne prime M_{2203} has 664 digits! This particular Mersenne prime was found by Raphael Robinson in the year 1952.

According to Wikipedia, as of October 2020, only 51 Mersenne primes are known. The largest Mersenne prime known, which has 24862048 digits, is

$$M_{82589933} = 2^{82589933} - 1$$

(by the time you are reading this hopefully it is larger!). See the Great Internet Mersenne Prime Search.

It has been conjectured by Lenstra, Pomerance, and Wagstaff that:

Conjecture 1.59. *There are an infinite number of Mersenne primes.*

This conjecture is at present unsolved, and fame and glory await the solver!

We have the following beautiful result of Euclid and Euler (see Theorem 6.90 below for a proof):

Theorem 1.60.

An even positive integer n is a perfect number

if and only if

$n = (2^p - 1)2^{p-1}$, *where* $2^p - 1$ *is a Mersenne prime!*

For example, the first few perfect numbers can be written as the products (where the first factor is a Mersenne prime):

$$\begin{aligned} 6 &= 3 \cdot 2, \\ 28 &= 7 \cdot 4, \\ 496 &= 31 \cdot 16, \\ 8128 &= 127 \cdot 64, \\ 33550336 &= 8191 \cdot 4096. \end{aligned}$$

1.9*. Twin primes: An excursion into the unknown

Question: What did one twin prime say to the other?
Answer: We are close, but we are not identical.

There are certainly lots of things that we do not know about primes. This includes the following topic, although much amazing progress has been made. A twin prime pair is a pair of primes, one of which is equal to 2 plus the other. For example, the twin prime pairs of primes under 100 are

$$(3,5),\ (5,7),\ (11,13),\ (17,19),\ (29,31),\ (41,43),\ (59,61),\ (71,73). \tag{1.87}$$

The largest twin prime pair under one thousand is

$$(881,883). \tag{1.88}$$

The largest twin prime pair under one million is

$$(999959,999961). \tag{1.89}$$

As of this writing, the largest known twin prime pair is

$$(2996863034895 \cdot 2^{1290000} - 1,\ 2996863034895 \cdot 2^{1290000} + 1). \tag{1.90}$$

This twin prime pair was found on September 14, 2016, by PrimeGrid using the Twin Prime Search.

The following is the Twin Prime Conjecture. As of this writing, this conjecture is still unsolved!

Conjecture 1.61. *There are an infinite number of twin prime pairs.*

A cousin prime pair is a pair of primes, one of which is equal to 4 plus the other. For example, the cousin prime pairs of primes under 100 are

$$(3,7),\ (7,11),\ (13,17),\ (19,23),\ (37,41),\ (43,47),\ (67,71),\ (79,83). \tag{1.91}$$

Currently, it is unknown whether there are an infinite number of cousin prime pairs.

Given a positive even integer k, let us say that a pair of primes is a **k-prime pair** if one of the primes is equal to k plus the other. Then twin and cousin prime pairs are the same as 2- and 4-prime pairs, respectively.

Yitang Zhang proved the following.[7]

---

[7] Zhang's work depends on a number of mathematicians' works including Bombieri, Deligne, Friedlander, Goldston, Heath-Brown, Iwaniec, Pintz, Vinogradov, and Yıldırım.

Theorem 1.62. *There exists a positive even integer k less than* 70 *million (yikes, that's a big number!) such that there are an infinite number of k-prime pairs.*

By the Polymath8 Project, this result has been improved to:

Theorem 1.63. *There exists a positive even integer $k \leq 246$ such that there are an infinite number of k-prime pairs.*

But, as of this writing, we do not know any *specific* such k, even though we know that such a k exists!

1.10*. Goldbach's conjecture

Primes greater than 2 are always odd. So the sum of two such primes is always even. This naturally leads to the following question:

Can we write each positive even integer as the sum of two prime numbers?

Of course, 2 cannot be written as such since although $2 = 0 + 2$ and $2 = 1 + 1$, neither 0 nor 1 are prime numbers. So we should start with the number 4. As the first few cases, we have

$$4 = 2 + 2,\ 6 = 3 + 3,\ 8 = 3 + 5,\ 10 = 3 + 7,\ 12 = 5 + 7,\ 14 = 7 + 7. \tag{1.92}$$

Here are a few more:

$$16 = 3 + 13,\ 18 = 5 + 13,\ 20 = 3 + 17,\ 22 = 3 + 19,\ 24 = 5 + 19. \tag{1.93}$$

Skipping a few to 100, we have

$$100 = 11 + 89,\ 102 = 5 + 97,\ 104 = 7 + 97,\ 106 = 17 + 89,\ 108 = 19 + 89. \tag{1.94}$$

A longstanding and tantalizing conjecture is the following.

Conjecture 1.64 (Goldbach's conjecture). *Every even integer greater than* 2 *is the sum of two primes.*

Here are some more examples of even integers which are the sums of two prime numbers:

$$1000 = 3 + 997 = 17 + 983 = 23 + 977 = \cdots = 491 + 509, \tag{1.95}$$

where there are in total 28 ways to write 1000 as the sum of two primes. For each of the next powers of 10, we just give one sum of primes, likely of very many:

$$
\begin{aligned}
10^4 &= 59 + 9941,\\
10^5 &= 11 + 99989,\\
10^6 &= 17 + 999983,\\
10^7 &= 29 + 9999971,\\
10^8 &= 11 + 99999989.
\end{aligned}
$$

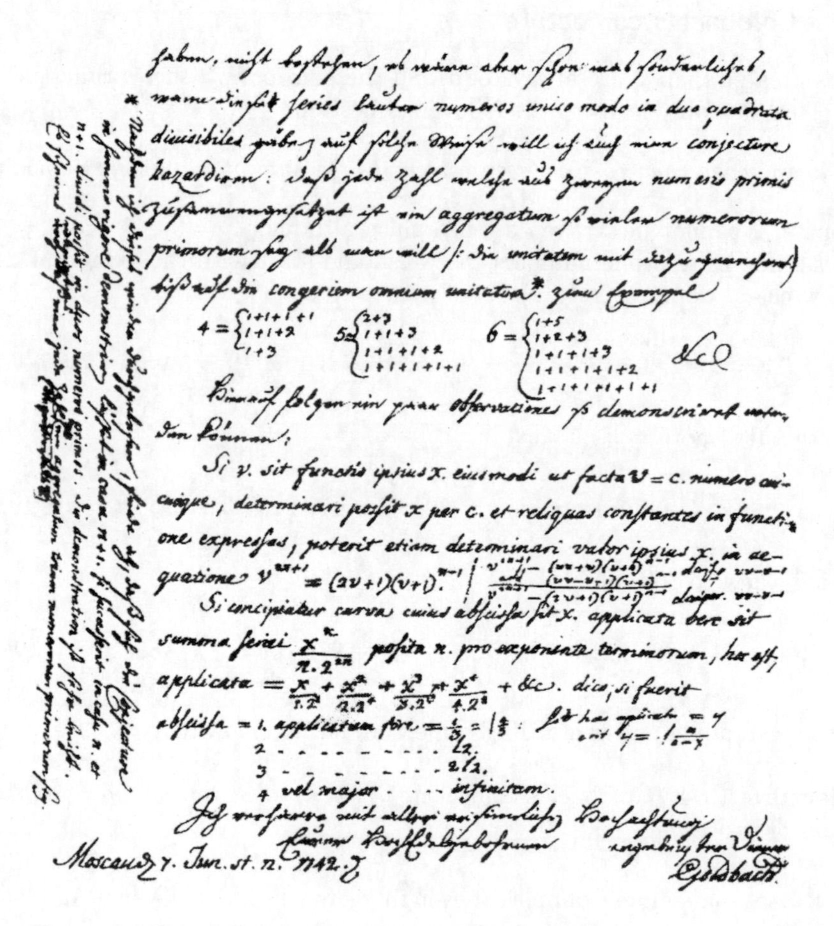

Figure 1.10*.1. Letter from Goldbach to Euler stating his conjecture, dated 1742. Wikimedia Commons, Public Domain.

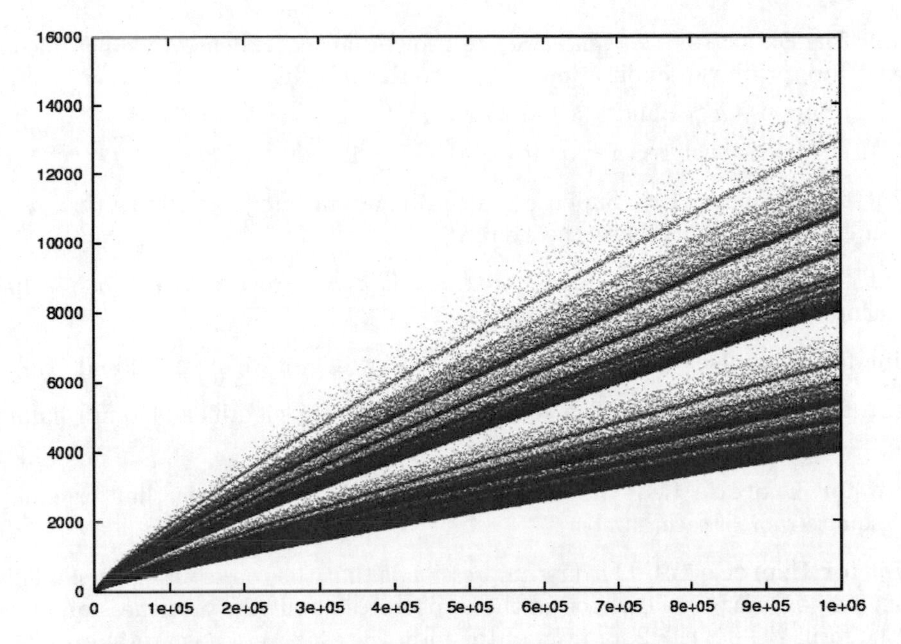

Figure 1.10*.2. A plot of points when the horizontal coordinate is a positive even integer n and the vertical coordinate is the number of ways n can be written as the sum of two primes. This plot gives us confidence that Goldbach's conjecture is true! Credit: Wikimedia Commons, author: Reddish at the English-language Wikipedia, licensed under the GNU Free Documentation license, version 1.2 or any later version, and the Creative Commons Attribution 3.0 Unported (https://creativecommons.org/licenses/by-sa/3.0/deed.en) license.

At the time of this writing, the following is known.

Theorem 1.65. *Every even integer greater than* 2 *and less that* 4×10^{18} *is the sum of two primes.*

So, unless you have access to a powerful computer and lots of time on your hands, we don't suggest that you look for counterexamples to Goldbach's conjecture!

1.11. Hints and partial solutions for the exercises

Hint for Exercise 1.1. The color coding in Figure 1.0.3 starts as follows. 1 is gray. 2 is orange. 3 is green. 4 since it is $2^2 = 2 \cdot 2$ is two oranges, 5 is blue, 6 since it is $2 \cdot 3$ is an orange and a green. 7 is purple. 8 since it is 2^3 is three oranges. Etc. See §1.6 on prime factorization for more food for thought on this.

Hint for Exercise 1.2. Suppose $a = n$. Then $n = a \cdot b = n \cdot b$. Now divide this equality by n.

Hint for Exercise 1.3. Since we are proving an equivalence (a.k.a. a biconditional), we prove two implications. Suppose that $n = ab$.

(1) Suppose $a > 1$ and $b > 1$. Prove that $1 < a < n$.

(2) Suppose that $1 < a < n$. Prove that $b > 1$.

Hint for Exercise 1.4. Suppose that 0 divides an integer n. Show that $n = 0$. Then explain why this solves the exercise.

Hint for Exercise 1.5. Suppose that $1 < a < n$. Prove that $1 < b < n$ using elementary properties about inequalities.

Hint for Exercise 1.6. The answer is given in each of Figure 1.1.2 and (1.6).

Hint for Exercise 1.7. The second, fourth, sixth, eighth, and tenth columns comprise the even integers.

Hint for Exercise 1.8. We just need to check for divisors less than or equal to the square root of each number.

Hint for Exercise 1.9. Our first impression is that the conjecture is false. This is from our knowing the falsehood of the conjectures about the primeness of integers of the form $2^n - 1$, $n \geq 2$, or even of the form $2^p - 1$, where p is a prime.

Hint for Exercise 1.10. Make the substitution $x = 2^a$. Then $2^n - 1 = x^b - 1$ by $2^a - 1 = x - 1$. Apply polynomial long division to dividing $x^b - 1$ by $x - 1$ (or you may already know what this ratio is having seen it somewhere).

Hint for Exercise 1.11. The key is to mimic Figures 1.2.1 and 1.2.2.

Hint for Exercise 1.12. Theorem 1.28 says that "The product of any integer and an even integer is even." Recast this statement as an "if-then" statement, and use the definition of even.

Hint for Exercise 1.13. Draw a $2k + 1$ by $2\ell + 1$ rectangle and consider the 4 "subrectangles" of dimensions $2k \times 2\ell$, $2k \times 1$, $1 \times 2\ell$, and 1×1.

Hint for Exercise 1.14. Let a be an even integer. Apply the result that the product of an even integer and any integer is even to the product $a^3 = a \cdot a^2$.

Hint for Exercise 1.15. Suppose that the product of two integers a and b is even. Assume for a contradiction that not at least one of the integers a, b is even. This means that both a and b are odd. Use a result in the book.

Hint for Exercise 1.16. Use that $\sqrt{11} < 4$.

Hint for Exercise 1.17. Assume for a contradiction that it is not true that $a \leq 100$ or $b \leq 100$. Then $a > 100$ and $b > 100$. Can you derive a contradiction from this assumption?

Hint for Exercise 1.18. Without loss of generality, we may assume that

$$0 < a \leq b \leq c. \tag{1.96}$$

Then show that

$$a^3 \leq n. \tag{1.97}$$

Hint for Exercise 1.19. We cross out all of the composite numbers from 2 to 100, inclusive. Any such composite number is the product of a prime p less than or equal to $\sqrt{100} = 10$ with an integer greater than 1. We call these products non-trivial multiples of p. These primes p are $2, 3, 5, 7$.

Observe that having crossed out all of the non-trivial multiples of primes, for primes less than or equal to p, the next uncrossed out integer is the next prime after p. For example, having crossed out all the non-trivial multiples of 2, 3, and 5, the next uncrossed out integer is 7. This in part explains why the process works.

Having crossed out all of the non-trivial multiples of $2, 3, 5, 7$, the remaining integers are 1 (which is not a prime) and the exact list of primes less than or equal to 100.

Hint for Exercise 1.20. We only need to consider primes less than or equal to $\sqrt{200}$, which is less than 15. So we cross out the (non-trivial) multiples of $2, 3, 5, 7, 11, 13$.

Hint for Exercise 1.21. Follow the advice of the exercise and check the fifth Fermat number. Use a computer and an online factorization applet.

Hint for Exercise 1.22. We simply apply Theorem 1.37 to the special case where $a = b = n$.

Hint for Exercise 1.23. Suppose a divides b. Then there exists an integer k such that $b = ka$. Square this equation.

Hint for Exercise 1.24. Suppose that 3 divides b. Then there exists an integer k such that $b = 3k$. Square this equation. For the second part of the exercise, replace 3 by a.

Hint for Exercise 1.25. Suppose that a divides b and that b divides c. Then there exist integers k and ℓ such that

$$b = ka, \qquad c = \ell b. \tag{1.98}$$

Continue.

Hint for Exercise 1.26. Let a be a positive divisor of a positive integer b with the property that $a \neq b$. Then $a < b$ (why?). There exists a positive integer k such that $ak = b$ (why?). We have $k \geq 2$ (why?). Continue.

Hint for Exercise 1.27. This is very similar to Solved Problem 1.44, so mimic the proof thereof.

Hint for Exercise 1.28. We have $a \geq 1$ and $b \geq 1$. Use a basic fact about inequalities, multiplication, and the maximum.

Hint for Exercise 1.29. For example, suppose that an integer n is equal to $p_1^{k_1} p_2^{k_2} p_3^{k_3}$, where p_1, p_2, p_3 are distinct primes and k_1, k_2, k_3 are positive integers. Then the positive divisors of n are given by

$$p_1^{j_1} p_2^{j_2} p_3^{j_3}, \quad \text{where } 0 \leq j_1 \leq k_1,\ 0 \leq j_2 \leq k_2,\ 0 \leq j_3 \leq k_3. \tag{1.99}$$

Hint for Exercise 1.30. Use that 5, 17, and 23 are prime to find all of the positive divisors of $5 \cdot 17$ and $23 \cdot 17$.

Hint for Exercise 1.31. Let c be a common divisor of d and b. Use that d divides a and the transitivity of division.

Hint for Exercise 1.32. Is 2 a common divisor of both a and b?

Hint for Exercise 1.33. We have $\frac{68}{51} = \frac{4}{3}$.

Hint for Exercise 1.34. Let p and q be non-equal prime numbers. Let $g := \gcd(p^2, q^2)$. Use that g is a positive integer dividing both p^2 and q^2 and that p and q are primes. You may use that the only positive divisors of p^2 are 1, p, and p^2, and similarly for p replaced by q.

Hint for Exercise 1.35. Let k be an integer. Substitute $m = 5 + 11k$ and $n = -4 - 9k$ into the expression $9m + 11n$ and verify that the answer is equal to 1.

Hint for Exercise 1.36. Counting by 13's yields

$$13, 26, 39, 52, 65, \ldots, \tag{1.100}$$

while counting by 19's yields

$$19, 38, 57, 76, 95, \ldots. \tag{1.101}$$

By comparing the integers on these two lists, can you find a solution to the linear Diophantine equation (1.72)?

For the second part of the exercise, multiply by 5.

For the third part of the exercise, let k be any integer and calculate $13(m_0 + 19k) + 19(n_0 - 13k)$.

Hint for Exercise 1.37. The answer is given by (1.75).

Hint for Exercise 1.38. One of the divisors is 53.

Hint for Exercise 1.39. (1) The only divisors of p are 1 and p.

(2) The only divisors of pq are $1, p, q, pq$. Note that it is hypothetically possible that $p = q$.

Hint for Exercise 1.40. By the previous exercise, we only need to show that the composite numbers strictly between 6 and 28 are all not perfect. To do this, one simply computes the sum of the positive divisors not equal to the composite number.

Hint for Exercise 1.41. We have that $2^{11} - 1 = 2047$. An online applet factors this as $2047 = 23 \cdot 89$.

Chapter 2

Mathematical Induction

Figure 2.0.1. As in the television series Manifest, we are all interconnected. If we tip over the bear, then the koala, penguin, marmot, hippo, panda, squirrel, cat , ... must all follow! Figure drawn using the Ti*k*Zlings package.

Goals of this chapter: To explain how to prove results by mathematical induction. For example, we will prove various interesting sequences of formulas and sequences of inequalities.

To show that there is a version of induction, called strong induction, that is more powerful in some important situations. To use strong induction to prove that any integer at least 2 is either a prime or a product of primes. To understand some properties of the amazing Fibonacci sequence.

Mathematical induction is a very useful method of proof!

2.1. Mathematical induction

Suppose that we want to prove an infinite number of statements $P(n)$, one statement for every positive integer n. If we prove these statements one at a time, then things will not go very well as we will need an infinite number of proofs, which is not humanly possible to achieve. However, for many interesting sequences of statements, there is a method of proof, called *mathematical induction*, which works. This method is based on an axiom, called the *axiom of induction*.

2.1.1. Example: An arithmetic sum.

Consider the question:

> Let n be any positive integer. What is the sum of the first n positive integers?

Firstly, we may write this sum as

$$1+2+\cdots+(n-1)+n=\sum_{i=1}^{n} i. \tag{2.1}$$

For example, by taking $n=3$, we have $1+2+3=6$. Apparently children learn the "dot dot dot" idea in grade school: Our daughter once taught us how to count to 100:

one, two, *skip a few*, ninety-nine, one hundred!

In mathematical notation, this is

$$1, 2, \ldots, 99, 100. ☺$$

Let us look at what the sums in (2.1) are for the first several positive integers n:

Table 2.1.1. Sums of the first n positive integers.

n	1	2	3	4	5	6	7	8	9	10
sum	1	3	6	10	15	21	28	36	45	55

Factoring these sums as the products of two integers helps us understand if there is a pattern. We observe that Table 2.1.1 may be rewritten as

Table 2.1.2. Sums of the first n positive integers rewritten.

n	1	2	3	4	5	6	7	8	9	10
sum	1	3	$2\cdot 3$	$2\cdot 5$	$3\cdot 5$	$3\cdot 7$	$4\cdot 7$	$4\cdot 9$	$5\cdot 9$	$5\cdot 11$

Notice that there is an interesting pattern in these factorizations. For n odd, the two factors of the sum are n and $\frac{n+1}{2}$. For n even, the two factors of the sum are $\frac{n}{2}$ and $n+1$. In both of these cases, this leads us to the following proposed formula:

Theorem 2.1. *If n is a positive integer, then*

$$1+2+\cdots+n=\frac{n(n+1)}{2}. \tag{2.2}$$

Since the right-hand side of (2.2) for $1 \leq n \leq 10$ agrees with Table 2.1.2, we are confident that this formula is true.

Presumably using this formula, as legend has it, young Carl Friedrich Gauss worked out on the spot in primary school:

$$1 + 2 + \cdots \textit{ don't skip any! } \cdots + 99 + 100 = \frac{100 \cdot 101}{2} = 5050. \tag{2.3}$$

Now the question is: How do we prove formula (2.2)? In fact, there are a number of ways. The way we will prove it here is a more formal way, called **mathematical induction**, or just **induction** for short.

We have observed that Table 2.1.2 verifies formula (2.2) for $1 \leq n \leq 10$. Let us pretend, for example, that we don't know that the formula is true for $n = 9$, but somehow we know that the formula is true for $n = 8$. Can we prove the formula for $n = 9$ from just knowing that the formula for $n = 8$ is true? Yes, here is a simple way, which is an example of the method of induction. We know that (the $n = 8$ case)

$$1 + 2 + 3 + 4 + 5 + 6 + 7 + 8 = \frac{8(8+1)}{2} = \frac{8 \cdot 9}{2}. \tag{2.4}$$

Thus, by adding 9 to both sides of this equation, we obtain

$$1+2+3+4+5+6+7+8+9 = (1+2+3+4+5+6+7+8)+9 = \frac{8 \cdot 9}{2}+9. \tag{2.5}$$

We can put the right-hand side over the common denominator 2 to get

$$1 + 2 + 3 + 4 + 5 + 6 + 7 + 8 + 9 = \frac{(8+2) \cdot 9}{2} = \frac{9 \cdot (9+1)}{2}. \tag{2.6}$$

This is the desired formula for $n = 9$! (That is, for $n = 9$; wow! Not $n = 9$ factorial; lol!) The upshot of all of this is that by just knowing that the formula is true for $n = 8$, we can prove that the formula is true for $n = 9$.

Exercise 2.1. *Let k be a positive integer. By just knowing formula (2.2) for $n = k$, can we prove (2.2) for $n = k + 1$? After all, we were able to do this for $k = 8$. Hint: The answer is given in §2.1.2. But see if you can work it out yourself!*

2.1.2. A proof by induction: Summing the positive integers. Firstly, we recall *implications*, which we briefly discussed in Chapter 1 and which is more formally treated in Chapter 3. So far in this book, we have been using "if-then" statements, which are statements of the form:

If P, then Q.

Another way to say this is

P implies Q.

And using symbols, we write this as

$$P \Rightarrow Q.$$

For our current purposes, regarding implications we will just use that *proving an implication $P \Rightarrow Q$* amounts to doing the following:

Suppose that P is true. Prove that then Q is true.

Our first proof by induction is a prototypical example for proving sequences of equalities.

Proof of Theorem 2.1. Let $P(n)$ be the statement $1+2+\cdots+n=\frac{n(n+1)}{2}$.

Base case ($n=1$)**:** Evidently, since $1=\frac{1(1+1)}{2}$, we have that $P(1)$ is true.

Inductive step ($P(k)\Rightarrow P(k+1)$)**:** Now, *assume* that k is a positive integer such that $P(k)$ is true. (In the base case we have just seen that $k=1$ is such an integer.[1]) We will now show that $P(k+1)$ is true.

Proof: By hypothesis, k is assumed to be such that

$$1+2+\cdots+k=\frac{k(k+1)}{2}. \tag{2.7}$$

Using this assumption, we calculate that (this is where the main work is)

$$\begin{aligned} 1+2+\cdots+(k+1) &= (1+2+\cdots+k)+(k+1) \\ &= \frac{k(k+1)}{2}+\frac{2}{2}\cdot(k+1) \qquad \text{(this is where we used (2.7))} \\ &= \frac{(k+2)(k+1)}{2} \qquad \text{(by the distributive law)} \\ &= \frac{(k+1)((k+1)+1)}{2}. \end{aligned} \tag{2.8}$$

This says that $P(k+1)$ is true. So we have proved for all positive integers k that the implication $P(k)\Rightarrow P(k+1)$ is true.

Now, the Axiom 2.5 of mathematical induction, which is formally stated below (see also Remark 2.3 below), is exactly the statement that in such a situation as this we can conclude that $P(n)$ is true *for all* positive integers n. □

Remark 2.2 (Thinking forward → and backward ←)**.** A nice, but less formal, way to see that formula (2.2) is true without using induction is to sum positive integers both "forward" and "backward" as follows:

$$\begin{array}{ccccccccccccc} 1 & + & 2 & + & 3 & + & \cdots & + & n-2 & + & n-1 & + & n \\ +\,n & + & n-1 & + & n-2 & + & \cdots & + & 3 & + & 2 & + & 1 \\ \hline n+1 & + & n+1 & + & n+1 & + & \cdots & + & n+1 & + & n+1 & + & n+1 \end{array}$$

Since there are n of the $n+1$'s in the last line, we obtain that

$$2(1+2+3+\cdots+(n-2)+(n-1)+n)=n(n+1), \tag{2.9}$$

which is equivalent to (2.2). We guess that this must have been young Gauss's trick in computing (2.3)! See Figure 2.3.1 below for a related visual proof of Theorem 2.1.

Remark 2.3. For the inductive step, we use the assumption that $P(k)$ is true, (2.7), to prove that $P(k+1)$ is true, (2.8). That is, in the proof above, for the inductive step, the key was to show the **implication**:

$$\textbf{If } (2.7) \text{ is true, } \textbf{then } (2.8) \text{ is true.} \tag{2.10}$$

[1] Note that in the base case, we didn't *assume* that $P(1)$ is true, we *proved* that $P(1)$ is true.

Metaphorically, we are building a bridge between the two statements. See Figure 2.1.1. Induction says that if the first statement $P(1)$ is true and if the implication (2.10) is true for all positive integers k, then $P(n)$ is true for all positive integers n.

Figure 2.1.1. A (Golden Gate) Bridge between the inductive hypothesis and the inductive conclusion. This bridge enables us to go from San Francisco ($P(k)$) to Sausalito ($P(k+1)$). Photo credit: Golden Gate Bridge, December 15, 2015, by D. Ramey Logan.jpg from Wikimedia Commons by D. Ramey Logan (https://don.logan.com), under Creative Commons CC-BY 4.0 (https://creativecommons.org/licenses/by/4.0/deed.en) license.

2.1.3. The general idea of induction. The concrete discussion above illustrates the fundamental idea of induction. Consider an infinite sequence of statements

$$P(1), P(2), P(3), \ldots. \tag{2.11}$$

Suppose we know that $P(1)$ is true. How can we prove that all of the statements are true?

In the examples above, firstly we observed that by just knowing $P(8)$ is true we can deduce that $P(9)$ is true. Secondly, we generalized this to show that, for every positive integer k, if $P(k)$ is true, then $P(k+1)$ is true.

Now, if we know that $P(1)$ implies $P(2)$, then since $P(1)$ is true, so is $P(2)$. Next, if we know that $P(2)$ implies $P(3)$, since $P(2)$ is true, we then know that $P(3)$ is true. Here, we need to know that $P(3)$ implies $P(4)$. If so, then $P(4)$ is true. And so on. We should be able to go on forever! This is the idea of induction.

Given a particular sequence of statements $P(n)$, $n \geq 1$, we can make a sweeping conjecture: No matter what positive integer k is, the implication

$$P(k) \text{ implies } P(k+1)$$

is true. This would mean that $P(1)$ implies $P(2)$, and $P(2)$ implies $P(3)$, ..., and $P(8)$ implies $P(9)$, and .... So all of the statements should be true since we assumed that $P(1)$ is true! This is what the axiom of mathematical induction is all about. Fortunately, we can often prove such a sweeping conjecture.

2.1.4. More proofs by induction: Arithmetic sums. Now, let us sum just the odd integers. Recall that $\mathbb{Z}^+$ denotes the set of positive integers.

Exercise 2.2. *Prove by induction that for all $n \in \mathbb{Z}^+$,*

$$\sum_{i=1}^{n}(2i-1) = 1+3+5+\cdots+(2n-1) = n^2. \tag{2.12}$$

Hint: Model your proof on the proof of Theorem 2.1.

See Figure 2.1.2 for a visual proof of (2.12)! Although in this book we are teaching formal proofs, some visual proofs such as this may be considered as superior to induction, easier to remember, and delightful!

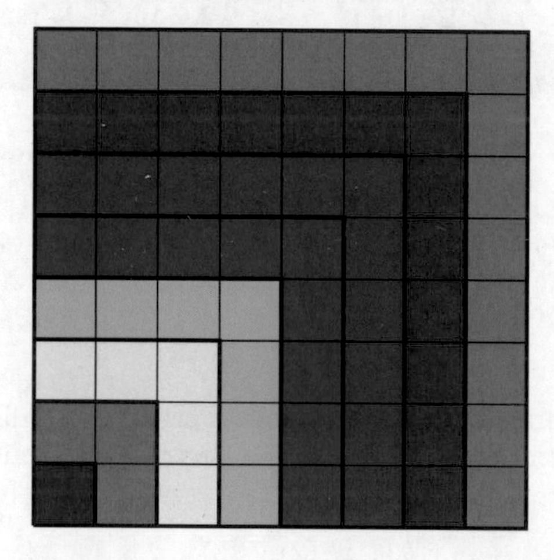

Figure 2.1.2. A visual proof that $1+3+5+7+9+11+13+15 = 8^2$. This is the $n = 8$ case of (2.12).

Let π be pi, and let e be Euler's number. So, approximately, $\pi \approx 3.14159$ and $e \approx 2.71828$.

Solved Problem 2.4. *Show by induction that for all positive integers n,*

$$\sum_{i=1}^{n}(\pi i + e) = \frac{n(n+1)}{2}\pi + ne. \tag{2.13}$$

Solution. *Base case*: $n = 1$. This is true because the formula says in this case that

$$\pi \cdot 1 + e = \frac{1(1+1)}{2}\pi + 1 \cdot e, \tag{2.14}$$

which is true since both sides equal $\pi + e$.

Inductive step: Suppose that k is a positive integer with the property that

$$\sum_{i=1}^{k}(\pi i + e) = \frac{k(k+1)}{2}\pi + ke. \tag{2.15}$$

We then compute that

$$\begin{aligned}\sum_{i=1}^{k+1}(\pi i+e) &= \sum_{i=1}^{k}(\pi i+e)+(\pi(k+1)+e)\\ &= \left(\frac{k(k+1)}{2}\pi+ke\right)+(\pi(k+1)+e)\\ &= \frac{k(k+1)}{2}\pi+(k+1)\pi+ke+e\\ &= \frac{(k+1)(k+2)}{2}\pi+(k+1)e.\end{aligned}$$

This is the desired inductive conclusion. By induction we are done.

Another way to solve this problem is to use the fact that $\sum_{i=1}^{n} i = \frac{n(n+1)}{2}$ by (2.2). Indeed, using this we calculate that

$$\sum_{i=1}^{n}(\pi i+e)=\pi\sum_{i=1}^{n} i+\sum_{i=1}^{n} e=\pi\frac{n(n+1)}{2}+ne. \tag{2.16}$$

Note that we used the commutativity of addition to split the sum into two sums. We also used the distributive law to pull out the factor of π in front of the first sum on the right-hand side.

More generally, we can sum any arithmetic progression.

Exercise 2.3. *Let a and b be real numbers. Prove that for all $n \in \mathbb{Z}^+$,*

$$\sum_{i=1}^{n}(b+ia)=(b+a)+\cdots+(b+na)=n\left(\frac{n+1}{2}a+b\right). \tag{2.17}$$

Observe that Exercise 2.2 is the special case where $a=2$ and $b=-1$ and Solved Problem 2.4 is the special case where $a=\pi$ and $b=e$.

2.1.5. Formal statement of induction.

Now, finally, we have the formal statement of mathematical induction as an axiom.

Axiom 2.5 (Mathematical Induction). *Let $P(1), P(2), P(3), \ldots$ be an infinite sequence of statements. Suppose that $P(1)$ is true, and suppose that $P(k)$ implies $P(k+1)$ for all $k \geq 1$. Then $P(n)$ is true for all $n \geq 1$.*

To wit, induction says that the following hypothesis implies the following conclusion:

Hypothesis:

(1) $P(1)$ is true.

(2) $P(k)$ implies $P(k+1)$ for all $k \geq 1$.

Conclusion: $P(n)$ is true for all $n \geq 1$.

Recall that to prove the implication $P(k)$ implies $P(k+1)$, we assume that $P(k)$ is true. Then, by logical deduction, we need to prove that $P(k+1)$ is true.

2.1.6. The domino effect. Mathematical induction can be visualized as the domino effect (not to be confused with the domino theory!). Imagine we have an infinite string of dominoes. We have carefully placed all of the dominoes so that for every k if the k-th domino in the line topples, then the $(k+1)$-st domino topples (this is the inductive step). Mathematical induction says that if we knock over the first domino, then *for every* n, the n-th domino (eventually) topples. To visualize induction, see Figure 2.1.3 for tipping dominoes instead of the tipping cute animals pictured at the beginning of this chapter.

Induction can be used to prove the well-known formula for geometric sums. Firstly, consider summing powers of 2:

$$1,\ 1+2,\ 1+2+4,\ 1+2+4+8,\ 1+2+4+8+16, \ldots.$$

We easily calculate that these sums are equal to

$$1,\ 3,\ 7,\ 15,\ 31, \ldots.$$

To see a pattern more clearly, we add 1 to each sum to obtain

$$2,\ 4,\ 8,\ 16,\ 32, \ldots,$$

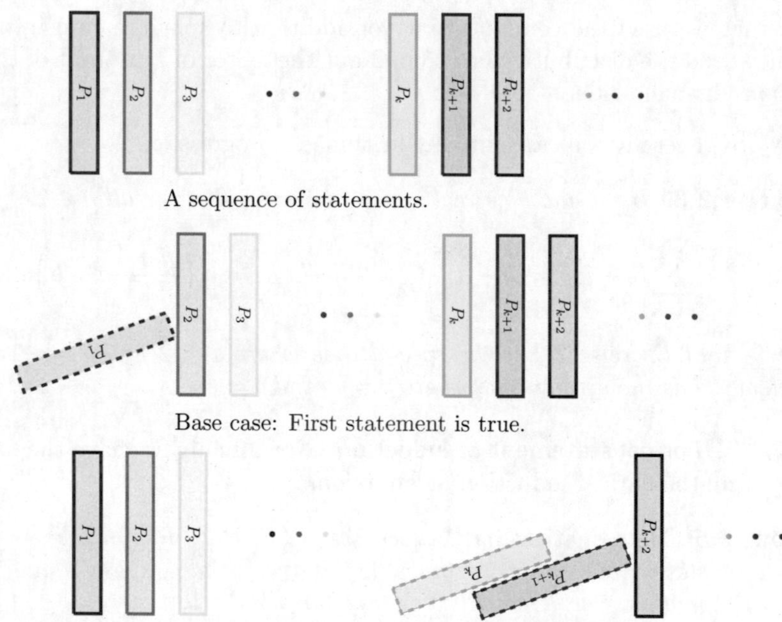

Inductive step: k-th statement implies $(k+1)$-st statement for every k.

Conclusion: all statements are true.

Figure 2.1.3. Dominoes illustrating the idea of mathematical induction. Given a line of dominoes $P_1, P_2, P_3, \ldots, P_k, P_{k+1}, P_{k+2}, \ldots$, if the first one P_1 falls and if the k-th one P_k falls *implies* the $(k+1)$-st one P_{k+1} falls for all k, then they all (P_n for all $n \geq 1$) fall.

which we can rewrite as powers of 2:

$$2^1,\ 2^2,\ 2^3,\ 2^4,\ 2^5, \dots,$$

Observe, e.g., that

$$1+2+4+8+16 = 2^0+2^1+2^2+2^3+2^4 = \sum_{i=0}^{4} 2^i$$

and that this sum is equal to

$$31 = 2^5 - 1 = 2^{4+1} - 1.$$

This leads us to guess the formula: For any non-negative integer n, we have

$$1+2+2^2+\cdots+2^n = \sum_{i=0}^{n} 2^i = 2^{n+1} - 1. \tag{2.18}$$

Secondly, consider summing powers of 3:

$$1,\ 1+3,\ 1+3+9,\ 1+3+9+27,\ 1+3+9+27+81, \dots.$$

We calculate that these sums are equal to

$$1,\ 4,\ 13,\ 40,\ 121, \dots.$$

If we multiply these sums by 2 and then add 1, then we obtain

$$3,\ 9,\ 27,\ 81,\ 243, \dots.$$

We can rewrite these integers as powers of 3:

$$3^1,\ 3^2,\ 3^3,\ 3^4,\ 3^5, \dots.$$

So we have, e.g., that

$$1 + 2\sum_{i=0}^{4} 3^i = 243 = 3^{4+1}.$$

Thus, we guess that for every non-negative integer n, we have

$$\sum_{i=0}^{n} 3^i = \frac{3^{n+1}-1}{2}. \tag{2.19}$$

In the above, we have guessed the formula for geometric sums with base equal to 2 and base equal to 3. We can actually prove, by induction, the formula for a geometric sum with any base r, so long as $r \neq 1$. Of course, on the other hand, if $r = 1$, then the sum is easy: $\sum_{i=0}^{n} 1^i = n+1$.

Exercise 2.4. *Let $r \in \mathbb{R}$, where $r \neq 1$. Prove that for all $n \in \mathbb{Z}^+$, we have for a finite geometric sum the formula*

$$\sum_{i=0}^{n} r^i = 1 + r + r^2 + \cdots + r^n = \frac{1-r^{n+1}}{1-r}. \tag{2.20}$$

In other words, for $n \in \mathbb{Z}^+$ and $r \in \mathbb{R}$,

$$r^{n+1} - 1 = (r-1)(r^n + r^{n-1} + \cdots + r^2 + r + 1). \tag{2.21}$$

We can go from finite series to infinite series by taking a limit. A formula which also has a geometric proof is the following.

Exercise 2.5. *Can you prove that the infinite geometric sum of the positive powers of one-half is equal to one-third:*

$$\sum_{n=1}^{\infty}\left(\frac{1}{2}\right)^{2n}=\frac{1}{3}? \tag{2.22}$$

This is an infinite sum (i.e., an infinite series), so you will have to take a limit of the (finite) partial sums

$$s_k := \sum_{n=1}^{k}\left(\frac{1}{2}\right)^{2n} \tag{2.23}$$

as $k \to \infty$.

Formula (2.22) may be visualized as the lovely Figure 2.1.4.

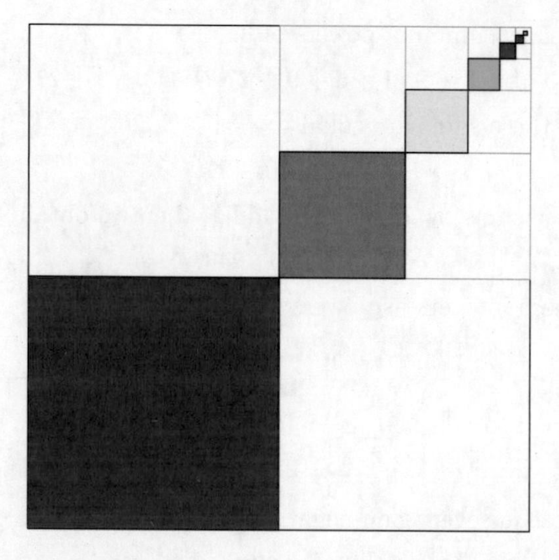

Figure 2.1.4. A visual proof of (2.22). For every rainbow square, there are two white squares of the same size. If the big square is 1-by-1, then the sum of the areas of the rainbow squares is equal to $\frac{1}{4}+\frac{1}{4^2}+\frac{1}{4^3}+\cdots=\frac{1}{3}$. The sum of the areas of the white squares is equal to $\frac{2}{3}$.

2.1.7. Caesar is a salad dressing. Let us think more about what induction means. This will help us understand the key ideas and clear up any misconceptions. Seemingly unrelated, part of a conversation with Socrates in the movie "Bill and Ted's Excellent Adventure" went like this:

Ted: What do I say?

Bill (shrugging): Philosophize with him.

Ted (shrugs; to Socrates): "*All we are is dust in the wind*", dude.[2]

However, let us imagine how this conversation might have continued:

Bill: Dude, like what is induction?

[2] "Dust in the Wind" is a song by Kansas.

Ted: It's like the song "Another One Bites the Dust" by Queen. (In the dialogue below, **Bill** and **Ted** try to stay true to the lyrics of the Queen song.)

Bill: Huh? What do you mean?

SOCRATES: Proving the base case $P(1)$ is like: "*Are you ready, hey, are you ready for this?*"

Bill: Yes, I'm hanging on the edge of my seat! Or, it's like "Start Me Up" (song by The Rolling Stones). Excellent!

Ted: Then we have the inductive hypothesis $P(k)$, which is like: "Another one bites the dust."

Bill: Dude, what do you do with this?

Ted: You work hard to prove the inductive conclusion $P(k+1)$, which is like: "*Hey, I'm gonna get you too. Another one bites the dust.*"

Bill: So "*I'm gonna get you too*" is like proving the implication $P(k) \Rightarrow P(k{+}1)$ in the inductive step.

SOCRATES: Right, dude (Socrates learns quickly how to use the word "dude"). Then, by induction we are done: "*Another one bites the dust. Another one bites the dust. And another one gone and another one gone.*" This is like the conclusion that $P(n)$ is true for all positive integers n.

Bill: That's pretty repetitive, dude.

Ted: It's infinitely repetitive, dude. That is the only way to get *everyone* to bite the dust. "*Are you happy, are you satisfied?*"

Bill: Yeah, I can stand the heat. You've convinced me! You are awesome. Anything you would like to add?

Ted (*as in the movie, concludes*): "Yes, and that thanks to great leaders, such as Genghis Khan, Julius Caesar, and Socratic Method, the world is ... full of history!"

With this Socratic understanding of induction, we return to the practical matter of proving statements by induction. Given that we have seen the induction proof of the formula for the sum of positive integers in §2.1.2, can you mimic this proof (in a broad sense) to prove formulas for sums of small powers of positive integers?

Exercise 2.6. *Prove that for all* $n \in \mathbb{Z}^+$,

$$\sum_{i=1}^{n} i^2 = 1^2 + 2^2 + 3^2 + \cdots + n^2 = \frac{n(n+1)(2n+1)}{6}. \tag{2.24}$$

Exercise 2.7. *Prove by induction that the sum of the squares of the first* n *positive odd integers is equal to*

$$\frac{n(2n-1)(2n+1)}{3} = \frac{4n^3 - n}{3}.$$

Exercise 2.8. *Prove that for all* $n \in \mathbb{Z}^+$,

$$\sum_{i=1}^{n} (-1)^i \, i^2 = -1^2 + 2^2 - 3^2 + \cdots + (-1)^n \, n^2 = (-1)^n \frac{n(n+1)}{2}. \tag{2.25}$$

Exercise 2.9. *Prove that for all $n \in \mathbb{Z}^+$,*

$$\sum_{i=1}^{n} i^3 = 1^3 + 2^3 + 3^3 + \cdots + n^3 = \frac{n^2(n+1)^2}{4}. \tag{2.26}$$

Furthermore, observe that this is equal to $(1+2+3+\cdots+n)^2$.

The following formula follows from a previous result and exercise, but can you prove it directly by induction?

Exercise 2.10. *Prove that for all $n \in \mathbb{Z}^+$,*

$$\sum_{i=1}^{n} i(i+1) = 1 \cdot 2 + 2 \cdot 3 + 3 \cdot 4 + \cdots + n \cdot (n+1) = \frac{n(n+1)(n+2)}{3}. \tag{2.27}$$

Let us consider sums of ratios, actually simple rational functions.

Exercise 2.11. *Prove that for all $n \in \mathbb{Z}^+$,*

$$\sum_{i=1}^{n} \frac{1}{i(i+1)} = \frac{1}{1 \cdot 2} + \frac{1}{2 \cdot 3} + \frac{1}{3 \cdot 4} + \cdots + \frac{1}{n \cdot (n+1)} = \frac{n}{n+1}. \tag{2.28}$$

Exercise 2.12. *Prove that for all $n \in \mathbb{Z}^+$,*

$$\begin{aligned} &\sum_{i=1}^{n} \frac{1}{(2i-1)(2i+1)} \\ &= \frac{1}{1 \cdot 3} + \frac{1}{3 \cdot 5} + \frac{1}{5 \cdot 7} + \cdots + \frac{1}{(2n-1)(2n+1)} = \frac{n}{2n+1}. \end{aligned} \tag{2.29}$$

Instead of sums, we can consider products.

Exercise 2.13. *Prove that if n is an even positive integer, then*

$$\left(1 - \frac{1}{2}\right)\left(1 + \frac{1}{3}\right)\left(1 - \frac{1}{4}\right) \cdots \left(1 - \frac{(-1)^n}{n}\right) = \frac{1}{2}. \tag{2.30}$$

Explain (not prove), without using induction, why the formula is true.

Exercise 2.14. *Prove that for all integers $n \geq 2$,*

$$\prod_{i=2}^{n} \left(1 - \frac{1}{i^2}\right) = \left(1 - \frac{1}{2^2}\right)\left(1 - \frac{1}{3^2}\right) \cdots \left(1 - \frac{1}{n^2}\right) = \frac{n+1}{2n}. \tag{2.31}$$

2.1.8. Flaws in attempts at inductive proofs.

We can still make mistakes in proving statements by induction if we are not careful. A classical flawed proof by induction (in this case a falsidical paradox) is the proof that

all horses are the same color.

The "fake proof" goes like this. (We will use some basic set theory notation. If you are not familiar with the notations, you may find them in Chapter 5 below.) We will grant ourselves the simplifying assumption that each horse has a single, specific color, so that we don't for example allow for striped horses (this is not where the proof breaks down).

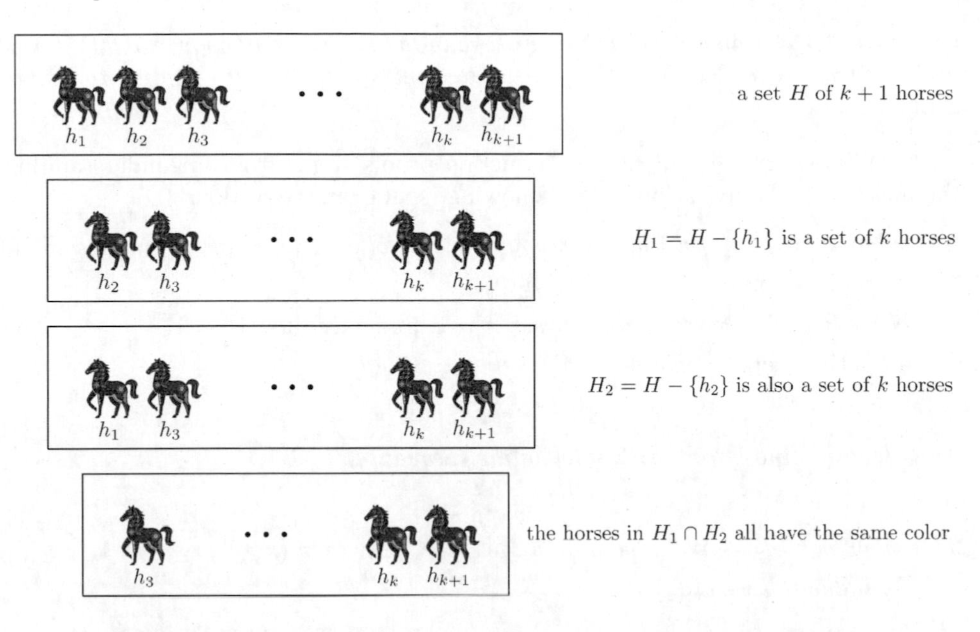

Figure 2.1.5. *Fake induction: the trouble is in getting started!* By the inductive hypothesis, each of the sets H_1 and H_2 (with k horses) consists of horses of a single color. So, since all of the horses in $H_1 \cap H_2$ are of the same color, we deduce that all of the horses in H are of the same color. But(!), for this argument to work, we need $k+1 \geq 3$, i.e., $k \geq 2$. In particular, this argument doesn't work for $k = 1$ since then $H_1 \cap H_2 = \emptyset$ is the empty set! Credit for the drawing of the red horse: Designed by tartila / Freepik.com.

Let $P(n)$ be the statement that the horses in any set of n horses all have the same color.

Base case. $P(1)$ is trivial since we have a collection of only 1 horse.

Inductive step. Suppose that k is a positive integer such that $P(k)$ is true. Let H be a set of $k+1$ horses. Take one horse h_1 away from H to get a set

$$H_1 := H - \{h_1\}$$

of k horses. By the inductive hypothesis, all the horses in H_1 have the same color C_1.

Next, choose a horse h_2 that is different than the horse h_1. Take away the horse h_2 to get the set

$$H_2 := H - \{h_2\}.$$

Since H_2 consists of k horses, by the inductive hypothesis again, all the horses in H_2 have the same color C_2. Now take a third horse h_3 different than both h_1 and h_2. We have that $h_3 \in H_1$ and $h_3 \in H_2$. This implies that h_3 has both the color C_1 and the color C_2. Therefore, C_1 and C_2 are the same color. We conclude that $h_1 \in H_2$ has the color C_2, which is the same as the color C_1. Therefore all of the horses in H have the same color C_1.

By induction, we are done!

Exercise 2.15. *Can you detect the mistake in the flawed proof above? Hint: Think about how to prove $P(2)$ from $P(1)$. A more detailed hint is in the caption to Figure* 2.1.5.

A different type of mistake in induction proofs stems from misunderstanding the meaning of "implication". The following is an example of this.

Exercise 2.16. *Explain what is wrong with the following "proof" by induction.*

Non-Theorem. *$n^2 \leq n$ for all $n \in \mathbb{Z}^+$.*

Non-Proof. *Clearly the* base case $n = 1$ *is true because* $1^2 = 1$.

Inductive step. *Suppose* $(k+1)^2 \leq k+1$. *Then*

$$k^2 + 2k + 1 \leq k + 1.$$

Thus (subtracting 1 *from both sides of the inequality)*

$$k^2 + 2k \leq k.$$

This implies $k^2 \leq k$. We have proved that if $k^2 \leq k$, then $(k+1)^2 \leq k+1$.

By induction we are done.

Exercise 2.17 (Be excellent to each other (and party on, dudes!)). *Explain what is not so excellent about the following statement and proof by induction (it's not the base case).*

Statement: All adventures have the same degree of excellency.

Proof by induction. *Let $P(n)$ be the statement:*

> *If A is any set of n adventures, then all the adventures in A have the same degree of excellency.*

Base case*: Let A be a set of* 1 *adventure. Since A only has one element, all adventures in A have the same degree of excellency (as the sole element).*

Inductive step*: Let k be a positive integer with the property that all the adventures in any set of k adventures have the same degree of excellency.*

Now let A be a set of $k+1$ adventures. Choose distinct adventures a_1 and a_2 in A. Then $B := A - \{a_1\}$ and $C := A - \{a_2\}$ each are sets of k adventures. So, by the inductive hypothesis, all of the adventures in B have the same degree of excellency, and all of the adventures in C have the same degree of excellency.

Let a_3 be any adventure in A distinct from both a_1 and a_2. Then a_3 is in both B and C. Take any adventure in A. Then the adventure is in both B and C and hence has the same degree of excellency as a_3. This proves that all of the adventures in A have the same degree of excellency.

This completes the proof of the inductive step.

By induction the result follows.

2.2. Rates of growth of functions

Just as we can prove formulas by induction, that is, equalities of different expressions, we can prove inequalities by induction. As you may have heard by now, exponential functions grow faster than polynomials. We will look at examples of this from the point of view of induction. Through this, we will get some experience working with inequalities.

2.2.1. Exponentials versus polynomials: Racing to infinity.

Question: An exponential function and a polynomial race to infinity. Who gets there first?
Answer: Neither. Both never get there!

A bit of calculus: Consider the limit

$$\lim_{x\to\infty} \frac{e^x}{x^2} = \infty. \tag{2.32}$$

Since $\lim_{x\to\infty} e^x = \infty$ and $\lim_{x\to\infty} x^2 = \infty$, the limit in (2.32) has the indeterminate form $\frac{\infty}{\infty}$. We can apply l'Hospital's rule (twice) to conclude that

$$\lim_{x\to\infty} \frac{e^x}{x^2} = \lim_{x\to\infty} \frac{e^x}{2x} = \lim_{x\to\infty} \frac{e^x}{2} = \infty. \tag{2.33}$$

So, certainly, we have for x sufficiently large that

$$e^x > x^2.$$

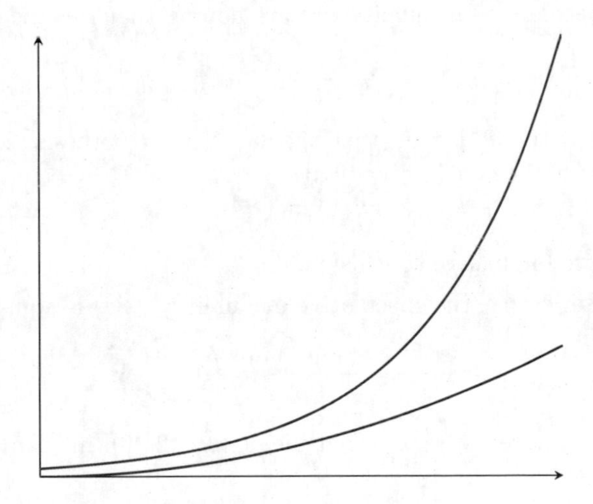

Figure 2.2.1. A race to infinity and beyond between the functions $x \mapsto x^2$ and $x \mapsto e^x$.

By induction, we can determine for which integer values of x we have this inequality. For this, we will introduce a slight generalization of the method of mathematical induction. For Mathematical Induction Axiom 2.5, the base case is $n = 1$, and we prove statements for all integers $n \geq 1$. More generally, we can assume that the base case is any given integer, for example, $n = 2$. In this case, for the inductive step we need to prove that $P(k) \Rightarrow P(k+1)$ for all integers $k \geq 2$. Then, the slight generalization of induction allows us to conclude that $P(n)$ is true for all integers $n \geq 2$. Formally, this version of induction is stated as Axiom 2.8 below.

Theorem 2.6. *For any $n \in \mathbb{Z}^+$, we have the inequality*

$$e^n > n^2. \tag{2.34}$$

Proof. There are various, roughly equivalent, approaches to proving this by induction. We choose to write the desired inequality in "ratio form":

$$\frac{e^n}{n^2} > 1. \tag{2.35}$$

That is, (2.34) and (2.35) are equivalent. We remark that two inequalities are *equivalent* if each inequality implies the other. That is, if either of the two inequalities is true, then the other inequality is true.

Clearly

$$\frac{e^1}{1^2} = e > 1, \tag{2.36}$$

so that the case $n = 1$ is true. (For reasons we will explain below, we will not call this the base case.) Now suppose that $k \geq 1$ is such that (the inductive hypothesis)

$$\frac{e^k}{k^2} > 1. \tag{2.37}$$

We observe that one way to ensure that the desired inductive conclusion is true is to prove the "speculative" inequality (we are hoping that this inequality is true)

$$\frac{e^{k+1}}{(k+1)^2} \geq \frac{e^k}{k^2}. \tag{2.38}$$

Indeed, if inequality (2.38) is true, then the inductive hypothesis (2.37) implies that

$$\frac{e^{k+1}}{(k+1)^2} > 1, \tag{2.39}$$

which is the desired inductive conclusion.

Now, we observe that the speculative inequality (2.38) is equivalent to

$$e = \frac{e^{k+1}}{e^k} \geq \frac{(k+1)^2}{k^2} = \left(1 + \frac{1}{k}\right)^2. \tag{2.40}$$

By taking square roots, we see that inequality (2.40), in turn, is equivalent to the inequality

$$\sqrt{e} \geq 1 + \frac{1}{k}. \tag{2.41}$$

By the aforestated equivalences and implication of inequalities, if (2.41) is true, then the inductive conclusion is true. Thus, it suffices to prove inequality (2.41).

If $k \geq 2$, then we have

$$1 + \frac{1}{k} \leq 1 + \frac{1}{2} = 1.5 = \sqrt{2.25} < \sqrt{e}. \tag{2.42}$$

Hence the desired inequality (2.41) is true provided $k \geq 2$. To summarize, we have proved for $k \geq 2$ the *implication*

$$\frac{e^k}{k^2} > 1 \quad \Rightarrow \quad \frac{e^{k+1}}{(k+1)^2} > 1. \tag{2.43}$$

Now since we could only prove (2.43) assuming that $k \geq 2$, we have to take the base case to be $k = 2$. In this case we have

$$\frac{e^2}{2^2} > 1, \tag{2.44}$$

so that the base case is true.

By the slight generalization of induction (Axiom 2.8 below), we conclude that for all $n \geq 2$,

$$\frac{e^n}{n^2} > 1. \tag{2.45}$$

But we have already observed that this inequality is also true for $n = 1$. Thus, we have completed the proof of the theorem. □

Remark 2.7 (Manipulating inequalities in proofs). The proof of Theorem 2.6 has the following general structure. We started with what we'll call "Inequality A", which we want to prove. We showed that this follows from "Inequality B" and that "Inequality B" is equivalent to "Inequality C". Then we prove "Inequality C", and we conclude that "Inequality A" is true. Regarding this, the equivalence of inequalities allows us to reason *backwards*, which is essential to our proof.

An example of where we cannot reason backwards is the following. Let x and y be real numbers. Suppose $x \geq y$. Then $0 \cdot x \geq 0 \cdot y$, which is equivalent to $0 \geq 0$. Although $0 \geq 0$ is true, in this case we cannot reason backwards and conclude that $x \geq y$ for all real numbers x and y.

Can you mimic the idea of the proof of Theorem 2.6 to solve the following?

Exercise 2.18. *Prove by induction that for each integer $n \geq 4$, $4^n \geq 16n^2$.*

2.2.2. Induction starting at a base case other than $n = 1$.

The proof of Theorem 2.6 is an example of a proof by induction where the base case is an integer different from 1. Here is the slightly more general version of induction we used above.

Axiom 2.8 (Mathematical induction slightly more generally). *Let b be an integer, and let $P(b), P(b+1), P(b+2), \ldots$ be an infinite sequence of statements. Suppose that $P(b)$ is true, and suppose that $P(k)$ implies $P(k+1)$ for all $k \geq b$. Then $P(n)$ is true for all $n \geq b$.*

This axiom, where the base case is an arbitrary fixed integer b, is equivalent to the usual induction axiom where the base case is the integer 1. We won't prove this equivalence. Readers looking for something fun to do may prove it for themselves.

We can also combine induction with a little bit of logical deduction to prove a fun result.

Exercise 2.19 (Kung Fu Panda meets The Penguin). *Prove the following statement is true for all integers $n \geq 2$. If a total of n pandas and penguins stand in a line to get gelato, where the first animal in the line is a panda and the last animal in the line is a penguin, then somewhere in the line there is a panda directly in front of a penguin.*

Remark 2.9. As usual, a hint for the exercise above is given at the end of the chapter. However, you may also wish to solve this exercise by reasoning (less formally) not using induction. Namely, in the gelato line there is always a last panda, and this panda is by assumption not the last animal in the line. What animal is behind the last panda?

2.2.3. Comparing exponentials and polynomials a little more generally. We can create many exercises similar to Exercise 2.18. Can you try your hand at one?

Exercise 2.20 (Create your own problem, but don't make it one of Jay-Z's 99). *Choose an interesting positive integer I and a boring positive integer B, where $B > I$. Using these diametrically opposite integers, find the least positive integer L such that: For each integer $n \geq L$,*

$$I^n \geq Bn^2. \tag{2.46}$$

Prove this statement by induction.

We can also change the order of the monomial from 2 to any positive real number p. To be more concrete, we choose the base of the exponential function to be 2.

Solved Problem 2.10 (Exponentials versus monomials: A general look). *Let p be a positive real number, and let $N \geq (2^{1/p} - 1)^{-1}$ be a positive integer. Prove that*

$$2^n \geq \frac{2^N}{N^p} n^p \quad \text{for all } n \geq N. \tag{2.47}$$

Solution. The proof is actually not difficult because although the inequalities may look complicated, they have been rigged to work.

The *base case* $n = N$ is true because in this case we have

$$2^N = \frac{2^N}{N^p} N^p. \tag{2.48}$$

Inductive step: Suppose that $k \geq N$ is such that $2^k \geq \frac{2^N}{N^p} k^p$. This inequality is equivalent to (we put the k's on one side and the N's on the other side)

$$\frac{k^p}{2^k} \leq \frac{N^p}{2^N}. \tag{2.49}$$

Thus, to prove the inductive step, it suffices to prove the inequality

$$\frac{(k+1)^p}{2^{k+1}} \leq \frac{k^p}{2^k}. \tag{2.50}$$

Indeed, since (2.49) is true, if (2.50) is true, then we have

$$\frac{(k+1)^p}{2^{k+1}} \leq \frac{N^p}{2^N}, \tag{2.51}$$

which is equivalent to the desired inductive conclusion $2^{k+1} \geq \frac{2^N}{N^p}(k+1)^p$.

Now, the inequality (2.50) we want to prove is equivalent to the inequality

$$\left(1 + \frac{1}{k}\right)^p = \left(\frac{k+1}{k}\right)^p = \frac{(k+1)^p}{k^p} \leq \frac{2^{k+1}}{2^k} = 2. \tag{2.52}$$

This, in turn, by taking p-th roots, is equivalent to

$$\frac{1}{k} \leq 2^{1/p} - 1,$$

which further in turn is equivalent to

$$k \geq \frac{1}{2^{1/p} - 1}. \tag{2.53}$$

Since $k \geq N$ and by our assumption $N \geq (2^{1/p} - 1)^{-1}$, we have that this last inequality is true. Thus (2.50) is true, and we conclude that the inductive conclusion is true. Finally, by induction (the slightly more general Axiom 2.8), we are done.

Remark 2.11. Let us see how we can use the problem to rig a special case of the general inequality so that we can prove it by induction.

For example, let $p = 3$. We compute that $(2^{1/3} - 1)^{-1} \approx 3.847$, which is less than 4. So we may choose $N = 4$. We then compute that $\frac{2^N}{N^p} = \frac{2^4}{4^3} = \frac{1}{4}$. Therefore a special case of (2.47) is

$$2^n \geq \frac{1}{4} n^3 \quad \text{for all } n \geq 4. \tag{2.54}$$

If Solved Problem 2.10 still seems a bit too abstract, we invite you to prove (2.54) in a more concrete way by substituting $p = 3$ and $N = 4$ into the proof of (2.47).

While we are looking at exponentials, we might as well prove the following fun fact.

Exercise 2.21. *Prove that for all $n \in \mathbb{Z}^+$,*

$$\sum_{i=1}^{n} (i+1) \cdot 2^i = 2 \cdot 2 + 3 \cdot 2^2 + 4 \cdot 2^3 + \cdots + (n+1) \cdot 2^n = n \cdot 2^{n+1}. \tag{2.55}$$

2.2.4. Factorials versus exponentials. As we have seen, in the race to infinity, exponentials beat polynomials. Now we will see that factorials beat exponentials! Recall that factorials are defined by $0! = 1$ and $n! = n(n-1) \cdots 2 \cdot 1$ for $n \in \mathbb{Z}^+$. We will discuss factorials more formally in Example 2.29 below.

The first several factorials and powers of 2 are listed in Table 2.2.1.

Table 2.2.1. Some positive integers and their factorials and powers of 2. It looks like the factorials may be winning the race to infinity. This is indeed true.

n	$n!$	2^n
1	1	2
2	2	4
3	6	8
4	24	16
5	120	32
6	720	64
7	5040	128

Theorem 2.12. *For any integer $n \geq 4$, we have*

$$n! > 2^n. \tag{2.56}$$

Proof. Let $P(n)$ be the statement $n! > 2^n$.

We observe that $4! = 24 > 16 = 2^4$, so that the base case $P(4)$ is true.

Observe that $P(n)$ is equivalent to the statement $Q(n)$ given by

$$\frac{n!}{2^n} > 1. \tag{2.57}$$

For the sequence of statements $Q(n)$, we can follow a similar pattern of proof as for Theorem 2.6, and we invite the reader to work this out.

For variety (after all, it is the spice of life!) we proceed slightly differently: Suppose that an integer $k \geq 4$ satisfies $k! > 2^k$; i.e., suppose that $P(k)$ is true. Then we calculate that

$$(k+1)! = (k+1) \cdot k! > (k+1) \cdot 2^k \geq 2 \cdot 2^k = 2^{k+1}, \tag{2.58}$$

where we used $k! > 2^k$ (our induction hypothesis) for the first (strict) inequality and that $k+1 \geq 2$ for the second (weak) inequality, which is trivially true since $k \geq 4$. Hence $P(k+1)$ is true.

By mathematical induction (Axiom 2.8), we conclude that $n! > 2^n$ for all $n \geq 4$. □

Alternate approach to the inductive step. By wishful thinking (which in this case works!), we speculate that the ratio

$$k \mapsto \frac{k!}{2^k} \tag{2.59}$$

is a monotonically non-decreasing (or perhaps, more strongly, increasing) function of k. So we consider trying to prove the speculative inequality

$$\frac{(k+1)!}{2^{k+1}} \geq \frac{k!}{2^k}. \tag{2.60}$$

By elementary algebraic manipulation, we see that this inequality is equivalent to the following inequality:

$$k+1 = \frac{(k+1)!}{k!} \geq \frac{2^{k+1}}{2^k} = 2,$$

which is plainly true for $k \geq 1$ (with strict inequality if $k > 1$). Thus, if $\frac{k!}{2^k} > 1$, then $\frac{(k+1)!}{2^{k+1}} > 1$, and so the inductive step is proved!

Commentary. Looking back at the proof, we see that inequality (2.58) holds for all $k \geq 1$. This is because we only used the induction hypothesis and the inequality $k+1 \geq 2$, which is true only assuming that $k \geq 1$. Therefore the following statement is true:

$$P(k) \text{ implies } P(k+1) \text{ for all } k \geq 1. \tag{2.61}$$

We can visualize all of this as

$$P(1) \Rightarrow P(2) \Rightarrow P(3) \Rightarrow P(4) \Rightarrow P(5) \Rightarrow \cdots. \tag{2.62}$$

Here, a green implication or statement is true, whereas a red statement is false. By (2.62), we have that if $P(1)$ is true, then $P(2)$ is true. But $P(1)$ is false, which means this implication doesn't tell us anything. It also says that if $P(2)$ is true, then $P(3)$ is true. But, similarly, $P(2)$ is false. Eventually, we get to $P(4)$ which we know is true! So, now we can conclude something, namely that $P(5)$ is true.

Exercise 2.22. *Give an intuitive explanation for why the factorial $n!$ grows faster than a^n for every fixed positive integer a.*

Let a be a positive integer. Let n_0 be an integer greater than a. Explain why

$$(2n_0)! > a^{n_0}. \tag{2.63}$$

Prove by induction that

$$(2n)! > a^n \quad \text{for all } n \geq n_0. \tag{2.64}$$

We remark that Stirling's formula gives the asymptotic behavior of $n!$:

$$n! \sim \sqrt{2\pi n}\left(\frac{n}{e}\right)^n. \tag{2.65}$$

Roughly speaking, $n!$ is more like n^n than a fixed base to the power n.

Here is a sum formula involving factorials.

Exercise 2.23. *Prove by induction on n, for all positive integers n,*

$$1 \cdot 1! + 2 \cdot 2! + 3 \cdot 3! + \cdots + n \cdot n! = (n+1)! - 1.$$

For more practice with inequalities, consider the following.

Exercise 2.24. *Prove by induction that for every integer $n \geq 2$,*

$$\frac{1}{\sqrt{1}} + \frac{1}{\sqrt{2}} + \frac{1}{\sqrt{3}} + \cdots + \frac{1}{\sqrt{n}} > \sqrt{n}. \tag{2.66}$$

Can you also give a simple explanation (without using induction or integration as in the integral test) for why the inequality is true?

2.3. Sums of powers of the first n positive integers

The idea of summing over positive integers can be replaced by summing over powers of positive integers. Let k be a positive integer. Consider the sum

$$\sum_{i=1}^{n} i^k = 1^k + 2^k + 3^k + \cdots + n^k. \tag{2.67}$$

In this section we will first give visual proofs of formulas for these sums for the powers $k = 1, 2, 3$. We then give a proof by induction of a formula for $k = 4$.

Firstly, consider again the $k = 1$ case, where the sum is

$$A_n := 1 + 2 + \cdots + n. \tag{2.68}$$

Recall that adding this sum to itself in the reverse order yields

$$\begin{array}{ccccccc} 1 & + & 2 & +\cdots+ & n \\ n & + & (n-1) & +\cdots+ & 1 \\ \hline (n+1) & + & (n+1) & +\cdots+ & (n+1) \end{array}, \tag{2.69}$$

so we obtain

(2.70) $$2A_n = 2(1+2+\cdots+n) = n(n+1).$$

By dividing by 2, we obtain (2.2), which is the formula given by Theorem 2.1. We can visualize this algebra by arranging squares suitably as in Figure 2.3.1.

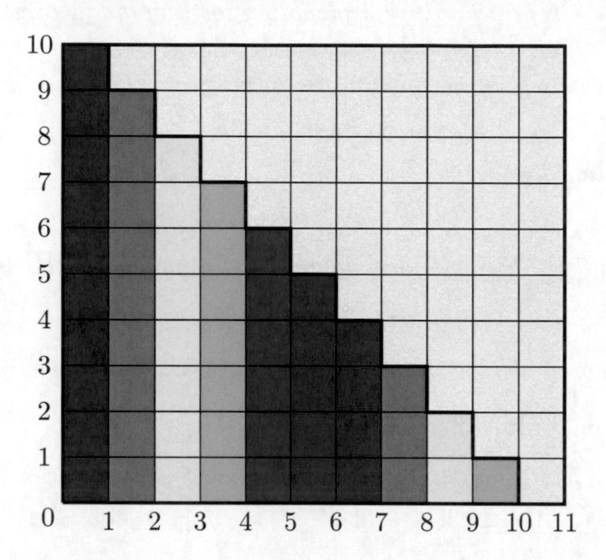

Figure 2.3.1. A picture is worth $2^3 5^3$ words: since A_{10} = rainbow area, which equals the gray area, we have $2A_{10} = 10 \cdot 11$ is the total area of the rectangle.

Secondly, for the case $k = 2$ we consider the marvelous formula (2.24) in Exercise 2.6:

(2.71) $$B_n := 1^2 + 2^2 + 3^2 + \cdots + n^2 = \frac{n(n+1)(2n+1)}{6}.$$

We picture k^2 as k^2 1-by-1-by-1 cubes combined to form a k-by-k-by-1 rectangular solid. We fit together a 1-by-1-by-1 cube, a 2-by-2-by-1 cube, and a 3-by-3-by-1 cube to form a **mini-sculpture** whose volume is equal to $1^2 + 2^2 + 3^2$; see Figure 2.3.2. We can now fit three mini-sculptures together to get a **mega-sculpture** which almost looks like a cube. Figure 2.3.2 shows us how to slice the top halves of the top level of the mega-scupture to obtain a rectangular solid whose dimensions are

(2.72) $$3\text{-by-}4\text{-by-}3.5.$$

Thus the volume of the mega-sculpture is $\frac{3\cdot 4\cdot(2\cdot 3+1)}{2}$. We conclude that the volume of each mini-sculpture is

(2.73) $$1^2 + 2^2 + 3^2 = \frac{3\cdot 4\cdot(2\cdot 3+1)}{6}.$$

This is precisely formula (2.71) for the case where $k = 2$ and $n = 3$.

Figure 2.3.2. A visual proof that $1^2+2^2+3^2+\cdots+n^2 = \frac{1}{3}n(n+1)(n+\frac{1}{2})$. Six times this sum is equal to $n(n+1)(2n+1)$. Credit: Wikimedia Commons, author: Salix alba, licensed under Creative Commons Attribution-Share Alike 4.0 International (https://creativecommons.org/licenses/by-sa/4.0/deed.en) license

Thirdly, for the case $k = 3$ the formula for the sum of the third powers as stated in Exercise 2.9 can be rewritten as

$$\sum_{i=1}^{n} i^3 = 1^3 + 2^3 + 3^3 + \cdots + n^3 = \left(\frac{n(n+1)}{2}\right)^2. \tag{2.74}$$

By using (2.2), we see that this is equivalent to the formula visualized in Figure 2.3.3.

Fourthly, for the case $k = 4$ we have the formula

$$\sum_{i=1}^{n} i^4 = 1^4 + 2^4 + 3^4 + \cdots + n^4 = \frac{1}{30}n(n+1)(2n+1)(3n^2+3n-1). \tag{2.75}$$

We prove this in Solved Problem 2.13 below.

Remarkably, there is a formula for the sum of the k-th powers for every positive integer k. We now make some comments about this, and we refer the interested reader to the literature for more discussion. Let B_j be the j-th Bernoulli number. We will not go into the details about the definitions of the Bernoulli numbers except to say that the interested reader may consult the Wikipedia link for "Bernoulli number" and other sources (so don't worry if you don't know what they are!).

Faulhaber's formula (which we will not prove) says the following:

$$\sum_{i=1}^{n} i^k = \frac{1}{k+1}n^{k+1} + \frac{1}{2}n^k + \sum_{j=2}^{k} \frac{k!\,B_j}{j!\,(k-j+1)!}\,n^{k-j+1}. \tag{2.76}$$

For example, for $k = 4$ we obtain the formula

$$\sum_{i=1}^{n} i^4 = \frac{1}{5}n^5 + \frac{1}{2}n^4 + B_2n^3 + B_3n^2 + B_4n. \tag{2.77}$$

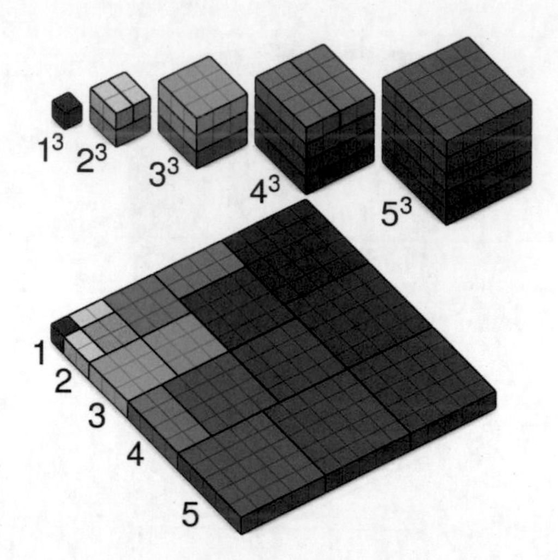

Figure 2.3.3. A visual proof that

$$1^3 + 2^3 + 3^3 + \cdots + n^3 = (1 + 2 + 3 + \cdots + n)^2.$$

Since $B_2 = \frac{1}{6}$, $B_3 = 0$, and $B_4 = -\frac{1}{30}$ (the Bernoulli numbers can be computed using recursive formulas), we obtain

$$\sum_{i=1}^{n} i^4 = \frac{1}{5}n^5 + \frac{1}{2}n^4 + \frac{1}{3}n^3 - \frac{1}{30}n. \tag{2.78}$$

This quintic polynomial in n is factored as the right-hand side of (2.75), so indeed (2.75) is a special case of Faulhaber's formula.

Solved Problem 2.13 (Proof of Faulhaber's formula for $k = 4$). *Prove* (2.78), *that is,*

$$\sum_{i=1}^{n} i^4 = \frac{1}{5}n^5 + \frac{1}{2}n^4 + \frac{1}{3}n^3 - \frac{1}{30}n,$$

by induction.

Solution. Here, it is convenient to use the Binomial Theorem 7.26 below, which says that

$$\begin{aligned}(a+b)^n &= \sum_{i=0}^{n} \binom{n}{i} a^{n-i} b^i \\ &= a^n + \cdots + \binom{n}{i} a^{n-i} b^i + \cdots + b^n.\end{aligned} \tag{2.79}$$

For example, taking $a = m$ and $b = 1$ we obtain for $n = 2, 3, 4, 5$ that

$$(m+1)^2 = m^2 + 2m + 1, \tag{2.80a}$$

$$(m+1)^3 = m^3 + 3m^2 + 3m + 1, \tag{2.80b}$$

$$(m+1)^4 = m^4 + 4m^3 + 6m^2 + 4m + 1, \tag{2.80c}$$

$$(m+1)^5 = m^5 + 5m^4 + 10m^3 + 10m^2 + 5m + 1. \tag{2.80d}$$

We will use the case $n = 4$. As background, in (2.79), $\binom{n}{i} = \frac{n!}{i!(n-i)!}$ denotes the *binomial coefficient* n choose i. If you are not familiar with binomial coefficients, you may just use the fact that the coefficients of the polynomials in (2.80) are given by Pascal's triangle.

We prove (2.78) by induction on n. We will just check the inductive step as the base is easy as usual. Let

$$p(n) := \frac{1}{5}n^5 + \frac{1}{2}n^4 + \frac{1}{3}n^3 - \frac{1}{30}n$$

denote the quintic polynomial on the right-hand side of (2.78). The inductive hypothesis is that m is a positive integer with the property that $\sum_{i=1}^{m} i^4 = p(m)$. The desired inductive conclusion is that the formula $\sum_{i=1}^{m+1} i^4 = p(m+1)$ is true. By taking differences, we see that this is equivalent to

$$(m+1)^4 = \sum_{i=1}^{m+1} i^4 - \sum_{i=1}^{m} i^4 = p(m+1) - p(m). \tag{2.81}$$

Therefore, to prove the inductive step, we just need to show that for every positive integer m, $(m+1)^4$ equals $p(m+1) - p(m)$. To verify this, we compute using (2.80) that

$$\begin{aligned}
&p(m+1) - p(m) \\
&\quad = \frac{1}{5}((m+1)^5 - m^5) + \frac{1}{2}((m+1)^4 - m^4) \\
&\quad\quad + \frac{1}{3}((m+1)^3 - m^3) - \frac{1}{30}((m+1) - m) \\
&\quad = \left(m^4 + 2m^3 + 2m^2 + m + \frac{1}{5}\right) + \left(2m^3 + 3m^2 + 2m + \frac{1}{2}\right) \\
&\quad\quad + \left(m^2 + m + \frac{1}{3}\right) - \frac{1}{30} \\
&\quad = m^4 + 4m^3 + 6m^2 + 4m + 1 \\
&\quad = (m+1)^4.
\end{aligned}$$

Since we have the desired equality, this proves the inductive step. By induction we are done. □

2.3.1. Relation of summing powers to integration. Let k be a positive integer. Define the n-th partial sum of the k-th power sum by

$$s_n := 1^k + 2^k + 3^k + \cdots + n^k. \tag{2.82}$$

Let

$$f(x) = x^k, \quad \text{for } x \in \mathbb{R}, \tag{2.83}$$

be the k-th power function. We observe that

$$\frac{s_n}{n^k} = \left(\frac{1}{n}\right)^k + \left(\frac{2}{n}\right)^k + \left(\frac{3}{n}\right)^k + \cdots + \left(\frac{n}{n}\right)^k = \sum_{i=1}^{n} f\left(\frac{i}{n}\right). \tag{2.84}$$

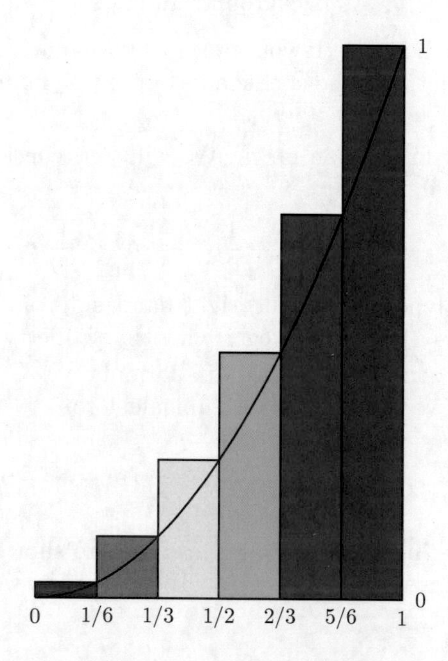

Figure 2.3.4. The graph of the squaring function $f(x) = x^2$ and a Riemann sum partitioning $[0, 1]$ into $n = 6$ equal subintervals and using right endpoints. The Riemann sum is approximately equal to the area under the graph, with equality in the limit as $n \to \infty$.

In general, for a continuous function $f : [0, 1] \to \mathbb{R}$, the **Riemann sum** of f over $[0, 1]$ using the right endpoints of a partition of $[0, 1]$ into n equal length subintervals is equal to

$$n\text{-th Riemann sum} = \sum_{i=1}^{n} f\left(\frac{i}{n}\right) \Delta x, \tag{2.85}$$

where $\Delta x := \frac{1}{n}$. By calculus, we know that

$$\lim_{n\to\infty} n\text{-th Riemann sum} = \int_0^1 f(x)\, dx = \int_0^1 x^k dx = \frac{1}{k+1}. \tag{2.86}$$

By combining all of this, we obtain

$$\frac{1}{k+1} = \lim_{n\to\infty} \frac{s_n}{n^k} \frac{1}{n} = \lim_{n\to\infty} \frac{s_n}{n^{k+1}}. \tag{2.87}$$

Equivalently, we have the limit formula

$$\lim_{n\to\infty} \frac{s_n}{\frac{n^{k+1}}{k+1}} = 1. \tag{2.88}$$

Since the ratio of s_n to $\frac{1}{k+1}n^{k+1}$ tends to 1 as n goes to infinity, we will express this as

$$\sum_{i=1}^{n} i^k = s_n \sim \frac{1}{k+1}n^{k+1} \quad \text{for } n \text{ large.} \tag{2.89}$$

This asymptotic formula gives us more confidence in Faulhaber's formula (2.76). That is, if we can prove a non-trivial consequence of a formula we are not sure about (e.g., a formula we haven't proved), then we have more confidence in the formula.

2.3.2. Revisiting visualizing the sum of the first n positive integers. Just like formal proofs, visual proofs are not unique. We often can come up with different visual proofs of the same result. For example, does Figure 2.3.5 convince you that

$$1+2+3+\cdots+n = \binom{n+1}{2}? \tag{2.90}$$

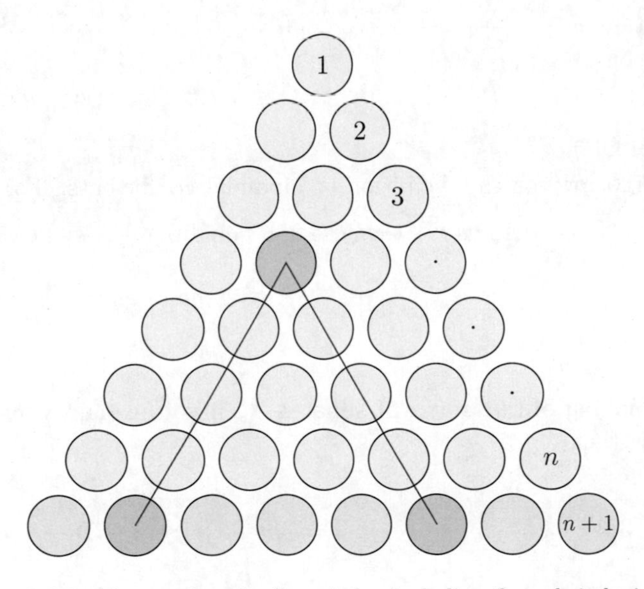

Figure 2.3.5. The number of yellow circles, including the red circle, is equal to $1+2+3+\cdots+n$. There are $n+1$ orange circles, including the two red circles, in the last row. Each yellow circle is in one-to-one correspondence with a choice of two of the $n+1$ orange circles.

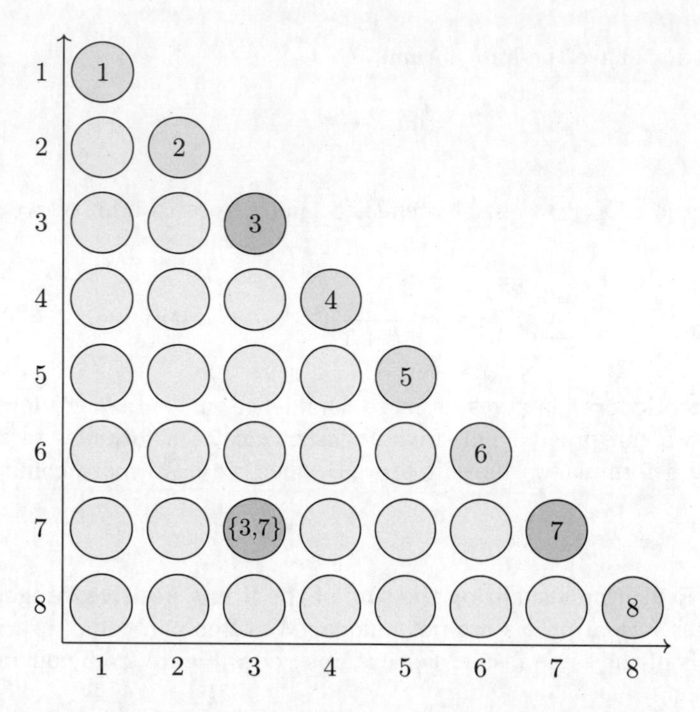

Figure 2.3.6. Another rendering for $n = 7$ of the same visualization as in Figure 2.3.5. Each yellow circle corresponds to a cardinality 2 subset of $\mathbb{N}_8$.

See Chapter 7 below for the definition of binomial coefficients. For now, we just remark that $\binom{n+1}{2}$, read as $n+1$ choose 2, is equal to $\frac{n(n+1)}{2}$, so we recover (2.2) from (2.90).

2.3.3. Thinking about sums of squares.

Can you explain why Figures 2.3.7 and 2.3.8 show that

$$3(1^2 + 2^2 + 3^2 + \cdots + n^2) = (2n+1)(1+2+3+\cdots+n)\,? \tag{2.91}$$

Observe that by applying (2.2) to this, we recover (2.24).

Exercise 2.25. *Can you complete from* (2.91) *a visual proof of the formula in Exercise* 2.6*?*

Figure 2.3.7. The sum of the integers in each of the three equilateral triangles equals $1^2 + 2^2 + 3^2 + \cdots + n^2$.

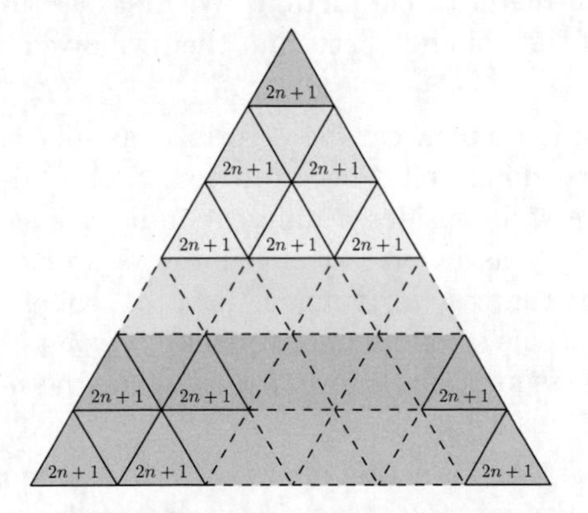

Figure 2.3.8. Each position in the three triangles sums to $2n+1$. Moreover, there are $1 + 2 + 3 + \cdots + n$ positions in each triangle.

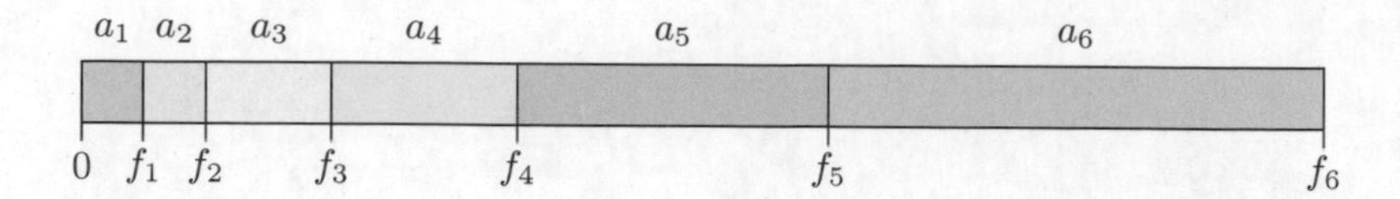

Figure 2.3.9. A visual explanation for why Exercise 2.26 is true.

2.3.4. Thinking about proving summation formulas by induction.

In the solution to Solved Problem 2.13 we employed a special case of the following general method for proving summation formulas.

Exercise 2.26. *Let $\{a_n\}_{n=1}^{\infty}$ be a sequence of real numbers, and let $f : \mathbb{Z}^+ \to \mathbb{R}$ be a function. Suppose that $a_1 = f(1)$ and that for all $k \in \mathbb{Z}^+$,*

$$a_{k+1} = f(k+1) - f(k). \tag{2.92}$$

Prove by induction that the formula

$$a_1 + a_2 + a_3 + \cdots + a_n = f(n) \tag{2.93}$$

is true for all $n \in \mathbb{Z}^+$.

Can you reprove some of the induction exercises we have seen thus far by this method?

2.4. Strong mathematical induction

Now that we are comfortable with mathematical induction, we can think about improvements. Strong mathematical induction is a form of induction, which in any circumstance is at least as powerful as the mathematical induction we have thus far considered.

2.4.1. Strong mathematical induction.

We first describe strong mathematical induction in terms of the domino effect, and then we give a more formal presentation.

The strong induction axiom can be described as follows. Firstly, we assume, as before, that the first domino has been toppled. So the idea of the base case has not changed. But we will consider a different inductive step. Suppose we have, perhaps not as carefully as before, arranged the dominoes so that we have the following property. If the first k dominoes have *all* toppled, then the $(k+1)$-st domino topples (this is the strong inductive step). Strong mathematical induction says that, under the assumptions above, for every n, the n-th domino eventually topples.

We can think of strong induction allowing us to use more dominoes to topple the "next" domino. For example, changing from dominoes to Dalmations, if we want to topple the 101-st Dalmation, then by the strong inductive hypothesis, we have all of the first 100 Dalmations at our disposal to use for toppling Dalmation number 101! Note that we don't always have to use all of the Dalmations at our disposal. For example, hypothetically, we might be in a situation where we only need to use Dalmation numbers 99 and 100 to topple Dalmation number 101. As we will see later, this is sometimes the case for results about the "Fibonacci numbers".

In analogy to Axiom 2.5, formally, we may state the mathematical induction axiom as:

Axiom 2.14 (Strong Mathematical Induction). *Let $P(1), P(2), P(3), \ldots$ be an infinite sequence of statements. Suppose:*

(1) *$P(1)$ is true.*

(2) *If $k \in \mathbb{Z}^+$ has the property that $P(1), P(2), \ldots, P(k)$ are all true, then $P(k+1)$ is true.*

Then $P(n)$ is true for all $n \geq 1$.

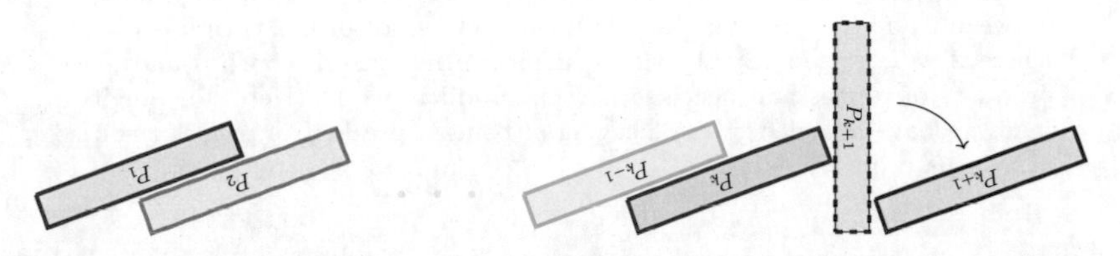

Figure 2.4.1. For the inductive step in strong induction, we suppose that all of the first k dominoes $P_1, P_2, \ldots, P_{k-1}, P_k$ have fallen and we need to prove that this implies that the $(k+1)$-st domino P_{k+1} falls.

2.4.2. Summary of strong induction. We can think of an infinite sequence of statements as a Stairway to Heaven. For mathematical induction, we are saying that if the first step exists and if we know the implication that the $(k+1)$-st step exists provided the k-th step exists, then the whole Stairway to Heaven exists.

For strong induction we are saying the following. If the first step exists and if we know the implication that the $(k+1)$-st step exists provided the first k steps all exist, then the whole Stairway to Heaven exists. Figure 2.4.1 gives a visualization of strong induction.

2.4.3. Existence part of the statement of the Prime Factorization Theorem. A wonderful application of strong induction is the following "existence" part of the Prime Factorization Theorem.

Theorem 2.15. *For any integer $n \geq 2$, n is prime or n is a product of primes.*

Proof. Let $P(n)$ be the following statement: n is prime or n is a product of primes.

Base case: $P(2)$ is true because 2 is a prime.

Strong inductive step: Suppose that $k \geq 2$ has the property that each of $2, 3, \ldots, k$ is a prime or is a product of primes (this is the strong inductive hypothesis).

Case 1. *$k+1$ is a prime.* Then we are done.

Case 2. *$k+1$ is not a prime.* By this assumption, there exist integers a and b satisfying $k+1 = a \cdot b$ and $1 < a, b < k+1$; that is, $2 \leq a, b \leq k$. By the strong inductive hypothesis, we know that a and b are each a prime or a product of primes. Hence their product $a \cdot b = k+1$ is a product of primes. Figure 2.4.2 gives a visualization of the inductive step for Case 2.

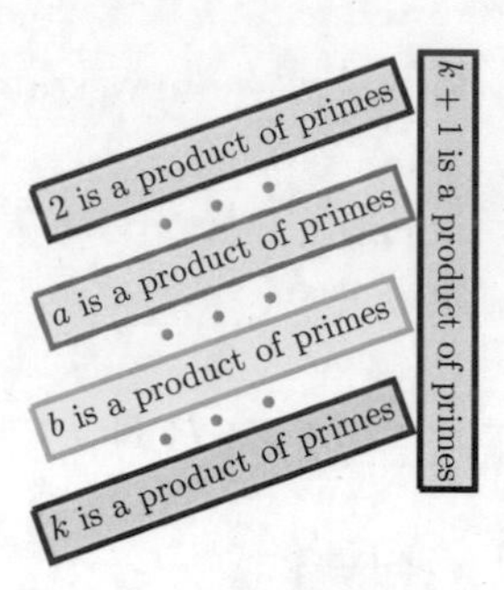

Figure 2.4.2. For the strong inductive step, we assume that all of the integers between 2 and k are products of primes. If $k+1$ is not prime, then $k+1 = ab$, where $2 \leq a, b \leq k$. By the strong inductive hypothesis, both a and b are products of primes and hence so is their product $k+1$. (Here, for brevity's sake, we have included in the meaning of being a product of primes the case of being a prime.)

In either case, we have proved the desired strong inductive conclusion that $k+1$ is a prime or a product of primes.

By strong mathematical induction (Axiom 2.14), we have proved that $P(n)$ is true for all $n \geq 2$. □

Remark 2.16. Another application of strong induction, to prime numbers, is Euclid's bound for the n-th prime number in Theorem 4.15 below.

2.4.4. Primes and the greatest common divisor. Primes have nice properties with respect to the greatest common divisor. We will be able to soup up properties by using induction. Firstly, we observe the following property.

Proposition 2.17 (Commonality with a prime is an all-or-nothing proposition). *If p is a prime and if a is not a multiple of p, then*

$$\gcd(a, p) = 1.$$

In other words, if p is a prime and if a is an integer, then

$$(a \text{ is a multiple of } p) \text{ or } \gcd(a, p) = 1.$$

In particular, if p and q are distinct primes, then $\gcd(p, q) = 1$ (we proved this last fact in §1.5.3).

Proof. Suppose that p is a prime and that a is not a multiple of p. Since p is a prime, its only positive divisors are 1 and p. Since $\gcd(a, p)$ is a positive divisor of p, we have $\gcd(a, p) = 1$ or $\gcd(a, p) = p$. Suppose $\gcd(a, p) = p$. Then, since $\gcd(a, p)$ divides a, we have p divides a, which contradicts our hypothesis. We conclude that $\gcd(a, p) = 1$.

Note that if p and q are distinct primes, then neither can be a multiple of the other. Thus $\gcd(p, q) = 1$. □

For example, if M is the set of multiples of, say, the prime 7, that is,

$$M = \{7k : k \in \mathbb{Z}\},$$

then $\gcd(n, 7) = 7$ for all $n \in M$. On the other hand, by Proposition 2.17 we have that $\gcd(n, 7) = 1$ for all $n \notin M$. ("$\in$" means "is an element of", and "$\notin$" means "is not an element of". See Chapter 5 below for every set theory notation you have not seen before.)

Can a prime have anything to do with the positive integers smaller than it, even if we multiply all of them together? The answer is provided by:

Corollary 2.18. *If p is a prime, then*

$$\gcd(p, (p-1)!) = 1. \tag{2.94}$$

Proof. By Proposition 2.17,

$$1 = \gcd(p, 1) = \gcd(p, 2) = \gcd(p, 3) = \cdots = \gcd(p, p-1).$$

Thus, by Exercise 2.27 right below (which is proved by induction), we have

$$\gcd(p, (p-1)!) = \gcd\left(p, 1 \cdot 2 \cdot 3 \cdots (p-1)\right) = 1. \tag{2.95}$$

□

For example, without having to do a calculation, we know that $\gcd(97, 96!) = 1$. By the way, $96! \approx 10^{150}$, so it would be hard to verify this fact without using the corollary!

Exercise 2.27. *Observe that Lemma 4.29 below says that if $\gcd(a, b) = 1$ and $\gcd(a, c) = 1$, then $\gcd(a, bc) = 1$. For this exercise, we will boldly assume this fact.*

Let a and $b_1, b_2, \ldots, b_k$ be non-zero integers such that $\gcd(a, b_i) = 1$ for $1 \leq i \leq k$. Prove by induction that

$$\gcd(a, b_1 b_2 \cdots b_k) = 1. \tag{2.96}$$

Hint: $b_1 \cdots b_{k+1} = (b_1 \cdots b_k) \cdot b_{k+1}$.

2.4.5. Numbers of tilings by dominoes of a rectangle. Dominoes, besides being useful for toppling, are also useful for tiling. Let $n \in \mathbb{Z}^+$. Consider the sequence of positive integers $\{t_n\}_{n=1}^{\infty}$ defined by:

> t_n is equal to the number of tilings of an $n \times 2$ rectangle by 1×2 rectangles (which we can think of as dominoes).

We'll call $\{t_n\}_{n=1}^{\infty}$ the sequence of **tiling numbers**. Clearly, $t_1 = 1$, and a moment's thought gives us that $t_2 = 2$. We can draw some pictures to compute t_n for n small. Figure 2.4.3 shows how to compute t_1, t_2, t_3, and t_4.

We have recursive formulas for t_3 and t_4:

Exercise 2.28. *Using Figure 2.4.4, explain why we have the following formulas:*

$$t_3 = t_1 + t_2 \quad \text{and} \quad t_4 = t_2 + t_3. \tag{2.97}$$

And we have recursive formula for t_5:

Exercise 2.29. *Verify that $t_5 = t_3 + t_4$. Hint: Mimic the idea presented for $t_3 = t_1 + t_2$ and $t_4 = t_2 + t_3$ in Figure 2.4.4. The secret to seeing all of these recursive formulas is to categorize tilings according to whether the rightmost tile(s) are vertical or horizontal!*

Exercise 2.30. *Can you guess a recursive formula for t_n for all integers $n \geq 3$? Hint: If you are not sure, the topic of the next section should give the answer away!*

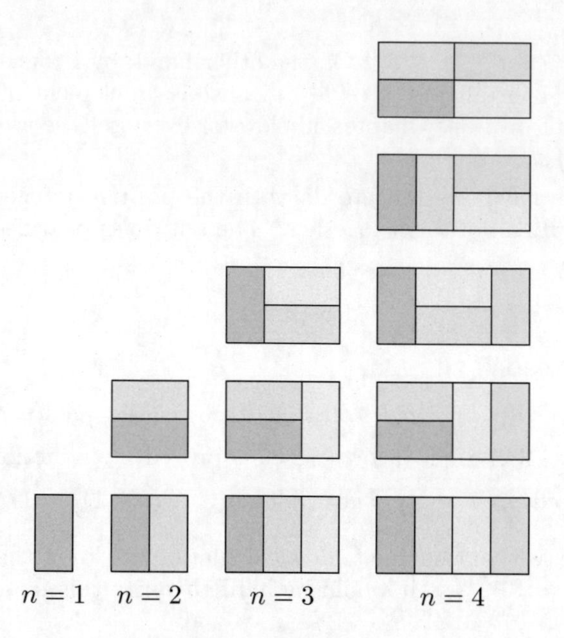

Figure 2.4.3. These arrangements convince us that $t_1 = 1$, $t_2 = 2$, $t_3 = 3$, $t_4 = 5$.

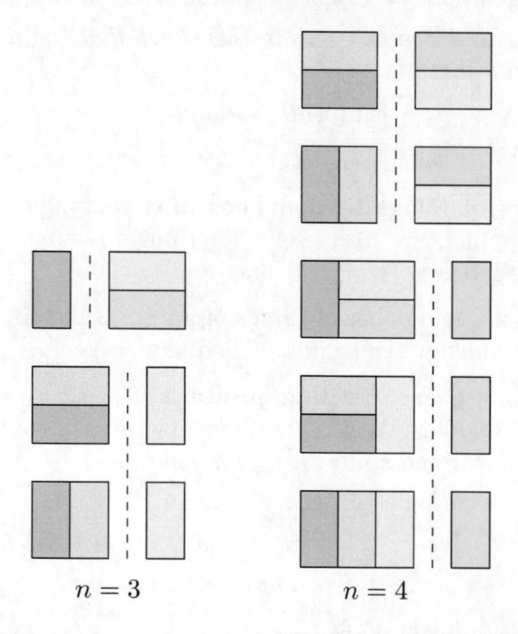

Figure 2.4.4. A visual proof that $t_3 = t_1 + t_2$ and $t_4 = t_2 + t_3$.

2.5. Fibonacci numbers

Fibonacci numbers have numerous applications, for example, the Fibonacci search technique in computer science. In this section, we'll learn about these important numbers.

Figure 2.5.1. Fibonacci (1170–1250). Wikimedia Commons, Public Domain.

The **Fibonacci numbers** form a sequence of positive integers. Let us denote the n-th Fibonacci number by f_n. We define the first two to be one:

$$f_1 = f_2 = 1. \tag{2.98}$$

The rest of the Fibonacci numbers can now be defined by induction:

$$f_n = f_{n-2} + f_{n-1} \quad \text{for } n \geq 3. \tag{2.99}$$

We see that $f_3 = f_1 + f_2 = 1 + 1 = 2$, $f_4 = f_2 + f_3 = 1 + 2 = 3$, and so forth.

The first several Fibonacci numbers (those less than 1000) are

$$1, 1, 2, 3, 5, 8, 13, 21, 34, 55, 89, 144, 233, 377, 610, 987. \tag{2.100}$$

The next several Fibonacci numbers (those less than 1000000) are

$$\begin{gathered} 1597, 2584, 4181, 6765, 10946, 17711, 28657, 46368, \\ 75025, 121393, 196418, 317811, 514229, 832040. \end{gathered}$$

Observe how quickly Fibonacci numbers grow!

For further discussion of inductive definitions, see §2.6.

If you have solved Exercise 2.30, then you will see that the Fibonacci numbers are equal to the tiling numbers: $f_n = t_{n-1}$ for $n \geq 2$.

Historical note. Fibonacci numbers appear in Indian mathematics and in particular in the work of Virahanka.

2.5.1. Patterns in the Fibonacci sequence.

The Fibonacci numbers exhibit many interesting patterns. Sometimes we can find these patterns in the Fibonacci sequence by methodically applying simple ideas. For example, let us start with a Fibonacci number and "push it down":

$$\begin{aligned} f_5 &= f_4 + f_3 \\ &= f_3 + f_3 + f_2 \\ &= f_3 + f_2 + f_2 + f_1 \\ &= f_3 + f_2 + f_1 + 1. \end{aligned}$$

Shall we see how to do this for the next Fibonacci number? Similarly, we compute that (using a different color coding of the Fibonacci numbers)

$$
\begin{aligned}
f_6 &= f_5 + f_4 \\
&= f_4 + f_4 + f_3 \\
&= f_4 + f_3 + f_3 + f_2 \\
&= f_4 + f_3 + f_2 + f_2 + f_1 \\
&= f_4 + f_3 + f_2 + f_1 + 1.
\end{aligned}
\tag{2.101}
$$

Figure 2.5.2 shows a way to see the calculation above visually.

f_6	=	f_5	f_4			
	=	f_4	f_3			
		f_4				
	=	f_4	f_3	f_2		
			f_3			
	=	f_4	f_3	f_2	f_1	
				f_2		
	=	f_4	f_3	f_2	f_1	1

Figure 2.5.2. A visual rendering of the calculation in equation (2.101).

As you may expect, the formulas above are part of a general pattern, given by the following.

Solved Problem 2.19. *Let f_n be the n-th Fibonacci number. Prove that for all positive integers n,*

$$\sum_{i=1}^{n} f_i = f_1 + f_2 + f_3 + \cdots + f_n = f_{n+2} - 1. \tag{2.102}$$

Solution. Base case: $n = 1$. We have

$$\sum_{i=1}^{1} f_i = f_1 = 1 = 2 - 1 = f_3 - 1 = f_{1+2} - 1.$$

So the base case is true.

Inductive step. Suppose that k is a positive integer satisfying

$$\sum_{i=1}^{k} f_i = f_{k+2} - 1. \tag{2.103}$$

Using $\sum_{i=1}^{k+1} f_i = \left(\sum_{i=1}^{k} f_i\right) + f_{k+1}$, the inductive hypothesis, and the Fibonacci recursive definition, we compute that

$$\begin{aligned}\sum_{i=1}^{k+1} f_i &= \left(\sum_{i=1}^{k} f_i\right) + f_{k+1} \\ &= f_{k+2} - 1 + f_{k+1} \\ &= f_{k+3} - 1 \\ &= f_{(k+1)+2} - 1.\end{aligned}$$

This proves the inductive step. By induction we are done. □

The Fibonacci numbers satisfy a plethora of identities. Here are two more:

Exercise 2.31. *Prove that for all positive integers n,*

$$f_1 + f_3 + f_5 + \cdots + f_{2n-1} = f_{2n}. \quad (2.104)$$

Exercise 2.32. *Prove that for all positive integers n,*

$$f_2 + f_4 + f_6 + \cdots + f_{2n} = f_{2n+1} - 1. \quad (2.105)$$

The Fibonacci numbers satisfy nice divisibility properties.

Exercise 2.33. *Prove that for every $k \in \mathbb{Z}^+$, the $3k$-th Fibonacci number f_{3k} is even. Hint: If you are having difficulty, see the following solved problem.*

Solved Problem 2.20. *Prove that for every $k \in \mathbb{Z}^+$, the $4k$-th Fibonacci number f_{4k} is divisible by 3.*

Solution. We have that $f_4 = 3$, which of course is divisible by 3, so the base case is true.

As the inductive hypothesis, suppose that k is a positive integer for which f_{4k} is divisible by 3. We then compute that (we go, step-by-step using the Fibonacci recursive definition, from $4k + 4$ back to $4k$)

$$\begin{aligned}f_{4(k+1)} &= f_{4k+4} \\ &= f_{4k+2} + f_{4k+3} \\ &= f_{4k+2} + (f_{4k+1} + f_{4k+2}) \\ &= f_{4k+1} + 2f_{4k+2} \\ &= f_{4k+1} + 2(f_{4k} + f_{4k+1}) \\ &= 2f_{4k} + 3f_{4k+1}.\end{aligned}$$

Now, since f_{4k} is divisible by 3 by our inductive hypothesis and certainly 3 is divisible by 3, their integral linear combination

$$(2)f_{4k} + (f_{4k+1})3 \quad (2.106)$$

is divisible by 3. We conclude that $f_{4(k+1)}$ is divisible by 3. This proves the inductive step.

By induction, we are done.

Exercise 2.34. *Prove that for every $k \in \mathbb{Z}^+$, the $5k$-th Fibonacci number f_{5k} is divisible by f_5.*

The following exercise, although a bit more difficult, is based on a similar idea.

Exercise 2.35. *Prove that for every $k, \ell \in \mathbb{Z}^+$, the $k\ell$-th Fibonacci number $f_{k\ell}$ is divisible by both f_k and f_ℓ.*

Hint: You may for example prove this by induction on ℓ. You may wish to write out explicitly the statements $P(\ell)$ that you will prove.

The Fibonacci sequence has interesting patterns regarding sums that are quadratic in the Fibonacci numbers.

Exercise 2.36. *Prove for all $n \in \mathbb{Z}^+$ that*

$$\sum_{i=1}^{n} f_i^2 = f_1^2 + f_2^2 + f_3^2 + \cdots + f_n^2 = f_n f_{n+1}. \tag{2.107}$$

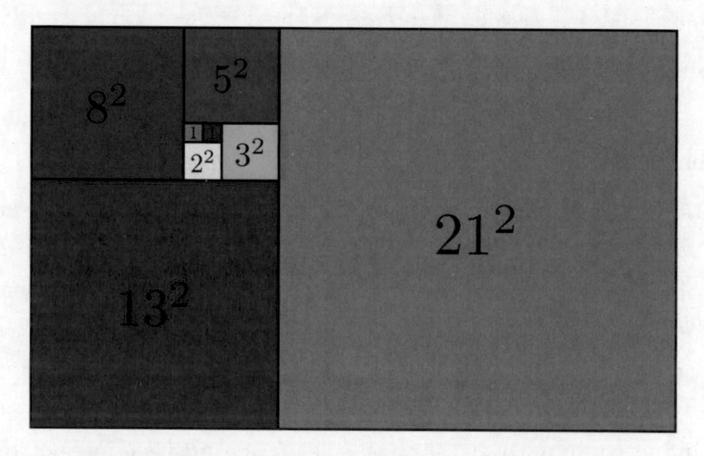

Figure 2.5.3. An illustration of formula (2.107) in the special case that $n = 8$:

$$1^2 + 1^2 + 2^2 + 3^2 + 5^2 + 8^2 + 13^2 + 21^2 = 21 \cdot 34.$$

Déjà vu all over again? This illustrates a rather convincing argument for (2.107), would you not say?

A more challenging formula, which is quadratic in the Fibonacci numbers, is:

Exercise 2.37. *Prove that for all $m, n \in \mathbb{Z}^+$,*

$$f_{m+n} = f_{m-1} f_n + f_m f_{n+1}. \tag{2.108}$$

Hint: You may for example prove this by strong induction on n.

Exercise 2.38. *Prove that for all positive integers n,*

$$f_n^2 + f_{n+1}^2 = f_{2n+1}. \tag{2.109}$$

Hint: You may use (2.129) below.

For sequences, the index is often a positive integer. For the Fibonacci sequence, we can naturally extend it so that the index is any integer, positive, zero, or negative!

Exercise 2.39. *Recall that we defined* $f_0 = 0$. *We can extend the Fibonacci sequence to negative indices by defining for negative integers* n,

$$f_n = f_{n+2} - f_{n+1}. \tag{2.110}$$

(Observe that this formula is actually true for all $n \in \mathbb{Z}$ *since it simply says that* $f_n + f_{n+1} = f_{n+2}$.*)*

We also obtain that

$$f_{-1} = 1,\ f_{-2} = -1,\ f_{-3} = 2,\ f_{-4} = -3,\ etc. \tag{2.111}$$

And we compute that f_{-11} *through* f_0 *are given by*

$$89,\ -55,\ 34,\ -21,\ 13,\ -8,\ 5,\ -3,\ 2,\ -1,\ 1,\ 0. \tag{2.112}$$

Prove the for $n \in \mathbb{Z}^+$,

$$f_{-n} = (-1)^{n+1} f_n. \tag{2.113}$$

Hint: Use strong induction.

The following two exercises also use strong induction.

Exercise 2.40 (Exponential bound for Fibonacci numbers). *Prove that*

$$f_n \leq 2^n \quad for\ n \geq 1. \tag{2.114}$$

Exercise 2.41. *Prove by strong induction that every positive integer is the sum of distinct Fibonacci numbers.*

For example, $1 = f_1$ *(or* f_2*)*, $2 = f_3$, $3 = f_4$, $4 = f_1 + f_4$, $5 = f_5$, $6 = f_1 + f_3 + f_4$.

Hint: If n *is not a Fibonacci number, then let* f *be the largest Fibonacci number less than* n, *and observe that*

$$n = f + (n - f). \tag{2.115}$$

Can you prove that $f > n - f$*? Apply the strong inductive hypothesis to* $n - f$.

2.5.2. Binet's formula. Miraculously, there is an explicit formula for the Fibonacci numbers: **Binet's formula**, which looks like it might be hard to prove, but, with a bit of care, turns out to be relatively easy.

Theorem 2.21. *Let*

$$\alpha = \frac{1+\sqrt{5}}{2} > 0, \qquad \beta = \frac{1-\sqrt{5}}{2} < 0. \tag{2.116}$$

For each $n \in \mathbb{Z}^+$, *the* n*-th Fibonacci number satisfies the formula*

$$f_n = \frac{\alpha^n - \beta^n}{\sqrt{5}}. \tag{2.117}$$

That is,

$$f_n = \frac{\left(\frac{1+\sqrt{5}}{2}\right)^n - \left(\frac{1-\sqrt{5}}{2}\right)^n}{\sqrt{5}}. \tag{2.118}$$

Proof. Observe that α and β are roots of the quadratic polynomial

$$p(x) = x^2 - x - 1 = \left(x - \frac{1}{2}\right)^2 - \frac{5}{4}; \tag{2.119}$$

that is,

$$\alpha^2 - \alpha - 1 = 0, \qquad \beta^2 - \beta - 1 = 0. \tag{2.120}$$

This explains why $\alpha + \beta = 1$ (i.e., the average of α and β is $\frac{1}{2}$) and also why α and β are special.

The number α is called the **golden ratio**.

Let the statement $P(n)$ be the formula $f_n = \frac{\alpha^n - \beta^n}{\sqrt{5}}$. A calculation yields

$$1 = \frac{\left(\frac{1+\sqrt{5}}{2}\right)^1 - \left(\frac{1-\sqrt{5}}{2}\right)^1}{\sqrt{5}} = \frac{\left(\frac{1+\sqrt{5}}{2}\right)^2 - \left(\frac{1-\sqrt{5}}{2}\right)^2}{\sqrt{5}}. \tag{2.121}$$

Thus $P(1)$ and $P(2)$ are true. These two statements form the base case.

Now suppose that $k \geq 2$ has the property that $P(k-1)$ and $P(k)$ are true. Then

$$f_{k-1} = \frac{\alpha^{k-1} - \beta^{k-1}}{\sqrt{5}}, \qquad f_k = \frac{\alpha^k - \beta^k}{\sqrt{5}}. \tag{2.122}$$

We want to prove that $P(k+1)$ is true. By these two formulas and the inductive definition of the Fibonacci sequence, we compute that

$$\begin{aligned} f_k &= f_{k-1} + f_k \\ &= \frac{\alpha^{k-1} - \beta^{k-1}}{\sqrt{5}} + \frac{\alpha^k - \beta^k}{\sqrt{5}} \\ &= \frac{1}{\sqrt{5}}\left(\alpha^{k-1}(1+\alpha) + \beta^{k-1}(1+\beta)\right) \\ &= \frac{1}{\sqrt{5}}\left(\alpha^{k-1}\alpha^2 + \beta^{k-1}\beta^2\right) \qquad \text{(by (2.120))} \\ &= \frac{1}{\sqrt{5}}\left(\alpha^{k+1} + \beta^{k+1}\right). \end{aligned} \tag{2.123}$$

This proves that $P(k+1)$ is true!

By strong mathematical induction, we conclude that $P(n)$ is true for all $n \in \mathbb{Z}^+$. □

2.5.3*. Another proof of Binet's formula. We may view the inductive formula $f_{k+2} = f_{k+1} + f_k$ for $k \geq 1$ for the Fibonacci numbers as a *vector equation* for column vectors of consecutive Fibonacci numbers: For $k \geq 1$,

$$\begin{pmatrix} f_{k+2} \\ f_{k+1} \end{pmatrix} = \begin{pmatrix} 1 & 1 \\ 1 & 0 \end{pmatrix} \begin{pmatrix} f_{k+1} \\ f_k \end{pmatrix}. \tag{2.124}$$

Indeed, the vector equation (2.124), using matrix multiplication on its right-hand side (take the dot products of the rows of the 2×2 matrix with the column vector), says that

$$f_{k+2} = f_{k+1} + f_k,$$
$$f_{k+1} = f_{k+1},$$

where the second equality is tautological (obvious).

We may group two consecutive pairs of consecutive Fibonacci numbers together to get the 2×2 *matrix equation*

$$\begin{pmatrix} f_{k+2} & f_{k+1} \\ f_{k+1} & f_k \end{pmatrix} = \begin{pmatrix} 1 & 1 \\ 1 & 0 \end{pmatrix} \begin{pmatrix} f_{k+1} & f_k \\ f_k & f_{k-1} \end{pmatrix} \tag{2.125}$$

for $k \geq 1$, where $f_0 = 0$ by definition (so that $f_2 = f_0 + f_1$). This matrix equation, like the vector equation (2.124), encodes the Fibonacci recursive definition.

For $k = 1$, (2.125) says that

$$\begin{pmatrix} f_3 & f_2 \\ f_2 & f_1 \end{pmatrix} = \begin{pmatrix} 1 & 1 \\ 1 & 0 \end{pmatrix} \begin{pmatrix} f_2 & f_1 \\ f_1 & f_0 \end{pmatrix} = \begin{pmatrix} 1 & 1 \\ 1 & 0 \end{pmatrix}^2.$$

For $k = 2$, (2.125) says that (we use the previous display)

$$\begin{pmatrix} f_4 & f_3 \\ f_3 & f_2 \end{pmatrix} = \begin{pmatrix} 1 & 1 \\ 1 & 0 \end{pmatrix} \begin{pmatrix} f_3 & f_2 \\ f_2 & f_1 \end{pmatrix} = \begin{pmatrix} 1 & 1 \\ 1 & 0 \end{pmatrix}^3.$$

So one expects that the following formula is true:

$$\begin{pmatrix} f_{k+1} & f_k \\ f_k & f_{k-1} \end{pmatrix} = \begin{pmatrix} 1 & 1 \\ 1 & 0 \end{pmatrix}^k. \tag{2.126}$$

Indeed, we can prove this formula by induction. We leave such a proof to the reader as an exercise.

Exercise 2.42. *Prove* (2.126) *for all positive integers k, where the* 0*-th Fibonacci number is defined to be $f_0 = 0$.*

With the help of some elementary linear algebra, we can use the matrix formula (2.126) as follows. The eigenvalues λ of the matrix $\begin{pmatrix} 1 & 1 \\ 1 & 0 \end{pmatrix}$ satisfy the characteristic equation

$$\begin{aligned} 0 &= \begin{vmatrix} 1-\lambda & 1 \\ 1 & -\lambda \end{vmatrix} \\ &= (1-\lambda)(-\lambda) - 1 \cdot 1 \\ &= \lambda^2 - \lambda - 1. \end{aligned}$$

This is the familiar equation (2.120)!

So $\alpha = \dfrac{1+\sqrt{5}}{2}$ and $\beta = \dfrac{1-\sqrt{5}}{2}$ are the eigenvalues of $\begin{pmatrix} 1 & 1 \\ 1 & 0 \end{pmatrix}$. The eigenvalue α has eigenvector $\begin{pmatrix} \alpha \\ 1 \end{pmatrix}$ and the eigenvalue $\beta = 1 - \alpha$ has eigenvector

$\begin{pmatrix} 1 \\ -\alpha \end{pmatrix}$. That is,

$$\begin{pmatrix} 1 & 1 \\ 1 & 0 \end{pmatrix}\begin{pmatrix} \alpha \\ 1 \end{pmatrix} = \alpha\begin{pmatrix} \alpha \\ 1 \end{pmatrix}, \qquad \begin{pmatrix} 1 & 1 \\ 1 & 0 \end{pmatrix}\begin{pmatrix} 1 \\ -\alpha \end{pmatrix} = \beta\begin{pmatrix} 1 \\ -\alpha \end{pmatrix}.$$

Thus, using (2.126), we calculate that

$$\begin{pmatrix} f_{k+1} & f_k \\ f_k & f_{k-1} \end{pmatrix}\begin{pmatrix} \alpha \\ 1 \end{pmatrix} = \begin{pmatrix} 1 & 1 \\ 1 & 0 \end{pmatrix}^k\begin{pmatrix} \alpha \\ 1 \end{pmatrix} = \alpha^k\begin{pmatrix} \alpha \\ 1 \end{pmatrix}.$$

The second row of this equation says

$$\alpha f_k + f_{k-1} = \alpha^k,$$

so that

$$\alpha^2 f_k + \alpha f_{k-1} = \alpha^{k+1}. \tag{2.127}$$

Similarly, we also have

$$\begin{pmatrix} f_{k+1} & f_k \\ f_k & f_{k-1} \end{pmatrix}\begin{pmatrix} 1 \\ -\alpha \end{pmatrix} = \begin{pmatrix} 1 & 1 \\ 1 & 0 \end{pmatrix}^k\begin{pmatrix} 1 \\ -\alpha \end{pmatrix} = \beta^k\begin{pmatrix} 1 \\ -\alpha \end{pmatrix}.$$

The second row of this equation says

$$f_k - \alpha f_{k-1} = -\beta^k\alpha. \tag{2.128}$$

So, by adding (2.127) and (2.128) together, we obtain

$$(\alpha^2 + 1)f_k = \alpha^{k+1} - \beta^k\alpha = \alpha(\alpha^k - \beta^k);$$

that is,

$$f_k = (\alpha^k - \beta^k)\frac{\alpha}{\alpha^2 + 1} = \frac{\alpha^k - \beta^k}{\sqrt{5}},$$

where for the second equality in the display above we used that

$$\alpha^2 + 1 = \frac{\left(1 + \sqrt{5}\right)^2 + 4}{4} = \frac{\sqrt{5} + 5}{2} = \sqrt{5}\alpha.$$

This reproves Binet's formula. □

Exercise 2.43. *Prove that for all integers $n \geq 2$,*

$$f_{n+1}f_{n-1} - f_n^2 = (-1)^n. \tag{2.129}$$

Hint: Determinant!

Recall that $\mathbb{Z}^{\geq}$ denotes the set of non-negative integers. We remark that, more generally, the ***Vajda–Everman identity*** *says that for $i, j, n \in \mathbb{Z}^{\geq}$,*

$$f_{n+i}f_{n+j} - f_n f_{n+i+j} = (-1)^n f_i f_j. \tag{2.130}$$

But we do not ask you to prove this!

Remark 2.22. Figure 2.5.4 is the basis for another solution to Exercise 2.43.

Exercise 2.44. *Prove for all $n \in \mathbb{Z}^+$ that*

$$\sum_{i=1}^{2n-1} f_i f_{i+1} = f_1 f_2 + f_2 f_3 + f_3 f_4 + \cdots + f_{2n-1}f_{2n} = f_{2n}^2. \tag{2.131}$$

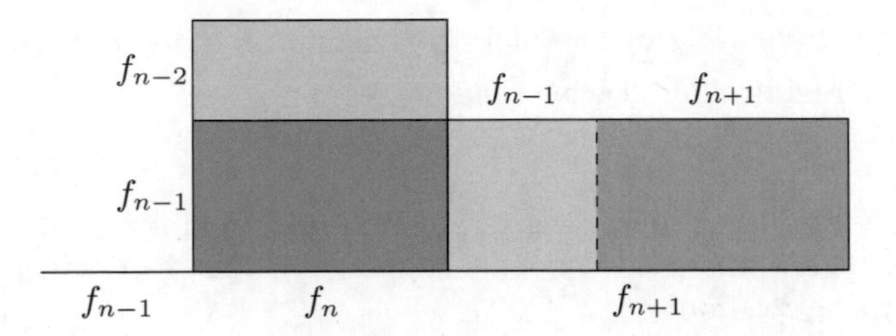

Figure 2.5.4. Can you deduce from considering the areas of the rectangles in this picture the recursive identity: $f_{n+1}f_{n-1} - f_n^2 = f_{n-1}^2 - f_n f_{n-2}$, or equivalently, $f_{n+1}f_{n-1} - f_{n-1}^2 = f_n^2 - f_n f_{n-2}$?

2.6. Recursive definitions

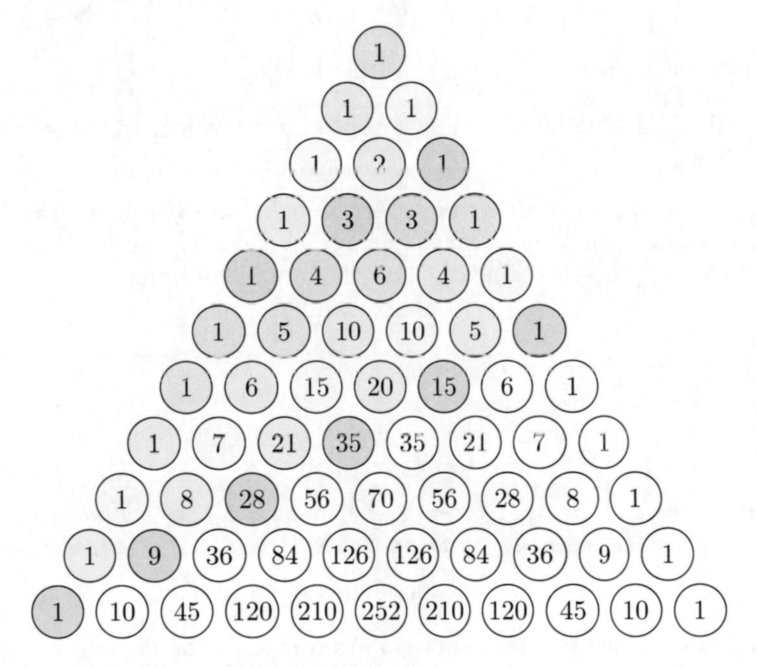

Figure 2.6.1. The diagonal sums of numbers in Pascal's triangle equal the Fibonacci numbers: Summing diagonally the numbers in circles of the same color yields: 1 (red), 1 (orange), 2 (yellow), 3 (green), 5 (blue), 8 (purple), 13 (red), 21 (orange), 34 (yellow), 55 (green), 89 (blue). See Chapter 7 for a discussion of binomial coefficients and Pascal's triangle. Can you find the proof, in your head or anywhere? Spoiler alert: see §7.4.5 below.

In this section we discuss recursive definitions (a.k.a. inductive definitions), of which the Fibonacci sequence is an example.

Example 2.23 (Non-negative powers of real numbers). Recall that the k-th power of a real number b is defined by

$$b^k := \underbrace{b \cdot b \cdots b}_{k \text{ times}}. \tag{2.132}$$

Using *inductive definition*, we can make this definition more formally rigorous.

Let b be a real number. Define the function $f : \mathbb{R} \to \mathbb{R}$ by

$$f(x) = bx.$$

This is simply the linear function given by multiplication by b. As we will see by Theorem 2.24 below, there exists a unique function $A : \mathbb{Z}^{\geq} \to \mathbb{R}$ satisfying $A(0) = 1$ and the recursion relation

$$A(k+1) = b \cdot A(k) \tag{2.133}$$

for all $k \in \mathbb{Z}^{\geq}$. The k-th power of b is formally defined by

$$b^k := A(k) \quad \text{for } k \in \mathbb{Z}^{\geq}. \tag{2.134}$$

From $A(0) = 1$ and (2.133), we see that

$$b^0 = 1, \quad b^1 = b \cdot b^0 = b \cdot 1 = b, \quad b^2 = b \cdot b^1 = b \cdot b, \quad b^3 = b \cdot b^2 = b \cdot b \cdot b,$$

etc., in agreement with (2.132).

By mathematical induction, we can prove the following, which we used in the example above.

Theorem 2.24 (Recursion theorem). *Let X be a set (if you prefer to be more concrete, you may take X to be the set of real numbers $\mathbb{R}$) and let $f : X \to X$ be a function. Then, for every element $a_1 \in X$, there exists a unique function*

$$A : \mathbb{Z}^+ \to X \tag{2.135}$$

satisfying

(1) $A(1) = a_1$,

(2) $A(k+1) = f(A(k))$ *for all* $k \in \mathbb{Z}^+$.

If we denote $a_n := A(n)$ for $n \in \mathbb{Z}^+$, then we have obtained by recursive definition a sequence (actually, by Definition 5.32 *below, A is already a sequence)*

$$\{a_n\}_{n=1}^{\infty} = \{a_1, a_2, a_3, \ldots\}. \tag{2.136}$$

Remark 2.25. In the theorem above, we can replace the domain $\mathbb{Z}^+$ by the set $\mathbb{Z}^{\geq}$, with the obvious resulting changes, e.g., $A(0) = a_0$.

Remark 2.26. Theorem 2.24 actually defines the *iterates* of the function $f : X \to X$. Namely, for each $k \in \mathbb{Z}^+$,

$$f^k : X \to X \tag{2.137}$$

is defined by $f^k(a_1) := A(k)$ for every $a_1 \in X$. In particular, $f^1(a_1) = f(a_1)$, $f^2(a_1) = f\big(f(a_1)\big)$, etc.

Example 2.27. Suppose that we choose instead $a_0 = A(0) = 3$, and we define $f : \mathbb{R} \to \mathbb{R}$ by $f(x) = \frac{x}{2}$. Let $a_k = f^k(a_0)$ for $k \in \mathbb{Z}^{\geq}$, where $f^0 := I_{\mathbb{R}}$ is the identity function. Then

$$a_k = \frac{3}{2^k} \quad \text{for } k \in \mathbb{Z}^{\geq}. \tag{2.138}$$

Proof of Theorem 2.24. Existence of A. Let $S \subset \mathbb{Z}^+$ be the set of n such that there exists a function $A_n : \mathbb{N}_n := \{1, 2, 3, \ldots, n\} \to \mathbb{R}$ satisfying

(1) $A_n(1) = a_1$,
(2) $A_n(k+1) = f(A_n(k))$ for all $1 \leq k \leq n-1$.

We will prove by induction that $S = \mathbb{Z}^+$.

Base case. Clearly $1 \in S$ since we may define $A_1 : \mathbb{N}_1 = \{1\} \to \mathbb{R}$ by $A_1(1) = a_1$.

Inductive step. Suppose that $m \in S$. We define $A_{m+1} : \mathbb{N}_{m+1} \to \mathbb{R}$ by:

(i) $A_{m+1}(k) = A_m(k)$ for $1 \leq k \leq m$.
(ii) $A_{m+1}(m+1) = f(A_m(m))$.

Since $m \in S$, we have

$$A_{m+1}(1) = A_m(1) = a_1. \tag{2.139}$$

Using also that $m \in S$ and (i), we have

$$A_{m+1}(k+1) = A_m(k+1) = f(A_m(k)) = f(A_{m+1}(k)) \tag{2.140}$$

for $1 \leq k \leq m-1$. Furthermore, by (ii) and corresponding to the $k = m$ case, we have that

$$A_{m+1}(m+1) = f(A_m(m)) = f(A_{m+1}(m)). \tag{2.141}$$

Thus, (2.140) is true for $1 \leq k \leq (m+1)-1$. By this and (2.139), the proof of the inductive step is complete. By induction, we conclude that $S = \mathbb{Z}^+$.

Now, since $S = \mathbb{Z}^+$, we can define the function $A : \mathbb{Z}^+ \to \mathbb{R}$ by

$$A(n) := A_n(n) \quad \text{for } n \in \mathbb{Z}^+. \tag{2.142}$$

This function satisfies (exercise!)

(a) $A(1) = a_1$,
(b) $A(k+1) = f(A(k))$ for all $k \in \mathbb{Z}^+$.

We have proved the existence of A.

Remark. We actually have $A|_{\mathbb{N}_n} = A_n$ (restriction) for $n \in \mathbb{Z}^+$ (see §5.3.5.2 below for the definition of the restriction of a function to a subset of its domain).

Uniqueness of A. Suppose that we have a second function $A' : \mathbb{Z}^+ \to X$ satisfying $A'(1) = a_1$ and $A'(k+1) = f(A'(k))$ for all $k \in \mathbb{Z}^+$.

We will now prove by induction that $A(n) = A'(n)$ for all $n \in \mathbb{Z}^+$.

Base case: $n = 1$. By assumption, $A(1) = a_1 = A'(1)$.

Inductive step. Suppose $k \in \mathbb{Z}^+$ is such that $A(k) = A'(k)$. Since f is a function and by our hypotheses, we have

$$A(k+1) = f(A(k)) = f(A'(k)) = A'(k+1). \tag{2.143}$$

By induction, we are done. This proves the uniqueness of A. □

A slightly more general form of recursive definition is as follows. We don't bother to prove this theorem. The interested readers are very welcome to prove it for themselves.

Theorem 2.28. *Let X be a set, let $a_0 \in X$, and let $f : X \times \mathbb{Z}^{\geq} \to X$ be a function. Then there exists a unique function $A : \mathbb{Z}^{\geq} \to X$ satisfying*

(1) $A(0) = a_0$,

(2) $A(k+1) := f(A(k), k)$ *for* $k \in \mathbb{Z}^{\geq}$.

Example 2.29. Recall that the factorial of a non-negative integer is defined by

(1) $0! := 1$,

(2) $n! = 1 \cdot 2 \cdot 3 \cdots (n-1) \cdot n$ for $n \in \mathbb{Z}^{+}$.

So we see the recursion relation

$$(k+1)! = (k+1) \cdot k!. \tag{2.144}$$

In terms of the general recursive definition in Theorem 2.28, we are taking $X = \mathbb{Z}^{+}$, $a_0 := 1$, and

$$f(m,k) := (k+1) \cdot m. \tag{2.145}$$

So $A(0) = 1$, and

$$A(k+1) = f(A(k), k) = (k+1) \cdot A(k)$$

for $k \in \mathbb{Z}^{\geq}$. We see that

$$A(k) = k!.$$

Another generalization of Theorem 2.24 is as follows (again, we don't prove this, although the proof is not difficult).

Theorem 2.30. *Let X be a set and let $f : X \times X \to X$ be a function. Then, for every element $a_1, a_2 \in X$, there exists a unique function $A : \mathbb{Z}^{+} \to X$ satisfying*

(1) $A(1) = a_1$, $A(2) = a_2$,

(2) $A(k+1) = f(A(k-1), A(k))$ *for integers* $k \geq 2$.

An example of this is the Fibonacci sequence $\{f_n\}_{n=1}^{\infty}$ discussed in §2.5, where we take

$$f(x,y) := x + y \tag{2.146}$$

in Theorem 2.30. Indeed, in this case we also take $a_1 = a_2 = 1$, and we compute that

$$\begin{aligned} A(3) &= f(A(1), A(2)) = A(1) + A(2) = 1 + 1 = 2, \\ A(4) &= A(2) + A(3) = 1 + 2 = 3, \end{aligned} \tag{2.147}$$

etc. We leave it to the reader to verify that $f_n = A(n)$ for all $n \in \mathbb{Z}^{+}$.

Of course, we can have more general versions of recursive definitions. The discussion above is just a sample.

Exercise 2.45. *Define inductively the sequence of numbers $\{g_n\}_{n=0}^{\infty}$ by $g_0 = 1$ and*

$$g_n = 2g_{n-1} + 2^n \quad \textit{for } n \geq 1. \tag{2.148}$$

Prove that $g_n = (n+1)2^n$ for all $n \geq 0$.

2.7. Arithmetic and algebraic equalities and inequalities

We can prove some easy but fundamental algebraic identities using induction. We start with something more fun.

2.7.1. A philatelic puzzle.

Exercise 2.46. *Prove that for every integer $n \geq 12$ there exist non-negative integers a and b such that*

$$n = 4a + 5b. \tag{2.149}$$

A philatelic way to describe this is that any postage amount which is at least 12 *cents (oh, remember the good old days!) can be made from* 4*-cent and* 5*-cent stamps.*

Hint: A case pertinent to the idea of the proof of the inductive step is $n = 15$. Also, split up the proof into two cases, where one of the cases is related to the aforestated hint.

2.7.2. Induction and even and odd. We consider some results regarding parity that can be proved by induction.

Solved Problem 2.31. *Prove by induction that the sum of any n even integers is even for all $n \in \mathbb{Z}^+$.*

Solution. The base case is obvious since the sum of one integer is itself, so we just prove the inductive step. Suppose that $k \in \mathbb{Z}^+$ has the property that the sum of any k even integers is even. Let $a_1, a_2, \ldots, a_{k+1}$ be even integers. Then

$$a_1 + a_2 + \cdots + a_{k+1} = (a_1 + a_2 + \cdots + a_k) + a_{k+1}.$$

Now both $a_1 + a_2 + \cdots + a_k$ and a_{k+1} are even, so their sum is even (by Theorem 1.24). This proves the inductive step.

By induction we are done.

Exercise 2.47. *Prove that for every positive integer n, if a is even, then a^n is even. Note that this exercise generalizes Exercise* 1.14*. Did you need to use induction?*

Exercise 2.48. *Prove by induction on n that for every positive integer n, the product of any n odd integers is odd.*

Exercise 2.49. *Prove that for every positive integer n, if the product of n integers is even, then at least one of the integers is even. This exercise generalizes Exercise* 1.15.

2.7.3. Induction and the distributive law. In this subsection we consider more general forms of the distributive law.

Exercise 2.50. *Prove that if $k \in \mathbb{Z}^+$ and if $b, a_1, a_2, \ldots, a_k$ are real numbers, then*

$$(a_1 + a_2 + \cdots + a_k)b = a_1 b + a_2 b + \cdots + a_k b. \tag{2.150}$$

For this exercise you may assume right distributivity for the "field" of real numbers with addition and multiplication (see §8.7). See Figure 2.7.1 *for a visualization of this formula.*

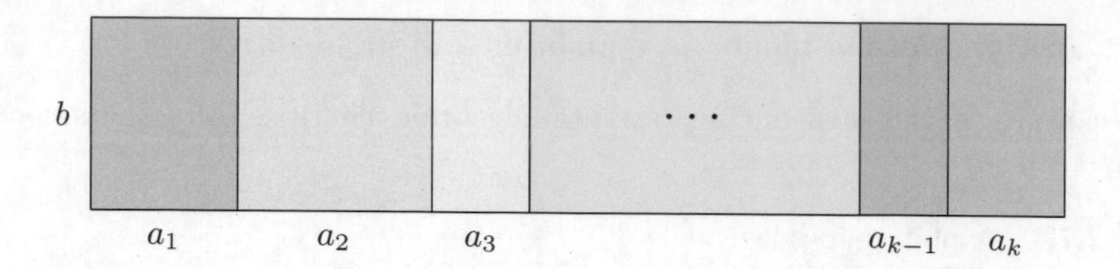

Figure 2.7.1. A visual proof that $(a_1+a_2+\cdots+a_k)b = a_1b+a_2b+\cdots+a_kb$. We show this proof since it is helpful for understanding different points of view.

Exercise 2.51. *Prove that if $k \in \mathbb{Z}^+$ and if $a_1, a_2, \ldots, a_k, b_1, b_2$ are real numbers, then*

$$\left(\sum_{i=1}^{k} a_i\right) \cdot (b_1 + b_2) = \sum_{i=1}^{k} a_i b_1 + \sum_{i=1}^{k} a_i b_2. \tag{2.151}$$

See Figure 2.7.2 for a visualization of this formula.

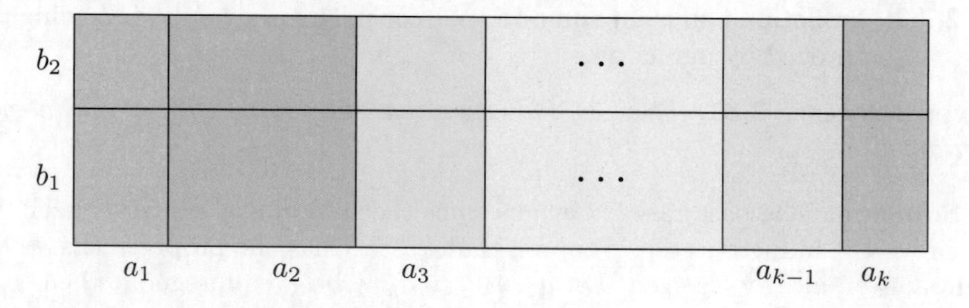

Figure 2.7.2. A visual proof of (2.151).

We have the following formula for the product of two sums.

Exercise 2.52. *Prove that if $k, \ell \in \mathbb{Z}^+$ and if $a_1, a_2, \ldots, a_k, b_1, b_2, \ldots, b_\ell$ are real numbers, then*

$$\left(\sum_{i=1}^{k} a_i\right) \cdot \left(\sum_{j=1}^{\ell} b_j\right) = \sum_{j=1}^{\ell}\left(\sum_{i=1}^{k} a_i b_j\right) =: \sum_{j=1}^{\ell}\sum_{i=1}^{k} a_i b_j. \tag{2.152}$$

For example, if $k = \ell = 2$, then this says that

$$(a_1 + a_2)(b_1 + b_2) = (a_1b_1 + a_2b_1) + (a_1b_2 + a_2b_2). \tag{2.153}$$

Hint: You may for example prove this by induction on ℓ.

The formula above generalizes to products of a finite collection of sums.

Exercise 2.53. *Without going through all the details, describe how you might go about proving the following by induction: If $r \in \mathbb{Z}^+$, $k_1, k_2, \ldots k_r \in \mathbb{Z}^+$, and $a_{ij} \in \mathbb{R}$, where $1 \leq i \leq r$ and $1 \leq j \leq k_i$, then*

$$\prod_{i=1}^{r}\left(\sum_{j_i=1}^{k_i} a_{ij_i}\right) = \sum_{j_r=1}^{k_r} \cdots \sum_{j_2=1}^{k_2}\sum_{j_1=1}^{k_1} a_{1j_1}a_{2j_2}\cdots a_{rj_r}. \tag{2.154}$$

For example, if $r = 3$ *and* $k_1 = k_2 = k_3 = 2$*, then this says that*

$$\begin{aligned}
&(a_{11} + a_{12})(a_{21} + a_{22})(a_{31} + a_{32}) \\
&= \sum_{k=1}^{2}\sum_{j=1}^{2}\sum_{i=1}^{2} a_{1i}a_{2j}a_{3k} \\
&= \sum_{j=1}^{2}\sum_{i=1}^{2} a_{1i}a_{2j}a_{31} + \sum_{j=1}^{2}\sum_{i=1}^{2} a_{1i}a_{2j}a_{32} \\
&= \left(\sum_{i=1}^{2} a_{1i}a_{21}a_{31} + \sum_{i=1}^{2} a_{1i}a_{22}a_{31}\right) + \left(\sum_{i=1}^{2} a_{1i}a_{21}a_{32} + \sum_{i=1}^{2} a_{1i}a_{22}a_{32}\right) \\
&= a_{11}a_{21}a_{31} + a_{12}a_{21}a_{31} + a_{11}a_{22}a_{31} + a_{12}a_{22}a_{31} \\
&\quad + a_{11}a_{21}a_{32} + a_{12}a_{21}a_{32} + a_{11}a_{22}a_{32} + a_{12}a_{22}a_{32}.
\end{aligned}$$

2.7.4. Induction and division. Here we consider powers and division. The following result may be interpreted as saying that the integers $(5+1)^n$ and 1^n have the same remainder 1 when divided by 5. For more general results regarding division, see Chapter 4.

Exercise 2.54. *Prove that for all* $n \in \mathbb{Z}^+$, $6^n - 1$ *is divisible by* 5. *See Figure* 2.7.3 *for a visualization of the inductive step.*

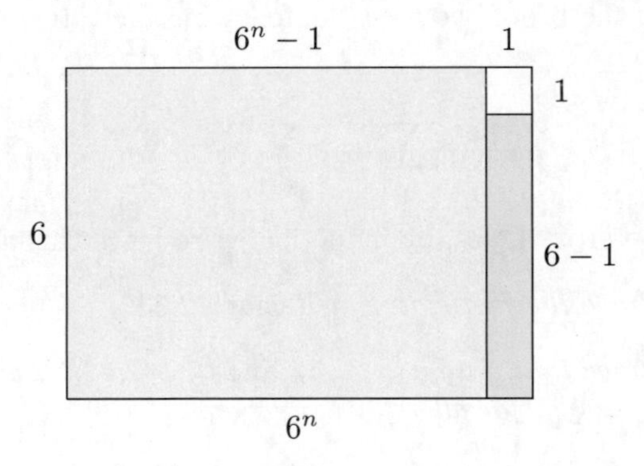

Figure 2.7.3. Visualizing the inductive step to prove Exercise 2.54 by considering the total area of the shaded region as the sum of the areas of the yellow and green rectangles. Note that the area of the big rectangle is $6^n \cdot 6 = 6^{n+1}$, whereas the area of the small white square is 1.

Exercise 2.55. *Generalize the previous exercise as follows. Let* k *be a positive integer. Prove that for all* $n \in \mathbb{Z}^+$, $(k+1)^n - 1$ *is divisible by* k.

2.7.5. Induction and algebraic inequalities. The following result, proved by induction, alternatively follows from the fact that for a finite set of real numbers, their minimum is at most their average, and their maximum is at least their average.

Solved Problem 2.32. *Let n be a positive integer. Prove that for every real number $b, a_1, a_2, \ldots, a_n$ such that*

(2.155) $$a_1 + a_2 + \cdots + a_n = b,$$

there exists $1 \leq i \leq n$ such that $a_i \leq \frac{b}{n}$ and there exists $1 \leq j \leq n$ such that $a_j \geq \frac{b}{n}$

Solution. We prove this by contradiction. Suppose there exist $n \in \mathbb{Z}^+$ and real numbers $b, a_1, a_2, \ldots, a_n$ such that

(2.156) $$a_1 + a_2 + \cdots + a_n = b,$$

but there does exist $1 \leq i \leq n$ such that $a_i \leq \frac{b}{n}$. Then we have $a_i > \frac{b}{n}$ for all $1 \leq i \leq n$. This implies that

(2.157) $$b = a_1 + a_2 + \cdots + a_n = \sum_{i=1}^{n} a_i > \sum_{i=1}^{n} \frac{b}{n} = n\frac{b}{n} = b,$$

which is a contradiction.

This proves the first statement of the problem. The second statement is proved similarly.

An alternative proof follows from the existence of a minimum element. Namely, by Theorem 4.46 below, there exists $1 \leq i \leq n$ such that $a_i \leq a_j$ for all $1 \leq j \leq n$. We then have that $na_i \leq \sum_{j=1}^{n} a_j = b$.

Remark 2.33. In the proof, we used the following fact. If $n \in \mathbb{Z}^+$ and if $a_i > b_i$ for $1 \leq i \leq n$, then

(2.158) $$\sum_{i=1}^{n} a_i > \sum_{i=1}^{n} b_i.$$

This fact we can be proved, yes, by induction! (We leave this as an exercise.)

Exercise 2.56. *Prove the fact stated in Remark* 2.33.

Exercise 2.57. *Prove that if $a_1, a_2, \ldots, a_n$ and $b_1, b_2, \ldots, b_n$ are non-negative real numbers satisfying $a_i \leq b_i$ for all $1 \leq i \leq n$, then*

(2.159) $$a_1 a_2 \cdots a_n \leq b_1 b_2 \cdots b_n.$$

Hint: Besides the base case $n = 1$, you may wish to prove the $n = 2$ case, which may then be used for the inductive step.

The following may be relevant to the next exercise: Is an analogous statement true where we replace weak inequalities $\leq$ by strong inequalities $<$?

Exercise 2.58. *Suppose that $n = a_1 a_2 \cdots a_k$, where all of the letters are positive integers. Prove that there exists $1 \leq i \leq k$ such that*

(2.160) $$a_i \leq \sqrt[k]{n}.$$

Hint: Prove this by contradiction. See Figure 2.7.4 *for a visualization when $k = 2$.*

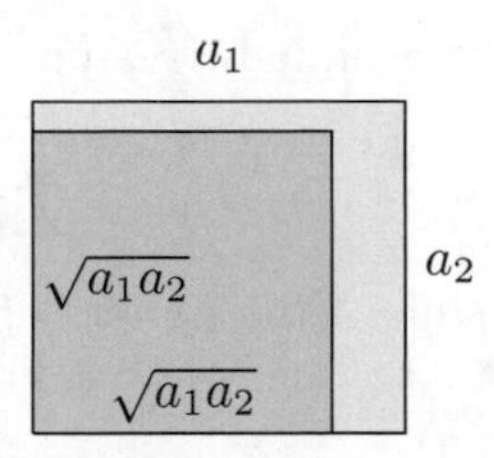

Figure 2.7.4. An impossible picture. Visualizing the $k = 2$ case of a proof by contradiction of (2.160). If $a_1, a_2 > \sqrt{a_1a_2}$, can the area of the purple square with side lengths $\sqrt{a_1a_2}$ equal the area of the orange a_1-by-a_2 rectangle? Do we need the a_i's to be integers, or can they be any positive real numbers?

2.7.6. Thinking about proving inequalities by induction. clco thru -part2 For some of the proofs by induction of inequalities we encountered earlier in this chapter, we proved a *monotonicity* property, which can be stated as follows.

Exercise 2.59. *Let $\{f_n\}_{n=1}^{\infty}$ and $\{g_n\}_{n=1}^{\infty}$ be sequences of positive real numbers. Suppose that $f_1 \geq g_1$ and that for all $k \geq 1$,*

$$\frac{f_{k+1}}{f_k} \geq \frac{g_{k+1}}{g_k}. \tag{2.161}$$

Prove by induction that for all $n \in \mathbb{Z}^+$,

$$f_n \geq g_n, \quad \textit{that is,} \quad \frac{f_n}{g_n} \geq 1. \tag{2.162}$$

Remark 2.34. Observe that (2.161) is equivalent to

$$\frac{f_{n+1}}{g_{n+1}} \geq \frac{f_n}{g_n}. \tag{2.163}$$

From this we see that

$$1 \leq \frac{f_1}{g_1} \leq \frac{f_2}{g_2} \leq \frac{f_3}{g_3} \leq \frac{f_4}{g_4} \leq \cdots. \tag{2.164}$$

This explains why the exercise is true, although the formal proof of it is by induction!

Note that we can interpret (2.161) as saying that, term-wise, the sequence f_k increases faster than g_k.

2.8. Hints and partial solutions for the exercises

Hint for Exercise 2.1. As the hint says, the answer is given in §2.1.2.

Hint for Exercise 2.2. For the inductive step, we need to show that **if** k is a positive integer with the property that the formula $1 + 3 + 5 + \cdots + (2k - 1) = k^2$ is true, **then** the formula $1 + 3 + 5 + \cdots + (2(k + 1) - 1) = (k + 1)^2$ is true. To prove this **implication**, use the inductive hypothesis and the easy fact that

$$1 + 3 + 5 + \cdots + (2(k + 1) - 1) = \big(1 + 3 + 5 + \cdots + (2k - 1)\big) + (2(k + 1) - 1).$$

Simplify the resulting quadratic expression in k.

Hint for Exercise 2.3. For the inductive step: Suppose that k is a positive integer such that $\sum_{i=1}^{k}(b+ia) = k\left(\frac{k+1}{2}a+b\right)$ is true. Show that $\sum_{i=1}^{k+1}(b+ia) = (k+1)\left(\frac{(k+1)+1}{2}a+b\right)$ is true. To do this, relate $\sum_{i=1}^{k+1}(b+ia)$ to $\sum_{i=1}^{k}(b+ia)$.

Hint for Exercise 2.4. The sum of the powers of r from 0 to $k+1$ is equal to the sum of the powers of r from 0 to k plus r^{k+1}. This will help with the inductive step.

Hint for Exercise 2.5. Use Exercise 2.4 with $r = 1/2$. Then take the limit of your formula for the partial sum s_n as $k \to \infty$.

Hint for Exercise 2.6. The inductive hypothesis is that k is a positive integer with the property that $\sum_{i=1}^{k} i^2 = \frac{k(k+1)(2k+1)}{6}$ is true.

Hint for Exercise 2.7. The inductive hypothesis is that k is a positive integer with the property that $1^2 + 3^2 + \cdots + (2k-1)^2 = \frac{k(2k-1)(2k+1)}{3}$ is true.

Hint for Exercise 2.8. The easy fact $\sum_{i=1}^{k+1}(-1)^i i^2 = \sum_{i=1}^{k}(-1)^i i^2 + (-1)^k k^2$ is useful for the inductive step. You may need to factor a quadratic polynomial in k.

Hint for Exercise 2.9. For the inductive step, the quartic polynomial you will encounter has factors of $k+1$ and $k+2$ (to what power(s)?).

Hint for Exercise 2.10. You can prove this directly by induction, or use the previous results for summing i and i^2.

Hint for Exercise 2.11. For the inductive step, as is often an algebraic step for such problems, put ratios over a common denominator.

Hint for Exercise 2.12. This is similar in spirit to the previous exercise.

Hint for Exercise 2.13. For the inductive step, showing that

$$\left(1 - \frac{(-1)^{2k+1}}{2k+1}\right)\left(1 - \frac{(-1)^{2(k+1)}}{2(k+1)}\right) = 1$$

will be helpful.

Hint for Exercise 2.14. For the inductive step, you may come across the expression $\left(1 - \frac{1}{(k+1)^2}\right)\frac{k+1}{2k}$. Simplify this.

Hint for Exercise 2.15. Although $P(k) \Rightarrow P(k+1)$ is true for all $k \geq 2$, it is not true for $k = 1$.

Hint for Exercise 2.16. The problem with this "proof by induction" is in the inductive step. The **converse** of an implication is not "logically equivalent" to the implication.

Hint for Exercise 2.17. This is the same falsidical paradox as the "all horses are the same color" falsidical paradox.

Hint for Exercise 2.18. Apply the inductive hypothesis to $4^{k+1} = 4 \cdot 4^k$. Then show that the inductive conclusion will follow from showing for $k \geq 4$ that $4k^2 \geq (k+1)^2$. By taking square roots, show that this is true whenever $k \geq 1$, which is certainly true if $k \geq 4$.

Hint for Exercise 2.19. The inductive hypothesis is: Suppose that a positive integer k has the property that for every line of k pandas and penguins, where the first animal in the line is a panda and the last animal in the line is a penguin, then somewhere in the line there is a panda directly in front of a penguin.

To prove the inductive conclusion, we assume the following: We have a line of $k+1$ pandas and penguins, where the first animal in the line is a panda and the last animal in the line is a penguin.

Now consider the first k animals and consider two cases: (1) The k-th animal in the line is a panda. (2) The k-th animal in the line is a penguin.

Hint for Exercise 2.20. Let us take $I = 7$ and $B = 23$ for example. (But you should choose different numbers.) Our first job is the find the least positive integer L such that: For each integer $n \geq L$,

$$7^n \geq 23n^2. \tag{2.165}$$

Show that this is false for $n = 2$ and true for $n = 3$. Let $n = 3$ be the base case. The proof is similar to the previous exercise, that is, the same general scheme, except with different numbers. So mimic the previous exercise.

Hint for Exercise 2.21. By now, you should have had a fair bit of practice proving formulas by induction. For this exercise, you may find the elementary exponential function fact that $2 \cdot 2^{k+1} = 2^{k+2}$ to be useful.

Hint for Exercise 2.22. Both a^n and $n!$ are products of n positive integers. In the former case all the n numbers are a, whereas in the latter case the numbers are the integers from 1 to n. Now suppose that n is much larger than a. We have

$$n! = \prod_{i=1}^{a-1} i \prod_{i=a}^{n} i = (a-1)! \prod_{i=a}^{n} i. \tag{2.166}$$

As n gets larger, we have more and more integers i between a and n where $i \gg a$ (informally, $\gg$ means much larger).

We have

$$\frac{n!}{a^n} = \frac{(a-1)!}{a^{a-1}} \cdot \frac{\prod_{i=a}^{n} i}{a^{n-a+1}}. \tag{2.167}$$

The first factor is a positive real number independent of n, but it is rather clear that the second factor limits to ∞ as $n \to \infty$. This is an intuitive explanation for why $\frac{n!}{a^n} \to \infty$ as $n \to \infty$.

Suppose that $k \geq n_0$ has the property that $(2k)! > a^k$. Use that $(2(k+1))! = (2k+2)(2k+1) \cdot (2k)!$ to prove the inductive conclusion.

Hint for Exercise 2.23. For the inductive step, you may find the elementary identity $(k+2) \cdot (k+1)! = (k+2)!$ to be useful.

Hint for Exercise 2.24. The inductive step boils down to proving the inequality $\sqrt{k} + \frac{1}{\sqrt{k+1}} \geq \sqrt{k+1}$. To see this, multiply everything by $\sqrt{k+1}$, and simplify.

The easier way (perhaps less rigorous) to solve the exercise is to observe that $\frac{1}{\sqrt{i}} > \frac{1}{\sqrt{n}}$ for $1 \leq i < n$ and that there are n terms in the sum.

Hint for Exercise 2.25. Use (2.2).

Hint for Exercise 2.26. This exercise is rigged to work since

$$f(k) + (f(k+1) - f(k)) = f(k+1).$$

Hint for Exercise 2.27. For this exercise, assume Lemma 4.29 below. This makes the cases $n = 1$ and $n = 2$ easy. The inductive hypothesis is: Suppose k is a positive integer with the property that if $a, b_1, b_2, \ldots, b_k$ are non-zero integers such that $\gcd(a, b_i) = 1$ for $1 \le i \le k$, then $\gcd(a, b_1 b_2 \cdots b_k) = 1$. To prove the inductive conclusion, use this, the fact that $b_1 \cdots b_{k+1} = (b_1 \cdots b_k) \cdot b_{k+1}$, and the $n = 2$ case (Lemma 4.29).

Hint for Exercise 2.28. For calculating t_3, if the rightmost tile is vertical, then by considering the two leftmost tiles, we see that there are $t_2 = 2$ such tilings. If the (two) rightmost tiles are horizontal, then there is just $t_1 = 1$ leftmost (vertical) tile.

A similar description may be made for calculating t_4.

Hint for Exercise 2.29. If the rightmost tile is vertical, then by considering the four leftmost tiles, we see that there are $t_4 = 5$ such tilings. If the (two) rightmost tiles are horizontal, then by considering the three leftmost tiles, we see that there are $t_3 = 3$ such tilings.

Hint for Exercise 2.30. We guess that $t_n = t_{n-2} + t_{n-1}$.

Hint for Exercise 2.31. You will find the easy identity $f_1 + f_3 + f_5 + \cdots + f_{2(k+1)-1} = (f_1 + f_3 + f_5 + \cdots + f_{2k-1}) + f_{2k+1}$ useful.

Hint for Exercise 2.32. This exercise is very similar to the previous exercise.

Hint for Exercise 2.33. Mimic the solution to Solved Problem 2.20. You'll find that the calculations are slightly shorter.

Hint for Exercise 2.34. Again mimic the solution to Solved Problem 2.20. The calculations are slightly longer, but still quite manageable.

Hint for Exercise 2.35. The base case $\ell = 1$ is easily seen to be true. As our inductive hypothesis, assume that m is a positive integer with the property that f_{km} is divisible by f_k. To prove the inductive conclusion, show that $f_{k(m+1)} = f_{km-1} f_k + f_{km} f_{k+1}$ and use the inductive hypothesis.

Hint for Exercise 2.36. Use the inductive hypothesis and the Fibonacci recursive definition to prove the inductive conclusion.

Hint for Exercise 2.37. We prove identity (2.108) by strong induction on n. The base case: $n = 1$ follows from the Fibonacci recursive definition. To prove the $n = 2$ case use the Fibonacci recursive definition and that $f_2 = 1$ and $f_3 = 2$.

For the inductive step: Suppose that identity (2.108) is true for $1 \le n \le k$. Start with $f_{m+(k+1)} = f_{m+k-1} + f_{m+k}$, and use that (2.108) is true for $n = k - 1$ and $n = k$.

Hint for Exercise 2.38. Use (2.129) (twice) to rewrite $f_n^2 + f_{n+1}^2$, and then use Exercise 2.37.

Hint for Exercise 2.39. As our strong inductive hypothesis, assume that $k \geq 2$ is an integer such that (2.113) is true for $1 \leq n \leq k$. Using this strong inductive hypothesis, compute $f_{-(k+1)}$.

Hint for Exercise 2.40. As our inductive hypothesis, assume that k is a positive integer satisfying

$$f_n \leq 2^n \quad \text{for } 1 \leq n \leq k. \tag{2.168}$$

Estimate f_{k+1}.

Hint for Exercise 2.41. For the strong inductive hypothesis: Suppose that the result is true for all positive integers up to and including $n-1$, where $n \geq 2$. Let f be the largest Fibonacci number less than n. Clearly,

$$n = f + (n - f), \tag{2.169}$$

where $n - f \in \mathbb{Z}^+$. Show by contradiction that $f > n - f$. Then apply the strong inductive hypothesis to prove the strong inductive conclusion.

Hint for Exercise 2.42. Assume the inductive hypothesis that k is a positive integer with the property that

$$\begin{pmatrix} f_{k+1} & f_k \\ f_k & f_{k-1} \end{pmatrix} = \begin{pmatrix} 1 & 1 \\ 1 & 0 \end{pmatrix}^k.$$

Compute $\begin{pmatrix} 1 & 1 \\ 1 & 0 \end{pmatrix}^{k+1}$.

Hint for Exercise 2.43. Use $f_{k+1}f_{k-1} - f_k^2 = \det \begin{pmatrix} f_{k+1} & f_k \\ f_k & f_{k-1} \end{pmatrix}$ and the previous exercise. You will need to know some basic facts about determinants of 2×2 matrices.

Hint for Exercise 2.44. As our inductive hypothesis, suppose that $k \in \mathbb{Z}^+$ satisfies $\sum_{i=1}^{2k-1} f_i f_{i+1} = f_{2k}^2$. Using this and the Fibonacci recursive definition, compute $\sum_{i=1}^{2(k+1)-1} f_i f_{i+1}$.

Hint for Exercise 2.45. The inductive step follows from a straightforward calculation.

Hint for Exercise 2.46. The base case is $n = 12$. As the inductive hypothesis, suppose $k \geq 12$ is an integer for which there exist $a, b \in \mathbb{Z}^{\geq}$ such that $k = 4a + 5b$. Split the proof into two cases: (1) $a \geq 1$. (2) $a = 0$.

How does the assumption that $a \geq 1$ help?

When $a = 0$, we have $k = 5b$. Explain why $b \geq 3$. Observe that if you replace three 5-cent stamps by four 4-cent stamps, the postage increases by 1 cent. How does this help?

Hint for Exercise 2.47. No need to use induction. Let n be a positive integer. Use that $a^n = a \cdot a^{n-1}$ and that an even integer times an integer is even.

Hint for Exercise 2.48. Use induction, the identity $a_1 a_2 \cdots a_{k+1} = (a_1 a_2 \cdots a_k) \cdot a_{k+1}$, and the fact that the product of two odd integers is odd.

Hint for Exercise 2.49. Take the contrapositive of the implication comprising the statement of the previous exercise.

Hint for Exercise 2.50. As the inductive hypothesis, suppose that n is a positive integer such that for every real number $b, a_1, a_2, \ldots, a_n$ we have

$$(a_1 + a_2 + \cdots + a_n)b = a_1 b + a_2 b + \cdots + a_n b. \tag{2.170}$$

Now suppose $b, a_1, a_2, \ldots, a_{n+1}$ are real numbers. Compute $(a_1 + a_2 + \cdots + a_{n+1})b$, while using the usual distributive law and the inductive hypothesis.

Hint for Exercise 2.51. For the inductive step, we assume that k is a positive integer with the property that for every real number $a_1, a_2, \ldots, a_k, b_1, b_2$ we have

$$\left(\sum_{i=1}^{k} a_i\right) \cdot (b_1 + b_2) = \sum_{i=1}^{k} a_i b_1 + \sum_{i=1}^{k} a_i b_2. \tag{2.171}$$

Now let $a_1, a_2, \ldots, a_{k+1}, b_1, b_2$ be real numbers. Compute $\left(\sum_{i=1}^{k+1} a_i\right) \cdot (b_1 + b_2)$.

Hint for Exercise 2.52. We prove the statement by induction on ℓ. The $\ell = 1$ case follows from the previous exercise. Suppose that m is a positive integer with the property that for every real number $a_1, a_2, \ldots, a_k, b_1, b_2, \ldots, b_m$ we have

$$\sum_{i=1}^{k} a_i \cdot \sum_{j=1}^{m} b_j = \sum_{j=1}^{m} \left(\sum_{i=1}^{k} a_i b_j \right). \tag{2.172}$$

Now let $a_1, a_2, \ldots, a_k, b_1, b_2, \ldots, b_{m+1}$ be real numbers. Compute $\sum_{i=1}^{k} a_i \cdot \sum_{j=1}^{m+1} b_j$ while using the previous exercise and the inductive hypothesis.

Hint for Exercise 2.53. We would prove this result by induction on r while using the result of Exercise 2.50.

Hint for Exercise 2.54. For the inductive step, assume that k is a positive integer with the property that $6^k - 1$ is divisible by 5. By definition, this means that there exists an integer a such that $6^k - 1 = 5a$. Show that $6^{k+1} - 1$ is divisible by 5.

Hint for Exercise 2.55. Use the same method as the previous exercise. The only difference is that this exercise is slightly more abstract.

Hint for Exercise 2.56. Use that

$$\sum_{i=1}^{n+1} a_i = \sum_{i=1}^{n} a_i + a_{n+1}.$$

Also use the general fact that if $a > b$ and $c > d$, then $a + c > b + d$. All of this will help you with the inductive step.

Hint for Exercise 2.57. Suppose that k is a positive integer with the property that if $a_1, a_2, \ldots, a_k$ and $b_1, b_2, \ldots, b_k$ are non-negative real numbers satisfying $a_i \leq b_i$ for all $1 \leq i \leq k$, then

$$a_1 a_2 \cdots a_k \leq b_1 b_2 \cdots b_k. \tag{2.173}$$

(That is, assume that the implication above is true for the positive integer k.) Now suppose that $a_1, a_2, \ldots, a_{k+1}$ and $b_1, b_2, \ldots, b_{k+1}$ are non-negative real numbers satisfying $a_i \leq b_i$ for all $1 \leq i \leq k + 1$. Continue (what do we want to prove?).

Hint for Exercise 2.58. This is not an induction proof per se. Since the set $\{a_1, a_2, \ldots, a_k\}$ is a finite set, it has a minimum element (this fact can be proved by induction and is discussed in a later chapter). How does $a_1 a_2 \cdots a_k$ compare to the k-th power of this minimum element?

Hint for Exercise 2.59. Since all of the numbers are positive, inequality (2.161) is equivalent to $\frac{f_{n+1}}{g_{n+1}} \geq \frac{f_n}{g_n}$. By hypothesis, we have $\frac{f_1}{g_1} \geq 1$. As the inductive hypothesis, suppose that k is a positive integer satisfying $\frac{f_k}{g_k} \geq 1$. Continue.

Chapter 3

Logic: Implications, Contrapositives, Contradictions, and Quantifiers

Figure 3.0.1. René Descartes (1596–1650). Portrait by Frans Hals in Louvre Museum. Wikimedia Commons, Public Domain.

Goals of this chapter: To get down to the business of proving theorems using logical reasoning. To give some fun examples of applying logic to solve riddles, which are toy models of theorems. To understand the basics of logic. To see proofs by contradiction, including the incredible fact that the square root of two is irrational. To discuss the key concepts of universal and existential quantifiers.

Cogito, ergo sum. – René Descartes

Question: Does the rock Descartes is sitting on exist?

Answer: We know it exists, but the rock doesn't.

Rebuttal: Those who believe in the simulation hypothesis doubt this.

We begin this chapter with some wrong proofs and some fun riddles and puzzles. The more rigorously minded reader may proceed directly to §3.5 (perhaps after skimming through the first few sections).

3.1. The need for rigor

Mathematics is based on logical deduction. In this book so far we have focused on presenting proofs with correct logical deduction. To expand the landscape of results we can prove, it is useful to lay down some more foundation for the logic we use to rigorously prove statements, which we do in this chapter. We have also seen that visual proofs are very effective and convincing, and so far we have presented correct visual proofs. But we present to you:

3.1.1. A cautionary tale: Not all "ducks" are ducks.

> If it looks like a duck, swims like a duck, and quacks like a duck, then it probably is a duck. See Figure 3.1.1.

We often use this type of reasoning in everyday life. And when we use this type of reasoning, we are usually correct! We may think of "superficial" as being what's on the surface, and "deep" as being what's beneath the surface. Sometimes what is deep agrees with what is superficial, but other times they are different. This is when we have to be careful with our reasoning. In the development of mathematics, by trial and error, humans have decided on how deeply and carefully we should think. So this decision of "rigor" is a social phenomenon. We will be conventional and present mathematics roughly from this point of view, which actually agrees with our personal viewpoint.

Figure 3.1.1. They look, swim, and quack like ducks. Are they ducks? Figure created using the LaTeX TikZducks package.

If true, the following superficial and false result might cause the universe to spontaneously combust (just kidding).

Non-Theorem 3.1.

$$\frac{63}{2} = \frac{65}{2}. \tag{3.1}$$

That is, by subtracting $\frac{63}{2}$ *from both sides, we get* $0 = 1$. ☺

Proof, not. The theorem follows immediately from the two claims:

Claim 1. The union of the red, orange, yellow, and green regions is a right triangle with non-hypotenuse side lengths 5 and 13. Thus the sum of their areas is equal to $\frac{1}{2} \cdot 5 \cdot 13 = \frac{65}{2}$.

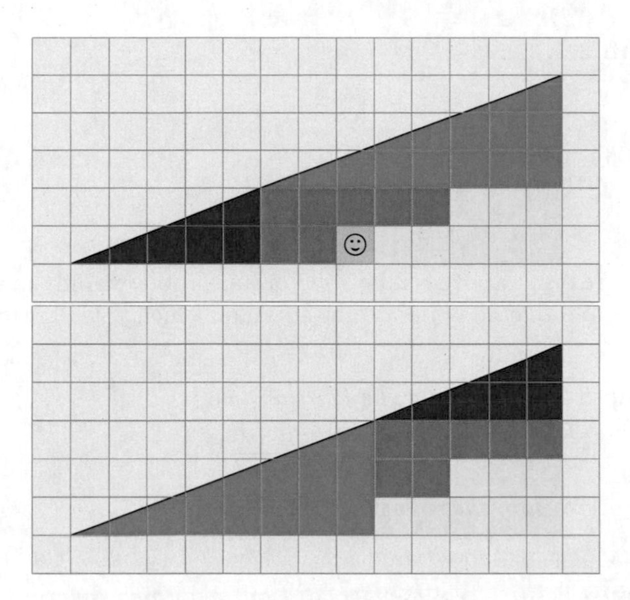

Figure 3.1.2. Two "triangles" that pass the duck test.

Claim 2. The union of the red, orange, yellow, and green regions is a right triangle with non-hypotenuse side lengths 5 and 13 *minus* a unit square ☺. Thus the sum of their areas is equal to $\frac{65}{2} - 1 = \frac{63}{2}$.

□

Where did we go wrong?

Well, both regions in Figure 3.1.2 look like triangles. But, as your mother told you, looks can be deceiving!

Let us look at the "triangle" in Claim 1, which is the top "triangle" pictured in Figure 3.1.2. The red triangle has non-hypotenuse side lengths 5 and 2, whereas the green triangle has non-hypotenuse side lengths 8 and 3. But for the "whole triangle" to really be a triangle, we need these two triangles to be *similar*. But they are not as 5/2 is not equal to 8/3.

In fact, the true sum of the areas of the colored triangles is equal to

$$\frac{8 \cdot 3}{2} + \frac{5 \cdot 2}{2} + 7 + 8 = 12 + 5 + 7 + 8 = 32 = \frac{64}{2}, \tag{3.2}$$

which happens to be the average of $\frac{63}{2}$ and $\frac{65}{2}$. So neither Claim 1 nor Claim 2 gives the correct area!

3.1.2. More fallacies: Let us make up some axioms.

Seen on a car bumper sticker: My karma ran over your dogma.
Seen on a dog leash: My dogma ran under your karma.

Besides wrong inductive or visual reasoning, we are attracted to wrong algebra rules because of their simplicity and beauty. For example, take the following.

Non-Theorem 3.2.

$$1 + 1 = 1. \tag{3.3}$$

Non-Proof. We compute

$$1 + 1 = \frac{1}{1} + \frac{1}{1} \overset{\text{☺}}{=} \frac{1+1}{1+1} = \frac{2}{2} = 1. \tag{3.4}$$ □

Of course, the result above is false. ☹ For its "proof" we only made one algebra mistake. So we see that it is important to know which rules are true and which rules are false.

Solved Problem 3.3. *Consider the fake formula*

$$\frac{a}{b} + \frac{c}{d} = \frac{a+c}{b+d}. \tag{3.5}$$

Prove or disprove each of the following statements:

(1) ***For all*** *positive real numbers* a, b, c, d, *equation* (3.5) *is true.*

(2) ***There exist*** *positive real numbers* a, b, c, d *for which equation* (3.5) *is true.*

Solution. (1) For statement (1) to be true, (3.5) must be true for all positive real numbers a, b, c, d. So if there exist positive real numbers a, b, c, d for which it is false, then the statement is false. That is, to prove that the statement is false, we just need to come up with one example of a, b, c, d for which (3.5) is false. This is easy: let $a = b = c = d = 1$. In this case, equation (3.5) says that (déjà vu)

$$1 + 1 = \frac{1}{1} + \frac{1}{1} \overset{\text{☹}}{=} \frac{1+1}{1+1} = \frac{2}{2} = 1,$$

which is false. We conclude that statement (1) is false.

(2) To prove that statement (2) is true, we would just need to find one example of a, b, c, d for which (3.5) is true. Unfortunately, we will show that there is no such example, and therefore statement (2) is also false. We can see this as follows. Equation (3.5) is equivalent to the equation obtained by multiplying it by $b + d$:

$$a\left(1 + \frac{d}{b}\right) + c\left(1 + \frac{b}{d}\right) = a + c. \tag{3.6}$$

This in turn is equivalent to

$$\frac{ad}{b} + \frac{cb}{d} = 0. \tag{3.7}$$

Since a, b, c, d are all positive, there are no solutions to this last equation. Therefore, there are no positive solutions to (3.5). We conclude that statement (2) is false. □

Exercise 3.1. *Prove or disprove each of the following statements (about elementary functions) that start with the following:*

For all *real numbers* x *and* y:

(1) $(x + y)^2 = x^2 + y^2$.

(2) $e^{x+y} = e^x + e^y$.

(3) $\ln(x + y) = \ln(x) + \ln(y)$, *where we assume in addition that* $x, y > 0$.

Hints (for this exercise and the next): Consider the quadratic case of the Binomial Theorem and the identities $e^{x+y} = e^x e^y$ *and* $\ln(xy) = \ln(x) + \ln(y)$.

Can rules that are not always true be sometimes true? The following shows that it depends on the types of functions we are considering.

Exercise 3.2. *Prove or disprove each of the following statements that start with the following:*

There exist positive *real numbers* x *and* y *such that:*

(1) $(x + y)^2 = x^2 + y^2$.
(2) $e^{x+y} = e^x + e^y$.
(3) $\ln(x + y) = \ln(x) + \ln(y)$.

3.1.3. Beauty and the beast: Mistakenly assuming linearity. Sometimes we think too fast by assuming a beautiful "formula". One way of slowing down our thinking is to check our "facts" against "reality". Knowing when to slow down our thinking comes from experience. If one is cautious, one checks one's calculations more.

The beast, a.k.a. the so-called freshman's dream, is

$$(x + y)^n \overset{☺}{=} x^n + y^n, \tag{3.8}$$

where $n > 1$ is an integer.

For example, the $n = 2$ case of the freshman's dream is

$$(x + y)^2 \overset{☺}{=} x^2 + y^2. \tag{3.9}$$

To check this dream against reality, we may consider various simple cases and see if they are formulas we know the truth value of. For example, if we take x arbitrary and $y = 0$, then we obtain the formula

$$(x + 0)^2 = x^2 + 0^2, \tag{3.10}$$

which is true since both sides are equal to x^2. However, if on the other hand we take $x = y = 1$, then we get the non-sensical (false) formula

$$(1 + 1)^2 = 1^2 + 1^2, \tag{3.11}$$

that is, $4 = 2$. So we conclude that the freshman's dream is only a false dream! This disproves the statement in Exercise 3.1(1).

Takeaway: The idea of checking a conjecture against special cases of the conjecture is very useful.

A similar misconception is: For all non-negative real numbers x and y we have

$$\sqrt{x + y} \overset{☺}{=} \sqrt{x} + \sqrt{y}. \tag{3.12}$$

Exercise 3.3. *Can you disprove formula* (3.12)*?*

Exercise 3.4. *In view of these examples, can you explain the title of this subsection about "mistakenly assuming linearity" in terms of properties of functions?*

3.1.4. A happy tale: The most famous theorem of plane geometry. Since we have seen how visual proofs can go wrong, to restore our faith in them, we consider one of the many geometric proofs of the Pythagorean theorem. Visual proofs of the Pythagorean theorem can be made formally rigorous. However, since they are "clearly" correct, we will not present the details of how to make them formally rigorous.

Consider a right triangle with side lengths a, b, c, where c is the length of the hypotenuse.

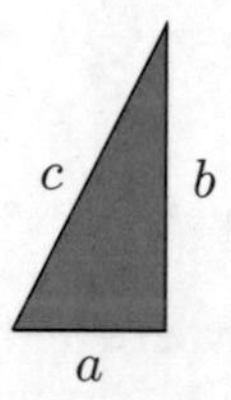

Figure 3.1.3. A right triangle.

Suppose we are asked to prove the Pythagorean theorem:

$$a^2 + b^2 = c^2. \tag{3.13}$$

A picture that may come to mind is the one in Figure 3.1.4. Here, we may think of our original triangle as a room. One of the proof ideas in geometry is to build a house around the room. Unfortunately, the house built in Figure 3.1.4 does not immediately suggest a proof, for why should the areas of the two smaller squares sum to the area of the bigger square?

Here is where we need to use our creativity. As an architect, given our triangular room, we can create any auxiliary structures attached to it that we like. In particular, if we mash up two of our right triangles together to form an $a \times b$ rectangle, it is not hard to imagine finally arriving at the diagram on the left of Figure 3.1.5.[1] The diagram on the right of Figure 3.1.5 is a square house of the same dimensions $(a + b) \times (a + b)$. The Pythagorean theorem follows from the fact that the two houses have the same area. Here is the algebra:

$$\text{Area of House One } = a^2 + b^2 + 2ab, \tag{3.14}$$

$$\text{Area of House Two } = c^2 + 2ab. \tag{3.15}$$

The Pythagorean theorem follows (just subtract $2ab$ "square feet" from both houses)!

In "The Matrix" movie series, the Architect is the "father" of The Matrix. Likewise, we are the "parents" of the proofs we create.

Theorem 3.4 (Pythagorean theorem)**.** *If a right triangle has hypotenuse of length c and the other side lengths are a and b, then*

$$a^2 + b^2 = c^2. \tag{3.16}$$

[1] Mashup of triangles is not to be confused with mashup of songs!

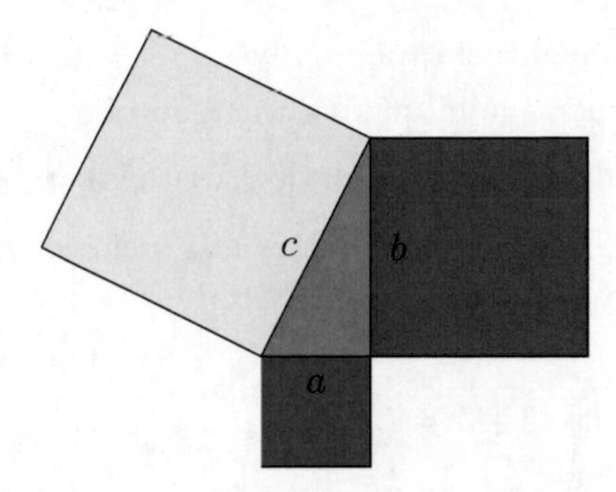

Figure 3.1.4. The right triangle and the squares of its sides.

Proof. Figure 3.1.5 is an illustration of a proof of the Pythagorean theorem. There are in fact many different proofs of the Pythagorean theorem; a google search will yield many examples. □

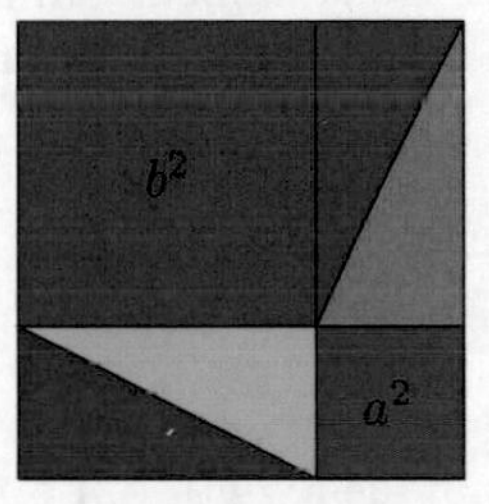

Figure 3.1.5. A visual proof of the Pythagorean theorem. The house that Jack built is on the left. The house that Jill built is on the right.

Exercise 3.5. *In your own words, write out an explanation of a proof of the Pythagorean theorem based on Figure* 3.1.5.

3.2. Statements

Although visual proofs are appealing and we often use *geometric* constructs to understand logical reasoning, it is important to learn how to *write* mathematical proofs. Before we learn how to write proofs, we need to learn how to correctly write the *statements* we are proving and/or learn how to understand and interpret statements. For example, we wish to shy away from *ambiguous* statements and questions.

Here are some examples of ambiguous sentences to put in a recommendation letter:

- A person like them is hard to find.
- They are an unbelievable worker.
- You will be lucky to get them to work for you.
- Their true ability is deceiving.

- We cannot recommend them too highly.
- You should waste no time in making them an offer.

Just as in any writing, bad mathematical writing can be ambiguous:

Question 3.5. *If The Hulk runs around the lake twice and Captain America runs around the lake 5 times more, how many times does Captain America run around the lake?*

Answer 1. 7, because $2 + 5 = 7$.

Answer 2. 12, because $2 + 5 \cdot 2 = 12$.

Which answer is correct? It depends on how one interprets the word "times" in the question. If each time is one time around the lake, Answer 1 is correct. If we mean 5 times the number of laps The Hulk has run, then Answer 2 is correct. Admittedly, this example of a communication breakdown was a cheap trick taking advantage of linguistic ambiguity! But, in general, we should be careful to make precise mathematical statements.

Now, what is a statement? In short, it is a sentence that is either true or false (but of course not both). For example,

"One and one and one is three"[2]

is a true statement, provided we interpret it as saying that $1+1+1 = 3$. Similarly,

"Two plus two is four. Minus one, that's three, quick maths"[3]

is also a true statement, at least in that $2 + 2 = 4$ and $4 - 1 = 3$. Whether this is quick maths may be a matter of opinion!

Some sentences cannot be either true or false, and hence these sentences are not statements. Consider a version of the liar paradox:

"This sentence is false."

Is this sentence a statement? That is, is it either true or false? Let's see: Suppose that the sentence is true. Then, by its very content, the sentence is false! On the other hand, suppose that the sentence is false. Then, it says that the sentence is not false; that is, the sentence is true. So, if the sentence is a statement, then we obtain a contradiction. We conclude that the sentence is not a statement!

Other versions of the liar paradox are: "I am lying" and "This sentence is a lie." Observe that these sentences are self-referential, a concept which occurs in language, logic, mathematics, and computing.

Remark 3.6. A formal way to present the liar paradox is as follows. Given a statement P, let $\mathrm{TV}(P)$ denote its **truth value**. That is, $\mathrm{TV}(P) = \mathrm{T}$ if P is true, and $\mathrm{TV}(P) = \mathrm{F}$ if P is false. Then the form of the liar paradox sentence P is

$$P = (\mathrm{TV}(P) = \mathrm{F}). \tag{3.17}$$

So, if $\mathrm{TV}(P) = \mathrm{T}$, then $\mathrm{TV}(P) = \mathrm{F}$. And, if $\mathrm{TV}(P) = \mathrm{F}$, then $\mathrm{TV}(P) = \mathrm{T}$. Each case leads to a contradiction.

[2] From the song "Come Together" by The Beatles.

[3] From the song "Man's Not Hot" by Big Shaq.

> Two professors of logic are in a classroom and they notice the statement "Only a doofus would believe a statement like this!" written on the blackboard. Puzzled, one of the professors asks the other, "Do you believe this statement?" Dismayed, the other replies, "Of course not! Only a doofus would believe a statement like this!"

3.3. Truth teller and liar riddle: Asking the right question

Mathematics is about asking and answering questions. We have seen the dangers of making statements or asking questions that are ambiguous, oxymoronic, or self-referential. Given a well-posed question, how do we answer it? Sometimes, it is by asking the right question in return!

Abbott and Costello have asked the question:

> Who's on first?

We instead ask the question:

> Who is telling the truth?

Lawrence of Arabia is at the officer's club in Cairo after a long walk in the desert. Two officers, named Beyoncé and Jay-Z, play a trick on him by hiding two glasses side by side behind the counter, one filled with hot coffee and the other filled with ice-cold lemonade. One of the officers always tells the truth and the other officer always lies, but Lawrence does not know who is who. Lawrence, who is in desperate need of a lemonade, is allowed only one question. What question should poor Lawrence ask, and how shall he subsequently choose?

The answer is that Lawrence should ask the question:

Beyoncé or Jay-Z, which glass (left or right) will the other of you two tell me is the glass with the lemonade?

Let us analyze this logically to determine how, given an answer, Lawrence should choose. Without loss of generality, we may assume that Beyoncé answers the question and that she says that Jay-Z will say that the lemonade glass is on the right.

Case 1: *Beyoncé is the truth teller.* Then Jay-Z is the liar, and so since Beyoncé is telling the truth, Jay-Z will say that the lemonade glass is on the right and he will be lying. So we conclude that the lemonade glass is on the left!

Case 2: *Beyoncé is the liar.* Then Jay-Z is the truth teller, and so since Beyoncé is lying, Jay-Z will say that the lemonade glass is on the left and he will be telling the truth. So, we again conclude that the lemonade glass is on the left!

Since either Case 1 or Case 2 must be true and they lead to the same conclusion, Lawrence should confidently and happily choose the glass on the left!

Exercise 3.6. *Justify why, without loss of generality, we may assume that Beyoncé answers the question and that she says that Jay-Z will say that the lemonade glass is on the right. Namely, why are all of the other cases essentially the same as this case by reason of symmetry?*

3.4*. Logic puzzles

Proving statements in mathematics, just like solving riddles, is about finding a logical argument to ascertain the truth or falsehood of a statement. Logic puzzles are toy examples of such statements. Solving logical puzzles gives us a taste of the logical reasoning we do in mathematics.

3.4.1. How many Terminators? Assume that at most one of the following three statements about Sarah Connor is true. What can you deduce from this assumption?

(1) Sarah owns at least two Terminators.[4]

(2) Sarah owns less than nine Terminators.

(3) Sarah owns at least nine Terminators.

Let us figure out the solution to this puzzle. Suppose that (3) is true. Then this implies that (1) is true, contradicting that at most one of the statements is true. Therefore (3) is false. This tells us that (2) is true. This in turn implies that (1) is false. So what we can conclude is that Sarah owns at most one Terminator.

Alternatively, since (3) is the negation of (2), we know that exactly one of (2) and (3) is true. Thus, if at most one of the three statements is true, then (1) is false.

More abstractly, in the situation above we have three statements of the form:

(1) P,

(2) not Q (this is the *negation* of Q, which we formally define in §3.5.3 below, but it is what you think it is),

(3) Q,

where $Q \Rightarrow P$; that is, (3) $\Rightarrow$ (1).

Thus, if (3) is true, then (1) is true, a contradiction since we cannot have more than one statement being true. So (3) is false; i.e., Q is false; that is, (2) is true. Since at most one of the statements is true, this implies that (1) is false; i.e., we conclude that P is false.

Remark 3.7. Observe that we actually proved that (1) and (3) are false, whereas (2) is true. In other words, we proved that "P is false" and "Q is false". Now, our assumption that $Q \Rightarrow P$ is equivalent to its *contrapositive* (not P) $\Rightarrow$ (not Q); see §3.7 below. So we see that "P is false" is the stronger of these two statements since it implies "Q is false".

3.4.2. Ride My Seesaw.

> "You can't always get it when you really want it."
> – From "Another Tricky Day" by The Who

Here is a rather tricky riddle (if you wish to avoid a lengthy analysis, you may skip it as it is not needed for any later discussion):

---

[4] A Terminator is a cybernetic organism: living tissue over a metal endoskeleton.

Solved Problem 3.8. *There are twelve bunnies on a beach in The Bahamas.*[5] *Eleven of them weigh exactly the same, and the twelfth bunny either weighs more or weighs less, but not the same as the other eleven. You have a seesaw that you can put any number of bunnies on. Using the seesaw only* three *times, find out who is the odd-bunny out weight-wise and whether that bunny weighs more or less than the others.*

This riddle looks like it will succumb to some sort of divide and conquer method. But, because we are allowed only 3 weighings, it is tricky to figure out how exactly we should proceed with this idea!

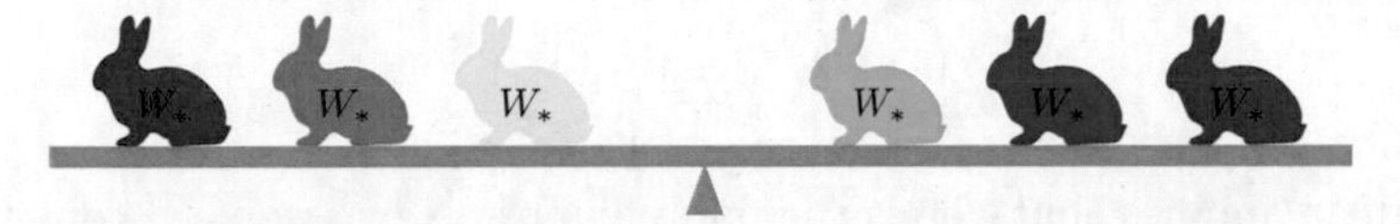

Figure 3.4.1. Six regular bunnies on a balanced seesaw. Tikz code adapted from the "improve the code for a rabbit" answer by samcarter_is_at_topanswers xyz at Stack Exchange. Licensed under Creative Commons Attribution-Share Alike 4.0 International License https://creativecommons.org/licenses/by-sa /4.0/deed.en).

Solution. Label the bunnies

$$B_1, B_2, B_3, \ldots, B_{12}. \tag{3.18}$$

Let their weights be

$$W_1, W_2, W_3, \ldots, W_{12}, \tag{3.19}$$

respectively. Call any one of the eleven bunnies weighing the same a **regular bunny**, and call the bunny weighing differently the **funny bunny**.

In the following, the color coding is as follows: a green bunny is a bunny we know is regular, an orange bunny is a bunny we don't know the type of, and a red bunny is a bunny we know is funny. So, hypothetically, if B_{12} is the funny bunny, then we have

$$W_1 = W_2 = \cdots = W_{11} \neq W_{12}.$$

A way to implement the aforementioned divide and conquer method is to divide a given group of bunnies of unknown type into three groups:

(1) ones on one side of the seesaw,

(2) ones on the other side of the seesaw,

(3) ones not on the seesaw.

A key observation is that since there is only one funny bunny, if we put *equal* numbers of bunnies on the two sides of the seesaw, then the seesaw is balanced *if and only if* all of the bunnies on the seesaw are regular. Here is how we can specifically proceed (this is not the only way):

First weighing. Place bunnies B_1, B_2, B_3, B_4 on one side of the seesaw and bunnies B_5, B_6, B_7, B_8 on the other side of the seesaw. The remaining bunnies $B_9, B_{10}, B_{11}, B_{12}$ are not weighed.

[5] One gets a rather different result if one uses 12 Monkeys. ☺

CASE 1: The *first weighing* is balanced:

$$W_1 + W_2 + W_3 + W_4 = W_5 + W_6 + W_7 + W_8. \tag{3.20}$$

In this case, the bunnies $B_1, B_2, \ldots, B_8$ are all regular and hence colored green. Thus, the funny bunny must be one of the bunnies $B_9, B_{10}, B_{11}, B_{12}$. (Observe that since the bunnies $B_1, B_2, \ldots, B_8$ all weigh the same, from the point of view of weight they are interchangeable.)

Case 1 second weighing. Place B_9 and B_{10} on one side of the seesaw, and place B_{11} and B_1 on the other side of the seesaw (so that B_{12} is not on the seesaw).

SUBCASE 1a: The *second weighing* is balanced:

$$W_9 + W_{10} = W_{11} + W_1. \tag{3.21}$$

So the bunnies B_9, B_{10}, B_{11} are regular. This, together with $B_1, B_2, \ldots, B_8$ all being regular, implies that B_{12} is the funny bunny. Next, we need to determine if the funny bunny weighs more or less than the regular bunnies.

Subcase 1a third weighing. So, for the *third weighing*, we put the funny bunny B_{12} on one side of the seesaw and your favorite regular bunny (of the other eleven) on the other side of the seesaw and see which of the two bunnies weighs more. This completes Subcase 1a.

SUBCASE 1b: The *second weighing* is unbalanced with

$$W_9 + W_{10} > W_{11} + W_1. \tag{3.22}$$

Subcase 1b third weighing. Now place B_9 on one side of the seesaw and B_{10} on the other side of the seesaw.

SUBSUBCASE 1b(i): The *third weighing* is balanced:

$$W_9 = W_{10}. \tag{3.23}$$

In this case we conclude that B_{11} is the funny bunny. By (3.22), we conclude that the funny bunny B_{11} weighs less than the regular bunnies. This completes Subsubcase 1b(i).

SUBSUBCASE 1b(ii):

$$W_9 > W_{10}. \tag{3.24}$$

We leave Subsubcase 1b(ii) as an exercise. Hint: Show that B_9 is the funny bunny.

SUBSUBCASE 1b(iii):

$$W_9 < W_{10}. \tag{3.25}$$

We leave Subsubcase 1b(iii) as an exercise. Hint: B_9 is a regular bunny.

This completes Subcase 1b.

SUBCASE 1c: The *second weighing* is unbalanced with

$$W_9 + W_{10} < W_{11} + W_1. \tag{3.26}$$

Subcase 1c third weighing. Again place B_9 on one side of the seesaw and B_{10} on the other side of the seesaw. Similarly to Subcase 1b, we consider the three subsubcases (i), (ii), and (iii) depending on the result of the third weighing. We leave this as an exercise.

This completes Subcase 1c and hence completes Case 1.

CASE 2: The *first weighing* is unbalanced with

$$W_1 + W_2 + W_3 + W_4 > W_5 + W_6 + W_7 + W_8. \tag{3.27}$$

Then $B_9, B_{10}, B_{11}, B_{12}$ are all regular bunnies.

Case 2 second weighing. Place B_1, B_2, B_5 on one side of the seesaw and place B_3, B_4, B_9 on the other side of the seesaw (so that B_6, B_7, B_8 are not on the seesaw).

SUBCASE 2a: The *second weighing* yields

$$W_1 + W_2 + W_5 = W_3 + W_4 + W_9. \tag{3.28}$$

In this case, the bunnies B_1, B_2, B_3, B_4, B_5 are all regular. Hence the funny bunny is one of

$$B_6, B_7, B_8. \tag{3.29}$$

And by (3.27), the funny bunny weighs less than the regular bunnies.

Subcase 2a third weighing. Place B_6 on one side of the seesaw and place B_7 on the other side of the seesaw.

If $W_6 = W_7$, then B_8 is the lighter funny bunny.

If $W_6 \neq W_7$, then the lighter of the two B_6 and B_7 is the funny bunny.

SUBCASE 2b: The *second weighing* yields

$$W_1 + W_2 + W_5 > W_3 + W_4 + W_9. \tag{3.30}$$

In this case, either B_1 or B_2 is the funny bunny—as, for every other possibility, (3.30) would contradict (3.27). So the funny bunny is heavier than the regular bunnies. The third step is to put B_1 and B_2 on the seesaw and see which bunny weighs more.

SUBCASE 2c: The *second weighing* yields

$$W_1 + W_2 + W_5 < W_3 + W_4 + W_9. \tag{3.31}$$

In this case, by comparing with (3.27), we see that either B_3 or B_4 is the funny bunny. And the funny bunny is heavier than the regular bunnies. Again, the third step is to put B_3 and B_4 on the seesaw and see which bunny weighs more.

CASE 3: The *first weighing* is unbalanced with

$$W_1 + W_2 + W_3 + W_4 < W_5 + W_6 + W_7 + W_8. \tag{3.32}$$

Exercise! This is very similar to Case 2. We can actually appeal to symmetry. □

The riddle above is a nice, albeit complicated, example of analyzing a problem by cases and subcases.

Exercise 3.7. *Fill in all the details that were omitted in the solution to Solved Problem* 3.8.

3.5. Logical connectives

As we have seen, riddles give a pleasant entryway to logical thinking in mathematics. In this section we begin our more formal discussion of the building blocks of logical thinking: logical connectives (or **connectives** for short). Since the reader no doubt already understands the words "or", "and", "not", and "implies" at least in the working way we have encountered them in the first two chapters, our present discussion is a bit non-linear in order to introduce interesting examples to start.

3.5.1. The logical connective "or". We begin with:

Definition 3.9 (The connective "or"). Let P and Q be statements. We say that the (compound) statement

$$P \text{ or } Q$$

is true if at least one of the two statements P and Q is true. The connective "or" is also referred to as a **disjunction**.

Example 3.10. The statement

$$(e > 3) \text{ or } (\pi > 3) \tag{3.33}$$

is true since, although $e > 3$ is false, we have that $\pi > 3$ is true.

In a formal mathematical sense, we can make the meaning of the logical connective "or" in Definition 3.9 more precise by defining it by a **truth table**. The truth table for the connective "or" is

Table 3.5.1. The truth table for the logical connective "or".

P	Q	P or Q
T	T	T
T	F	T
F	T	T
F	F	F

For example, the third row says that if P is false and Q is true, then "P or Q" is true. This is actually the case for Example 3.10 with P being the statement "$e > 3$" and Q being the statement "$\pi > 3$".

We also denote "or" by the symbol $\vee$, so that "P or Q" is written as

$$P \vee Q. \tag{3.34}$$

In fact, we see from the truth table that the statement "P or Q" is false exactly when both the statement "P" and the statement "Q" are false.

Observe that, given the statements P and Q, there are $4 = 2^2$ possible truth values in total since for each of the 2 statements P and Q there are 2 possible truth values. Each of the rows in the truth table above represents 1 of these 4 possibilities.

Note also that "Q or P" has the same truth values as "P or Q". The essential reason for this is that the truth values of "P or Q" in the second and third rows of Table 3.5.1 are the same.

Here is an example of a true statement involving the connective "or":

Lemma 3.11. *Let a and b be real numbers satisfying $ab = 0$. Then $a = 0$ or $b = 0$.*

Solved Problem 3.12. *Prove Lemma* 3.11.

Solution. Let a and b be real numbers satisfying $ab = 0$.

Case 1: $a = 0$. In this case we are done, since $a = 0$ implies $a = 0$ or $b = 0$.[6]

Case 2: $a \neq 0$. In this case, $\frac{1}{a}$ is a well-defined real number and we may multiply the equation $0 = ab$ by it to obtain

$$0 = \frac{1}{a} \cdot 0 = \frac{1}{a} \cdot ab = b. \tag{3.35}$$

Now, since $b = 0$, we also conclude that $a = 0$ or $b = 0$.

Since, for either Case 1 or Case 2 we have $a = 0$ or $b = 0$, we have proved the result. □

Exercise 3.8. *Observe that the proof above is according to the following template.*

Statement: *Suppose P. Then Q or R.*

Proof. *Suppose P.*

Case 1: *Assume Q. Then we are done.*

Case 2: *Assume not Q. Prove R.*

Explain why this template (provided you can accomplish the goal of Case 2) proves the statement.

The connective "or" is sometimes referred to as the "**inclusive or**". On the other hand, an example from everyday life of the '**exclusive or**', in the form of a question, is:

"Would you like tea **or** coffee?"

Most people would answer either "Tea, please" or "Coffee, please", using the "exclusive or", but a mathematician might answer "Yes, please!" using the "inclusive or".

So, to avoid funny answers, one should instead ask a mathematician the following question:

> "Would you like only coffee, only tea, neither, or both? Please choose exactly one of these mutually exclusive options. Thank you!" ☺

In this book, we will use: **Either** P **or** Q to denote the exclusive or.

The properties of inequalities, addition, and multiplication provide nice examples to understand logical connectives. Recall the following.

[6] We take for granted this last implication, which is formally stated in Example 3.26(1) below.

Axiom 3.13 (Trichotomy law for inequalities). *Let a and b be real numbers. Then exactly one of the following is true:*

$$a < b, \quad b < a, \quad a = b. \tag{3.36}$$

By definition, $a > b$ if and only if $b < a$. Also by definition, $a \leq b$ if and only if $a < b$ or $a = b$. Similarly, by definition, $a \geq b$ if and only if $a > b$ or $a = b$.

We have the following properties.

Axiom 3.14 (Inequalities, addition, and multiplication). *Let a, b, c be real numbers.*

(1) (Addition) *If $a < b$, then $a + c < b + c$.*

(2) (Multiplication by a positive number) *If $a < b$ and $c > 0$, then $ac < bc$.*

(3) (Multiplication by a negative number) *If $a < b$ and $c < 0$, then $ac > bc$.*

For example, if $a < b$, then $2a < 2b$, but $-2a > -2b$.

The following shows us how the multiplication axiom changes when we replace $<$ and $>$ by $\leq$ and $\geq$, respectively.

Solved Problem 3.15. *Prove that if $a \leq b$ and $c \geq 0$, then $ac \leq bc$.*

Solution. Suppose that $a \leq b$ and $c \geq 0$.

Case 1: $a = b$. In this case we have $ac = bc$, which implies that $ac \leq bc$, so we are done.

Case 2: $a \neq b$. Since $a \leq b$, by definition we have $a = b$ or $a < b$. Since $a \neq b$, we have that $a = b$ is false, so that $a < b$ is true.

Subcase 2a: $c = 0$. In this case we have $ac = 0 = bc$, and we are done.

Subcase 2b: $c \neq 0$. Since $c \geq 0$, we have $c > 0$. Since we also have $a < b$, we can apply Axiom 3.14(2) to conclude that $ac < bc$. Since this implies that $ac \leq bc$, we are done.

In conclusion, we have proved the result by cases. □

Exercise 3.9. *Let a, b, c be real numbers. Prove: If $a \leq b$, then $a + c \leq b + c$.*

Exercise 3.10. *Let a, b, c be real numbers. Prove: If $c > 0$ and $ac < bc$, then $a < b$. Hint: If c is a positive real number, then $\frac{1}{c}$ is a well-defined positive real number.*

3.5.2. The connective "and".

We begin with:

Definition 3.16 (The connective "and"). Given two statements P and Q, we say that the statement

$$P \text{ and } Q$$

is true if both of the two statements P and Q are true. The connective "and" is also referred to as a **conjunction**.

If $a \geq 0$, then we say that a is **non-negative**. If $a \leq 0$, then we say that a is **non-positive**. Let us prove the following result with a conjunction in the hypothesis.

Lemma 3.17. *Let a be a real number. If a is non-negative and a is non-positive, then $a = 0$.*

Proof. By the hypothesis, we have

$$(a > 0 \text{ or } a = 0) \text{ and } (a < 0 \text{ or } a = 0). \tag{3.37}$$

Suppose for a contradiction that $a > 0$. Then, by ($a < 0$ or $a = 0$) and the trichotomy law (Axiom 3.13), we have a contradiction. Therefore, the statement $a > 0$ is false. This and ($a > 0$ or $a = 0$) being true imply that $a = 0$. □

So, being simultaneously non-negative and non-positive is not contradictory. This reminds us of Schrödinger's cat.

Wanted, dead *and* alive: Schrödinger's cat.

All the difference a logical connective makes. Switch it and you get a Bon Jovi song: Wanted Dead *or* Alive.

We can summarize the definition of "and" by the **truth table**:

(3.38)

P	Q	P and Q
T	T	T
T	F	F
F	T	F
F	F	F

This truth table gives us a formal definition of the connective "and".

For example, the second row of the truth table says that if P is true and if Q is false, then "P and Q" is false. We also denote "and" by the symbol $\wedge$, so that "P and Q" is written as

$$P \wedge Q. \tag{3.39}$$

We see from the truth table that the statement "P and Q" is true exactly when both the statement "P" and the statement "Q" are true.

Here is another simple example of a true statement involving the connective "and":

Lemma 3.18. *Let a and b be non-negative real numbers such that $a + b = 0$. Then $a = 0$ and $b = 0$.*

Exercise 3.11. *Prove Lemma 3.18. Hint: After stating the hypotheses, one can, for example, start the proof by "Suppose for a contradiction that $a \neq 0$."*

As a visual device to think about some logical statements, we represent these statements by Venn diagrams. In Figure 3.5.1, each statement is represented by a set. The logical connective "and" is represented by "intersection", and the logical connective "or" is represented by "union". The motivations for these representations are given in Chapter 5 on set theory.

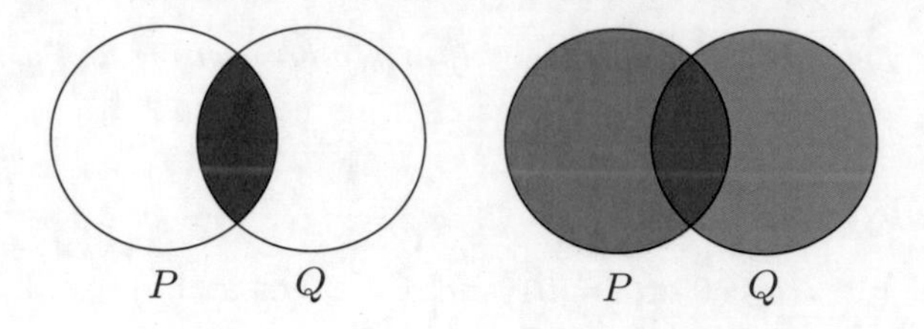

Figure 3.5.1. Left: Visualizing P and Q. Right: Visualizing P or Q.

3.5.3. The logical connective "not". Given a statement P, we say that the statement

$$\text{not } P$$

is true if P is false.

For example, since the statement "π is an integer" is false, we have that the statement "not (π is an integer)" is true.

The connective "not" is also referred to as a **negation**.

The truth table for "not" is

(3.40)

P	not P
T	F
F	T

We also denote "not" by the symbol $\neg$, so that "not P" is written as

$$\neg\, P. \tag{3.41}$$

From the truth table we see that "P" is true exactly when "not P" is false. We also see that "P" is false exactly when "not P" is true.

The negation of a negation is equivalent to the original statement. We can see this easily by the following truth table:

(3.42)

P	not P	not (not P)
T	F	T
F	T	F

To see why the truth values in the third column are so, we may rename the statement "not P" as "Q". Then we see that the second and third columns of (3.42) are the same as the truth table (3.40) with P replaced by Q. That is, in (3.42) the third column follows from the second column.

Since the truth values of the first and third columns are the same, we have that the statement P is **logically equivalent** to the statement not (not P). That is,

$$P \text{ is true} \quad \text{exactly when} \quad \text{not}\,(\text{not } P) \text{ is true.} \tag{3.43}$$

For example, if we say, "That's not not true", we mean "That's true." Perhaps, a more common version of this is: "That's not untrue!"

Example 3.19. Consider the statement: not $(x < y)$. By the trichtomy law, this is equivalent to $y < x$ or $x = y$; that is, $y \leq x$ (equivalently, $x \geq y$).

Exercise 3.12. *Let x be a real number. Prove: If $x \neq 0$, then $x^2 > 0$. Hint: Since x is non-zero by assumption, we have $x > 0$ or $x < 0$. Now use Axiom* 3.14.

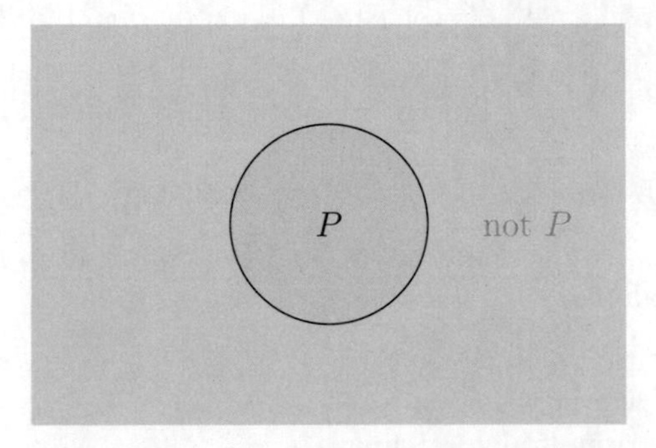

Figure 3.5.2. Visualizing the negation.

In Figure 3.5.2, a statement P is represented by a set, and its negation (not P) is represented by the "complement" of the set. See Chapter 5 for the motivation for this representation.

3.5.4. Tautologics. Now that we have discussed the logical connectives "or", "and", and "not", we can consider statements consisting of two or more statements joined by these logical connectives. We start with, no pun intended, trivialities.

Consider the statement

$$\text{“}2B \text{ or (not } 2B\text{)”}.$$

To most people, this is a famous poetic sentence of Shakespeare, but we will rather think of it as a mathematical statement! In fact, this sentence is Much Ado About Nothing: it is a *tautology*, which we will now explain.

A **tautology** is a compound statement in the form of a "formula" of statements which is always true independent of the truth values of the statements it is comprised of.

We can see that the compound statement "$2B$ or (not $2B$)" is a tautology by considering its truth table:

$2B$	not $2B$	$2B$ or (not $2B$)
T	F	T
F	T	T

So, no matter what the truth value of $2B$ is, the statement "$2B$ or (not $2B$)" is true, and hence it is a tautology! So Shakespeare was correct after all!

Example 3.20. The compound statement

$$(P \text{ or } Q) \text{ or } ((\text{not } P) \text{ or } (\text{not } Q))$$

is a tautology by the following truth table:

P	Q	P or Q	not P	not Q	(not P) or (not Q)	(P or Q) or ((not P) or (not Q))
T	T	T	F	F	F	T
T	F	T	F	T	T	T
F	T	T	T	F	T	T
F	F	F	T	T	T	T

Some Yogi-isms, i.e., statements by Yogi Berra, are tautologies when suitably interpreted. For example,

It ain't over 'til it's over.
It's déjà vu all over again.
You can observe a lot by watching.

Of course, this is all tongue in cheek. We've often watched while not observing.

Many Yogi-isms are *contradictions* (see the next subsection):

Nobody goes there anymore. It's too crowded.
I really didn't say everything I said.

What do you make of the following Yogi-isms? Are they tautologies, contradictions, or neither?

A nickel ain't worth a dime anymore.
Always go to other people's funerals; otherwise they won't go to yours.
If you can't imitate him, don't copy him.

So, what is the goal of mathematics? Starting with your premises, use a bunch of *tautologies* to logically deduce your desired conclusion!

Example 3.21. Let P and Q be statements. We have the following tautologies:

(1) P or (not P).
(2) (P or Q) or ((not P) or Q).
(3) (P and Q) or $\big($(not P) or (not Q)$\big)$.

Exercise 3.13. *Prove that the statements in parts* (2) *and* (3) *of Example* 3.21 *are tautologies.*

3.5.5. Contradictions. On the other hand, some trivialities can be oxymoronic. A **contradiction** is a compound statement in the form of a "formula" of statements which is always false independent of the truth values of the statements it is comprised of.

For example:

Proposition 3.22. The non-Shakespearean statement

"$2C$ and (not $2C$)"

is a contradiction; i.e., it is false (independent of what the statement "$2C$" is).

Proof. We prove this by constructing a truth table. Let $P = 2C$. We have

(3.44)

P	not P	P and (not P)
T	F	F
F	T	F

Here, we used the truth tables for "and" in (3.38) and for "not" in (3.40). In particular, we used the truth values in the second and third rows of (3.38). □

Most of the time, when we derive a contradiction in a proof, it is a statement of the form "P and (not P)".

Historical note: The 1992 movie "Wayne's World" made popular the use of the word "not" at the end of sentence. Here is an example:

"I find this lecture interesting, not!"

Here is an interesting self-contradictory quote:

"We learn from history that we do not learn from history." – Georg W. F. Hegel

Example 3.23. The compound statement

$$(P \text{ or } Q) \text{ and } ((\text{not } P) \text{ and } (\text{not } Q))$$

is a contradiction by the following truth table:

P	Q	P or Q	not P	not Q	(not P) and (not Q)	(P or Q) and ((not P) and (not Q))
T	T	T	F	F	F	F
T	F	T	F	T	F	F
F	T	T	T	F	F	F
F	F	F	T	T	T	F

Example 3.24. Let P, Q, and R be statements. We have the following contradictions:

(1) P and (not P).

(2) (P and Q) and ((not P) and R).

(3) (P and Q) and ((not P) or (not Q)).

Exercise 3.14. *Prove that the statements in parts* (2) *and* (3) *of Example* 3.24 *are contradictions.*

Exercise 3.15. *Is*

(3.45) $$(P \text{ or } Q) \text{ and } ((\text{not } P) \text{ or } (\text{not } Q))$$

a contradiction? Hint: What happens for example if P is true and Q is false?

In §3.8 below we will discuss proof by contradiction! Namely, suppose we are trying to prove the following statement: "*quod erat demonstrandum*". We begin by assuming that *quod erat demonstrandum* is false. If, from this assumption, we can prove a *contradiction*, then we are done. That is, we will have proved that *quod erat demonstrandum* is true!

3.6. Implications

> "In mathematics you don't understand things. You just get used to them." – John von Neumann

In this section we discuss the familiar notion of *implication*, which is a logical connective and which we have been using throughout the book up to now, in a more formal way.

3.6.1. A literary introduction to implications.

In the folk tale Cinderella, Cinderella's stepmother tells her:

> If you complete your chores, then you can go to the ball.

This is a **conditional statement** or **implication**, that is, an if-then statement!

Such a statement has the general form:

$$P \Rightarrow Q, \tag{3.46}$$

where P and Q themselves are statements. Recall that a statement is either true or false.

What is the mathematical *meaning* of an implication? Firstly, here is what it is not: causation. For example, consider the sentence:

If it rains, then my porch gets wet.

In this sentence we see causation: the rain is what causes the porch to get wet. On the other hand, in mathematics, we view implications differently:

An implication does not imply (pun intended) *cause and effect.*

Given two statements P and Q, we say that:

(1) The implication $P \Rightarrow Q$ is false if P is true and Q is false.
(2) The implication $P \Rightarrow Q$ is true if any one of the following cases is true:
  (a) P and Q are true.
  (b) P is false and Q is true.
  (c) P and Q are false.

Let us first see if such a definition of implication is reasonable. For fun, consider first the implication above from the story of Cinderella. Let us suppose, according the stepmother's wishes, that Cinderella will never "P" = complete her chores and will never "Q" = go to the ball. In this case, both P and Q are false, but by rule 2(c) above, the implication $P \Rightarrow Q$ is true!

Example 3.25. Consider the following implication:

If the moon is made of green cheese, then the earth is flat.

Here, again, both the hypothesis and conclusion are false (let us make the assumption that we are stating facts and not opinions!), and hence the implication is true!

It is useful to illustrate the meaning of *implication* by the truth table defining it as a logical connective:

(3.47)

P	Q	$P \Rightarrow Q$
T	T	T
T	F	F
F	T	T
F	F	T

This truth table defines in a formal sense the implication $\Rightarrow$. From this truth table we see that the truth value of $P \Rightarrow Q$ depends only on the truth values of P and Q, as is the case with the other logical connectives that we have discussed.

Observe that if Q is true, then the implication $P \Rightarrow Q$ is true independent of whether P is true or false. For example, the statement

$$(1^2 = -1) \Rightarrow (1^2 = 1) \tag{3.48}$$

is true simply because $1^2 = 1$ is true. In a sense, the interesting aspect of an implication $P \Rightarrow Q$ is when P is true, and in this case when the implication is true, we then know that Q is true.

In a logical argument, when we write

"If P, then Q",

we mean that if P is true, then Q is true. In particular, if P is false, then nothing is said about whether Q is true. Other ways of writing this implication are:

P implies Q.

Q if P.

P only if Q (we will not use this much).

Example 3.26. We have the following examples of tautologies involving implications:

(1) $P \Rightarrow (P \text{ or } Q)$.

(2) $(P \text{ and } Q) \Rightarrow P$.

(3) $(P \text{ and } Q) \Rightarrow (P \text{ or } Q)$.

(4) $(P \text{ and } Q) \Leftrightarrow (Q \text{ and } P)$.

(5) $(P \text{ or } Q) \Leftrightarrow (Q \text{ or } P)$.

Solved Problem 3.27. *Prove that the statement in part* (1) *of Example* 3.26 *is a tautology.*

Solution. We give two proofs.

(1) *Proof by logical inference.* Suppose P is true. Then P or Q is true, since we know that at least P is true. Therefore the implication $P \Rightarrow (P \text{ or } Q)$ is true.

(2) *Proof by truth table.* We have the following truth table:

(3.49)

P	Q	P or Q	$P \Rightarrow (P \text{ or } Q)$
T	T	T	T
T	F	T	T
F	T	T	T
F	F	F	T

Since the truth values in the last column are all T's, we have proved the result.

Exercise 3.16. *Prove that each of the statements in parts* (2)–(5) *of Example* 3.26 *is a tautology.*

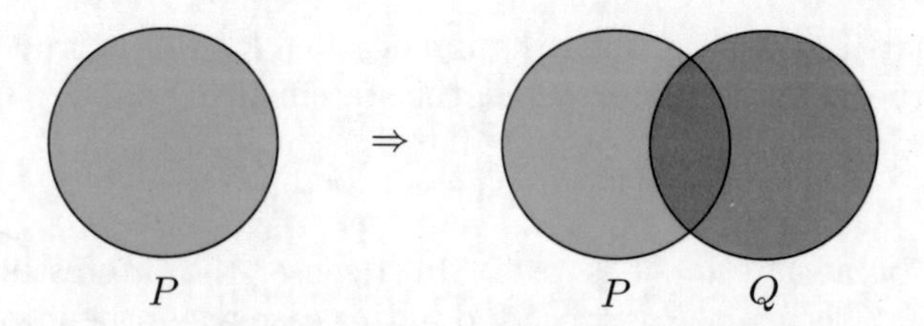

Figure 3.6.1. A Venn diagram visualization of Example 3.26(1).

The truth of an implication $P \Rightarrow R$ is represented by a Venn diagram as the set P being a *subset* of the set R.[7] In Figure 3.6.1, since the set P is a subset of the union of P and Q, this represents the implication $P \Rightarrow (P \text{ or } Q)$ being true. So this gives a visualization of part (1) of Example 3.26.

Exercise 3.17. *Can you draw Venn diagrams to illustrate the tautologies in parts* (2)–(5) *of Example* 3.26*?*

3.6.2. The converse. The title of this subsection does not refer to the manufacturer of a shoe!

Given an implication

$$P \Rightarrow Q, \tag{3.50}$$

its **converse** implication is

$$Q \Rightarrow P. \tag{3.51}$$

In other words, the converse of the statement "If P, then Q" is the statement

"If Q, then P."

It is a rather easy fact that if an implication is true, then its converse is not necessarily true. So, in general, as in this case, an implication does not have the same meaning as its converse. A simple example is the following implication: *If n is an even integer, then n is an integer.* Clearly this implication is true, whereas its converse is false.

[7]See Chapter 5 for a formal definition of the notion of *subset*.

The converse of “If an integer a is even, then a^2 is even” is

$$\text{“If } a^2 \text{ is even, then } a \text{ is even”}$$

(being understood that a is an integer). For this implication, both it and its converse are true.

Example 3.28. Consider the following true statement:

$$\text{If } x \geq 0, \text{ then } x^2 \geq 0.$$

Its converse is

$$\text{If } x^2 \geq 0, \text{ then } x \geq 0.$$

However, the converse is false since $-1 < 0$ and $(-1)^2 \geq 0$.

The following is the converse of the implication in Exercise 3.12.

Exercise 3.18. *Prove the following: Let $x \in \mathbb{R}$. If $x^2 > 0$, then $x \neq 0$.*

We have the following implication:

Exercise 3.19. *Let a and b be real numbers. Prove that if ($a > 0$ and $b > 0$) or ($a < 0$ and $b < 0$), then $ab > 0$.*

And its converse:

Exercise 3.20. *Let a and b be real numbers. Prove that if $ab > 0$, then ($a > 0$ and $b > 0$) or ($a < 0$ and $b < 0$).*

3.6.3. Transitivity of implication.

Having seen some examples of implications, we now consider one of its fundamental properties.

Lemma 3.29. *Suppose $P \Rightarrow Q$ and $Q \Rightarrow R$ are both true. Then $P \Rightarrow R$ is true.*

See Figure 3.6.2 for a visualization of the transitivity property of implication.

Exercise 3.21. *Prove Lemma 3.29.*

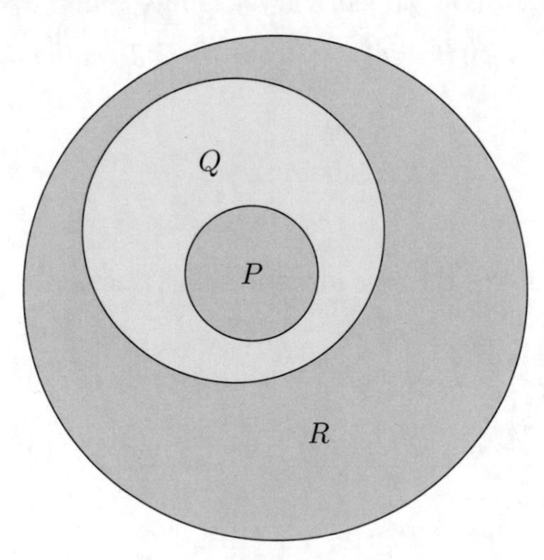

Figure 3.6.2. Visualizing the transitivity of implication: small $\Rightarrow$ medium $\Rightarrow$ large.

3.6.4. The biconditional. By combining an implication and its converse using the logical connective "and", we obtain the following.

Definition 3.30. Let P and Q be statements. The statement $P \Leftrightarrow Q$ means that $(P \Rightarrow Q)$ and $(Q \Rightarrow P)$. That is, the implication $(P \Rightarrow Q)$ and its converse are true. This is called a **biconditional.**

In other words, we have both:

"If P is true, then Q is true" and "If Q is true, then P is true."

In this case, we also say that "P if and only if Q" or "P is equivalent to Q."

Example 3.31. The following true statements are each biconditional:

(1) For any integer n, n is even if and only if n^2 is even.
(2) For any real number x, $x^2 > 0$ if and only if $x \neq 0$.
(3) For any prime p, p is odd if and only if $p > 2$.

Exercise 3.22. *Prove, using either a truth table or a logical argument, that each of the following statements are tautologies:*

(1) P *or* (*not* P).
(2) $(P \Rightarrow Q) \Leftrightarrow$ (*not* $Q \Rightarrow$ *not* P).
(3) ((*not* $P \Rightarrow Q$) *and* (*not* $P \Rightarrow$ *not* Q)) $\Rightarrow P$.
(4) *not* (P *and* Q) $\Leftrightarrow$ (*not* P *or not* Q)
(5) ((P *or* Q) *and* $(P \Rightarrow R)$ *and* $(Q \Rightarrow R)$) $\Rightarrow R$.

3.6.5. De Morgan's laws. Particularly useful equivalences of compound statements are the following. In a sense, negating an "or" yields an "and", and vice versa.

Solved Problem 3.32 (De Morgan's laws). *Prove that*

not (P *or* Q) $\Leftrightarrow$ (*not* P) *and* (*not* Q)

and

not (P *and* Q) $\Leftrightarrow$ (*not* P) *or* (*not* Q)

are tautologies.

Solution. We can see that the first statement is a tautology by considering its truth table in Table 3.6.1.

Table 3.6.1. Truth table for not (P or Q) $\Leftrightarrow$ (not P) and (not Q).

P	Q	P and Q	not (P and Q)	not P	not Q	(not P) or (not Q)
T	T	T	F	F	F	F
T	F	F	T	F	T	T
F	T	F	T	T	F	T
F	F	F	T	T	T	T

Evidently, we see that the truth values under the statements "not (P and Q)" and "(not P) or (not Q)" match (indeed, they are from top to bottom: F, T, T, T), so the statement that they are equivalent is a tautology, i.e., always true.

Similarly, one can prove the second statement. We leave this to you, dear reader!

We can alternatively prove that the first statement is a tautology by a logical argument as follows.

Suppose: not (P and Q). This is equivalent to: P and Q is false. By the definition of the logical connective "and", this means that at least one of P and Q is false. Equivalently, P is false or Q is false (we always use the inclusive "or"). That is, (not P) or (not Q) is true. We have established the desired equivalence.

See Figure 3.6.3 for a visualization.

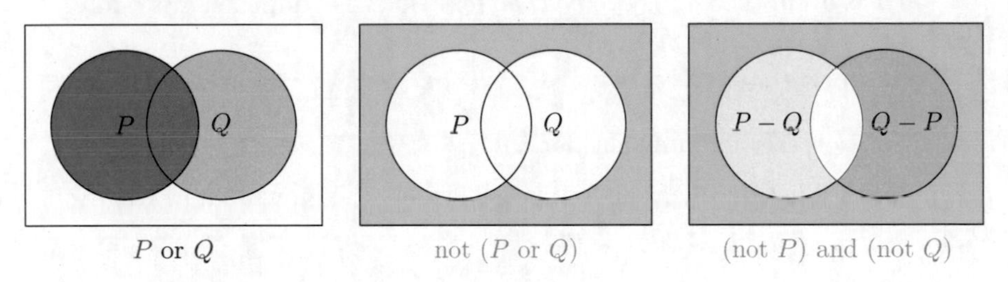

Figure 3.6.3. Visualizing one of De Morgan's laws. On the far right: (not P) consists of the union of $P - Q$ and (not (P or Q)), and (not Q) consists of the union of $Q - P$ and (not (P or Q)). We see that their intersection (corresponding to "and") is (not (P or Q)).

Exercise 3.23. *Let a and b be real numbers with $a < b$. Let x be a real number such that $a < x < b$ is false. Prove that $x \leq a$ or $x \geq b$. Hint: $a < x < b$ means $x > a$ and $x < b$.*

3.6.6. Rules of inference. In this subsection we discuss slightly more formally some of the logical reasoning concepts we have encountered thus far. However, the overall style of this book is to consider logic more informally. So we do not go into formal logic in any detail.

Propositional logic is the study of statements (a.k.a. propositions), relations between them, and logical arguments involving these statements.

Modus ponens is the following rule of inference:

> Suppose that P is true and that $P \Rightarrow Q$ is true. Then we can deduce that Q is true.

Example 3.33. Here is an example of *modus ponens*:

> 2 divides x^2. If 2 divides x^2, then 2 divides x. *Therefore*: 2 divides x.

Here, we assume that x is an integer whose square is divisible by 2. Then we state the fact that if x^2 is divisible by 2, then so is x. By *modus ponens*, we can conclude that x is divisible by 2.

Modus tollens is the following rule of inference:

If P, then Q. Not Q is true. *Therefore*: not P is true.

Example 3.34. Here is an example of *modus tollens*:

If 3 divides both a and b, then 3 divides $5a+7b$. 3 does not divide $5a+7b$. *Therefore*: 3 does not divide both a and b.

Modus tollens is related to contraposition, which is the following rule of inference (compare with the contrapositive defined in §3.7 below):

$P \Rightarrow Q$ is true. *Therefore*: not $Q \Rightarrow$ not P is true.

Example 3.35. Here is an example of *contraposition*:

If x is even, then x^2 is even. *Therefore*: If x^2 is odd, then x is odd.

3.7. Contrapositive

I am not, therefore I do not think.

Goodness! The statement above must be true! ☺ You will know why from this section.

3.7.1. The definition of contrapositive and examples. As we have seen, for an implication, its converse is not logically equivalent. However, we can add negation into the mix to rectify this.

Definition 3.36. The **contrapositive** of the statement $P \Rightarrow Q$ is the statement

$$(\text{not } Q) \Rightarrow (\text{not } P). \tag{3.52}$$

Example 3.37. Let the domain of discourse be $\mathbb{R}$. Consider the implication:

If $x > 0$, then $x^2 > 0$.

The contrapositive of this statement is the implication:

If $x^2 \leq 0$, then $x \leq 0$.

Observe that the following true implication is a stronger statement than the contrapositive above:

If $x^2 \leq 0$, then $x = 0$.

Equivalent to this implication is its contrapositive:

If $x \neq 0$, then $x^2 > 0$.

3.7.2. Equivalence of an implication with its contrapositive. This is a consequence of Truth Table 3.7.1. Since the truth values of the third and sixth columns therein are the same, we have the logical equivalence of an implication $P \Rightarrow Q$ with its contrapositive (not Q) $\Rightarrow$ (not P).

Exercise 3.24. *Let x be a real number, and let k be a positive integer. Prove that if $x^k > 0$, then $x \neq 0$.*

Table 3.7.1. The truth table for an implication and its contrapositive.

P	Q	$P \Rightarrow Q$	not P	not Q	(not Q) $\Rightarrow$ (not P)
T	T	T	F	F	T
T	F	F	T	F	F
F	T	T	F	T	T
F	F	T	T	T	T

3.7.3. Contrapositives of some results that we have proven. In Chapter 1 we actually considered a number of implications. We now consider some of their contrapositives. Recall from Corollary 1.30 that the square of an odd integer is odd and the square of an even integer is even.

By taking the contrapositives of these implications, we obtain:

Corollary 3.38. *We have the following implications regarding squares of integers:*

(1) *If a is an integer such that a^2 is even, then a is even.*

(2) *If b is an integer such that b^2 is odd, then b is odd.*

By combining this with Corollary 1.30 itself, we get the biconditional statements:

Corollary 3.39. *Let a be an integer. Then:*

(1) *a is even if and only if a^2 is even.*

(2) *a is odd if and only if a^2 is odd.*

Exercise 3.25. *Using results from Chapter* 1*, give short proofs of Corollaries* 3.38 *and* 3.39.

Exercise 3.26. *Prove that if the sum of two integers is even, then either both integers are even or both integers are odd.*

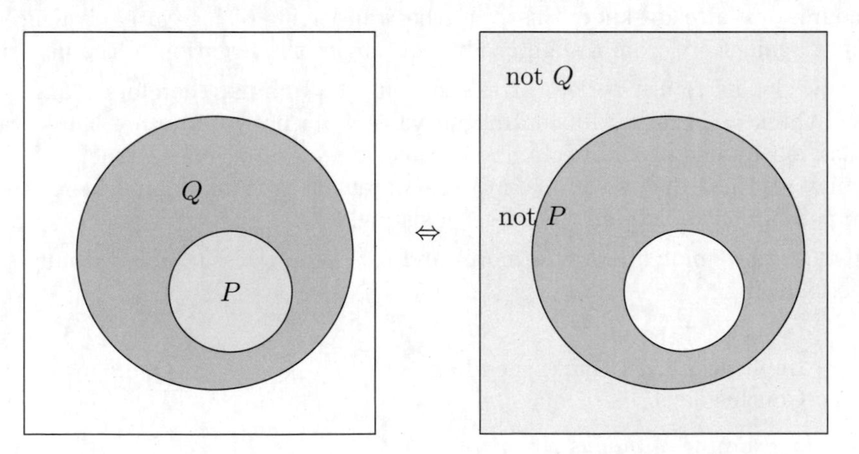

Figure 3.7.1. Visualizing the contrapositive (not Q) $\Rightarrow$ (not P) of the implication $P \Rightarrow Q$. The blue and green regions together represent Q, and the red and yellow regions represent (not P).

Exercise 3.27. *Prove that if the product of two integers is odd, then both integers are odd.*

Recall that Theorem 1.27 says that the sum of two odd integers is even. So, the contrapositive of this statement is:

If the sum of two integers is not even, then they are both not odd.

That is:

If the sum of two integers is odd, then they are both even.

Oops (ahem), where did we go wrong?! When negating the implication, we incorrectly negated the hypothesis. The contrapositive should read:

If the sum of two integers is not even, then they are not both odd.

What a difference in meaning interchanging the words "both" and "not" makes!

In other words:

If the sum of two integers is odd, then at least one of them is even.

Although this is a nice and true statement, we can do better (we leave this as an exercise):

If the sum of two integers is odd, then exactly one of them is even.

3.7.4. The contrapositive of the contrapositive. Given a conditional statement $P \Rightarrow Q$, its contrapositive is (not Q) $\Rightarrow$ (not P). In turn, the contrapositive of this second implication is

$$\big(\text{not (not } P)\big) \Rightarrow \big(\text{not (not } Q)\big). \tag{3.53}$$

Since the negation of the negation of a statement is the statement itself, we have that this implication is equivalent to the original implication

$$P \Rightarrow Q. \tag{3.54}$$

Of course, we already knew this from the transitivity of logical equivalence, but here we again see it from a slightly different angle (the negation of the negation).

Now, let us return to Descartes's statement "I think, therefore I am". This says: I think $\Rightarrow$ I am. So its contrapositive is: I am not $\Rightarrow$ I do not think. So the implication quoted at the beginning of this subsection is indeed true! Of course, you may say that the person making this statement is thinking and hence there is an apparent contradiction to them not existing!

Let us now think about this more carefully. What Descartes is actually saying is the following:

Premise: I think.
Implication: If I think, then I am.
Conclusion: I am.

This is an example of *modus ponens*.

However, for brevity Descartes's formatted the statement as

Premise, therefore Conclusion.

Indeed, the Implication is *implicit* in his statement.

The contrapositive of the implication in Descartes's argument is:

Contrapositive: If I am not, then I do not think.

In order to obtain the conclusion "I do not think", we need to assume the premise "I am not." However, by the very making of this premise, we are existing. ☺

On the other hand, the *converse* of Descartes's statement is:

I am, therefore I think.

We like to think of this as a false implication. But we can debate this. For example, this statement is self-referential. So, is the entity making this statement thinking? However, the entity might be a computer. Do computers think? Also, are self-referential sentences necessarily statements? And we can go on. In any case, our noodle is baked! Fortunately, we will not consider such sentences as mathematical statements.

3.7.5. Powers in Mersenne primes are prime. The following result, whose proof is formulated in terms of contraposition, certainly helps in the search for Mersenne primes. If you did Exercise 1.10, then you have already seen the proof, which is accomplished by proving the contrapositive implication.

Theorem 3.40. *If n is a positive integer such that $2^n - 1$ is a prime, then n itself is prime.*

Proof. Firstly, if $n = 1$, then $2^n - 1 = 1$ is not prime. Therefore we may assume that $n \geq 2$.

We prove the contrapositive. Suppose that $n \geq 2$ is a composite number; that is, $n = ab$, where $a, b > 1$ are integers. We have

$$2^n - 1 = (2^a)^b - 1. \tag{3.55}$$

Now recall the general formula (see (2.21))

$$x^b - 1 = (x - 1)(x^{b-1} + x^{b-2} + \cdots + x + 1). \tag{3.56}$$

Taking $x = 2^a$, which is an integer, we obtain

$$2^n - 1 = (2^a)^b - 1 = (2^a - 1)\big(2^{a(b-1)} + 2^{a(b-2)} + \cdots + 2^a + 1\big). \tag{3.57}$$

Observe that the factor $2^a - 1 > 1$ since $a > 1$. Similarly, the factor

$$2^{a(b-1)} + 2^{a(b-2)} + \cdots + 2^a + 1 > 1$$

since $a, b > 1$. We conclude that $2^n - 1$ is a composite number. □

3.8. Proof by contradiction

Ted: What if we were lying?
Bill: Why would we lie to ourselves?

Checkmate! Thus describes the essence of *proof by contradiction*, a.k.a. *reductio ad absurdum* (Latin for "reduction to absurdity"). We will consider absurdity as a synonym for contradiction. So for reductio ad absurdum, we start with assuming what we want to prove is false and deduce a contradiction from this assumption. Our only conclusion can be that what we want to prove is true!

In other words:

Mathematician Ted: Suppose that we were lying.
Mathematician Bill: This leads to a contradiction.
Dear reader: Therefore Bill and Ted were not lying!

We have the following lovely quote:

> *Reductio ad absurdum*, which Euclid loved so much, is one of a mathematician's finest weapons. It is a far finer gambit than any chess play: a chess player may offer the sacrifice of a pawn or even a piece, but a mathematician offers the game.
>
> – G. H. Hardy, A Mathematician's Apology

3.8.1. Proofs by contradiction that we have seen.

For example, the following proofs which we have already encountered in Chapter 1 are proofs by contradiction. We encourage the reader to review that these proofs are indeed by contradiction.

(1) The proof of Lemma 1.5 characterizing when an integer at least 2 is not a prime.

(2) The proof of Lemma 1.8 that any divisor of a positive integer is at most that integer.

(3) The proof of Lemma 1.21 that 3 is an odd integer, and its generalization to Lemma 1.22.

(4) Theorem 1.25 that the sum of an even integer and an odd integer is odd.

(5) The proof of Theorem 1.37 that any factorization of 1 by positive integers is trivial.

(6) In the proof of Corollary 1.38, we proved the statement that if $n^2 = 1$, then $n \neq 0$.

(7) After Definition 1.51 we proved that two distinct primes are coprime.

You've probably heard the expression (paraphrasing):

> There are many ways to be wrong, but there is only one way to be right.

Similarly, there are usually many ways to prove something wrong whereas there are usually few ways to prove something right. This explains why *proof by contradiction* is so effective:

> There are few ways to prove a statement, but there are many ways to disprove the negation of the statement.

Maybe we are serious; maybe we are not! ☺

In any case:

> *Proof by contradiction is one of the most important methods of proof in mathematics!*

Solved Problem 3.41. *Prove that if a divides b and if a does not divide c, then b does not divide c.*

Solution. Suppose for a contradiction that a divides b and that a does not divide c, but b divides c. Then, since a divides b and b divides c, by the transitivity of division, we obtain that a divides c. However, this contradicts a does not divide c. □

Mini-analysis of the solution. The template of the statement is:

$$\text{If } P \text{ and (not } Q\text{), then (not } R\text{).}$$

For a contradiction, we assumed that P, (not Q), and R are true. We derived from this that Q is true. Since this contradicts (not Q), the statement is proved.

Exercise 3.28. *Prove Corollary* 1.23 *from Theorem* 1.19. *(We didn't write out the detailed proof earlier.)*

Exercise 3.29. *Prove that if $a > b > 0$ are integers, then a is not a divisor of b.*

Exercise 3.30. *Let a, b, c, m, n be non-zero integers. Prove that if c does not divide $ma + nb$, then c does not divide a, or c does not divide b.*

Exercise 3.31. *Generalize Proposition* 3.42 *to the following statement: Suppose that d divides both a and b. Prove that if d does not divide c, then there do not exist integers m and n that solve the equation*

$$am + bn = c. \tag{3.58}$$

3.8.2. Confounded by conflating contradictions with contrapositives.

Proof by contrapositive is a special case of proof by contradiction. The template for a proof by contradiction of an implication $P \Rightarrow Q$ is:

> Suppose P and (not Q). *Yada, yada, yada.* Then R and (not R). □

Here, R is a suitable statement. On the other hand, the template for a proof by contrapositive is:

> Suppose (not Q). *Yada, yada, yada.* Then (not P). □

So, we can think of proof by contrapositive as the special case where $R = P$ of proof by contradiction. So, when one proves something by contrapositive, it may seem natural to think of it in terms of proving by the more general method of contradiction, although characterizing the proof as by contrapositive is more apt.

3.8.3. Solving, or not solving(!), linear equations over the integers.

The following is an example of a **linear Diophantine equation**, which we will study in earnest in §4.5. For this example, we prove the non-existence of solutions by *contradiction.*

Proposition 3.42. There do not exist integers m and n that solve the equation

$$15m + 18n = 1024. \tag{3.59}$$

Proof. Here is the idea of the proof: Suppose, for a contradiction, that the proposition is false. Then there exist integers m and n that satisfy the equation

$$15m + 18n = 1024. \tag{3.60}$$

What is wrong with this picture?

We observe (here, if you don't know your powers of two, is where we may have made it difficult for you!) that

$$1024 = 2^{10}. \tag{3.61}$$

In particular, the only prime divisor of the right-hand side of equation (3.60) is 2. In particular, 1024 is not divisible by the prime 3. However, the left-hand side is divisible by 3 since

$$15m + 18n = 3 \cdot (5m + 6n). \tag{3.62}$$

So, *if* a solution exists, we find that there exists a number, namely the number 1024, which is at the same time divisible by 3 and not divisible by 3, which is a contradiction! Thus, no solution to (3.60) exists. □

The proposition leads to the following (easy) generalization, which we leave as an exercise to the reader: Suppose that integers m and n satisfy an equation of the form

$$Am + Bn = C, \tag{3.63}$$

where A, B, C are integers. If d is a common divisor of A and B, then d divides C. In particular, the greatest common divisor

$$g := \gcd(A, B) \text{ divides } C.$$

We conclude that if g does not divide C, then (3.63) has no integer solutions.

Example 3.43. The linear Diophantine equation $51m + 68n = 1699$ has no solutions. Indeed, $\gcd(51, 68) = 17$, whereas 17 does not divide 1699.

3.8.4. Results about the greatest common divisor.

Solved Problem 3.44. *Prove that if an integer $n \geq 2$ is not a prime number, then $\gcd(n, (n-1)!) > 1$. In other words, its contrapositive: If an integer $n \geq 2$ satisfies $\gcd(n, (n-1)!) = 1$, then n is a prime number.*

Solution. Suppose that $n \geq 2$ is not a prime number. Then there exist integers $1 < a, b < n$ such that $n = ab$. It is easy to see that a divides both n and $(n-1)!$. Thus,

$$\gcd(n, (n-1)!) \geq a > 1.$$

□

The interested reader may wish to prove that if $n \geq 2$ is a composite number, then $\gcd(n, (n-1)!) = n$.

Exercise 3.32. *Let p be a prime and let a be an integer. Prove that if $\gcd(a, p) > 1$, then p divides a.*

How does this exercise compare to Exercise 1.34*?*

Exercise 3.33. *Let $n \geq 2$ be an integer. Prove the converse of Solved Problem* 3.44*: If $\gcd(n, (n-1)!) > 1$, then n is not a prime. Hint: You may quote a result about the case where n is a prime.*

3.8.5. The square root of 2 is irrational!

Question: What is an irrational number?
Answer: A number that thinks that two plus two is five!

Sadly, we've tried over and over again to teach the square root of two to think logically, but alas:

Theorem 3.45. *$\sqrt{2}$ is an irrational number.*

Proof. For a contradiction(!), suppose that $\sqrt{2}$ is a rational number. Then there exist integers a and b such that

$$\sqrt{2} = \frac{a}{b}.$$

By Theorem 1.50, we may assume that $\gcd(a, b) = 1$. In particular, a and b cannot both be even.

Now $2 = \frac{a^2}{b^2}$, so that

$$2b^2 = a^2. \tag{3.64}$$

By Corollary 3.38, since a^2 is even, a is even. Now, since a is even, there exists $q \in \mathbb{Z}$ such that $a = 2q$. Hence $2b^2 = (2q)^2 = 4q^2$, so that

$$b^2 = 2q^2. \tag{3.65}$$

Since b^2 is even, b is even. That both a and b are even contradicts that $\gcd(a, b) = 1$. We conclude that $\sqrt{2}$ is irrational. □

This is an iconic proof.

Remark 3.46. If we do not assume that a and b are coprime, we can prove the irrationality of $\sqrt{2}$ by the method of infinite descent. The idea is that a and b have to be even, and hence we can replace them by the integers $a_1 = a/2$ and $b_1 = b/2$. But then a_1 and b_1 are also both even, so in turn we can replace each of them by its half, retaining the property of being integers. We would then be able to continue this process of dividing by two ad infimum (forevermore), yielding a contradiction (why?)! Hint: Look for the well-ordering principle later in the book.

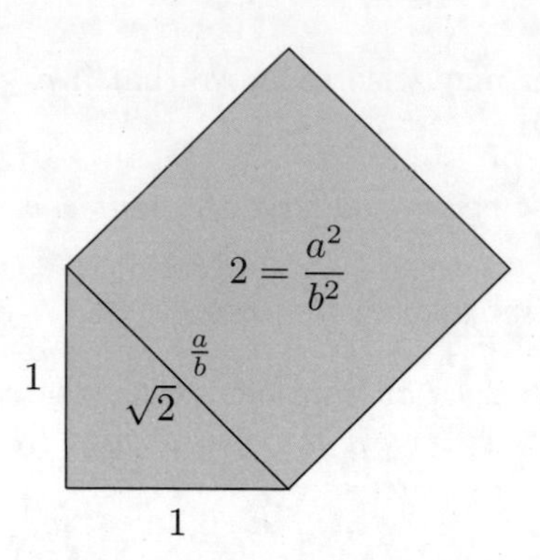

Figure 3.8.1. For a contradiction we assume that $\sqrt{2}$ is rational.

Exercise 3.34. *Prove that if x is a positive real number satisfying $x^3 = 2$, then x is irrational.*

3.8.6. The square root of any prime number is irrational. With the same logic as in the proof by contradiction in the previous section, we can prove the following.

Theorem 3.47. *If p is a prime, then $\sqrt{p}$ is irrational.*

The key to this is the following divisibility fact (see Corollary 4.24 below).

Lemma 3.48. *If a prime p divides the square a^2 of an integer, then p divides a.*

Given this fact, we leave it to the reader to prove Theorem 3.47.

Exercise 3.35. *Prove that if p is a prime, then $\sqrt{p}$ is irrational. Hint: Use Lemma* 3.48. *Note that Theorem* 3.51 *is a generalization of this exercise, but we would like you to go through the logic directly instead of appealing to the theorem.*

Just like the fact that square roots of primes are irrational, we have that m-th roots of primes are irrational. We consider the case of the prime 3.

Solved Problem 3.49. *Let $m \geq 2$ be an integer. Prove that $3^{1/m}$ is irrational.*

Solution. Suppose for a contradiction that $3^{1/m}$ is rational. Then there exist coprime integers a and b such that

$$3^{1/m} = \frac{a}{b}, \quad \text{so that} \quad 3 = \frac{a^m}{b^m}. \tag{3.66}$$

Thus, $b^m 3 = a^m$, and hence 3 divides a^m. Since 3 is a prime, this implies that 3 divides a (this will be proved in the next chapter). Thus, there exists an integer k such that $a = 3k$. Substituting this into the previous equation, we obtain

$$b^m 3 = a^m = 3^m k^m, \quad \text{so that} \quad b^m = 3^{m-1} k^m. \tag{3.67}$$

Since, $m - 1 \geq 1$, this implies that 3 divides b^m. Therefore, 3 divides b, which contradicts that a and b are coprime. □

3.8.7. When square roots are irrational. A **semiprime** is the product of distinct primes. Their square roots are also irrational.

Solved Problem 3.50. *Let p and q be distinct primes. Prove that $\sqrt{pq}$ is irrational.*

Solution. Suppose for a contradiction that $\sqrt{pq}$ is rational. Then there exist coprime integers a and b such that

$$\sqrt{pq} = \frac{a}{b}, \quad \text{so that} \quad pq = \frac{a^2}{b^2}. \tag{3.68}$$

Thus $a^2 = b^2pq$. In particular, p divides a^2. Since p is a prime, this implies that p divides a. Thus, there exists an integer k such that $a = pk$. Substituting this into the previous equation, we obtain

$$p^2k^2a^2 = b^2pq, \quad \text{so that} \quad pk^2a^2 = b^2q. \tag{3.69}$$

In particular, p divides b^2q. Since p and q are distinct primes, we have that p and q are coprime. By Theorem 4.23 below, this implies that p divides b^2, which in turn implies that p divides b. Since p divides both a and b, we have $\gcd(a,b) \geq p > 1$, which contradicts that a and b are coprime. □

The following (completely general) result can actually be proved with an argument similar to the previous ones. It says that the square root of a positive integer is either an integer or irrational.

Theorem 3.51 (Square roots are integers or irrational). *Let n be integer of the form*

$$n = p_1^{k_1} p_2^{k_2} \cdots p_r^{k_r}, \tag{3.70}$$

where $p_1 < p_2 < \cdots < p_r$ are primes and $k_1, k_2, \ldots, k_r$ are positive integers. If at least one of the integers $k_1, k_2, \ldots, k_r$ is odd, then $\sqrt{n}$ is irrational.

Otherwise, if $k_1, k_2, \ldots, k_r$ are all even, then $\sqrt{n}$ is an integer.

Proof. Let i be such that k_i is odd. Write $k_i = 2\ell_i + 1$, where ℓ_i is a non-negative integer. We have

$$n = (p_i^{\ell_i})^2 m, \quad \text{that is,} \quad \sqrt{n} = p_i^{\ell_i}\sqrt{m}, \tag{3.71}$$

where

$$m := p_1^{k_1} \cdots p_{i-1}^{k_{i-1}}\, p_i\, p_{i+1}^{k_{i+1}} \cdots p_r^{k_r}. \tag{3.72}$$

It suffices to prove that $\sqrt{m}$ is irrational. Suppose, for a contradiction, there exists a rational number in lowest terms $\frac{a}{b}$ satisfying

$$\left(\frac{a}{b}\right)^2 = m. \tag{3.73}$$

Then

$$a^2 = b^2 p_1^{k_1} \cdots p_{i-1}^{k_{i-1}}\, p_i\, p_{i+1}^{k_{i+1}} \cdots p_r^{k_r}. \tag{3.74}$$

This implies that p_i divides a^2. Now, by Lemma 3.48, we have that p_i divides a. We write

$$a = cp_i. \tag{3.75}$$

By substituting this into equation (3.74) and dividing both sides by p_i, we obtain

$$c^2 p_i = b^2 p_1^{k_1} \cdots p_{i-1}^{k_{i-1}} p_{i+1}^{k_{i+1}} \cdots p_r^{k_r}. \tag{3.76}$$

Here, we have to use a believable result from down the road (see Theorem 4.23 and Lemma 4.29 below): Since p_i divides b^2 times a product of primes all of which are distinct from p_i, we have that p_i divides b^2. Now we can apply Lemma 3.48 again to obtain that p_i divides b. We have proved that $p_i > 1$ is a common divisor of a and b, contradicting the assumption that the fraction $\frac{a}{b}$ is in lowest terms. □

3.8.8. Thinking irrationally about roots ☺. Suppose a stranger comes up to you and blurts out:

> Mesdames et Messieurs, there exist positive real numbers x and y that are both irrational and such that x^y is rational.

How do you react to l'Étranger's assertion? Are they saying gibberish (maybe they Forgot About Dre?) or is there something to it?

Since there are lots of irrational numbers, we might hope that the stranger's statement is true. Perhaps we can let x be the unique positive real number satisfying $x^{\sqrt{2}} = 2$. Then, all we have to do is prove that x is irrational. Now, $x = 2^{1/\sqrt{2}}$, so maybe it isn't that difficult. But, alas at this point we seem to be stuck. So, apparently, we need to explore another path.

Since, by now, we have been brainwashed into thinking that proof by contradiction is a powerful method, let us explore this avenue. Suppose for a contradiction that for all positive irrational real numbers x and y, the number x^y is irrational.

Let $x = \sqrt{2}^{\sqrt{2}}$. Then, by our assumption, x is irrational. Now what can we say about

$$x^{\sqrt{2}}\,? \tag{3.77}$$

Have you figured out the proof of the stranger's assertion?

Is this related to the original path we tried to take?

Here is a way to think about the preceding logical argument visually. Let $\mathbb{Q}$ denote the rational numbers, and let $\mathbb{I}$ denote the irrational numbers. Define the function $f : \mathbb{R}^+ \to \mathbb{R}^+$ by

$$f(x) = x^{\sqrt{2}}. \tag{3.78}$$

Observe that

$$f(f(x)) = f(x^{\sqrt{2}}) = (x^{\sqrt{2}})^{\sqrt{2}} = x^{\sqrt{2}\cdot\sqrt{2}} = x^2. \tag{3.79}$$

Now let x be any positive irrational number whose square x^2 is rational. By Theorem 3.45, $x = \sqrt{2}$ is such a number. Consider the schematic picture in Figure 3.8.2. Schematically, the blue interval represents the rationals and the red interval represents the irrationals.

Exercise 3.36. *Give a (formal) proof of* l'Étranger*'s assertion at the beginning of Subsection* 3.8.8.

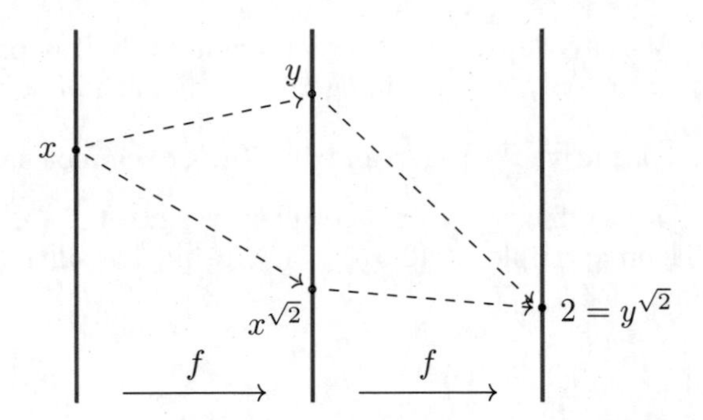

Figure 3.8.2. x is irrational. If x maps by f to a rational number, then we have that $x^{\sqrt{2}}$ is rational. If x maps by f to an irrational number y, then we use that $y^{\sqrt{2}}$ is rational.

Goal:	Case 1:	Case 2:
$\exists x, y,\ x^y \in \mathbb{Q}$	$\sqrt{2}^{\sqrt{2}} \in \mathbb{Q}$	$\sqrt{2}^{\sqrt{2}} \notin \mathbb{Q},\ \left(\sqrt{2}^{\sqrt{2}}\right)^{\sqrt{2}} = 2 \in \mathbb{Q}$

Figure 3.8.3. In Case 1, $x = \sqrt{2}$. In Case 2, $x = \sqrt{2}^{\sqrt{2}}$. In both cases, $y = \sqrt{2}$. In both cases we also have that x and y are irrational, whereas x^y is rational.

3.9. Pythagorean triples

Baseball fans know:

> These are better than the Pythagorean doubles, but not as good as the Pythagorean homers.

And now we discuss a topic (not baseball) that has been of mathematical interest since antiquity.

3.9.1. Right triangles. Suppose that we are given a right triangle, that is, a triangle where one of the (interior) angles is a right angle of 90°, i.e., $\pi/2$ radians. Call the lengths of the two sides (called **legs**) incident to the right angle a and b. Call the remaining side, the **hypotenuse**, c. The Pythagorean Theorem 3.4 then says that

$$a^2 + b^2 = c^2. \tag{3.80}$$

Definition 3.52. If a, b, and c are all positive integers, then we say that (a, b, c) is a Pythagorean triple. We call a and b **leg lengths** and c the **hypotenuse length** of the Pythagorean triple.

Example 3.53. Since $3^2 + 4^2 = 5^2$, we have that $(3, 4, 5)$ is a Pythagorean triple. Another Pythagorean triple is $(5, 12, 13)$. For a few more examples, see Table 3.9.1.

If (a, b, c) is a Pythagorean triple, then clearly we have

$$0 < a, b < c. \tag{3.81}$$

Definition 3.54. We say that a Pythagorean triple (a, b, c) is **primitive** if a, b, c are coprime; that is, the only positive integer dividing each of a, b, c is 1.

In particular, a primitive Pythagorean triple (a, b, c) cannot have a, b, c all even.

Example 3.55. The Pythagorean triple of positive integers $(3, 4, 5)$ is primitive, whereas the Pythagorean triple $(9, 12, 15)$ is not primitive since 3 divides each of $9, 12, 15$. See Figure 3.9.1.

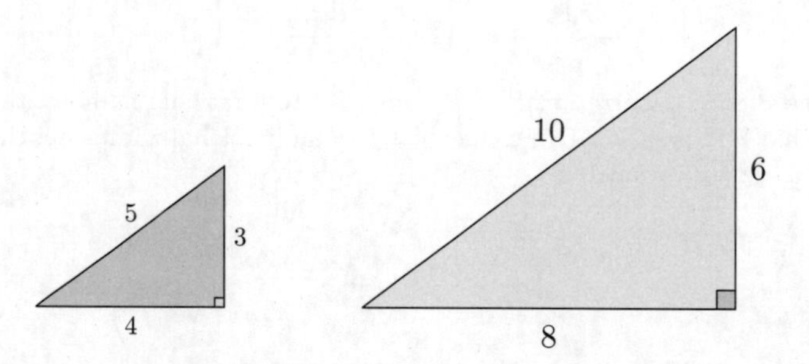

Figure 3.9.1. Pythagorean triples. The one on the left is primitive, and the one on the right is not primitive.

3.9.2. Pythagorean triples and parity.

The following result gives constraints on the parities of integers in Pythagorean triples.

Lemma 3.56. *If (a, b, c) is a Pythagorean triple, then at least one of a and b is even.*

Furthermore, if (a, b, c) is primitive, then exactly one of a and b is even (and one is odd), and hence c is odd.

Proof. For a contradiction, suppose that both a and b are odd. Then there exist non-negative (actually positive) integers k and ℓ such that $a = 2k+1$ and $b = 2\ell+1$. We have

$$c^2 = a^2 + b^2 = (2k+1)^2 + (2\ell+1)^2 = 4(k^2 + k + \ell^2 + \ell) + 2. \tag{3.82}$$

This implies that c^2 is even, which in turn implies that c is even. Thus there exists an integer m such that $c = 2m$. Hence $c^2 = 4m^2$ is a multiple of 4, which contradicts (3.82). We conclude that at least one of a and b is even.

Now suppose that (a, b, c) is primitive. If a and b are both even, then c is even, which is a contradiction. Thus, in this case exactly one of a and b is even, and exact one of them is odd. Finally, this implies that c^2 is odd, which implies that c is odd. □

Exercise 3.37. *Prove that if (a, b, c) is a Pythagorean triple, then $a+b-c$ is even.*

Exercise 3.38. *Let (a, b, c) be a primitive Pythagorean triple. Prove:*

(1) *Exactly one of (a, b, c) is divisible by 3.*
(2) *Exactly one of (a, b, c) is divisible by 4.*

(3) *Exactly one of* (a, b, c) *is divisible by* 5.

In particular, abc *is divisible by* 60.

Exercise 3.39. *Prove that if* (a, b, c) *is a Pythagorean triple, then*

$$\frac{(c-a)(c-b)}{2} \tag{3.83}$$

is a perfect square. (You may first check a few examples to give you confidence that this statement is true.)

3.9.3. Euclid's formula.

Theorem 3.57. *There are an infinite number of primitive Pythagorean triples.*

This follows from:

Lemma 3.58 (Euclid's formula). *If* m *and* n *are positive integers with* $m > n$, *then*

$$a = m^2 - n^2, \quad b = 2mn, \quad c = m^2 + n^2 \tag{3.84}$$

yields a Pythagorean triple.

Observe that

$$c - b = (m - n)^2 \tag{3.85}$$

is a perfect square, as well as

$$\frac{c-a}{2} = n^2 \tag{3.86}$$

being a perfect square.

For example, for every positive integer n, *by taking* $m = n + 1$ *we obtain the Pythagorean triple*

$$a = 2n + 1, \quad b = 2n^2 + 2n, \quad c = 2n^2 + 2n + 1. \tag{3.87}$$

Note that, here, $a < b = c - 1 < c$. *In particular, any odd integer at least* 3 *is the leg length of a Pythagorean triple.*

Examples of Euclid's formula are given in Table 3.9.1.

Table 3.9.1. Some instances of Euclid's formula for Pythagorean triples. Primitive triples are in red.

m	n	a	b	c
2	1	3	4	5
3	1	8	6	10
3	2	5	12	13
4	1	15	8	17
4	2	12	16	20
4	3	7	24	25
5	1	24	10	26
5	2	21	20	29

Exercise 3.40. *The Pythagorean triples of positive integers less than* 100 *after Table* 3.9.1 *are*

$$(12, 35, 37),\ (9, 40, 41),\ (28, 45, 53),\ (11, 60, 61),\ (16, 63, 65),\ (33, 56, 65),$$
$$(48, 55, 73),\ (13, 84, 85),\ (36, 77, 85),\ (39, 80, 89),\ (65, 72, 97).$$

For each of these Pythagorean triples, find the corresponding pair of positive integers (m, n) *satisfying* $m > n$ *and Euclid's formula* (3.84). *(You may need to switch* a *and* b *to get the form* (3.84) *in the same order.)*

Exercise 3.41. *Prove that Theorem* 3.57 *follows from Lemma* 3.58.

Exercise 3.42. *Prove Lemma* 3.58. *Hint: This is just a calculation.*

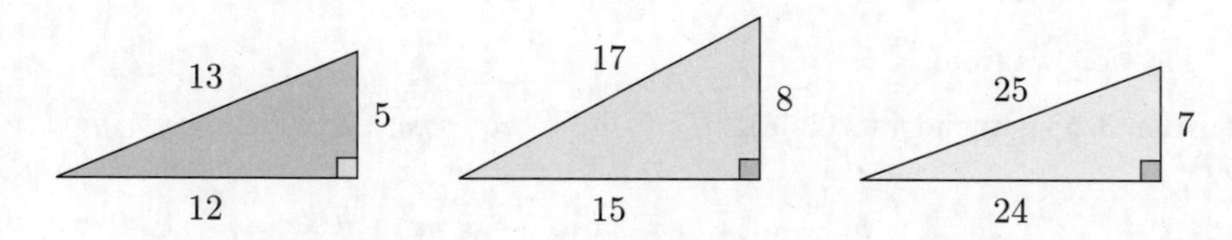

Figure 3.9.2. Visualization of three primitive Pythagorean triples drawn to (different) scale(s).

3.9.4*. Primitive Pythagorean triples.

Conversely, we have the following result.

Lemma 3.59 (Converse of Euclid's Lemma). *If* (a, b, c) *is a primitive Pythagorean triple with* a *odd and* b *even, then there exist positive integers* m *and* n *with* $m > n$ *such that* (a, b, c) *is of the form* (3.84)*:*

$$a = m^2 - n^2, \quad b = 2mn, \quad c = m^2 + n^2.$$

Moreover, m *and* n *are coprime, and one is even and one is odd.*

Proof. Denote $P = \left(\frac{b}{c}, \frac{a}{c}\right)$. Since $\left(\frac{b}{c}\right)^2 + \left(\frac{a}{c}\right)^2 = 1$, the point P lies on the unit circle in $\mathbb{R}^2$, and also P lies in the interior of the first quadrant of the plane since a, b, c are all positive.

Let $Q = (0, -1)$ be the "south pole" of the unit circle, and let R be the intersection of the line segment $\overline{PQ}$ with the (positive) x-axis. We easily find that (for example, by similar right triangles; see Figure 3.9.3.)

$$R = \left(\frac{\frac{b}{c}}{1 + \frac{a}{c}}, 0\right) = \left(\frac{b}{a + c}, 0\right). \tag{3.88}$$

On the other hand, we can write P as an affine combination of Q and R:

$$\left(\frac{b}{c}, \frac{a}{c}\right) = P = (1 - t)(0, -1) + t\left(\frac{b}{a + c}, 0\right) = \left(t\frac{b}{a + c}, t - 1\right), \tag{3.89}$$

where the parameter $t \neq 0$ is determined by the condition that P lies on the unit circle. So we obtain

$$1 = t^2 \frac{b^2}{(a + c)^2} + (t - 1)^2; \tag{3.90}$$

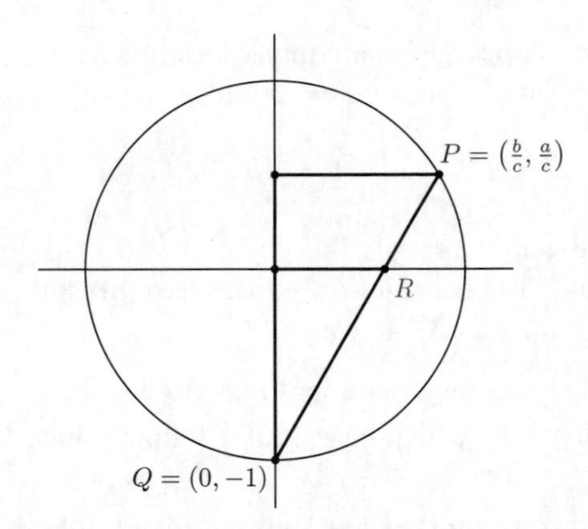

Figure 3.9.3. Similar triangles used to derive formula (3.88) for R.

that is, after simplifying and factoring out the non-zero variable t,

$$t\left(\frac{b^2}{(a+c)^2} + 1\right) - 2 = 0. \tag{3.91}$$

Therefore,

$$t = \frac{2}{\frac{b^2}{(a+c)^2} + 1} = \frac{2(a+c)^2}{b^2 + (a+c)^2}. \tag{3.92}$$

By substituting this into (3.89), we obtain the identities

$$\frac{b}{c} = \frac{2b(a+c)}{b^2 + (a+c)^2}, \qquad \frac{a}{c} = \frac{(a+c)^2 - b^2}{b^2 + (a+c)^2}. \tag{3.93}$$

Now let $g = \gcd(b, a+c)$, and define $m := \frac{a+c}{g}$ and $n := \frac{b}{g}$. We then obtain

$$\frac{b}{c} = \frac{2mn}{m^2 + n^2}, \qquad \frac{a}{c} = \frac{m^2 - n^2}{m^2 + n^2}. \tag{3.94}$$

Finally, we claim that (3.84) holds; that is,

$$a = m^2 - n^2 =: \tilde{a}, \quad b = 2mn =: \tilde{b}, \quad c = m^2 + n^2 =: \tilde{c}. \tag{3.95}$$

To see this, first observe that (3.94) says that

$$\frac{b}{c} = \frac{\tilde{b}}{\tilde{c}}, \qquad \frac{a}{c} = \frac{\tilde{a}}{\tilde{c}}, \tag{3.96}$$

so that

$$\frac{\tilde{a}}{a} = \frac{\tilde{b}}{b} = \frac{\tilde{c}}{c} =: k. \tag{3.97}$$

We also write this as $(\tilde{a}, \tilde{b}, \tilde{c}) = k(a, b, c)$.

Clearly $k \in \mathbb{Q}$ since it is equal to a quotient of integers.

Claim. k is a (positive) integer.

Proof of the claim. Suppose for a contradiction that k is not an integer. Then $k = \frac{r}{s}$, where $r, s \in \mathbb{Z}^+$ satisfy $\gcd(r, s) = 1$ and $s > 1$. We have that

$$\tilde{a} = \frac{r}{s}a, \quad \tilde{b} = \frac{r}{s}b, \quad \tilde{c} = \frac{r}{s}c \tag{3.98}$$

are all integers. Since $\gcd(r, s) = 1$, by Theorem 4.23 this implies that s divides each of a, b, c. We have a contradiction to the assumption that the Pythagorean triple (a, b, c) is primitive. q.e.d. the claim.

Note that $\gcd(m, n) = 1$ by Theorem 1.50. We will show that the Pythagorean triple $(\tilde{a}, \tilde{b}, \tilde{c})$ is primitive, in which case (3.95) follows since k is a positive integer which then must be equal to 1.

Suppose for a contradiction that the Pythagorean triple $(\tilde{a}, \tilde{b}, \tilde{c})$ is not primitive. Then the integer k in (3.97) satisfies $k > 1$.

Let p be an odd prime dividing k. Then p divides each of $2mn$, $m^2 - n^2$, and $m^2 + n^2$.

Claim. p divides both m and n, a contradiction to $\gcd(m, n) = 1$.

Case 1: Since the prime $p > 2$ divides both $2m^2$ and $2n^2$ and since $\gcd(p, 2) = 1$, by Theorem 4.23 p divides both m^2 and n^2. By Corollary 4.24, this implies that p divides both m and n. q.e.d. the claim.

By the claim, k is equal to a power of 2; that is, $k = 2^t$ for some positive integer t. By (3.97), we have

$$m^2 - n^2 = \tilde{a} = 2^t a, \quad 2mn = \tilde{b} = 2^t b, \quad m^2 + n^2 = \tilde{c} = 2^t c. \tag{3.99}$$

Since $\gcd(m, n) = 1$, we cannot have both m and n even. Since 2 divides $m^2 - n^2$, we have that m and n have the same parity and hence m and n are both odd. However, $2mn = 2^t b$ now yields a contradiction since $t > 0$ and b is even.

We conclude that the Pythagorean triple $(\tilde{a}, \tilde{b}, \tilde{c})$ is primitive and hence equal to (a, b, c). The lemma follows. □

So, in conclusion, we have also proved Theorem 3.57 by Exercise 3.41.

3.10. Quantifiers

So far in this book we have had a smattering of existential and universal quantifiers in our discussions. In this section we give a more "conscious" discussion of these quantifiers.

3.10.1. Existential quantifiers.

Question: What is an existential crisis?
Answer: When you forget the meaning of $\exists$.

An existential quantifier is about the existence of something. That is, there *is* something of some sort.

3.10.1.1. *Existence.* Existentialism (existence):

$$\exists \text{ means "there exists".}$$

We call $\exists$ an **existential quantifier**.

For example:

Theorem 3.60. *There* exists *an irrational real number.*

Proof. By Theorem 3.45, $\sqrt{2}$ is an irrational real number. □

Since we are asked to prove the existence of an "animal", in this case an irrational number, all we need to do is produce just one of this animal. So we are done!

So, *if it comes easy*, the template for the proof of the existential statement: $\exists x \in X$, $P(x)$ is

$$\text{Jean} \in X \text{ and } P(\text{Jean}) \text{ is true.}$$

3.10.1.2. *Non-existence.* But, in general and according to Ringo Starr, "*It Don't Come Easy*". Consider the nihilistic statement (non-existence):

There is no free lunch.

Mathematically, as long as a "free lunch" is a well-defined term, this says:

There does not exist a free lunch. I.e., $\nexists$ free lunch.

3.10.1.3. *Proving existence and non-existence.* The following comes easy:

Solved Problem 3.61. *Prove that there exists a rational number x satisfying $x > 1$ and $x < 2$.*

Solution. Let $x = \frac{3}{2}$. Then x is a rational number, $x > 1$, and $x < 2$. □

And it can be easy when it don't come easy:

Solved Problem 3.62. *Prove that there does not exist a real number x satisfying $x < 1$ and $x > 2$.*

Solution. Suppose for a contradiction that there does exist a real number x satisfying $x < 1$ and $x > 2$; i.e., $2 < x$. We then have that $2 < 1$ by the transitivity of $<$. This is a contradiction. □

Exercise 3.43. *Prove that there exists $x \in \mathbb{R}$ satisfying both $x > a$ and $x < b$ if and only if $a < b$.*

Exercise 3.44. *Prove that for every positive integer n, there exists a prime p such that $p > n$.*

Exercise 3.45. *Prove that for every positive rational number r, there exists a positive rational number s satisfying $s < r/100$.*

3.10.2. Universal quantifiers. A universal quantifier is about *everything* of some sort having some property. Namely, ubiquity. Not much ado about everything:

$$\forall \text{ means "for all".}$$

We call $\forall$ a **universal quantifier**.

For example, consider the statement: "For all $x \in \mathbb{R}$, $x^2 \geq 0$." This means that the square of *every* real number is non-negative.

How not to prove a universal statement:

$$\text{Joan} \in X \text{ and } P(\text{Joan}) \text{ is true}$$

... unless: Joan is the only element of X!

On the other hand, X may consist of "Jean and Joan and a who knows who", according to Carlos Santana. In particular, if X has an infinite number of elements, then you would have to check that $P(x)$ for all x in X, which is an infinite number of x's.

For example, we cannot just say that $\pi \in \mathbb{R}$ and $\pi^2 \geq 0$; therefore, $\forall x \in \mathbb{R}$, $x^2 \geq 0$. But we can argue: Let x be (*any*) real number. Blah, blah, blah.[8] Therefore $x^2 \geq 0$.

Solved Problem 3.63. *Prove that for every rational number r, r^2 is a rational number.*

Solution. Let r be any rational number. Then there exist integers a and b such that $r = \frac{a}{b}$ (so $b \neq 0$). We then compute that $r^2 = \frac{a^2}{b^2}$. Since a^2 and b^2 are integers (and $b^2 \neq 0$), we conclude that r^2 is a (well-defined) rational number. □

Exercise 3.46. *Prove that for every rational number r and s, $r + s$ and rs are rational numbers.*

3.10.3. Disproving existential statements. Suppose that we are asked to disprove that there exists a unicorn on Earth. Assuming that any unicorn on Earth, if it exists, is an animal, this is equivalent to proving the following universal statement: For every animal on Earth, the animal is not a unicorn.

More abstractly, consider the existential statement:

$$\exists x_0 \in X,\ P(x_0).$$

To disprove this statement, we need to prove the following universal statement:

$$\forall x \in X,\ \text{not } P(x).$$

In other words, this last statement is equivalent to the negation of the statement $\exists x_0 \in X$, $P(x_0)$, which we may write as $\nexists x \in X$, $P(x)$.

To disprove an existence statement, i.e., to prove a non-existence statement, one method is proof by contradiction, as in Solved Problem 3.62.

To summarize, the negation of an existential statement is equivalent to a universal statement. For example, if we flippantly think of a solution x to the inequalities $2 < x < 1$ as being evil, then we are saying that the statement "there is no evil" is

[8] We leave it to you, dear reader, to recall the details of the proof, where the idea is to consider the two cases of squares of (i) non-negatives and (ii) negatives.

equivalent to the statement "it's all good (from Diego to the Bay)", assuming that evil and good are negations of each other! On the other hand, if there exists evil, then not all is good! In general, by taking negations in the first statement of this paragraph, we see that an existential statement is equivalent to the negation of a universal statement.

3.10.4. Disproving universal statements. Now suppose we are asked to disprove the following statement: *For every fish on Earth, the fish can swim.* Disproving this universal statement is equivalent to proving the following existential statement: *There exists a fish on Earth that cannot swim.*

More abstractly, consider the statement

$$\forall y \in Y,\, Q(y).$$

To disprove this statement, all we need to do is provide an element Shaq $\in Y$ such that $Q(\text{Shaq})$ is false. This is because the *negation* of the universal statement above is equivalent to the existential statement

$$\exists y_0 \in Y, \text{ not } Q(y_0).$$

3.10.5. Examples of earlier statements as universal or existential statements. In this subsection we rephrase statements of the results we have encountered in Chapter 1 more formally in terms of existential and universal quantifiers, and implications.

Recall that $\mathbb{E}$ is the set of even integers and $\mathbb{O}$ is the set of odd integers.

Theorem 1.24 says: $\forall a, b \in \mathbb{Z},\, a, b \in \mathbb{E} \Rightarrow a + b \in \mathbb{E}$. In other words, the sum of any two even integers is even.

Theorem 1.27 says: $\forall a, b \in \mathbb{Z},\, a, b \in \mathbb{O} \Rightarrow a + b \in \mathbb{E}$.

Corollary 1.30 says:

(1) $\forall a \in \mathbb{Z},\, a \in \mathbb{E} \Rightarrow a^2 \in \mathbb{E}$. In other words, the square of any even integer is even.

(2) $\forall a \in \mathbb{Z},\, a \in \mathbb{O} \Rightarrow a^2 \in \mathbb{O}$.

Theorem 1.52 says: $\forall a, b \in \mathbb{Z}$,

$$c|a \text{ and } c|b \quad \Rightarrow \quad \forall m, n \in \mathbb{Z},\, c|(ma + nb). \tag{3.100}$$

In other words, for any integers a, b, c, if c divides both a and b, then c divides any integral linear combination of a and b.

Exercise 1.23 says:

(1) $\forall a, b, c \in \mathbb{Z},\, a|b \Rightarrow a|bc$.

(2) $\forall a, b \in \mathbb{Z},\, a|b \Rightarrow a|b^2$.

3.10.6. More mathematical implications: Implications with free variables. Regarding "more", the title of this subsection is punny (but not puny!). We look at some implications with a free variable. These are implications of which we have already considered many examples. For example, the implication with the free variable $x \in \mathbb{R}$: For all $x \in \mathbb{R}$, if $x \neq 0$, then $x^2 > 0$.

Consider the general form of an implication in one free variable:

$$\forall x \in X,\ P(x) \Rightarrow Q(x). \tag{3.101}$$

In terms of sets, we can think of this statement as saying the following. Let

$$\mathfrak{P} := \{x \in X : P(x) \text{ is true}\},$$
$$\mathfrak{Q} := \{x \in X : Q(x) \text{ is true}\}$$

be the **truth sets** of P and Q, respectively. That is, $\mathfrak{P}$ and $\mathfrak{Q}$ are the sets of all x where $P(x)$ and $Q(x)$, respectively, are true.

We have that (3.101) is equivalent to the set containment of truth sets (see Figure 3.10.1):

$$\mathfrak{P} \subset \mathfrak{Q}. \tag{3.102}$$

Indeed, (3.101) says that if $P(x)$ is true, then $Q(x)$ is true. On the other hand, (3.102) says that if $x \in \mathfrak{P}$, then $x \in \mathfrak{Q}$. By the definition of the sets $\mathfrak{P}$ and $\mathfrak{Q}$, this says that if $P(x)$ is true, then $Q(x)$ is true. This proves the aforestated equivalence.

As a device to remember this equivalence better, let us define for the moment that x is a *P-true element* if $P(x)$ is true and x is a *Q-true element* if $Q(x)$ is true. So $\mathfrak{P}$ and $\mathfrak{Q}$ are the sets of P-true and Q-true elements, respectively. So, in this language, (3.102) says that any P-true element is a Q-true element.

Figure 3.10.1. Visualizing an implication: The rare occurrence of a duck $\mathfrak{P}$ in a pond $\mathfrak{Q}$ surrounded by sand X.

Consider, as an example, the following implication:

Let $x \in \mathbb{R}$. If $x > 0$, then $x^2 > 0$.

This is a true statement and it illustrates all three "true" rows of the four rows in the truth table for $\Rightarrow$. Let $P(x) = (x > 0)$ and let $Q(x) = (x^2 > 0)$. Let us consider the three cases:

(1) $x > 0$. Then $P(x)$ is true and $Q(x)$ is true.

(2) $x = 0$. Then $P(x)$ is false and $Q(x)$ is false.

(3) $x < 0$. Then $P(x)$ is false and $Q(x)$ is true.

In all three cases, the statement $P(x) \Rightarrow Q(x)$ is true. Since, for every real number x we have the trichotomy $x > 0$, $x = 0$, or $x < 0$, we conclude that the statement that $P(x) \Rightarrow Q(x)$ for all $x \in \mathbb{R}$ is true.

The example above was given to illustrate a point about implications. You may have noticed that we earlier proved the following stronger statement:

Let $x \in \mathbb{R}$. If $x \neq 0$, then $x^2 > 0$.

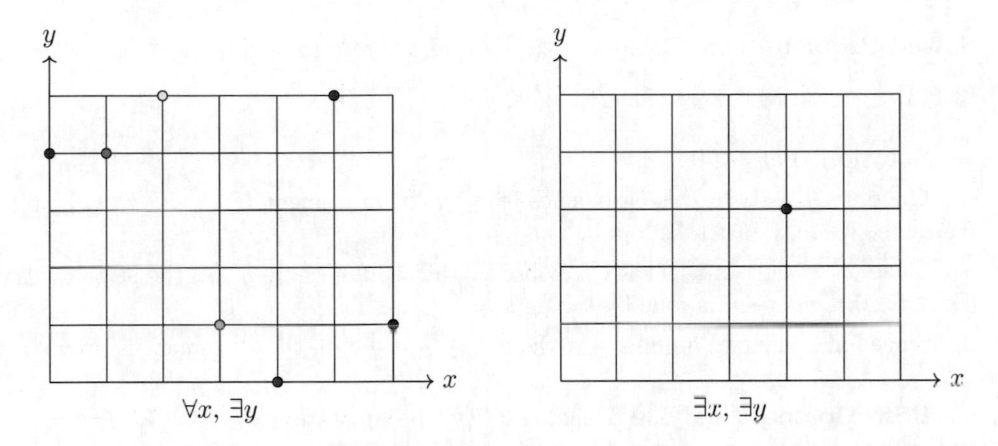

Figure 3.10.2. Visualizing universal and existential quantifiers of two variables. The right and left grids visualize parts (1) and (2) of Example 3.64, respectively.

3.10.7. Mixing universal and existential quantifiers of two variables. Let us start out very simply.

Example 3.64 (Fun with Dick (existential) and Jane (universal)).

(1) $\exists x \in \mathbb{R}, \exists y \in \mathbb{R}, x > y$. Here, we just have to find two real numbers where one is greater than the other. So, randomly, we let $x = 1$ and $y = 0$. Since $1 > 0$ and hence $x > y$, we have proved the statement.

(2) $\forall x \in \mathbb{R}, \exists y \in \mathbb{R}, x > y$. Here we have to show that no matter what x is, there is some y greater than x. So we: Let $x \in \mathbb{R}$. (Given this x,) we then let $y = x - 1$, say. Since $x > x - q$ and hence $x > y$, we have proved the statement.

(3) $\exists x \in \mathbb{R}, \forall y \in \mathbb{R}, x > y$. This statement is false. For: Let $x \in \mathbb{R}$. Then let $y = x$. Then $x \not> y$. This proves that for all $x \in \mathbb{R}$, there exists $y \in \mathbb{R}$ such that (not $x > y$).

(4) $\exists x \in \mathbb{R}, \forall y \in \mathbb{R}, x > y$. We leave it as an exercise to prove that this statement is false!

Exercise 3.47. *Prove each of the following statements.*

(1) $\exists x \in \mathbb{R}, \exists y \in \mathbb{R}, x^2 = -y^2$.

(2) $\forall x \in \mathbb{R}, \exists y \in \mathbb{R}, x^2 \leq y - 100$.

(3) $\exists x \in \mathbb{R}, \forall y \in \mathbb{R}, x \leq y^2 - 100$.

(4) $\forall x \in \mathbb{R}, \forall y \in \mathbb{R}, x^2 \geq -y^2$.

Exercise 3.48. *Disprove each of the following statements.*

(1) $\exists x \in \mathbb{R}, \exists y \in \mathbb{R}, x^2 = -y^2 - 1$.

(2) $\forall x \in \mathbb{R}, \exists y \in \mathbb{R}, x \geq y^2 - 100$.

(3) $\exists x \in \mathbb{R}, \forall y \in \mathbb{R}, x^2 \leq y + 100$.

(4) $\forall x \in \mathbb{R}, \forall y \in \mathbb{R}, x^2 > -y^2$.

Solved Problem 3.65 (Bananas *split*). *Prove the following statement:*

$$\forall y \in \mathbb{R}, \quad (\exists x \in \mathbb{R},\ x^2 \leq y) \quad \textit{or} \quad (\exists x \in \mathbb{R},\ 0 < x < -y). \tag{3.103}$$

Solution. Let $y \in \mathbb{R}$.

Case 1: $y \geq 0$. In this case, since $0^2 \leq y$, the statement $(\exists x \in \mathbb{R},\ x^2 \leq y)$ is true since we may just take $x = 0$.

Case 2: $y < 0$. In this case, we observe that $0 < -\frac{y}{2} < -y$. So the statement $(\exists x \in \mathbb{R},\ 0 < x < -y)$ is true by taking $x = -\frac{y}{2}$.

Since for every real number y we have $y \geq 0$ or $y < 0$, the statement is proved.

□

Post-Mortem Analysis. Denote by $P(y)$ the substatement $(\exists x \in \mathbb{R},\ x^2 \leq y)$, and denote by $Q(y)$ the substatement $(\exists x \in \mathbb{R},\ 0 < x < -y)$. Then we see that statement (3.103) takes the general form

$$\forall y \in \mathbb{R},\ P(y) \text{ or } Q(y). \tag{3.104}$$

In our solution above, we showed that:

(1) If $y \geq 0$, then $P(y)$ is true.

(2) If $y < 0$, then $Q(y)$ is true.

In fact, we have the stronger facts

$$P(y) \Leftrightarrow y \geq 0 \quad \text{and} \quad Q(y) \Leftrightarrow y < 0. \tag{3.105}$$

Exercise 3.49. (1) *Prove* (3.105).

(2) *Explain why to prove a statement of the general form* (3.104), *it suffices to show that union of the set of y for which $P(y)$ is true and the set of y for which $Q(y)$ is true is the set of all real numbers.*

Exercise 3.50. *Prove the following statement:*

$$\forall x \in \mathbb{R}, \quad (\forall y \in \mathbb{R},\ -y^2 \leq x - 2) \quad \textit{or} \quad (\exists y \in \mathbb{R},\ x^3 < y < 8). \tag{3.106}$$

3.11. Hints and partial solutions for the exercises

Hint for Exercise 3.1. All three statements are false. Use the identities in the hint to help find in each case real numbers x and y where the stated identity is false.

Hint for Exercise 3.2. We remark that a calculator often does not give exact answers (never if the number is irrational). So we do not recommend using a calculator.

(1) is false, whereas (2) and (3) are true.

Hint for Exercise 3.3. Consider the special case where $x = y = 1$.

Hint for Exercise 3.4. Take for example the mistaken formula (3.12). Let $f(x) = \sqrt{x}$. Then (3.12) says that $f(x+y) = f(x) + f(y)$, which would be true for a linear function but isn't true for this function.

Hint for Exercise 3.5. The area of our right triangle is equal to $\frac{1}{2}ab$. The same is true for every rotation of our triangle.

Hint for Exercise 3.6. Regarding the symmetry in left and right, the point is that there are two locations for the glasses, one of which we can call "L" and one of which we can call "R". The structure of the argument doesn't change if we switch L with R.

Hint for Exercise 3.7. Details for Subsubcase 1b(ii): By (3.24), either B_9 or B_{10} is the funny bunny. Suppose for a contradiction that B_{10} is the funny bunny. Then B_9 and B_{11} are regular bunnies, and thus (3.24) contradicts (3.22). We conclude that B_9 is the funny bunny, and it weighs more than the regular bunnies.

Hint for Exercise 3.8. In Case 1 we are done since Q implies Q or R. In Case 2 we are done since R implies Q or R.

Hint for Exercise 3.9. Separate the proof into two cases: (1) $a = b$, (2) $a < b$.

Hint for Exercise 3.10. Use the hint and multiply the inequality by $\frac{1}{c}$.

Hint for Exercise 3.11. Let a and b be non-negative real numbers such that $a + b = 0$. Then we have

$$0 \leq a = -b \leq 0. \tag{3.107}$$

Continue.

Hint for Exercise 3.12. If $x < 0$, then $x \cdot x > 0 \cdot x$.

Hint for Exercise 3.13. One way to solve the exercise is by truth tables.

Will illustrate another way for part (2). Let P and Q be statements.

Case 1. Suppose P is true. Then (P or Q) is true, which implies that (P or Q) or ((not P) or Q) is true.

Case 2. Suppose P is false. Then ((not P) or Q) is true, which implies that (P or Q) or ((not P) or Q) is true.

Since Case 1 or Case 2 is true and in each case we have proved the desired conclusion, we are done.

Hint for Exercise 3.14. We prove part (2).

Case 1. Suppose P is true. Then $\big($(not P) and $R\big)$ is false. Hence the statement (P and Q) and $\big($(not P) and $R\big)$ is false.

Case 2. Suppose P is false. Then (P and Q) is false. Hence the statement (P and Q) and $\big($(not P) and $R\big)$ is false.

Since Case 1 or Case 2 is true and in each case we have shown that the statement (P and Q) and ((not P) and R) is false, we are done.

Hint for Exercise 3.15. No, it is not a contradiction by the hint.

Hint for Exercise 3.16. Proof of (3). Suppose (P and Q). Then P is true. This implies that (P or Q) is true. We conclude (P and Q) $\Rightarrow$ (P or Q).

Hint for Exercise 3.17. The Venn diagram proof of part (1) is given by Figure 3.6.1.

Hint for Exercise 3.18. Suppose $x^2 > 0$. Suppose for a contradiction that $x = 0$. Continue.

Hint for Exercise 3.19. We prove this by cases. Let a and b be real numbers satisfying ($a > 0$ and $b > 0$) or ($a < 0$ and $b < 0$). Let *Case 1* be ($a > 0$ and $b > 0$). Let *Case 2* be ($a < 0$ and $b < 0$). In each of these cases, use elementary properties of inequalities and multiplication to obtain the desired conclusion.

Hint for Exercise 3.20. Suppose $ab > 0$. Firstly, show that $a \neq 0$. Secondly, consider two cases: (1) $a > 0$, (2) $a < 0$. In case (1) prove that $b > 0$, and in case (2) prove that $b < 0$.

Hint for Exercise 3.21. Suppose $P \Rightarrow Q$ and $Q \Rightarrow R$ are both true. Let P. Then since $P \Rightarrow Q$, we have Q. Continue.

Hint for Exercise 3.22. (1) *Case 1*: P is true. Then P or (not P) is true (since P or R is true for every statement R).

Case 2: P is false. Then (not P) is true, so P or (not P) is true (since R or (not P) is true for every statement R).

Here is the truth table method of the proof:

(3.108)

P	not P	P or (not P)
T	F	T
T	T	T

(2) This result is saying that an implication is equivalent to its contrapositive.

($\Rightarrow$) Suppose $P \Rightarrow Q$. Then suppose not Q. Now suppose P. Since $P \Rightarrow Q$, we have Q, which is a contradiction (to not Q). Therefore, P is false; i.e., not P is true. We have proved that not $Q \Rightarrow$ not P.

($\Leftarrow$) Suppose: not $Q \Rightarrow$ not P. Then suppose P. Now suppose not Q. Since not $Q \Rightarrow$ not P, we have that not P is true, i.e., P is false, which is a contradiction. Therefore not Q is false; i.e., Q is true. We have proved $P \Rightarrow Q$.

Hint for Exercise 3.23. The real number x has the property that

$$\text{not } (x > a \text{ and } x < b)$$

is true.

Hint for Exercise 3.24. Suppose for a contradiction that $x = 0$.

Hint for Exercise 3.25. *Corollary* 3.38(1): The contrapositive is: If a is odd, then a^2 is odd. This is the second statement of Corollary 1.30.

Corollary 3.38(2): The contrapositive is: If b is even, then b^2 is even. This is the first statement of Corollary 1.30.

Hint for Exercise 3.26. Prove the contrapositive using Theorem 1.25.

Hint for Exercise 3.27. Use Theorem 1.28.

Hint for Exercise 3.28. 2 is a prime. Now suppose that there exists an even prime p besides 2. Derive a contradiction.

Hint for Exercise 3.29. Suppose that a positive integer a is a divisor of a positive integer b. Then there exists an integer k such that

$$b = ka. \tag{3.109}$$

Explain why $k \geq 1$ and why this yields a contradiction if we assume that $a > b > 0$.

Hint for Exercise 3.30. The contrapositive of this implication is Theorem 1.52.

Hint for Exercise 3.31. Suppose that d divides both a and b but d does not divide c. Suppose that m and n are a solution to the equation

$$am + bn = c. \tag{3.110}$$

Derive a contradiction.

Hint for Exercise 3.32. The gcd must be equal to p.

Hint for Exercise 3.33. If n is a prime, then $\gcd(n, (n-1)!) = 1$. This is the contrapositive of the statement we want to prove.

Hint for Exercise 3.34. For a contradiction, suppose that $\sqrt[3]{2}$ is a rational number. Then there exist integers a and b such that $\sqrt[3]{2} = \frac{a}{b}$. By Theorem 1.50, we may assume that $\gcd(a, b) = 1$. Now $2 = \frac{a^3}{b^3}$, so that $2b^3 = a^3$. Continue.

Hint for Exercise 3.35. For a contradiction, suppose that $\sqrt{p}$ is a rational number. Then there exist integers a and b such that $\sqrt{p} = \frac{a}{b}$. Use Theorem 1.50 and Lemma 3.48.

Hint for Exercise 3.36. Let $x = \sqrt{2}^{\sqrt{2}}$. If x is rational, then we are done. So we may assume that x is irrational. Consider $x^{\sqrt{2}}$. Does this provide an example of l'Étranger's assertion?

Hint for Exercise 3.37. Let (a, b, c) be a Pythagorean triple. By Lemma 3.56, at least one of a and b is even. Consider the two cases: (1) both a and b are even and (2) exactly one of a and b is even.

Hint for Exercise 3.38. Proof of part (1): The remainder of a perfect square after dividing by 3 is either 0 or 1. If c^2 is divisible by 3, this implies that both a^2 and b^2 are divisible by 3. This in turn implies that each of a, b, c is divisible by 3, which contradicts the primitiveness assumption. Thus c^2 has remainder 1 after dividing by 3. This implies that one of a^2 and b^2 has remainder 0 after dividing by 3 and the other has remainder 1 after dividing by 3. Of the two integers a and b, the one whose square is divisible by 3 is the one which is divisible by 3, and the other isn't.

Hint for Exercise 3.39. Firstly, assume that (a, b, c) is primitive, and use Lemma 3.59. Secondly, reduce the general case to the primitive case. E.g., if (a, b, c) is not primitive, then there exists an integer $k > 1$ such that

$$\left(\frac{a}{k}, \frac{b}{k}, \frac{c}{k}\right) \tag{3.111}$$

is a primitive Pythagorean triple.

Hint for Exercise 3.40. $(6, 1)$ gives $(35, 12, 37)$, and $(5, 4)$ gives $(9, 40, 41)$.

Hint for Exercise 3.41. By Lemma 3.58, any odd integer at least 3 is the leg length of some Pythagorean triple. In particular, any odd prime number is the leg length of some Pythagorean triple. Continue.

Hint for Exercise 3.42. Let m and n be positive integers with $m > n$, and let

$$a = m^2 - n^2, \quad b = 2mn, \quad c = m^2 + n^2. \tag{3.112}$$

Show that $a^2 + b^2 = c^2$.

Hint for Exercise 3.43. ($\Rightarrow$): Use the transitivity of inequality. ($\Leftarrow$): Consider the average of a and b.

Hint for Exercise 3.44. There are an infinite number of primes.

Hint for Exercise 3.45. Divide r by an integer greater than 100.

Hint for Exercise 3.46. Since r and s are rational, there exist integers a, b, c, d such that $r = \frac{a}{b}$ and $s = \frac{c}{d}$. Calculate.

Hint for Exercise 3.47. (1) Let $x = 0$. What should we let y equal?

(2) Given x, define y appropriately in terms of x.

Hint for Exercise 3.48. (3) Prove that $\forall x \in \mathbb{R}, \exists y \in \mathbb{R}, x^2 > y + 100$.

(4) Let $x = 0$. What should we let y equal?

Hint for Exercise 3.49. For the first biconditional: Use that if $x \in \mathbb{R}$, then $x^2 \geq 0$.

For one of the directions of the second biconditional: Let $x = -\frac{y}{2}$.

Chapter 4

The Euclidean Algorithm and Its Consequences

$$\frac{277}{5} = 55 + \frac{2}{5} \Leftrightarrow 277 = 5 \cdot 55 + 2$$

Figure 4.0.1. A numismatic way to think of this is that \$2.77 equals 55 nickels plus two pennies for your thoughts (inflation)!

Goals of this chapter: Now we start to get serious about elementary number theory. To show that being serious does not mean being difficult ☺. To exhibit the power of the Division Theorem. To understand the greatest common divisor, a powerful concept. To solve linear Diophantine equations. To prove the Fundamental Theorem of Arithmetic.

4.1. The Division Theorem

We have seen the statement of the Division Theorem in §1.4.3. Now we return to this fundamental result and discuss its proof as well as a number of important applications to arithmetic.

4.1.1. Division in everyday life. When we divide the integer 277 by 5 to obtain a remainder of 2, we really mean that there exists an integer q (called the quotient) such that

$$277 = 5q + 2; \tag{4.1}$$

namely $q = 55$. The significance of 55 is that this is the unique integer that produces a remainder r with $0 \leq r < 5$; namely $r = 2$.

If for example we instead divide -277 by 5, then to get a remainder r with $0 \leq r < 5$, we would write

$$-277 = 5(-56) + 3;$$

namely in this case the quotient is $q = -56$ and the remainder is $r = 3$.

Now, given any integer a, such as $a = 277$ or $a = -277$, and given any integer q, for dividing a by 5 with quotient q, we may define the remainder r by

$$a = 5q + r. \tag{4.2}$$

For the Division Theorem, we are looking for the unique integer q that produces the smallest positive remainder r.

Here's another, more practical, example: The Superbowl, which is on a Sunday, is 50 days from now. What day of the week is it today? Without batting an eyelash, we announce that today is Saturday! How did we know this? Modular arithmetic! Namely, as an example of modular arithmetic, since there are seven days in a week, we did the calculation $50 = 7 \cdot 7 + 1$ and then used this to subtract 1 day from Sunday.

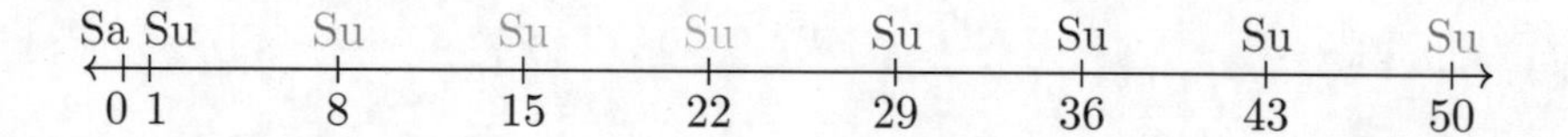

Figure 4.1.1. Counting backwards from the day of the Superbowl to today!

Now suppose that we are closer to the big day. The Superbowl, which is at 3 pm Pacific Time, is 100 hours from now. What time of day is it? Again, we use modular arithmetic and calculate that $100 = 4 \cdot 24 + 4$. Since there are 24 hours in a day, we conclude that it is 11 am Pacific Time. Time to get chips and salsa!

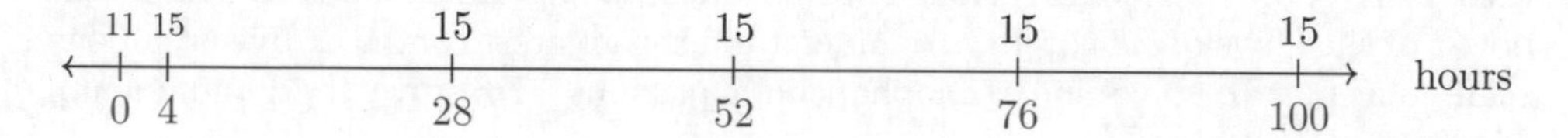

Figure 4.1.2. Using division to find what time it is (does anybody really know?). 3 pm is 15:00 hours in the 24-hour time notation. $96 = 4 \cdot 24$ hours before then is still 15:00 hours. Since $100 = 96 + 4$, the current time is 15:00 − 4:00 = 11:00 hours, i.e., 11 am.

4.1.2. Statement of the Division Theorem. In general, we may uniquely divide any integer a by any positive integer m to obtain a remainder r with $0 \leq r < m$. The **Division Theorem** says:

Theorem 4.1. *Let a be an integer and let m be a positive integer. Then there are unique integers q and r such that*

$$a = m\,q + r \quad \textit{and} \quad 0 \leq r < m. \tag{4.3}$$

So, for example, when we write

$$277 = 5q + r, \quad \text{with } 0 \leq r < 5,$$

the *only* possibility is that $q = 55$ and $r = 2$.

Similarly, there is only one solution (q, r) to

$$1234567890123456789 = 5q + r, \quad \text{with } 0 \leq r < 5.$$

It is rather easy to see that $r = 4$ (thanks to our base 10 number system), even though we cannot see quite as quickly what q is (but we know that q exists and it is unique).

4.1.3. Even and odd integers revisited. Before we proceed to prove the Division Theorem, we consider the simplest non-trivial special case. In particular, if we take $m = 2$ in the Division Theorem, then we obtain the statement:

Theorem 4.2. *Let a be an integer. Then there are unique integers q and r such that*

$$a = 2q + r \quad \text{and} \quad 0 \leq r < 2; \text{ that is, } r = 0 \text{ or } r = 1. \tag{4.4}$$

Observe that if a is even, then r must equal 0, while if a is odd, then r must equal 1. We leave the proofs of these two facts as easy exercises. Consequently,

$$a \text{ is even} \quad \Leftrightarrow \quad a = 2k \text{ for some } k \in \mathbb{Z}, \tag{4.5a}$$

$$a \text{ is odd} \quad \Leftrightarrow \quad a = 2k + 1 \text{ for some } k \in \mathbb{Z}. \tag{4.5b}$$

Solved Problem 4.3. *Let n and $n + 1$ be consecutive integers. Prove that one of them is divisible by 2 (even) and one of them has remainder 1 when divided by 2 (odd).*

Solution. Case 1: n is even. Then n is divisible by 2, and so $n = 2k$ for some integer k. Hence, $n+1 = 2k+1$, so that $n+1$ has remainder 1 when divided by 2.

Case 2: n is odd. Then, by Theorem 4.2, n has remainder 1 when divided by 2, and so $n = 2k + 1$ for some integer k. Hence, $n + 1 = 2k + 2 = 2(k + 1)$, so that $n + 1$ is divisible by 2. □

Exercise 4.1. *Let $n, n+1, n+2$ be consecutive integers. Prove that exactly one of these 3 integers is divisible by 3.*

4.1.4. The well-ordering principle. We will need the following axiom for the proof of the Division Theorem.

Axiom 4.4. *The **well-ordering principle** is the statement:*

Every non-empty set of positive integers has a least element.

In other words:

If X is a non-empty subset of $\mathbb{Z}^+$, then X has a minimum element.

We give a proof of the following in §4.11 (the appendix to this chapter) below.

Theorem 4.5. *The Axiom 2.5 of mathematical induction and the well-ordering principle are equivalent. That is, each implies the other.*

We can slightly generalize the well-ordering principle by the following.

Exercise 4.2. *Let b be an integer. Let Y be a non-empty subset of $\mathbb{Z}$ such that $y \geq b$ for all $y \in Y$. Prove that Y has a minimum element. Hint: Define*

$$X := \{y - b + 1 : y \in Y\}. \tag{4.6}$$

4.1.5. Proof of the Division Theorem. In this subsection we prove the Division Theorem 4.1. If you prefer to see the proof in a more concrete setting, first see the example after the proof.

Step 1. *Proof in the case where* $a \geq 0$. Note that if $a = mq + r$, then $r = a - mq$.

Existence (when $a \geq 0$**).** Consider the **set** S **of all non-negative remainders**; that is, define

$$S = \{a - m\bar{q} \mid \bar{q} \in \mathbb{Z}\} \cap \mathbb{Z}^{\geq}. \tag{4.7}$$

The remainder we seek is the smallest one:

(1) Since $a - m \cdot 0 = a \geq 0$, we have $a \in S$, so that S is non-empty.

(2) Since $S \subset \mathbb{Z}^{\geq}$ and $S \neq \emptyset$ (the empty set), by the well-ordering principle (Exercise 4.2 with $b = 0$), there exists a **smallest element of** S, which we call r.

Since $r \geq 0$ and since by the definition of $r \in S$ there exists $q \in \mathbb{Z}$ such that

$$r = a - mq, \quad \text{i.e.,} \quad a = mq + r,$$

we just need to show that $r < m$ in order to prove the *existence part* of the Division Theorem.

Observe that $r - m$ is a remainder since

$$r - m = (a - mq) - m = a - m(q + 1)$$

and $q + 1 \in \mathbb{Z}$. On the other hand, since $r - m < r$ and since r is the smallest element of S, we must have $r - m \notin S$. By the definition of S, this implies $r - m < 0$; i.e., $r < m$.[1] This completes the proof of the existence of q and r satisfying (4.3).

Uniqueness. (Here we do not use the condition that $a \geq 0$**.)** Suppose that q, r and $\tilde{q}, \tilde{r}$ are both pairs of integers such that we have both

$$a = mq + r \quad \text{and} \quad 0 \leq r < m$$

and

$$a = m\tilde{q} + \tilde{r} \quad \text{and} \quad 0 \leq \tilde{r} < m.$$

Then $mq + r = m\tilde{q} + \tilde{r}$ and hence

$$\tilde{r} - r = m\,(q - \tilde{q})\,.$$

Observe that since $\tilde{r} < m$ and $r \geq 0$, we have that $\tilde{r} - r < m - 0 = m$. Similarly, since $\tilde{r} \geq 0$ and $r < m$, we have that $\tilde{r} - r > 0 - m = -m$. Thus

$$-m < \tilde{r} - r < m,$$

and equivalently,

$$-m < m\,(q - \tilde{q}) < m; \quad \text{i.e.,} \quad -1 < q - \tilde{q} < 1.$$

Since $q - \tilde{q}$ is an integer and since it is strictly between -1 and 1, we conclude that

$$q - \tilde{q} = 0,$$

[1] To wit, $r - m$ is a remainder smaller than r (the smallest non-negative remainder), so it must be negative.

which in turn by $\tilde{r} - r = m(q - \tilde{q})$ implies that

$$\tilde{r} - r = 0.$$

Hence $q = \tilde{q}$ and $r = \tilde{r}$. This completes the proof of uniqueness. □

Step 2. (Existence when $a < 0$.) It only remains to prove existence in the case where $a < 0$. Again we let

$$S = \{a - m\bar{q} \mid \bar{q} \in \mathbb{Z}\} \cap \mathbb{Z}^{\geq}$$

be the set of non-negative remainders.

Observe that $a(1 - m) \in S$, so that S is non-empty. Indeed, we have that $a(1 - m) = a - ma = a - m\bar{q}$, where $\bar{q} = a$, and that $a(1 - m) \geq 0$ since $a < 0$ and $1 - m \leq 0$ (from $m > 0$). This says that $a(1 - m)$ is a non-negative remainder and hence is in S.

This is the only place in the proof where we previously used the condition that $a \geq 0$. Thus the rest of the proof of *existence* is exactly the same and we are done with the proof of the Division Theorem. □

Example 4.6. Consider the division in (4.1), where the solution is given by

$$277 = 5 \cdot 55 + 2. \tag{4.8}$$

We can obtain the quotient $q = 55$ and remainder $r = 2$ as follows. Let

$$S = \{277 - 5\bar{q} : \bar{q} \in \mathbb{Z}\} \cap \mathbb{Z}^{\geq}, \tag{4.9}$$

where $\mathbb{Z}^{\geq}$ denotes the set of non-negative integers. By inspection, we see that (see also Figure 4.1.3)

$$S = \{2, 7, 12, 17, 22, 27, \ldots\} = \{r + 5p : p \in \mathbb{Z}^{\geq}\}.$$

The remainder $r = 2$ is the minimum element of S. One then recovers the quotient q by $2 = 277 - 5q$; that is,

$$q = \frac{277 - 2}{5} = 55.$$

To reprove in this more concrete case that these are the unique remainder and quotient, we suppose that q and r satisfy

$$5q + r = 5 \cdot 55 + 2, \quad \text{where } 0 \leq r < 5.$$

By elementary algebra, we have

$$r - 2 = 5(55 - q).$$

Since $0 \leq r \leq 4$, this implies that

$$-2 \leq 5(55 - q) \leq 2;$$

that is,

$$-1 < -\frac{2}{5} \leq 55 - q \leq \frac{2}{5} < 1.$$

Since q is an integer, we conclude that $55 - q = 0$; that is, $q = 55$. From this we then easily recover that r must be equal to 2.

The reader may wish to compare this example with the general proof of the Division Theorem.

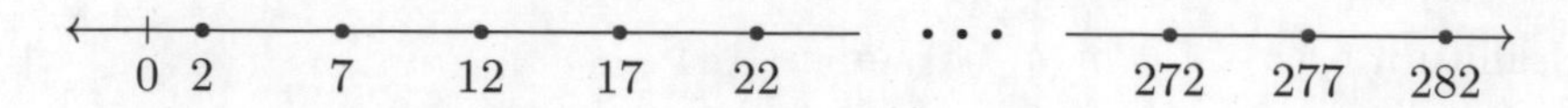

Figure 4.1.3. The Division Theorem illustrated: The set S of non-negative remainders in (4.9) consists of the red point and the (infinitely many) blue points, where the red point is the minimum element of S and is the smallest non-negative remainder.

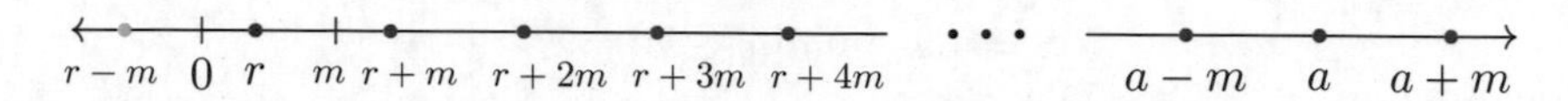

Figure 4.1.4. The general version of Figure 4.1.3. The smallest remainder r by definition satisfies $r \geq 0$. Since r is the smallest, $r - m < 0$, which implies $r < m$. By virtue of $r \in S$, we have $r = a - mq$ for some integer q. That is, $a = mq + r$, where $0 \leq r < m$.

Exercise 4.3. *Remarkably, the integer $n^2 + n + 41$ is prime for $0 \leq n \leq 39$. Prove that it is not a prime for both $n = 40$ and $n = 41$. (So the highest level we can reach is level 39. We proved that levels 40 and 41 are unattainable, not to mention Level 42!)*

Does there exist a positive integer k such that n^2+n+k is prime for all $n \in \mathbb{Z}^{\geq}$?

The following generalizes Exercise 4.1.

Exercise 4.4. *Let s be an integer and let m be a positive integer. Show that exactly one of the m integers $s+1, s+2, s+3, \ldots, s+m$ is divisible by m. Hint: Let r be the remainder of $s+m$. Note that $1 \leq m - r \leq m$. Consider $s + (m - r)$.*

Exercise 4.5. *Let a be an integer. Prove that there exist unique integers q' and r' satisfying*

$$a = 5q' + r' \quad \text{and} \quad 15 \leq r' \leq 19. \tag{4.10}$$

Hint: Use the Division Theorem.

4.1.6. The remainder and quotient functions. The Division Theorem in fact defines two functions. Let m be a positive integer. Define

$$R_m = \{0, 1, 2, \ldots, m-1\}. \tag{4.11}$$

Another way to describe this set is

$$R_m = \{a \in \mathbb{Z} : 0 \leq a < m\}. \tag{4.12}$$

This is the same as the **set of remainders**. Define the **remainder function**

$$\mathbf{r} : \mathbb{Z} \to R_m \tag{4.13}$$

by

$$\mathbf{r}(a) = r, \tag{4.14}$$

where r is the unique remainder given by the Division Theorem. Define the **quotient function**

$$\mathbf{q} : \mathbb{Z} \to \mathbb{Z} \tag{4.15}$$

by

$$\mathbf{q}(a) = q, \tag{4.16}$$

where q is the unique quotient given by the Division Theorem. So we have

$$a = m\,\mathbf{q}(a) + \mathbf{r}(a), \qquad 0 \leq \mathbf{r}(a) < m. \tag{4.17}$$

Example 4.7. For $m = 5$ we have for example that

$$2 = \mathbf{r}(2) = \mathbf{r}(-3) = \mathbf{r}(7) = \mathbf{r}(-8) = \mathbf{r}(12) = \mathbf{r}(-13) = \mathbf{r}(17) = \mathbf{r}(-18) = \cdots,$$
$$\mathbf{q}(2) = 0,\ \mathbf{q}(-3) = -1,\ \mathbf{q}(7) = 1,\ \mathbf{q}(-8) = -2,\ \mathbf{q}(12) = 2,\ \mathbf{q}(-13) = -3, \cdots.$$

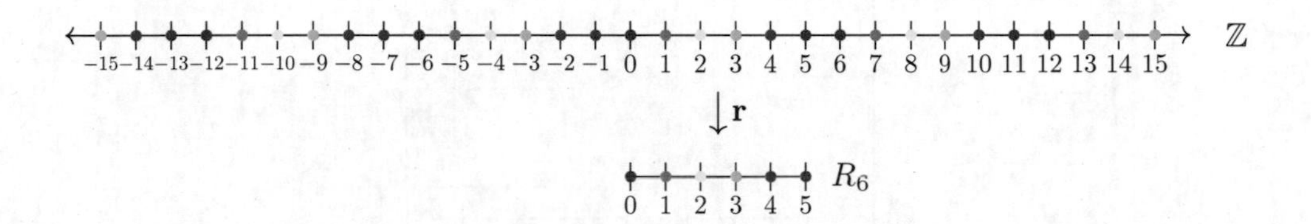

Figure 4.1.5. Visualization of the remainder map $\mathbf{r} : \mathbb{Z} \to R_6$ $(m = 6)$.

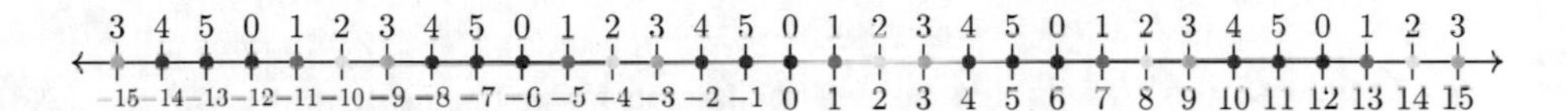

Figure 4.1.6. Alternatively, we can picture the remainder map $\mathbf{r} : \mathbb{Z} \to R_6$ by placing the inputs below the dots and the corresponding outputs above the dots. Note that there is a one-to-one correspondence between the outputs and the colors.

Solved Problem 4.8. *Show that*

$$\mathbf{r}(a) = 0 \quad \textit{if and only if} \quad m \textit{ divides } a. \tag{4.18}$$

Solution. Suppose that $\mathbf{r}(a) = 0$. Then $a = m\,\mathbf{q}(a) + \mathbf{r}(a) = m\,\mathbf{q}(a)$. Thus, m divides a. On the other hand, suppose that m divides a. Then there exists an integer q such that $a = mq$. Thus, the remainder for this q is $r = 0$. By the uniqueness of the remainder, $\mathbf{r}(a) = 0$. □

Exercise 4.6. *Let a and b be integers such that $\mathbf{r}(a) = \mathbf{r}(b)$ and $\mathbf{q}(a) = \mathbf{q}(b)$. Prove that $a = b$.*

Exercise 4.7. *Let n be an even integer. Prove that n or $n + 2$ is divisible by 4.*

4.1.7. Division and squares. What can we say about the remainders of perfect squares? After all, perfect squares aren't just any numbers. They're perfect (but not *perfect numbers*) and they're square numbers (but not *square*)! In this subsection we consider this question with respect to division by m, where m is a small positive integer. Besides squares, we also consider sums of two squares.

Firstly, we recall the following from Corollary 3.39: Let a be an integer. Then

$$2 \text{ divides } a \quad \text{if and only if} \quad 2 \text{ divides } a^2. \tag{4.19}$$

Theorem 4.9. *Let a be an integer. Then*

$$3 \text{ divides } a \quad \Leftrightarrow \quad 3 \text{ divides } a^2. \tag{4.20}$$

Proof. ($\Rightarrow$) This is a special case of Exercise 1.23 (note that a divides a^2).

($\Leftarrow$) We prove the contrapositive! Namely, we prove that if 3 does not divide a, then 3 does not divide a^2.

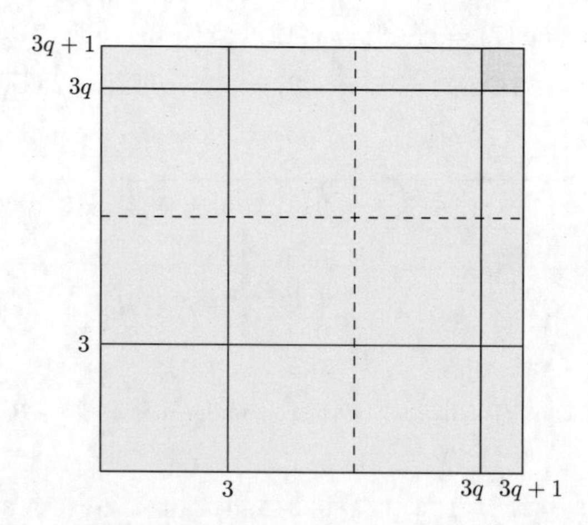

Figure 4.1.7. Visualizing that $(3q+1)^2$ has remainder 1 when divided by 3. Can you visualize the remainder of $(3q+2)^2$?

Suppose 3 does not divide a. Now, when we divide a by 3, by the Division Theorem we obtain a unique remainder r that is equal to 0, 1, or 2. Since 3 does not divide a, $r \neq 0$. Thus we have two cases:

Case 1: $r = 1$. In this case there exists an integer q such that $a = 3q + 1$. We then compute that

$$a^2 = (3q+1)^2 = 9q^2 + 6q + 1 = 3(3q^2 + 2q) + 1,$$

which by (4.18) shows that 3 does not divide a^2 (since our remainder is equal to 1).

Case 2: $r = 2$. In this case there exists an integer q such that $a = 3q + 2$. We then compute that

$$a^2 = (3q+2)^2 = 9q^2 + 12q + 4 = 3(3q^2 + 4q + 1) + 1,$$

which shows that 3 does not divide a^2 (since again our remainder is equal to 1).

This proves the desired contrapositive statement. □

Exercise 4.8. *Prove that if n is the square of an integer, then $n = 3q$ or $n = 3q+1$ for some $q \in \mathbb{Z}$. In particular, if m is an integer of the form $m = 3q + 2$ for some $q \in \mathbb{Z}$, then m is not the square of an integer. So, for example, we know immediately that* 999632 *is not a square, neither is* 1234567892.

When we consider arithmetic from the point of view of dividing by a given positive integer m, we say that we are doing arithmetic *modulo m*.

Solved Problem 4.10 (Sums of two squares modulo 4). (1) *Let m and n be integers such that 4 divides $m^2 + n^2$. Prove that both m and n are even.*

(2) *Let x and y be integers such that 4 divides $x^2 + y^2 + 2$. Prove that both x and y are odd.*

Solution. (1) If m is even, then $m = 2k$ for some integer k. In this case, $m^2 = 4k^2$ is divisible by 4. Since 4 divides $m^2 + n^2$ by hypothesis, we have that 4 divides $m^2 + n^2 - m^2 = n^2$. Thus, n is even.

Now, suppose that m is odd. Then $m = 2\ell + 1$ for some integer ℓ. This implies that $m^2 = 4(\ell^2 + \ell) + 1$. Thus 4 divides $m^2 - 1$. Since 4 divides $m^2 + n^2$ by hypothesis, we have that 4 divides $m^2 + n^2 - (m^2 - 1) = n^2 + 1$. Hence n^2 is odd, which implies that n is odd. This implies that 4 divides $n^2 - 1$. However, this contradicts that 4 divides $n^2 + 1$. We conclude that m is not odd. By the preceding paragraph, this implies that both m and n are even.

(2) If both x and y are even, then 4 divides $x^2 + y^2$, which contradicts our hypothesis. Thus, at least one of x and y is odd. Without loss of generality, we may assume that x is odd. By the proof of part (1) we have that 4 divides $x^2 - 1$. Since 4 divides $x^2 + y^2 + 2$ by hypothesis, we obtain that 4 divides $x^2 + y^2 + 2 - (x^2 - 1) = y^2 + 3$. Thus y^2 is odd, which implies that y is odd. □

Remark. For a sum of perfect squares $a^2 + b^2$, we have that after dividing by 4 the remainder is exactly one of the following:

(1) 0, in which case both a and b are even,

(2) 1, in which case for a and b one is even and one is odd,

(3) 2, in which case both a and b are odd.

It is not possible for the remainder to be equal to 3. We leave it as an exercise for the reader to verify this!

Exercise 4.9. *Prove that if m and n are integers such that 3 divides $m^2 + n^2$, then 3 divides both m and n.*

Exercise 4.10. *Prove that if n is the square of an integer, then $n = 4q$ or $n = 4q + 1$ for some $q \in \mathbb{Z}$.*

Exercise 4.11. *Prove that if n is the square of an integer, then $n = 5q$, $n = 5q + 1$, or $n = 5q + 4$ for some $q \in \mathbb{Z}$.*

The remainders of perfect squares are called *quadratic residues*. We will return to this topic in a later chapter.

Solved Problem 4.11 (Quadratic residues modulo 7). *Let n be a perfect square. Prove that there exists an integer q such that $n = 7q + r$, where r is equal to 0, 1, 2, or 4.*

Solution. Since n is a perfect square, there exists an integer k such that $n = k^2$. By the Division Theorem, there exist (unique) integers p and s such that

$$k = 7p + s, \quad \text{where } 0 \le s < 7. \tag{4.21}$$

We compute that

$$\begin{aligned} n = k^2 &= (7p+s)^2 \\ &= 7(7p^2 + 2ps) + s^2. \end{aligned}$$

So, if $s^2 = 7m + r$, where $0 \leq r < 7$, then we have that

$$n = 7q + r, \quad \text{where } q = 7p^2 + 2ps + m. \tag{4.22}$$

That is, the remainder of $n = k^2$ after dividing by 7 is equal to the remainder of s^2 after dividing by 7. Since $0 \leq s < 7$, we only need to consider the remainders of the squares of the first 7 non-negative integers. They are

$$\begin{aligned} \mathbf{r}(0^2) &= 0, \\ \mathbf{r}(1^2) = \mathbf{r}(6^2) &= 1, \\ \mathbf{r}(2^2) = \mathbf{r}(5^2) &= 4, \\ \mathbf{r}(3^2) = \mathbf{r}(4^2) &= 2. \end{aligned}$$

Here, we used that $\mathbf{r}(s^2) = \mathbf{r}((7-s)^2)$. □

Exercise 4.12. *Regarding perfect squares and remainders, we can summarize what we have seen so far in the following table:*

Table 4.1.1. Remainders of perfect squares after dividing by m for $2 \leq m \leq 7$.

m	*remainders of perfect squares after dividing by* m
2	$0, 1$
3	$0, 1$
4	$0, 1$
5	$0, 1, 4$
7	$0, 1, 2, 4$

Complete Table 4.1.1 *for* $m \leq 10$ *by considering the cases* $m = 6, 8, 9, 10$.

Exercise 4.13. *From the earlier exercises we conclude that* n *is not a perfect square* if *any of the following conditions hold:*

1. *dividing* n *by* 3 *yields a remainder of* 2,
2. *dividing* n *by* 4 *yields a remainder of* 2 *or* 3,
3. *dividing* n *by* 5 *yields a remainder of* 2 *or* 3.
4. *dividing* n *by* 7 *yields a remainder of* 3, 5, *or* 6.

Using the previous exercise, add to this list of conditions the cases of dividing n *by* m *for* $m = 6, 8, 9, 10$.

Exercise 4.14. *Apply the previous exercise to rule out as many integers between* 1 *and* 100 *that you can from being perfect squares.*

For example we can rule out 2 *since dividing* 2 *by* 4 *yields a remainder of* 2. *We can rule out* 3 *since dividing* 3 *by* 4 *yields a remainder of* 3. *We can rule out* 5 *since dividing* 5 *by* 3 *yields a remainder of* 2.

Exercise 4.15. *Prove by induction on* n *that for every positive integer* n *and integers* x *and* y, *we have that* $x^n - y^n$ *is divisible by* $x - y$.

4.2. There are an infinite number of primes

Recall the marvelous Theorem 1.56, which was proved using the Division Theorem.

Theorem 4.12 (Euclid's Theorem). *There are infinitely many primes.*

As with many fundamental results, there are multiple proofs of the infinitude of the number of primes. The next pair of exercises combine to present one such proof.

Exercise 4.16. *Suppose that there exists a strictly increasing sequence of integers at least 2, $\{a_n\}_{n=1}^{\infty}$, such that any two distinct terms of the sequence are coprime. That is, for all $n \neq k$, $\gcd(a_n, a_k) = 1$. Prove that this implies that there are infinitely many primes. You may use the Prime Factorization Theorem, which is the part of the Fundamental Theorem of Arithmetic that says that any integer at least 2 can be written as a product of primes (including the case of being a prime).*

Exercise 4.17. *Let*

$$F_n := 2^{2^n} + 1 \tag{4.23}$$

be the n-th Fermat number, where $n \in \mathbb{Z}^{\geq}$. For example, the Fermat numbers F_n for $0 \leq n \leq 6$ are

$$3, 5, 17, 257, 65537, 4294967297, 18446744073709551617. \tag{4.24}$$

Prove that the Fermat numbers are all coprime to each other; that is, $\gcd(F_n, F_k) = 1$ for all $n \neq k$. By the previous exercise, this implies that there are infinitely many primes.

Hint: First prove the formula

$$\prod_{i=0}^{n-1} F_i = F_n - 2 \quad \textit{for } n \geq 1. \tag{4.25}$$

Exercise 4.18. *Prove that for $n \geq 2$, the last digit of the n-th Fermat number F_n is 7.*

Remark 4.13. The Euclid–Mullin sequence of prime numbers a_n, $a \in \mathbb{Z}^+$, is defined as follows:

$$a_1 := 2,$$
$$a_n := \text{least prime factor of } a_1 a_2 \cdots a_{n-1} + 1 \text{ for } n \geq 2.$$

So, for example,

$$a_2 = 3,$$
$$a_3 = 7 \text{ since } 2 \cdot 3 + 1 = 7 \text{ is prime},$$
$$a_4 = 43 \text{ since } 2 \cdot 3 \cdot 7 + 1 = 43 \text{ is prime},$$
$$a_5 = 13 \text{ since 13 is the smallest prime factor of } 2 \cdot 3 \cdot 7 \cdot 43 + 1 = 1807.$$

As of September 2012, only the first 51 terms of the Euclid–Mullin sequence are known. For example,

$$a_{28} = 6436797949634662230815098 57.$$

We leave it to you, dear reader, to prove that the terms of the Euclid–Mullin sequence are distinct; that is, $a_i \neq a_j$ for $i \neq j$.

4.2.1. Euclid's bound for the n-th prime number. We can obtain a crude (a.k.a. coarse) bound for the size of the n-th prime using an idea related to the proof that there are infinitely many primes.

For $n \in \mathbb{Z}^+$, let p_n denote the n-th smallest prime number. That is, the set of primes less than p_n is a finite set with cardinality $n-1$. For example, $p_1 = 2$, $p_2 = 3$, $p_3 = 5$, etc.

Lemma 4.14. *We have*

$$p_{n+1} \leq 1 + \prod_{i=1}^{n} p_i = 1 + p_1 p_2 \cdots p_n. \tag{4.26}$$

Proof. Since none of $p_1, p_2, \ldots, p_n$ divide $1 + \prod_{i=1}^{n} p_i$, by the Prime Factorization Theorem there must exist a prime p_k, where $k \geq n+1$, that divides $1 + \prod_{i=1}^{n} p_i$. We conclude that

$$p_{n+1} \leq p_k \leq 1 + \prod_{i=1}^{n} p_i. \tag{4.27}$$

□

Exercise 4.19. *What is the smallest positive integer n for which $2^{(2^n)} > 1$ googol?*

Using the lemma, we can prove the following (ridiculously large by Exercise 4.19 ☹) upper bound for the n-th prime number.

Theorem 4.15 (Euclid). *For all $n \in \mathbb{Z}^+$, the n-th prime satisfies the bound*

$$p_n < 2^{2^n} := 2^{(2^n)}. \tag{4.28}$$

Proof. We prove this result by strong induction.

Base case. We have

$$p_1 = 2 < 2^{2^1} = 4. \tag{4.29}$$

So the base case is true.

Strong inductive step. Suppose that $k \in \mathbb{Z}^+$ is such that

$$p_i < 2^{2^i} \quad \text{for } 1 \leq i \leq k. \tag{4.30}$$

Then, using Lemma 4.14, we estimate as follows:

$$\begin{aligned} p_{k+1} &\leq 1 + \prod_{i=1}^{k} p_i \\ &< 1 + \prod_{i=1}^{k} 2^{2^i} \\ &= 1 + 2^{\sum_{i=1}^{k} 2^i} \\ &= 1 + 2^{(2^{k+1}-2)} \\ &< 2^{2^{k+1}}. \end{aligned}$$

By strong mathematical induction, we are done. □

We leave it to the reader to read about (the more fancy) Euler's proof of Euclid's Theorem that there are infinitely many prime numbers. Can you think about how to make Euler's proof rigorous?

4.3. The Euclidean algorithm

Division and the greatest common divisor go hand in hand. The Euclidean algorithm simplifies step by step the calculation of the greatest common divisor.

4.3.1. Basic gcd properties.

The key to the Euclidean algorithm is the following observation.

Theorem 4.16. *For non-zero integers a and b suppose that*

$$a = bq + r \quad \textit{where} \quad q, r \in \mathbb{Z}.$$

Then

$$\gcd(a, b) = \gcd(b, r). \tag{4.31}$$

This is particularly useful when $a > b > r \geq 0$. For in this case, among the three integers, effectively the largest integer is replaced by the smallest integer.

Proof. Suppose that d is a common divisor of a and b. Then, by Theorem 1.52 we have that d divides the following integral linear combination of a and b:

$$a \cdot 1 + b \cdot (-q) = a - bq = r,$$

which is the remainder. Hence d is a common divisor of b and r. This implies that $\gcd(a, b) \leq \gcd(b, r)$.

Conversely, if d is a common divisor of b and r, then d divides the following integral linear combination of b and r:

$$b \cdot q + r \cdot 1 = bq + r = a.$$

Hence d is a common divisor of b and a. This implies that $\gcd(b, r) \leq \gcd(a, b)$.

We conclude that $\gcd(a, b) = \gcd(b, r)$. □

Given non-zero integers $a > b$, if we use the Division Theorem to divide a by b to get a remainder r with $0 \leq r < b$, then Theorem 4.16 says that the greatest common divisor of a and b remains unchanged if we replace a by the remainder r. This makes the problem of finding the gcd easier since $r < b < a$.

Lemma 4.17. *Suppose that $a \in \mathbb{Z}$ and $b \in \mathbb{Z}^+$ are such that b divides a. Then* $\gcd(a, b) = b$.

Proof. Since b divides itself and since b divides a by hypothesis, b is a common divisor of a and b. So $b \leq \gcd(a, b)$. On the other hand, by Lemma 1.8 and since $b > 0$, we have $\gcd(a, b) \leq b$. We conclude that $\gcd(a, b) = b$. □

Example 4.18. (1) We have that 19 divides 114. Thus $\gcd(114, 19) = 19$.

(2) We have $114 = 36 \cdot 3 + 6$. Therefore $\gcd(114, 36) = \gcd(36, 6) = 6$.

4.3.2. The Euclidean algorithm.

Now we can proceed to demonstrate the Euclidean algorithm.

4.3.2.1. *An example of the Euclidean algorithm.* Consider the problem of finding the gcd of 936 and 324. By applying the Division Theorem repeatedly, we calculate that

$$936 = 324 \cdot 2 + 288, \tag{4.32a}$$

$$324 = 288 \cdot 1 + 36, \tag{4.32b}$$

$$288 = 36 \cdot 8 + 0. \tag{4.32c}$$

By Theorem 4.16, we conclude that

$$\gcd(936, 324) = \gcd(324, 288) = \gcd(288, 36) = 36, \tag{4.33}$$

where the last equality is because 36 divides 288.

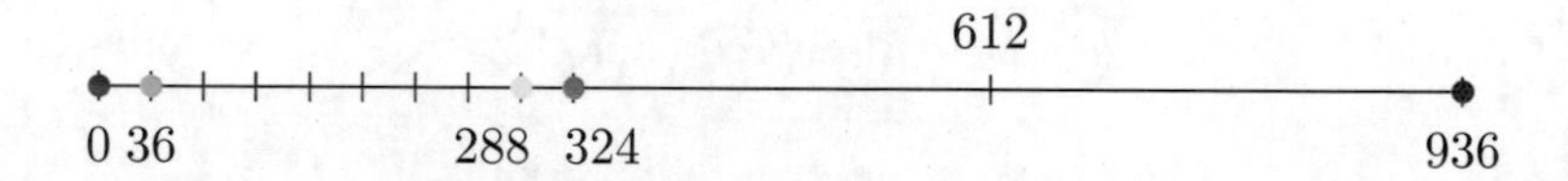

Figure 4.3.1. How is this diagram a visualization of the Euclidean algorithm? Note that counting backwards from 936 by 324's yields 612 and 288. Counting backwards from 324 by 288 yields 36.

Exercise 4.20. *Use the Euclidean algorithm to compute the greatest common divisor of* 812 *and* 188.

Exercise 4.21. *Use the Euclidean algorithm to compute the greatest common divisor of* 744 *and* 258.

4.3.2.2. *How the Euclidean algorithm is proved.* Now let us consider the Euclidean algorithm for a general pair of positive integers $a > b$. From the example above, we see that we successively find remainders of dividing a larger positive number by a smaller positive number. After each step, the larger of the two integers is replaced by the remainder, which is smaller than the smaller of the two integers. When does this all stop? We claim that eventually we get a remainder of zero!

Since the Euclidean algorithm produces a *sequence* of integers, we are going to put *subscripts* on the letters denoting the integers. Call, for the fun of it (okay, maybe this is not so fun, but you will see the pattern!), $a =: r_{-1}$, $b =: r_0$, and their remainder $r =: r_1$. So $a = bq + r$, $0 \leq r < b$, becomes

$$r_{-1} = r_0\, q_1 + r_1, \quad 0 \leq r_1 < r_0, \tag{4.34}$$

where we also called $q =: q_1$. If $r_1 = 0$, then we stop and celebrate that $\gcd(a, b) = b$. Otherwise, assuming $r_1 > 0$, the next application of the Division Theorem yields the existence of integers q_2 and r_2 satisfying

$$r_0 = r_1\, q_2 + r_2, \quad 0 \leq r_2 < r_1. \tag{4.35}$$

If $r_2 = 0$, we celebrate, congratulate ourselves, and observe that

$$\gcd(a, b) = \gcd(r_0, r_1) = r_1. \tag{4.36}$$

It is rather clear that if we continue in this way, we obtain a sequence of non-negative integers

$$a = r_{-1} > b = r_0 > r = r_1 > r_2 > r_3 > \cdots, \tag{4.37}$$

where we stop when $r_k = 0$ for some k. For such a k, we have $r_{k-2} = r_{k-1}q_k$ and hence

$$\gcd(a,b) = \gcd(r_0, r_1) = \gcd(r_1, r_2) = \cdots = \gcd(r_{k-2}, r_{k-1}) = r_{k-1} > 0. \quad (4.38)$$

This is the Euclidean algorithm for finding the gcd of a and b.

Now the 20 million dollar question (inflation!) is: How do we prove that there exists a positive integer k such that $r_k = 0$? That is, we actually do stop.

To see that we must stop, we think intuitively: A priori (with some observation we can do better in terms of the number of steps), the worst case scenario (the slowest sequence arriving at zero) is when the sequence of remainders is the arithmetic sequence

$$a = r_{-1} > a-1 = r_0 > a-2 = r_1 > a-3 = r_2 > \cdots > 1 = r_{a-2} > 0 = r_{a-1}, \quad (4.39)$$

so we must have the existence of such a k with $k \leq a-1$. Now, as long as r_i is not 0, the algorithm produces a strictly smaller, but still positive, r_{i+1}. So, a list of such r_i has to reach 0 within a steps by the argument above.

Summary. For the gcd of two distinct positive integers, we can divide the larger integer by the smaller integer to get a remainder. The gcd is equal to the gcd of the smaller integer and the remainder. In this way we can step by step reduce the size of the integers. Eventually we get a zero remainder. In this last case, the smaller number divides the larger number, and hence the gcd equals the smaller integer.

4.3.3. Reversing the Euclidean algorithm. An additional benefit of the Euclidean algorithm is that by reversing it we can write the gcd of two integers a and b as an integral linear combination of a and b.

4.3.3.1. *Example of the gcd as an integral linear combination.* Let us see how this works by considering the example in the previous subsection.

By (4.32b), we have

$$36 = 324 - 288. \quad (4.40)$$

By (4.32a), we have

$$288 = 936 - 324 \cdot 2. \quad (4.41)$$

Substituting this into the previous equation yields

$$\begin{aligned} 36 &= 324 - 288 \\ &= 324 - (936 - 324 \cdot 2) \\ &= (-1) \cdot 936 + 3 \cdot 324. \end{aligned} \quad (4.42)$$

That is, $\gcd(936, 324) = 36$, and

$$(-1)936 + (3)324 = 36. \quad (4.43)$$

4.3.3.2. *The gcd is always an integral linear combination.* In general, given two positive integers $a > b$, reversing the Euclidean algorithm finds us two integers m and n satisfying the **linear Diophantine equation**

$$ma + nb = g, \tag{4.44}$$

where g is the gcd of a and b. In our example, where $a = 936$ and $b = 324$, we found the solution given by the pair $m = -1$ and $n = 3$.

Exercise 4.22. *Write* $\gcd(812, 188)$ *as an integral linear combination of* 812 *and* 188.

Exercise 4.23. *Write* $\gcd(744, 258)$ *as an integral linear combination of* 744 *and* 258.

Recall from Definition 1.51 that integers a and b, not both zero, are **coprime** if $\gcd(a, b) = 1$. Breaking down what we have stated above, we have:

Theorem 4.19 (The gcd is an integral linear combination). *Let a and b be integers, not both zero. Then there exist integers m and n satisfying*

$$ma + nb = \gcd(a, b). \tag{4.45}$$

In particular, if a and b are coprime, then there exist integers m and n satisfying

$$ma + nb = 1. \tag{4.46}$$

Proof. We use the notation $r_0 = a$ and $r_1 = b$ as above. Let $r_{k-1} = g$ be the last non-zero remainder as above, where g is the greatest common divisor of a and b. By reversing the Euclidean algorithm step by step, we firstly have that r_{k-1} is an integral linear combination (**i.l.c.**) of r_{k-3} and r_{k-2}. Secondly, r_{k-2} is an i.l.c. of r_{k-4} and r_{k-3}. Thus, r_{k-1} is an i.l.c. of r_{k-4} and r_{k-3}. Continuing in this way, we finally see that $r_{k-1} = g$ is an i.l.c. of $r_0 = a$ and $r_1 = b$ (a rigorous proof of this uses induction). □

Exercise 4.24. *By reversing the Euclidean algorithm, find integers m and n satisfying*

$$83m + 17n = 1. \tag{4.47}$$

4.4. Consequences of the Division Theorem

As one might expect, the Division Theorem is fundamentally important in the study of (positive) integers from the point of view of multiplication. In particular, it is important in the study of prime numbers, the greatest common divisor, the property of two integers being coprime, divisors, etc. In this section we discuss some of the further consequences of the Division Theorem.

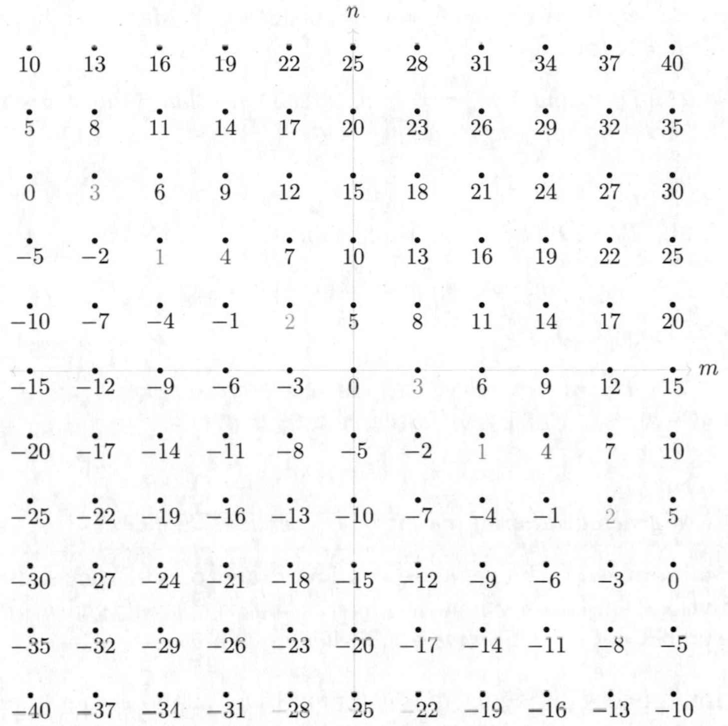

n

10	13	16	19	22	25	28	31	34	37	40
5	8	11	14	17	20	23	26	29	32	35
0	3	6	9	12	15	18	21	24	27	30
−5	−2	1	4	7	10	13	16	19	22	25
−10	−7	−4	−1	2	5	8	11	14	17	20
−15	−12	−9	−6	−3	0	3	6	9	12	15
−20	−17	−14	−11	−8	−5	−2	1	4	7	10
−25	−22	−19	−16	−13	−10	−7	−4	−1	2	5
−30	−27	−24	−21	−18	−15	−12	−9	−6	−3	0
−35	−32	−29	−26	−23	−20	−17	−14	−11	−8	−5
−40	−37	−34	−31	−28	−25	−22	−19	−16	−13	−10

m

Figure 4.3.2. At each point (m, n) the integral linear combination $3m+5n$ is displayed right below it. For example, -19 is displayed right below the point $(-3, -2)$. Since 3 and 5 are coprime, every integer may be represented as an integral linear combination.

4.4.1. Characterization of two integers being coprime.

One characterization of a and b being coprime is that there does not exist a prime that divides both a and b. Here is another important characterization, which is mainly because of Theorem 4.19 on being able to express the gcd as an integral linear combination.

Theorem 4.20 (Characterizing coprime integers). *Non-zero integers a and b are coprime $\Leftrightarrow$ the integer 1 can be written as an integral linear combination of a and b.*

Proof. ($\Rightarrow$) This follows immediately from Theorem 4.19.

($\Leftarrow$) Suppose there exist $m, n \in \mathbb{Z}$ such that

$$ma + nb = 1.$$

Let c be any positive common divisor of a and b. Since c divides a and b, we have c divides $ma + nb = 1$. This and $c > 0$ imply $c = 1$. We conclude that $\gcd(a, b) = 1$. □

Summary. In other words, if $c = g$ is the gcd of a and b, then there exist integers m, n such that $(m)a + (n)b = c$. Conversely if there exist integers m, n

such that $(m)a + (n)b = c$, where $c > 0$, then since g divides c, we have $g \leq c$. In particular, if $c = 1$, then $g = 1$.

Exercise 4.25. *Let p and q be prime numbers. Prove that p and q are not equal if and only if there exist integers m, n such that*

$$mp + nq = 1. \tag{4.48}$$

Exercise 4.26. *Prove that for every integer n,*

$$\gcd(5n + 2, 12n + 5) = 1. \tag{4.49}$$

Hint: Use Theorem 4.20.

Exercise 4.27. *Let us generalize the previous exercise. Let a, b, c, d be integers satisfying $ad - bc = 1$. Prove that for every integer n,*

$$\gcd(an + b, cn + d) = 1. \tag{4.50}$$

Is this really a generalization of the previous exercise? Explain.

Solved Problem 4.21 (Integers are coprime if and only if they have no prime common divisors). *Prove that non-zero integers a and b satisfy* $\gcd(a, b) = 1$ *if and only if there does not exist a prime p that divides both a and b.*

Solution. ($\Rightarrow$) Considering the contrapositive, suppose that there exists a prime p that divides both a and b. Then $\gcd(a, b) \geq p > 1$, so that $\gcd(a, b) \neq 1$.

($\Leftarrow$) Considering the contrapositive (again!), suppose that $g := \gcd(a, b) \neq 1$. Then $g > 1$. By Theorem 2.15, there exists a prime p dividing g. Since g divides both a and b, by the transitivity of division this implies that p divides both a and b. □

Solved Problem 4.22. *Suppose that non-zero integers a, b, c, d satisfy*

$$\gcd(a, b) = \gcd(a, d) = \gcd(c, b) = \gcd(c, d) = 1. \tag{4.51}$$

(We can think of this hypothesis as saying that the two integers in ac have nothing in common multiplicatively with the two integers in bd.) Prove that

$$\gcd(ac, bd) = 1. \tag{4.52}$$

Solution. Suppose for a contradiction that $\gcd(ac, bd) \neq 1$. Then, by the previous problem, there exists a prime p that divides both ac and bd. Thus, by Lemma 4.26,

$$(p \text{ divides } a \text{ or } p \text{ divides } c) \quad \text{and} \quad (p \text{ divides } b \text{ or } p \text{ divides } d)\,.$$

By considering the four resulting cases, we see that this contradicts (4.51). □

Here is an alternative proof. By hypothesis, a and b are coprime, and a and d are coprime. Hence, by Lemma 4.29, we have that a and bd are coprime. Similarly, we see that c and bd are coprime.

Now, since bd and a are coprime and since bd and c are coprime, by Lemma 4.29 we have that bd and ac are coprime. □

4.4.2. Division of products and coprimeness. Consider the situation of an integer dividing the product of two integers. The following gives us a criterion for when the integer will divide one of the two factors.

Theorem 4.23. *Suppose $a, b, c \in \mathbb{Z}^+$ and a and b are coprime. If a divides bc, then a divides c.*

Note that the converse of this implication is true by Exercise 1.23.

Proof. The idea is to use that a and b are coprime. By Theorem 4.20, there exist $m, n \in \mathbb{Z}$ such that

$$1 = am + bn.$$

The key is to link this to c. The idea is simply to multiply this equation by c to get

$$c = cam + cbn.$$

Now a clearly divides cam since $a \cdot cm = cam$. On the other hand, since a divides bc by hypothesis, we have that a divides cbn since $cbn = bc \cdot n$. Hence a divides the sum $cam + cbn$. Since $cam + cbn = c$, we conclude that a divides c. □

This is an elegant and important proof.

Interpretation. Intuitively the statement says that if a "goes completely into the product" bc while a has "nothing to do with" b, then a must "go completely into" c.

Taking advantage of the fact that primes have either all or nothing in common multiplicatively with other integers, we can deduce the following.

Corollary 4.24. *If a prime p divides the square a^2 of an integer, then p divides a.*

Proof. Since p is a prime, we have $\gcd(a, p)$ is equal to 1 or p.

Case 1: $\gcd(a, p) = p$. This implies that p divides a, and we are done.

Case 2: $\gcd(a, p) = 1$. We can apply Theorem 4.23 to the hypothesis that p divides $a \cdot a$ to conclude that p divides a, where this a is the second a in $a \cdot a$ (and we used $\gcd(a, p) = 1$ for the first a in $a \cdot a$).

In either case, we have p divides a. □

Remark 4.25. Here is another argument to prove Corollary 4.24. Can you fill in the details? We can write a as a product of primes: $a = p_1 p_2 \cdots p_r$, where the primes p_i might not be distinct. Then $a^2 = p_1^2 p_2^2 \cdots p_r^2$. Since p divides a^2, we must have that $p = p_i$ for some i. Therefore p divides a.

We have the following generalization of Corollary 4.24.

Lemma 4.26. *If a prime p divides a product ab, then p divides a or p divides b.*

Exercise 4.28. *Prove Lemma* 4.26.

Solved Problem 4.27. *For each part of this problem, if you use a result about division, state the result.*

(1) *Suppose* 12 *divides* $15k$, *where k is an integer. Prove that* 4 *divides* k.

(2) *Suppose that* $\gcd(9, ab) > 1$. *Prove that (3 divides a) or (3 divides b).*

Solution. (1) Division fact (Theorem 4.23): *If a divides bc and if* $\gcd(a, b) = 1$, *then a divides c.* Since 12 divides $15k$, we have that 4 divides $5k$. Since $\gcd(4, 5) = 1$ and by the aforestated division fact, we conclude that 4 divides k.

(2) Division fact (Lemma 4.26): *If a prime p divides a product of integers ab, then (p divides a) or (p divides b).* We apply this fact with $p = 3$. Since $g := \gcd(9, ab) > 1$ and since g divides 9, we have that g equals 3 or 9. In particular, 3 divides g. Since g divides ab, by the transitivity of division this implies that 3 divides ab. Finally, by the aforestated division fact, we conclude that (3 divides a) or (3 divides b). □

Solved Problem 4.28 (If a prime divides a positive power of an integer, then it divides the integer). *If a prime p divides a^n, where a is an integer and n is a positive integer, then p divides a.*

Solution. We prove this by induction on n. To be more formal this time, we explicitly mention that the statement $P(n)$ that we want to prove for all $n \geq 1$ is:

For any prime p and integer a, if p divides a^n, then p divides a.

Clearly the base case $P(1)$ is true.

As the inductive hypothesis, suppose that k is a positive integer such that $P(k)$ is true; that is, for every prime p and integer a, if p divides a^k, then p divides a.

Now let p be a prime and let a be a positive integer such that p divides a^{k+1}. Since we may think of this as saying that p divides the product $a^k a$, since p is a prime and by Lemma 4.26, we have that p divides a^k or p divides a. By the inductive hypothesis, we have that p divides a^k implies that p divides a. Thus, we conclude that p divides a. This proves $P(k + 1)$ and hence proves the inductive step.

By induction, we have proved $P(n)$ for all $n \geq 1$. □

The following says that if an integer has nothing in common multiplicatively with each of two integers, then that integer has nothing in common multiplicatively with the product of those two integers.

Lemma 4.29. *If a and b are coprime and if a and c are coprime, then a and bc are coprime.*

Proof. Suppose $\gcd(a, b) = 1$ and $\gcd(a, c) = 1$. Let d be a positive common divisor of a and bc. Then, by Exercise 1.31, we have

$$\gcd(d, b) \leq \gcd(a, b) = 1, \tag{4.53}$$

which implies that $\gcd(d, b) = 1$. Now, by Theorem 4.23, since d divides bc and since $\gcd(d, b) = 1$, we have that d divides c. This and using that $d > 0$ divides a implies that $d = 1$ since a and c are coprime. Finally, $d = 1$ implies that a and bc are coprime. □

Remark 4.30. An alternative, elegant proof of Lemma 4.29 uses the characterization of Theorem 4.20 for being coprime. Namely, by the hypotheses, there exist integers x, y, z, w such that

$$1 = ax + by, \qquad 1 = az + cw.$$

By multiplying(!) these two equations together, we obtain that

$$\begin{aligned} 1 &= (ax+by)(az+cw) \\ &= aaxz + acxw + abyz + bcyw \\ &= a(axz + cxw + byz) + bc(yw). \end{aligned}$$

By Theorem 4.20, this implies that a and bc are coprime.

Remark 4.31. Here is yet another argument to prove Lemma 4.29. Can you fill in the details? Assuming the hypothesis, suppose for a contradiction that a and bc are not coprime. Then there exists a prime p that divides both a and bc. Since p divides bc, we have that p divides b or that p divides c.

Exercise 4.29. *Let a, b, c be non-zero integers. Prove that if a and bc are not coprime, then a and b are not coprime or a and c are not coprime.*

4.4.3. Division of products and the gcd. On the other hand, suppose we have the same situation as above, except that a has "multiplicatively something to do with" b, such as $a = 6$ and $b = 15$. That is, suppose that a and b are not coprime. Then we can take for example $c = 14$ (an even number) and we get that

$$a = 6 \text{ divides } b \cdot c = 15 \cdot 14 = 210$$

but $a = 6$ does not divide $c = 14$. What happens is that part of 6 (namely 3) goes into $b = 15$ whereas the other part of 6 (namely 2) goes into $c = 14$. More generally, we have the following result.

Theorem 4.32. *With $\mathbb{Z}$ as the universal set, suppose that a non-zero integer c divides a product ab. Let $g = \gcd(c, a)$. Then*

$$\frac{c}{g} \text{ divides } b. \tag{4.54}$$

The converse of this implication is also true.

Observe that Theorem 4.23 is the special case where $g = 1$.

Proof. By the hypothesis, it follows that the integer $\frac{c}{g}$ divides the product of integers $\frac{a}{g} \cdot b$. Since $\gcd\left(\frac{c}{g}, \frac{a}{g}\right) = 1$ by Theorem 1.50, we conclude by Theorem 4.23 that $\frac{c}{g}$ divides b.

To see the converse, suppose that $\frac{c}{g}$ divides b. Then c divides gb. Since g divides a, this implies that c divides ab. □

Exercise 4.30. *Prove that for every $n \in \mathbb{Z}^+$,*

$$\gcd(a^n, b^n) = \gcd(a, b)^n. \tag{4.55}$$

Hint: For showing that the assumption that $\gcd(a,b)^n < \gcd(a^n, b^n)$ leads to a contradiction, use Solved Problem 4.28.

4.4.4. Coprime integers dividing an integer. Suppose that 5 and 6 both divide an integer x. Then we know that 30 divides x. The reason is the following result.

Proposition 4.33. Suppose that m_1 and m_2 are coprime positive integers. If m_1 and m_2 both divide an integer x, then their product $m_1 m_2$ divides x.

Proof. Since m_1 divides x, we have that $x = m_1 y$ for some integer y. Now, since m_2 divides the product $m_1 y$ and $\gcd(m_1, m_2)$, we have by Theorem 4.23 that m_2 divides y. Therefore there exists an integer z such that $y = m_2 z$. We conclude that

$$x = m_1 y = m_1 m_2 z,$$

so that $m_1 m_2$ divides x. □

On the other hand, if 6 and 10 both divide an integer x, then 30 divides x. The reason is now the following.

Exercise 4.31. *Generalize Proposition 4.33 by proving that if positive integers m_1 and m_2 both divide an integer x, then $m_1 m_2 / g$ divides x, where $g = \gcd(m_1, m_2)$.*

We remark that $m_1 m_2 / g$ is the least common multiple *of m_1 and m_2; see §4.9 below.*

By induction, we can extend the proposition to the following.

Corollary 4.34. *Suppose that $m_1, m_2, \ldots, m_k$ are pairwise coprime positive integers. If each of $m_1, m_2, \ldots, m_k$ divides an integer x, then their product $m_1 m_2 \cdots m_k$ divides x.*

Exercise 4.32. *Prove Corollary 4.34 by induction.*

Example 4.35. If $7, 11, 13$ each divide an integer x, then 1001 divides x.

4.5. Solving linear Diophantine equations

Figure 4.5.1. Diophantus (~207–~291).

In general, given integers a, b, and c, we want to find all integer solutions m and n to the equation

$$am + bn = c. \tag{4.56}$$

We call this a **linear Diophantine equation**. This is the simplest non-trivial type of equation over the integers.

For example, consider the equation

$$4199m + 1748n = 95, \tag{4.57}$$

where we would like to find all integer solutions m and n to this equation. We have

$$\gcd(4199, 1748) = 19, \tag{4.58}$$

and by *reversing the Euclidean algorithm*, one sees that a solution to

$$4199m + 1748n = 19 \tag{4.59}$$

is given by $m = 5$ and $n = -12$. Multiplying all of this by 5, we find that a particular solution to (4.57) is given by

$$m = 25 \quad \text{and} \quad n = -60. \tag{4.60}$$

4.5.1. Necessary and sufficient condition for the existence of a solution to a linear Diophantine equation. In general, the consequence above of reversing the Euclidean algorithm gives us the following existence result for solving linear Diophantine equations.

Theorem 4.36. *Let $a, b, c \in \mathbb{Z}^+$. Then there exist $m, n \in \mathbb{Z}$ such that*

$$am + bn = c \tag{4.61}$$

if and only if $\gcd(a, b)$ *divides* c.

That is, given positive integers a, b, c, we can solve the equation $am + bn = c$ for some pair of integers m, n if and only if c is divisible by $\gcd(a, b)$.

For example, we can solve the equation

$$4199m + 1748n = c$$

for integers m and n if and only if c is divisible by 19. This is why we were able to find a solution when $c = 95$. On the other hand if $77 \leq c \leq 94$ (76 is the next multiple of 19), then the equation $4199m + 1748n = c$ has no integer solutions m and n.

Proof of Theorem 4.36. ($\Leftarrow$) Suppose $\gcd(a, b)$ divides c. By Theorem 4.19, there exist $\bar{m}, \bar{n} \in \mathbb{Z}$ such that

$$\gcd(a, b) = a\bar{m} + b\bar{n}.$$

Since $\gcd(a, b)$ divides c, there exists $q \in \mathbb{Z}$ such that

$$c = \gcd(a, b) \cdot q.$$

Thus

$$\begin{aligned} c &= \gcd(a, b) \cdot q \\ &= (a\bar{m} + b\bar{n})\, q \\ &= a(\bar{m}q) + b(\bar{n}q). \end{aligned}$$

Taking $m = \bar{m}q$ and $n = \bar{n}q$ yields $am + bn = c$.

($\Rightarrow$) Suppose there exist $m, n \in \mathbb{Z}$ such that $am + bn = c$. Since $\gcd(a, b)$ divides both m and n, it divides the integral linear combination $am + bn$. Now, since $am + bn = c$, we conclude that $\gcd(a, b)$ divides c. $\square$

So we know when the equation $am + bn = c$ has at least *one* solution. Now, when it has a solution, how do we describe *all* solutions?

Example 4.37. We have that $m_0 = 1$ and $n_0 = 2$ solve the linear Diophantine equation $60 = 24m_0 + 18n_0$; that is,

$$60 = 24 \cdot 1 + 18 \cdot 2.$$

We can now describe all solutions. Suppose that m and n also solve

$$60 = 24m + 18n.$$

Then, by taking the difference of the two displayed equations above, we get

$$\begin{aligned} 0 &= 24m + 18n - (24 \cdot 1 + 18 \cdot 2) \\ &= 24(m - 1) + 18(n - 2). \end{aligned}$$

This equation is equivalent to the one obtained by dividing by $\gcd(24, 18) = 6$; namely

$$-3(n - 2) = 4(m - 1).$$

Since $\gcd(3, 4) = 1$ and since 3 divides $4(m - 1)$, by Theorem 4.23 we conclude that 3 divides $m - 1$. Hence there exists an integer k such that

$$m - 1 = 3k.$$

By substituting this display into the previous display, we obtain

$$-3(n - 2) = 4(m - 1) = 4 \cdot 3k,$$

so that (by cancelling the 3 factors)

$$n - 2 = -4k.$$

We have shown that if m, n is any solution, then we must have that

$$\begin{aligned} m &= 1 + 3k, \\ n &= 2 - 4k, \end{aligned}$$

where k is an integer.

Now we have to check that m and n of the form above are actually solutions. That is, *conversely*, $m = 1 + 3k$ and $n = 2 - 4k$ is a solution for every integer k. Indeed, we check that for all integers k:

$$24(1 + 3k) + 18(2 - 4k) = 24 + 72k + 36 - 72k = 60.$$

Summarizing, we have shown that m and n are integer solutions to $60 = 24m + 18n$ *if and only if*

$$\begin{aligned} m &= 1 + 3k, \\ n &= 2 - 4k, \end{aligned}$$

where k is an integer.

Exercise 4.33. (1) *Find a particular solution to the linear Diophantine equation* $812m + 188n = 28$.

(2) *Find a particular solution to the linear Diophantine equation* $744m + 258n = 18$.

4.5.2. How to solve the general linear Diophantine equation. By following the method in Example 4.37, we can now describe how to solve any linear Diophantine equation. Given positive integers a, b, and c, consider the linear Diophantine equation

$$a\,m + b\,n = c.$$

Let $g = \gcd(a, b)$. We assume that g divides c; otherwise, there are no solutions. By Theorem 4.19, there exist $\bar{m}, \bar{n} \in \mathbb{Z}$ such that

$$g = a\,\bar{m} + b\,\bar{n}.$$

Since $\gcd(a, b)$ divides c, there exists $q \in \mathbb{Z}$ such that $c = gq$. Thus

$$c = a\,(\bar{m}q) + b\,(\bar{n}q)\,.$$

Taking $m_0 = \bar{m}q$ and $n_0 = \bar{n}q$ yields a "particular solution":

$$a\,m_0 + b\,n_0 = c.$$

Now we find the general solution m and n. Suppose that $am + bn = c$. Then subtracting the display above from this, we obtain the "homogeneous equation"

$$a\,(m - m_0) + b\,(n - n_0) = 0.$$

We divide this equation by the gcd g to get

$$\frac{a}{g}\,(m - m_0) = -\frac{b}{g}\,(n - n_0).$$

Since $\gcd\left(\frac{a}{g}, \frac{b}{g}\right) = 1$ and since $\frac{a}{g}$ divides $-\frac{b}{g}\,(n - n_0)$, conclude from Theorem 4.23 that $\frac{a}{g}$ divides $n - n_0$. So there exists an integer k such that

$$n - n_0 = \frac{a}{g}\,k.$$

Plugging this display into the previous display, we obtain

$$\frac{a}{g}\,(m - m_0) = -\frac{b}{g}\frac{a}{g}\,k,$$

so that

$$m - m_0 = -\frac{b}{g}\,k.$$

We conclude that:

Theorem 4.38. *If $m = m_0$ and $n = n_0$ is a particular solution to the linear Diophantine equation $am + bn = c$, then the general solution is given by*

$$n = n_0 + \frac{a}{g}\,k,$$

$$m = m_0 - \frac{b}{g}\,k,$$

where k is an integer.

Conversely, if m and n are given as above, then we calculate that

$$\begin{aligned} a\,m + b\,n &= a\left(m_0 - \frac{b}{g}\,k\right) + b\left(n_0 + \frac{a}{g}\,k\right) \\ &= a\,m_0 + b\,n_0 - a\,\frac{b}{g}\,k + b\,\frac{a}{g}\,k \\ &= a\,m_0 + b\,n_0 \\ &= c. \end{aligned}$$

Therefore, the general solution to the linear Diophantine equation $am + bn = c$ is given by the display above the previous display.

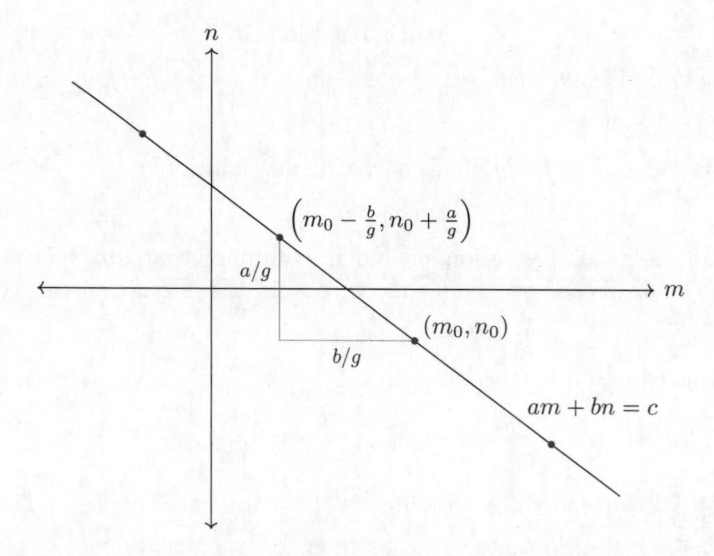

Figure 4.5.2. Visualizing the general solution to a linear Diophantine equation. Starting with a particular solution (m_0, n_0) (the red dot), the general solution is obtained by adding integer multiples of $\left(-\frac{b}{g}, \frac{a}{g}\right)$ to get the blue dots. Note that the slope of the line is equal to $-a/b$.

Solved Problem 4.39. *Find all solutions to the linear Diophantine equation*

$$140\,m + 63\,n = 35. \tag{4.62}$$

Solution. We first find $\gcd(140, 63)$ by the Euclidean algorithm:

$$\begin{aligned} 140 &= 63 \cdot 2 + 14, \\ 63 &= 14 \cdot 4 + 7, \\ 14 &= 7 \cdot 2 + 0. \end{aligned}$$

Thus $\gcd(140, 63) = 7$. Since 7 divides 35, we know that the linear Diophantine equation has a solution. We now proceed to find all solutions.

By reversing the Euclidean algorithm, we have

$$\begin{aligned} 7 &= 63 + 14 \cdot (-4) \\ &= 63 + (140 + 63 \cdot (-2)) \cdot (-4) \\ &= 140 \cdot (-4) + 63 \cdot 9. \end{aligned}$$

Thus a particular solution to our linear Diophantine equation is given by $m_0 = -20$ and $n_0 = 45$. Since $\frac{a}{g} = 20$ and $\frac{b}{g} = -9$, the general solution to (4.62) is given by

$$m = -20 - 9k, \quad n = 45 + 20k,$$

where $k \in \mathbb{Z}$. □

Exercise 4.34. (1) *Find the general solution to the linear Diophantine equation* $812m + 188n = 28$.

(2) *Find the general solution to the linear Diophantine equation* $744m+258n = 18$.

4.6. "Practical" applications of solving linear Diophantine equations (wink ☺)

In the movie *Die Hard with a Vengeance*, Simon asks John McClane and Zeus Carver to solve the following linear Diophantine equation in the guise of a water pouring puzzle:

> All you have is a 3-gallon jug, a 5-gallon jug, and an unlimited supply of water. Engineer it so that you have exactly 4 gallons of water in the 5-gallon jug. See Figure 4.6.1.

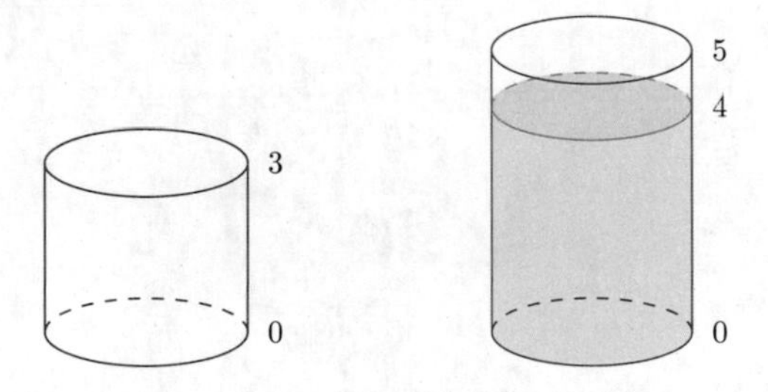

Figure 4.6.1. A 3-gallon jug and a 5-gallon jug. We wish to fill the 5-gallon jug with 4 gallons of water.

Here is a solution:

Fill the 5-gallon jug with water. Pour water from the 5-gallon jug into the 3-gallon jug until it is full. This leaves 2 gallons of water in the 5-gallon jug. Empty the 3-gallon jug. Pour the 2 gallons of water from the 5-gallon jug into the 3-gallon jug. Now fill the 5-gallon jug with water again. Pour water from the 5-gallon jug into the 3-gallon jug until it is full. Since the 3-gallon jug had exactly 2 gallons of water, exactly 1 gallon of water was poured from the 5-gallon jug, so that exactly 4 gallons of water remains in the 5-gallon jug. Sigh of relief!

Now suppose that the supply of water is unlimited, but it costs \$10/gallon. How do you minimize the number of gallons you need to buy? Hint: The solution above uses 10 gallons, which costs \$100! Can you think of a (not much) cheaper way to do this by first filling the 3-gallon jug with water?

By working backward, here is a solution to a linear Diophantine equation corresponding to the first solution above:

$$\begin{aligned} 4 &= 5 - 1 \\ &= 5 - (3 - 2) \\ &= 5 - (3 - (5 - 3)) \\ &= 2 \cdot 5 + (-2) \cdot 3. \end{aligned}$$

4.7*. (Polynomial) Diophantine equations

Generalizing linear Diophantine equations, a Diophantine equation is a polynomial equation where we are looking only for integer solutions.

For example, recall from (3.80) that a **Pythagorean triple** is a positive integer solution (x, y, z) to the Diophantine equation

$$x^2 + y^2 = z^2. \tag{4.63}$$

A famous example of a Diophantine equation is the equation of Fermat's Last Theorem:

$$x^n + y^n = z^n, \tag{4.64}$$

where $n \geq 3$, x, y, and z are positive integers.

Figure 4.7.1. Marie-Sophie Germain (1776–1831). Wikimedia Commons, Public Domain.

Over the span of a few hundred years, many mathematicians contributed to the solution of Fermat's Last Theorem, including, just to name a few: Fermat, Germain, Klein, Fricke, Hurwitz, Hecke, Dirichlet, Dedekind, Langlands, Tunnell, Deligne, Rapoport, Katz, Mazur, Igusa, Eichler, Shimura, Taniyama, Frey, Bloch, Kato, and Selmer.

The conjecture of Fermat was finally solved by Andrew Wiles and Richard Taylor. Namely, they proved:

Theorem 4.40 (Fermat's Last Theorem). *There do not exist positive integers $n \geq 3$, x, y, and z satisfying the equation*

$$x^n + y^n = z^n. \tag{4.65}$$

This is a beautiful and simple statement, but it was remarkably difficult to prove.

Exercise 4.35. *Using a calculator or computer, calculate*

$$(3987^{12} + 4365^{12})^{1/12}. \tag{4.66}$$

Is your answer an integer?

Without relying on Fermat's Last Theorem (unless you know its proof!), prove that

$$3987^{12} + 4365^{12} \neq 4472^{12}. \tag{4.67}$$

Hint: Consider the remainders after division by 4.

One can look at special cases of Fermat's Last Theorem and hope that they are easier to prove. For example, Fermat himself proved the following.

Theorem 4.41. *There do not exist positive integers x, y, and z satisfying the equation*

$$x^4 + y^4 = z^4. \tag{4.68}$$

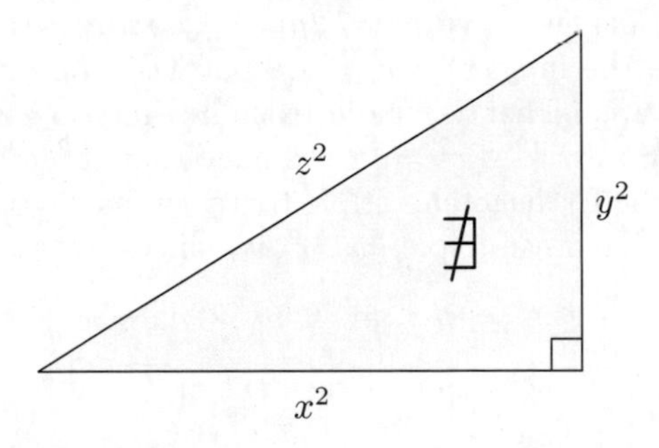

Figure 4.7.2. Theorem 4.41 says that there cannot be a right triangle with leg lengths x^2 and y^2 and hypotenuse length z^2, where x, y, z are positive integers. If such a triangle existed, the Pythagorean theorem would say that $(x^2)^2 + (y^2)^2 = (z^2)^2$. Replacing x^2 by w corresponds to Proposition 4.42.

This result is a corollary of the following result (for simply take $w = z^2$ in the display below).

Proposition 4.42. There do not exist positive integers x, y, and w satisfying the equation

$$x^4 + y^4 = w^2. \tag{4.69}$$

Proof. Let S be the set of all Pythagorean triples where each of the legs is a perfect square:

$$S := \{(x^2, y^2, w) \,|\, x^4 + y^4 = w^2\}.$$

Let $T \subset \mathbb{Z}^+$ be the set of all hypotenuses of Pythagorean triples in the set S:

$$T := \{w \,|\, (x^2, y^2, w) \in S\}.$$

Suppose for a contradiction that the proposition is false. Then S is non-empty, and hence T is a non-empty subset of $\mathbb{Z}^+$. By the well-ordering principle (Axiom 4.4), T has a minimum element w. By the definition of T, there exist positive integers x and y such that (x^2, y^2, w) is a Pythagorean triple.

We have that x^2 or y^2 is odd, since otherwise w could not be the minimum element of T. Thus, without loss of generality, we may assume that x^2 is odd; that is, x is odd. By Lemma 3.56, this implies that y is even. Furthermore, by minimality of w in T, it is easy to show that the Pythagorean triple (x^2, y^2, w) is primitive. So, by Lemma 3.59 we have that there exist positive integers $m > n$ such that (Euclid's formula)

$$x^2 = m^2 - n^2, \quad y^2 = 2mn, \quad w = m^2 + n^2. \tag{4.70}$$

In particular, $x^2 + n^2 = m^2$, so that (x, n, m) is a Pythagorean triple. Now we may apply Lemma 3.59 again to obtain that there exist positive integers $q > r$ such that

$$x = q^2 - r^2, \quad n = 2qr, \quad m = q^2 + r^2, \tag{4.71}$$

where q and r are coprime and one is even and one is odd (since x is odd). Furthermore, by $m = q^2 + r^2$, we have that m and q are coprime and m and r are coprime.

By (4.70) and (4.71), we have $y^2 = 2mn = 4qrm$, so that $qrm = \left(\frac{y}{2}\right)^2$ is a perfect square. Since the integers q, r, m are pairwise coprime, we conclude that each is a perfect square. That is, there exist integers a, b, c such that $q = a^2$, $r = b^2$, and $m = c^2$. So, by $q^2 + r^2 = m$, we have that (a^2, b^2, c) is a Pythagorean triple. Observe that the Pythagorean triple (a^2, b^2, c) has hypotenuse less than the hypotenuse of the Pythagorean triple (x^2, y^2, w) since

$$c \leq c^2 = m \leq m^2 < m^2 + n^2 = w.$$

This contradicts that w is the minimum element of T. □

4.8. The Fundamental Theorem of Arithmetic

Recall that Theorem 2.15 says that any integer $n \geq 2$ is prime or is a product of primes. We upgrade this result by adding to it the *uniqueness* of the prime factorization:

Theorem 4.43 (Fundamental Theorem of Arithmetic)**.** *Any integer n greater than 1 can be uniquely written as the product of primes in non-decreasing order.*

That is, there exist primes $p_1 \leq p_2 \leq \cdots \leq p_k$ such that

$$n = p_1 p_2 \cdots p_k. \tag{4.72}$$

Moreover, if

$$n = q_1 q_2 \cdots q_\ell, \tag{4.73}$$

where $q_1 \le q_2 \le \cdots \le q_\ell$ *are primes, then* $k = \ell$ *and* $p_i = q_i$ *for* $1 \le i \le k$.

Proof. The Prime Factorization Theorem 2.15 yields the existence part of the statement, so we are just left to prove the uniqueness part.

A way to simplify things is to cancel all p's and q's that are equal.[2] In view of this, one argues that if uniqueness fails, then one can assume that the p's and q's are distinct from each other; that is, for all i, j, $p_i \neq q_j$. We then obtain a contradiction because taking, say, p_1, since p_1 divides the product of the q's (exercise: explain why), it must divide one of the q's, which means that p_1 must equal one of the q's.

We expressed the proof in this way for conceptual clarity. A formal proof would take more effort. □

4.9. The least common multiple

On the flip side of the greatest common divisor, we have the least common multiple.

Definition 4.44. Let a and b be non-zero integers. The **least common multiple** $\ell = \operatorname{lcm}[a, b]$ of a and b is the unique positive integer ℓ satisfying the following properties:

(1) ℓ is a common multiple of a and b; that is, a divides ℓ and b divides ℓ.

(2) ℓ is less than or equal to any positive common multiple of a and b; that is, if m is a positive common multiple of a and b, then $\ell \le m$.

Let S be the set of positive integers that are common multiples of a and b. That is,

$$S = \{m \in \mathbb{Z}^+ : a \text{ divides } m \text{ and } b \text{ divides } m\}.$$

We have that $|ab| \in S$, so that S is non-empty. By the well-ordering principle, we have that S has a least element. This element is the lcm of a and b. Since the least element of S is unique, the least common multiple $\ell = \operatorname{lcm}[a, b]$ of a and b is unique.

We now sketch a proof of the formula: For any positive integers a and b,

$$\gcd(a, b) \cdot \operatorname{lcm}[a, b] = ab. \tag{4.74}$$

Can you fill in the details?

Denote $\ell := \operatorname{lcm}[a, b]$ and $g := \gcd(a, b)$.

Assertion *As Small*:

$$\frac{ab}{\ell} \le g. \tag{4.75}$$

Assertion *As Big*:

$$\frac{ab}{\ell} \ge g. \tag{4.76}$$

[2] For example, if $p_{17} = q_{23} = q_{24}$, then we choose one of q_{23}, q_{24} to cancel with p_{17}.

Why it's as small:

- For any common multiple m of a and b, ℓ divides m, for if ℓ does not divide m, then by the Division Theorem there exists $q, r \in \mathbb{Z}$ such that
$$m = q \cdot \ell + r, \text{ where } 0 < r < \ell.$$
Then, r is a common multiple of a and b, and hence $r \geq \ell$, a contradiction.
- ℓ divides ab. Thus, $\frac{ab}{\ell}$ is a positive integer.
- $\frac{ab}{\ell}$ is a common divisor of a and b. E.g., explain why ab divides $a\ell$ is true and why that helps.

Why it's as big:
$$\frac{ab}{g} \text{ is a common multiple of } a \text{ and } b.$$

Explain why this is easy to prove and why this is all that we need!

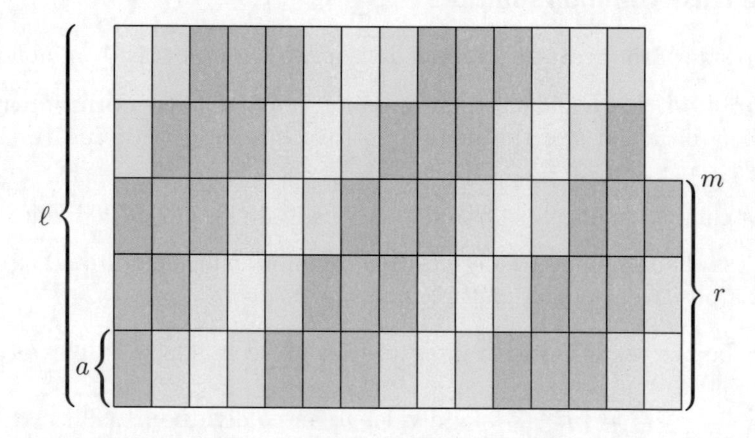

Figure 4.9.1. We divide m by $\ell = \operatorname{lcm}[a, b]$ to obtain a remainder r. Since a divides both ℓ and m, we have that a divides r. Similarly, b divides r. Since $r < \ell$, we conclude that $r = 0$. That is, ℓ divides m. This shows that the least common multiple divides any common multiple.

4.9.1. The gcd and lcm via the prime decomposition. We can see the relationship between the greatest common divisor and the least common multiple of two integers by consideration of the prime decompositions of the integers. Let $N \geq 2$ be an integer. By the Fundamental Theorem of Arithmetic, we may (uniquely) write

$$n = p_1^{k_1} p_2^{k_2} \cdots p_r^{k_r}, \tag{4.77}$$

where $p_1 < p_2 < \cdots < p_r$ are prime and $k_1, k_2, \ldots, k_r$ are positive integers.

Now, let m and n be integers, both of which are at least 2. Then there exist primes $q_1 < q_2 < \cdots < q_s$ and non-negative integers $k_1, k_2, \ldots, k_s$ and $\ell_1, \ell_2, \ldots, \ell_s$ such that

$$n = q_1^{k_1} q_2^{k_2} \cdots q_s^{k_s} \quad \text{and} \quad m = q_1^{\ell_1} q_2^{\ell_2} \cdots q_s^{\ell_s}. \tag{4.78}$$

Can you prove the following formulas for the greatest common divisor and the least common multiple?

(4.79a) $$\gcd(n,m) = q_1^{\min\{k_1,\ell_1\}} q_2^{\min\{k_2,\ell_2\}} \cdots q_s^{\min\{k_s,\ell_s\}},$$

(4.79b) $$\operatorname{lcm}[n,m] = q_1^{\max\{k_1,\ell_1\}} q_2^{\max\{k_2,\ell_2\}} \cdots q_s^{\max\{k_s,\ell_s\}}.$$

Using this, can you explain again why

$$\gcd(n,m) \cdot \operatorname{lcm}[n,m] = nm\,?$$

You may need to prove a little lemma about max and min!

4.10. Residues modulo an odd prime

We are particularly interested in division by primes. In this section we consider elementary consequences of the Division Theorem in this setting.

Residues are siblings of remainders. Let p be an odd prime, and define

$$b := \frac{p-1}{2},$$

which is an integer.

Observe that, corresponding to $p = 7$, we have that for any integer a there exist unique integers q and r such that

$$a = 7q + r, \qquad -3 \le r \le 3.$$

We call r the residue of a modulo 7. Generalizing the fact above, we have the following.

Exercise 4.36. *Prove that for every integer a, there exist unique integers q and r such that*

(4.80) $$a = pq + r, \qquad -b \le r \le b.$$

Hint: Apply the proof of the Division Theorem. You only will need that p is odd (it is not necessary for p to be a prime).

Note that if we use the Division Theorem verbatim, then we obtain a quotient q' and a remainder r' with $0 \le r' \le p-1 = 2b$. This exercise says that we can effectively shift the remainder interval from $[0, 2b]$ to $[-b, b]$.

Let a be an integer which is not a multiple of p. Consider the first b positive multiples of a:

$$1a, 2a, 3a, \ldots, ba.$$

For $1 \le i \le b$, let $-b \le r_i \le b$ be the unique remainder of ia given by (4.80), so that

(4.81) $$p \text{ divides } ia - r_i.$$

We call r_i the **residue** of ia (modulo p).

Exercise 4.37. (1) *Explain why all the residues of ia are non-zero: $r_i \ne 0$ for all $1 \le i \le b$.*

(2) *Prove that if $r_i < 0$ and $r_j > 0$, then $r_i + r_j$ is not divisible by p. Hint: Can $(i+j)a$ be a multiple of p?*

Exercise 4.38. *Prove that*

$$\prod_{r_i<0}(-r_i)\prod_{r_j>0} r_j = b!. \tag{4.82}$$

Here, the exclamation mark means factorial, so that it is b! instead of us getting really excited about b. ☺

Example 4.45. Let $p = 11$, so that $b = 5$. Let $a = 8$. Then, the first 5 positive multiples of a are $8, 16, 24, 32, 40$. The corresponding residues are

$$-3,\ 5,\ 2,\ -1,\ -4. \tag{4.83}$$

And, indeed, $3 \cdot 5 \cdot 2 \cdot 1 \cdot 4 = 5!$.

4.11. Appendix

4.11.1. Equivalence of induction and the well-ordering principle. Firstly, we show that mathematical induction and the well-ordering principle are equivalent:

Proof of Theorem 4.5. We want to prove:

$$\text{mathematical induction} \quad \Leftrightarrow \quad \text{well-ordering principle.}$$

($\Rightarrow$) Assume mathematical induction. Let A be a subset of $\mathbb{Z}^+$ that has no least element. It suffices to prove that A is empty.

Let $P(n)$ be the statement that $\mathbb{N}_n \subset \mathbb{Z}^+ - A$; that is, $A \cap \mathbb{N}_n = \emptyset$.

Base case: Suppose for a contradiction that $1 \in A$. Since 1 is the least element of $\mathbb{Z}^+$ and since A is a subset of $\mathbb{Z}^+$, we conclude that 1 is the least element of A, a contradiction to our assumption on A. Therefore $1 \notin A$, which is equivalent to saying $\mathbb{N}_1 \subset \mathbb{Z}^+ - A = A^c$ (taking $\mathbb{Z}^+$ to be the universal set).

Inductive step: Suppose that k is a positive integer such that $\mathbb{N}_k \subset A^c$. Suppose for a contradiction that $k+1 \in A$. Suppose $m \in \mathbb{Z}^+$ satisfies $m < k+1$. Then $m \in \mathbb{N}_k$, which implies that $m \notin A$. This proves that $k+1$ is the least element of A, a contradiction. Therefore $k+1 \notin A$. This and $\mathbb{N}_k \subset A^c$ imply that $\mathbb{N}_{k+1} \subset A^c$.

Since we have assumed that mathematical induction is true, we can now conclude that $P(n)$ is true for all $n \in \mathbb{Z}^+$. In particular, for all $n \in \mathbb{Z}^+$ we have $n \in \mathbb{N}_n \subset A^c$, which says that $n \notin A$. That is, no positive integer is in A. Since A is a subset of the set of positive integers, we conclude that A is empty.

($\Leftarrow$) Assume that the well-ordering principle is true. We want to prove mathematical induction.

To this end, let A be a subset of $\mathbb{Z}^+$ which satisfies the following properties:

(1) $1 \in A$.

(2) For all $k \in \mathbb{Z}^+$, if $k \in A$, then $k+1 \in A$.

We need to show that $A = \mathbb{Z}^+$; that is, $A^c = \emptyset$.

Suppose for a contradiction that A^c is non-empty. Since we have assumed that the well-ordering principle is true, we have that A^c has a least element. Call this least element ℓ. Since $\ell \in A^c$, we have $\ell \notin A$.

Since property (1) says that $1 \in A$, we have that $\ell \neq 1$. Since $\ell \in \mathbb{Z}^+$, this tells us that $\ell \geq 2$.

Suppose for a contradiction that $\ell - 1 \in A$. Since $\ell \geq 2$, $\ell - 1 \in \mathbb{Z}^+$. So property (2) and $\ell - 1 \in A$ tell us that $\ell = (\ell - 1) + 1 \in A$, which contradicts $\ell \notin A$. We conclude from this that $\ell - 1 \notin A$; that is, $\ell - 1 \in A^c$.

But since $\ell - 1 < \ell$, this contradicts that ℓ is the least element of A^c. From all of this we conclude that A^c is empty, as desired. □

4.11.2. Finite sets have extrema. Secondly, we consider the minimum and maximum for finite sets of real numbers. The following result is proved by induction.

Theorem 4.46. *If X is a non-empty finite set of real numbers, then X has both a minimum element and a maximum element.*

Proof. We prove that X has a minimum by induction on the cardinality of X. The existence of a maximum is proved similarly.

Base case: $n = 1$. In this case X consists of a single real number, which must be (both the maximum and) the minimum of the set.

Inductive step: Suppose that $k \in \mathbb{Z}^+$ has the property that the theorem is true for all sets of real numbers of cardinality k. Let X be a set of real numbers with cardinality $k + 1$. Choose an element of X and call it x_0.

Case 1: x_0 is the minimum element of X. Then we are done.

Case 2: x_0 is not the minimum element of X. Let $Y = X - \{x_0\}$. Then Y is a finite set of real numbers with cardinality k. Hence we can apply the inductive hypothesis to obtain that Y has a minimum element y_0.

We claim that y_0 is the minimum element of X. Actually this is pretty easy to see: Let $x \in X$.

If $x = x_0$, then since x_0 is not the minimum element of X, there exists $x_1 \in X$ such that $x = x_0 > x_1 \geq y_0$, where the last equality follows from $x_1 \in X - \{x_0\} = Y$ and y_0 being the minimum element of Y.

On the other hand, if $x \neq x_0$, then $x \in X - \{x_0\} = Y$. Since y_0 is the minimum element of Y, we have that $x \geq y_0$.

In either case, we have proved that $x \geq y_0$. We conclude that y_0 is the minimum element of X. This completes the inductive step.

By induction we are done. □

4.11.3. An alternate proof of the well-ordering principle. Using Theorem 4.46, we can give an alternate proof of the well-ordering principle.

Proof of Axiom 4.4. Let X be a non-empty subset of $\mathbb{Z}^+$. Since X is non-empty, there exists an integer n_0 in X. Let

$$Y := \{x \in X : 1 \leq x \leq n_0\}. \tag{4.84}$$

Then Y is non-empty since $n_0 \in Y$, and Y is finite since $Y \subset \mathbb{N}_{n_0}$. Hence, by Theorem 4.46, Y has a minimum element, which we call m.

We will now prove that m is the minimum element of X. Let $x \in X$.

Case 1: $x \leq n_0$. Then $x \in Y$. Since m is the minimum element of Y, we have $m \leq x$.

Case 2: $x > n_0$. Then $m \leq n_0 < x$, so that $m \leq x$.

This proves that m is the minimum element of X. □

4.12. Hints and partial solutions for the exercises

Hint for Exercise 4.1. Let $i := \mathbf{r}(n)$. Consider two cases: (1) $i = 0$ and (2) $0 < i < 3$. In case (1) explain why n is the only integer divisible by 3. In case (2) explain why $n + 3 - i$ is the only integer divisible by 3.

Hint for Exercise 4.2. By the well-ordering principle, X has a minimum element m. Using this, what should the minimum element of Y be? Prove it.

Hint for Exercise 4.3. Let $f(n) = n^2 + n + 41$. We have

$$f(40) = 40^2 + 40 + 41 = 40 \cdot 41 + 41 = 41^2, \tag{4.85}$$

which is not prime.

Hint for Exercise 4.4. Use the hint given. You may compare with the hint for Exercise 4.1 if you like.

Hint for Exercise 4.5. To prove existence, you can take the q and r from the Division Theorem and subtract 3 from q, which adds 15 to r.

To prove uniqueness, you may "transfer" the problem to the case of the Division Theorem, or alternatively you may mimic the proof of uniqueness in the Division Theorem.

Hint for Exercise 4.6. We have $a = m\,\mathbf{q}(a) + \mathbf{r}(a)$ and similarly for b.

Hint for Exercise 4.7. Let n be an even integer. Then there exist integers q, r such that $n = 4q + r$ where $r \in R_4$ is even; that is, r equals 0 or 2. Consider these two cases.

Hint for Exercise 4.8. There exist integers $k \in \mathbb{Z}$ and $r \in R_3 = \{0, 1, 2\}$ such that

$$n = 3k + r. \tag{4.86}$$

We compute that

$$n^2 = 3(3k^2 + 2kr) + r^2. \tag{4.87}$$

Consider the following three cases: (1) $r = 0$, (2) $r = 1$, and (3) $r = 2$. We have for example in case (2) that $n^2 = 3q + 1$, where $q = 3k^2 + 2k$.

Hint for Exercise 4.9. We have that $m^2 = 3q_1 + r_1$ and $n^2 = 3q_2 + r_2$, where $0 \le r_1, r_2 \le 1$. Thus

$$m^2 + n^2 = 3(q_1 + q_2) + r_1 + r_2, \tag{4.88}$$

where

$$0 \le r_1 + r_2 \le 2. \tag{4.89}$$

Continue.

Hint for Exercise 4.10. Consider the two case where n is even or odd.

Hint for Exercise 4.11. Let n be a square of an integer. Then there exist integers $k \in \mathbb{Z}$ and $r \in R_5 = \{0, 1, 2, 3, 4\}$ such that

$$n = 5k + r. \tag{4.90}$$

If, for example, $r = 4$, then $n^2 = 5q + 1$, where $q = 5k^2 + 8r + 3$.

Hint for Exercise 4.12. The completed table follows:

m	remainders of perfect squares after dividing by m
2	0, 1
3	0, 1
4	0, 1
5	0, 1, 4
6	0, 1, 3, 4
7	0, 1, 2, 4
8	0, 1, 4
9	0, 1, 4, 7
10	0, 1, 4, 5, 6, 9

Hint for Exercise 4.13.

We have the following table of elements of R_m that are not remainders of perfect squares:

m	elements of R_m that are not remainders of perfect squares
2	none
3	2
4	2, 3
5	2, 3
6	2, 5
7	3, 5, 6
8	2, 3, 5, 6, 7
9	2, 3, 5, 6, 8
10	2, 3, 7, 8

Hint for Exercise 4.14.

Condition	m	Integers ruled out from being perfect squares
(1)	3	2, 5, 8, 11, 14, 17, 20, 23, 26, 29, 32, 35, 38, 41, 44, 47, 50, 53, 56, 59, 62, 65, 68, 71, 74, 77, 80, 83, 86, 89, 92, 95, 98
(2)	4	2, 3, 6, 7, 10, 11, 14, 15, 18, 19, 22, 23, 26, 27, 30, 31, 34, 35, 38, 39, 42, 43, 46, 47, 50, 51, 54, 55, 58, 59, 62, 63, 66, 67, 70, 71, 74, 75, 78, 79, 82, 83, 86, 87, 90, 91, 94, 95, 98, 99
(3)	5	2, 3, 7, 8, 12, 13, 17, 18, 22, 23, 27, 28, 32, 33, 37, 38, 42, 43, 47, 48, 52, 53, 57, 58, 62, 63, 67, 68, 72, 73, 77, 78, 82, 83, 87, 88, 92, 93, 97, 98
(4)	6	2, 5, 8, 11, 14, 17, 20, 23, 26, 29, 32, 35, 38, 41, 44, 47, 50, 53, 56, 59, 62, 65, 68, 71, 74, 77, 80, 83, 86, 89, 92, 95, 98
(5)	7	3, 5, 6, 10, 12, 13, 17, 19, 20, 24, 26, 27, 31, 33, 34, 38, 40, 41, 45, 47, 48, 52, 54, 55, 59, 61, 62, 66, 68, 69, 73, 75, 76, 80, 82, 83, 87, 89, 90, 94, 96, 97
(6)	8	2, 3, 5, 6, 7, 10, 11, 13, 14, 15, 18, 19, 21, 22, 23, 26, 27, 29, 30, 31, 34, 35, 37, 38, 39, 42, 43, 45, 46, 47, 50, 51, 53, 54, 55, 58, 59, 61, 62, 63, 66, 67, 69, 70, 71, 74, 75, 77, 78, 79, 82, 83, 85, 86, 87, 90, 91, 93, 94, 95, 98, 99
(7)	9	2, 3, 5, 6, 8, 11, 12, 14, 15, 17, 20, 21, 23, 24, 26, 29, 30, 32, 33, 35, 38, 39, 41, 42, 44, 47, 48, 50, 51, 53, 56, 57, 59, 60, 62, 65, 66, 68, 69, 71, 74, 75, 77, 78, 80, 83, 84, 86, 87, 89, 92, 93, 95, 96, 98
(8)	10	12, 13, 17, 18, 22, 23, 27, 28, 32, 33, 37, 38, 42, 43, 47, 48, 52, 53, 57, 58, 62, 63, 67, 68, 72, 73, 77, 78, 82, 83, 87, 88, 92, 93, 97, 98

The integers from 1 to 100 that are not ruled out as perfect squares by considering division by m for $3 \leq m \leq 10$ are

$$1, 4, 9, 16, 25, 36, 49, 64, 81, 100. \tag{4.91}$$

Each of these integers is a perfect square! So, for the first 100 positive integers, we can rule out all non-perfect square by considering division by m for $3 \leq m \leq 10$.

Hint for Exercise 4.15. Let x and y be integers. Let $P(n)$ be the statement that $x^n - y^n$ is divisible by $x - y$.

Clearly $P(1)$ is true since $x - y$ is divisible by $x - y$. Suppose that $P(k)$ is true. Then there exists an integer a such that

$$x^k - y^k = a(x - y). \tag{4.92}$$

We compute that

$$\begin{aligned} x^{k+1} - y^{k+1} &= (x^{k+1} - x^k y) + (x^k y - y^{k+1}) \\ &= x^k(x - y) + (x^k - y^k)y \\ &= x^k(x - y) + a(x - y)y \\ &= (x^k + ay)(x - y). \end{aligned}$$

Since $x^k + ay$ is an integer, this proves that $x^{k+1} - y^{k+1}$ is divisible by $x - y$. That is, $P(k+1)$ is true.

The result follows from induction.

Hint for Exercise 4.16. For each $n \in \mathbb{Z}^+$, let $S(n)$ be the set of primes that divide at least one of the integers $a_1, a_2, \dots, a_n$. Each $S(n)$ is a finite set. By the Prime Factorization Theorem, $S(1)$ is non-empty. Clearly, $S(n) \subset S(n+1)$. We will prove that $S(n)$ is a proper subset of $S(n+1)$ for each n. That is, there exists an element of $S(n+1)$ that is not an element of $S(n)$. Suppose for a contradiction that $S(n) = S(n+1)$. Let p be a prime that divides a_{n+1}. Then $p \in S(n+1) = S(n)$. Since $p \in S(n)$, p divides a_i for some $1 \leq i \leq n$. Therefore $\gcd(a_{n+1}, a_i) \geq p > 1$ for some $1 \leq i \leq n$, which is a contradiction.

Now we have $|S(n+1)| \geq |S(n)| + 1$ for all $n \in \mathbb{Z}^+$. This implies that there are an infinite number of primes.

Hint for Exercise 4.17. We follow the hint and we first prove by induction that

$$\prod_{i=0}^{n-1} F_i = F_n - 2 \quad \text{for } n \geq 1. \tag{4.93}$$

Check the base case.

As our inductive hypothesis, suppose that $k \in \mathbb{Z}^+$ has the property that

$$\prod_{i=0}^{k-1} F_i = F_k - 2. \tag{4.94}$$

Show that

$$\begin{aligned} F_{k+1} &= (2^{2^k})^2 + 1 \\ &= (F_k)^2 - 2F_k + 2. \end{aligned}$$

By multiplying (4.94) by F_k, we obtain

$$\prod_{i=0}^{k} F_i = (F_k)^2 - 2F_k. \tag{4.95}$$

Finish the proof by induction of (4.93).

Now we prove that the Fermat numbers are pairwise coprime. Let $k < \ell$ be non-negative integers. We have

$$F_\ell = \prod_{i=0}^{\ell-1} F_i + 2 = 2 + F_k \cdot \prod_{0 \leq i \leq \ell-1,\, i \neq k} F_i.$$

Suppose for a contradiction that F_k and F_ℓ are not coprime. Since F_k and F_ℓ are both odd, this means that there exists an odd prime p dividing both F_k and F_ℓ. However, this is a contradiction since then the previous display implies that p divides 2.

Hint for Exercise 4.18. We prove this by induction. Check the base case. As our inductive hypothesis, suppose that $k \geq 2$ has the property that F_k has last digit equal to 7. By the calculation in the solution to the previous problem, we have that

$$F_{k+1} = (F_k)^2 - 2F_k + 2. \tag{4.96}$$

Moreover, by our inductive hypothesis there exists an integer q such that

$$F_k = 10q + 7. \tag{4.97}$$

Show that

$$F_{k+1} = 10(10q^2 + 12q + 3) + 7.$$

Hint for Exercise 4.19. As a crude estimate, since $2^3 < 10 < 2^4$, we have $2^{300} < 10^{100} < 2^{400}$.

Hint for Exercise 4.20. The gcd is 4.

Hint for Exercise 4.21. The gcd is 6.

Hint for Exercise 4.22. $4 = -95 \cdot 188 + 22 \cdot 812$.

Hint for Exercise 4.23. $6 = -49 \cdot 258 + 17 \cdot 744$.

Hint for Exercise 4.24. $m = 8$ and $n = -39$.

Hint for Exercise 4.25. Since p and q are distinct and prime, they are coprime.

Hint for Exercise 4.26. Observing that $12 \cdot 2 = 24$ and $5 \cdot 5 = 25$, we compute that

$$(-12)(5n + 2) + (5)(12n + 5) = 1. \tag{4.98}$$

By Theorem 4.20, we conclude that $\gcd(5n + 2, 12n + 5) = 1$.

Hint for Exercise 4.27. We compute that

$$(-c)(an + b) + (a)(cn + d) = -bc + ad = 1. \tag{4.99}$$

Hint for Exercise 4.28. Suppose that a prime p divides the product ab. We have $g := \gcd(p, a)$ is a positive integer dividing p. Hence $g = 1$ or $g = p$.

If $g = p$, then since g divides a, we have that p divides a.

On the other hand, if $g = 1$, then by Theorem 4.23 we have that p divides b.

We have proved that p divides a or p divides b.

Hint for Exercise 4.29. Let a, b, c be non-zero integers. The statement we want to prove is the contrapositive of the implication: If (a and b are coprime) and (a and c are coprime), then a and bc are coprime. This statement is true by Lemma 4.29.

Hint for Exercise 4.30. Let $g = \gcd(a, b)$. Then g divides both a and b. This implies that g^n divides both a^n and b^n. We conclude from this that

$$g^n \leq \gcd(a^n, b^n). \tag{4.100}$$

Now suppose for a contradiction that

$$g^n < \gcd(a^n, b^n). \tag{4.101}$$

We have that g^n divides $\gcd(a^n, b^n)$. Thus there exists an integer $k > 1$ such that

$$\gcd(a^n, b^n) = kg^n. \tag{4.102}$$

Let p be a prime that divides k. Then $g^n p$ divides both a^n and b^n, which implies that p divides both $\left(\frac{a}{g}\right)^n$ and $\left(\frac{b}{g}\right)^n$. This in turn implies that p divides both $\frac{a}{g}$ and $\frac{b}{g}$. This proves that g is not the gcd of a and b, which is a contradiction.

Hint for Exercise 4.31. Show that m_1 and m_2/g are coprime.

Hint for Exercise 4.32. We prove Corollary 4.34 by induction on k.

The base case $k = 1$ is obvious.

Suppose that $k \in \mathbb{Z}^+$ has the property that for every pairwise coprime positive integer $m_1, m_2, \ldots, m_k$, if each of $m_1, m_2, \ldots, m_k$ divides an integer x, then their product $m_1 m_2 \cdots m_k$ divides x.

Now suppose that $m_1, m_2, \ldots, m_{k+1}$ are pairwise coprime positive integers such that each of $m_1, m_2, \ldots, m_{k+1}$ divides an integer x. Since then each of $m_1, m_2, \ldots, m_k$ divides x, we have that $m_1 m_2 \cdots m_k$ divides x. Now since the $m_1, m_2, \ldots, m_{k+1}$ are pairwise coprime, we have

$$\gcd(m_1 m_2 \cdots m_k, m_{k+1}) = 1. \tag{4.103}$$

Since m_{k+1} also divides x, by Proposition 4.33, we have that the product

$$m_1 m_2 \cdots m_k \cdot m_{k+1} = m_1 m_2 \cdots m_{k+1} \tag{4.104}$$

divides x. This proves the inductive step.

By induction, we are done.

Finally, to be complete, we prove (4.103) by induction.

Suppose as an inductive hypothesis that k is a positive integer with the property that if $m_1, m_2, \ldots, m_k, b$ are pairwise coprime positive integers. Then $\gcd(m_1 m_2 \cdots m_k, b) = 1$.

Let $m_1, m_2, \ldots, m_{k+1}, b$ be pairwise coprime positive integers. By the inductive hypothesis, $\gcd(m_1 m_2 \cdots m_k, b) = 1$. This and $\gcd(m_{k+1}, b) = 1$ imply that

$$1 = \gcd((m_1 m_2 \cdots m_k) m_{k+1}, b) = \gcd(m_1 m_2 \cdots m_{k+1}, b). \tag{4.105}$$

This proves the inductive step, so by induction we have proved (4.103).

Note that for (4.105) we used the fact that if $\gcd(a, c) = \gcd(b, c) = 1$, then $\gcd(ab, c) = 1$. This is one of the lemmas in the book.

Hint for Exercise 4.33. Multiply the solution to Exercise 4.22 by 7.

Multiply the solution to Exercise 4.23 by 3.

Hint for Exercise 4.34. Use Theorem 4.38.

Hint for Exercise 4.35. We compute using a "big number calculator" that

$$3987^{12} = 16134474609751291283496491970515151715346481 \tag{4.106}$$

and

$$4365^{12} = 47842181739947321332739738982639336181640625. \tag{4.107}$$

Therefore

$$\begin{aligned}(3987^{12}+4365^{12})^{1/12} &= 63976656349698612616236230953154487896987106^{1/12}\\ &= 4472.0000000071.\end{aligned}$$

The answer is very close to, but not equal to, an integer!

Next, we prove that

$$3987^{12}+4365^{12}\neq 4472^{12}. \tag{4.108}$$

Let n be an odd integer. Then there exist integers q and r such that

$$n = 4q + r, \tag{4.109}$$

where r equals 1 or 3.

Case 1: $r = 1$. In this case,

$$n^{12} = 4s + 1 \tag{4.110}$$

for some integer s.

Case 2: $r = 3$. In this case, there is an integer s such that

$$r^4 = 4s + 3^4 = 4(s + 20) + 1.$$

Thus, there is an integer t such that

$$r^{12} = (r^4)^3 = 4t + 1.$$

We have just proved that any odd integer to the power 12 when divided by 4 has a remainder of 1.

From all of this, we conclude that the integer $3987^{12}+4365^{12}$ when divided by 4 has a remainder of 2. However, 4472^{12} when divided by 4 has a remainder of 0. Thus $3987^{12}+4365^{12}$ cannot be equal to 4472^{12}.

Hint for Exercise 4.36. Firstly, apply the Division Theorem to get a quotient q' and a remainder r' with $0 \le r' < p$. (1) If $r' \le b$, let $r = r'$. (2) If $r' > b$, let $r = r' - p$. Continue to prove existence. To prove uniqueness, mimic the proof of the uniqueness part of the Division Theorem.

Hint for Exercise 4.37. (1) Show that p cannot divide ia for $1 \le i \le b$.

(2) We have $0 \le i + j \le 2b < p$. So $(i + j)a$ cannot be a multiple of p.

Hint for Exercise 4.38. When $r_i < 0$, we have $0 < -r_i \le b$. When $r_j > 0$, we have $0 < r_j \le b$. None of the $-r_i$'s and r_j's are congruent to each other, and therefore they are distinct.

Chapter 5

Sets and Functions

$$\begin{aligned}
\text{The } \clubsuit\text{'s} &= \{A\clubsuit, K\clubsuit, Q\clubsuit, J\clubsuit, 10\clubsuit, 9\clubsuit, 8\clubsuit, 7\clubsuit, 6\clubsuit, 5\clubsuit, 4\clubsuit, 3\clubsuit, 2\clubsuit\}.\\
\text{The } \diamondsuit\text{'s} &= \{A\diamondsuit, K\diamondsuit, Q\diamondsuit, J\diamondsuit, 10\diamondsuit, 9\diamondsuit, 8\diamondsuit, 7\diamondsuit, 6\diamondsuit, 5\diamondsuit, 4\diamondsuit, 3\diamondsuit, 2\diamondsuit\}.\\
\text{The } \heartsuit\text{'s} &= \{A\heartsuit, K\heartsuit, Q\heartsuit, J\heartsuit, 10\heartsuit, 9\heartsuit, 8\heartsuit, 7\heartsuit, 6\heartsuit, 5\heartsuit, 4\heartsuit, 3\heartsuit, 2\heartsuit\}.\\
\text{The } \spadesuit\text{'s} &= \{A\spadesuit, K\spadesuit, Q\spadesuit, J\spadesuit, 10\spadesuit, 9\spadesuit, 8\spadesuit, 7\spadesuit, 6\spadesuit, 5\spadesuit, 4\spadesuit, 3\spadesuit, 2\spadesuit\}.
\end{aligned}$$

Figure 5.0.1. Four 13-element sets whose union is a deck of cards, without Jokers (or thieves)!

Figure 5.0.2. A lonely joker. This work has been released into the public domain by its author, Trocche100 at Italian Wikipedia.

Goals of this chapter: To introduce the language of mathematics: set theory. To understand general properties of functions. To relate functions to counting the number of elements in finite sets. To understand injections, surjections, and bijections.

We begin with some fun examples of sets: a set of suits, with four elements:

$$\{\clubsuit, \diamondsuit, \heartsuit, \spadesuit\}. \tag{5.1}$$

A set of suites, with three elements:

{Bach Suite No. 3 in D major, Holst The Planets Suite, Grieg Peer Gynt Suites}.

A set of sweets, with four elements:

{Creme tangerine, Coffee dessert, Cool cherry cream, Coconut fudge},

not to mention a ginger sling with a pineapple heart, all after the Savoy Truffle.[1]

[1] Rock trivia question: Why is this Beatles song, written by George Harrison, about Slowhand, a.k.a. Eric Clapton?

5.1. Basics of set theory

Per its namesake, we start with:

5.1.1. The notion of a set. By definition, a **set** is a collection of objects.

For example, $\mathbb{Z}$, $\mathbb{Q}$, $\mathbb{R}$ denote the sets of integers, rational numbers, and real numbers, respectively. $\mathbb{Z}^+$ denotes the set of positive integers. We have also considered the set E of even integers and the set O of odd integers, as well as many other sets.

Sets occur in everyday life, especially when we organize or categorize stuff.

5.1.2. Elements of sets. We call the objects in sets **elements**. If x is an element of a set A, then we denote this by

$$x \in A.$$

So $x \in \mathbb{R}$ is just another way of saying that x is a real number. For example, $\pi \in \mathbb{R}$ since π is a real number.

If an object x is not an element of A, then we denote this by

$$x \notin A.$$

That is, $x \notin A$ if and only if not $(x \in A)$. For example, $\frac{3}{2} \notin \mathbb{Z}$ since $\frac{3}{2}$ is not an integer. We also have $\pi \notin \mathbb{Z}$. But $\frac{3}{2} \in \mathbb{Q}$ since $\frac{3}{2}$ is a rational number. And $\pi \notin \mathbb{Q}$, although this last fact is not so obvious.

Figure 5.1.1. The element x is a "hare" inside A; i.e., $x \in A$. Also, y is not an element of A; i.e., $y \notin A$.

5.1.3. Subsets and equality of sets. Now that we know what a set is, we want to understand relations between sets.

We say that a set A is a **subset** of a set B, read as $A \subset B$, if $x \in A$ implies $x \in B$. In other words, every element of A is an element of B.

We say that two sets A and B are **equal**, denoted $A = B$, if and only if $A \subset B$ and $B \subset A$. Equivalently, $A = B$ if and only if $x \in A \Leftrightarrow x \in B$. In other words, the elements of A are the same as the elements of B.

Example 5.1. The set of positive integers is equal to the set of integers greater than or equal to 1. This is because there is no integer n satisfying $0 < n < 1$.

Observe that we have the inclusion of sets

$$\mathbb{Z} \subset \mathbb{Q} \quad \text{and} \quad \mathbb{Q} \subset \mathbb{R}. \tag{5.2}$$

This is because every integer is a rational number and every rational number is a real number.

We say that a set A is a **proper subset** of a set B if A is a subset of B and $A \neq B$.

It is easy to see that $\mathbb{Z}$ is a proper subset of $\mathbb{Q}$ since $\frac{1}{2}$ is rational but is not an integer.

By Theorem 3.45, which states that the square root of 2 is an irrational real number, we have that $\mathbb{Q}$ is a proper subset of $\mathbb{R}$.

We have the following basic property of the subset relation.

Lemma 5.2 (Transitivity of the subset relation). *If $A \subset B$ and $B \subset C$, then $A \subset C$.*

Proof. Suppose that $A \subset B$ and $B \subset C$. Let $x \in A$. Since $A \subset B$, this implies $x \in B$. Now, since $B \subset C$, this implies $x \in C$. We have proved that if $x \in A$, then $x \in C$. That is, $A \subset C$. □

Do you notice any similarity between the proof of this lemma and the proof of Lemma 3.29 (Exercise 3.21)?

5.1.4. Sets of sets: A primer. A *set* itself can be an *element* of another set. Consider the following set, which is different than the set in (5.1):

$$\mathcal{S} := \{\text{The } \clubsuit\text{'s, The } \diamondsuit\text{'s, The } \heartsuit\text{'s, The } \spadesuit\text{'s}\}, \tag{5.3}$$

where the elements The ♣'s, The ♢'s, The ♡'s, and The ♠'s are given by Figure 5.0.1 at the beginning of this chapter! The set $\mathcal{S}$ is a set with 4 elements, where each element is itself a set with 13 elements! For example, we have

$$\text{Q}\heartsuit \in \text{ The } \heartsuit\text{'s} \quad \text{and} \quad \text{The } \heartsuit\text{'s} \in \mathcal{S}.$$

So, in set theory, it is perfectly fine to consider *sets of sets*, although there are some paradoxes lurking around which require any rigorous treatment to be well thought out. We will mention some of the issues in set theory, but for the most part we will sidestep a detailed treatment of the (commonly accepted) axioms of set theory.

Exercise 5.1. *Explain why $\text{Q}\heartsuit \notin \mathcal{S}$.*

Example 5.3. Let $2 = \{2\spadesuit, 2\heartsuit, 2\diamondsuit, 2\clubsuit\}, \ldots, 2 = \{A\spadesuit, A\heartsuit, A\diamondsuit, A\clubsuit\}$. Then we have the set of sets $\mathcal{T} = \{2, 3, \ldots, K, A\}$, which has 13 elements, each of which is a set of 4 elements since there are 4 suits.

5.1.5. Much ado about nothing: The empty set.

Zero, zip, zilch, nada ... yada yada yada.

A fundamental, but somewhat trivial, set is an **empty set** $\emptyset$. This is a set with no elements. In contrast to its name, the empty set appears ubiquitously in set theory.

For the following, let us assume that a *zebra* is a well-defined mathematical notion.

Proposition 5.4 ($x \in \emptyset$ implies your favorite statement, true or false). Let $\emptyset$ be an empty set. If $x \in \emptyset$, then x is a zebra.

Proof. This is a true statement! This is because the hypothesis of the implication, $x \in \emptyset$, is false since $\emptyset$ has no elements. Any implication with a false hypothesis is a true statement. □

The idea of the proposition above implies that there is only one empty set:

Theorem 5.5 (Uniqueness of nothing). *The empty set is unique.*

Proof. Suppose $\emptyset$ and $\emptyset'$ are two empty sets; i.e., each of these two sets has no elements. Then the implication $x \in \emptyset$ implies $x \in \emptyset'$ is true since $x \in \emptyset$ is false (because $\emptyset$ has no elements). Similarly, the implication $x \in \emptyset'$ implies $x \in \emptyset$ is true since $\emptyset'$ has no elements. Hence

$$\emptyset = \emptyset'. \tag{5.4}$$

□

Exercise 5.2. (1) *Let $\emptyset$ be an empty set. Prove that for every set A, we have that*

$$\emptyset \subset A. \tag{5.5}$$

(2) *Use this to give another proof of Theorem* 5.5.

Since an empty set is unique, from now on we will refer to it as *the* empty set.

Exercise 5.3 (Part of nothing is nothing). *Prove that if a set A is a subset of $\emptyset$, then $A = \emptyset$.*

Question: What did the empty set say during an existential crisis?
Answer: Why is there nothing rather than something?

We say that a set A is **non-empty** if $A \neq \emptyset$. In particular, if A contains some element, call it x_0, then A is non-empty. We also observe that if B is a non-empty set, then $\emptyset$ is a proper subset of B.

Given the empty set $\emptyset$, we can define the set

$$\{\emptyset\}. \tag{5.6}$$

This set is non-empty! Indeed, it has a single element, the empty set, which is something! That is, in this case, nothing is something. ☺ We can also consider $\{\{\emptyset\}\}$, which is the set whose sole element *is* the set $\{\emptyset\}$.

To make sure we distinguish between similarly appearing notations, we consider the following.

Solved Problem 5.6. *For each of the following statements, either prove or disprove:*

(1) $\emptyset \in \emptyset$.
(2) $\emptyset \in \{\emptyset\}$.
(3) $\emptyset \subset \{\emptyset\}$.

Solution. (1) This statement is false. The empty set has no elements, but the statement $\emptyset \in \emptyset$ says that $\emptyset$ is an element of $\emptyset$.

(2) This statement is true. The set $\{\emptyset\}$ is the set with one element, where the one element is $\emptyset$. So, $\emptyset \in \{\emptyset\}$.

(3) This statement is true. This is because $\{\emptyset\}$ is a set and the empty set $\emptyset$ is a subset of any set.

By the set $\{\{\emptyset\}\}$ we mean the set with one element, where the one element is $\{\emptyset\}$. □

To try to further confound matters, we present:

Exercise 5.4 (Out of the void). *For each of the following statements, either prove or disprove:*

(i) $\{\emptyset\} \subset \{\emptyset\}$.
(ii) $\{\emptyset\} \in \{\emptyset\}$.
(iii) $\{\emptyset\} \in \{\{\emptyset\}\}$.
(iv) $\{\emptyset\} \subset \{\{\emptyset\}\}$.
(v) $\emptyset \subset \{\{\emptyset\}\}$.
(vi) $\emptyset \in \{\{\emptyset\}\}$.

5.1.6. Russell's Paradox.

We briefly discuss one of the paradoxes in the unrestrained consideration of sets of sets, as in §5.1.4. Russell's Paradox, which mildly boggles the mind, is the following. Let

R be the set of all sets that are not elements of themselves,

Figure 5.1.2. Bertrand Russell (1872–1970). Dutch National Archives, Author: Anefo, Creative Commons CC0 1.0 Universal Public Domain Dedication.

known as the "**Russell set**". In other words, we define

$$R := \{A : A \notin A\}, \tag{5.7}$$

where the object A to the left of the colon is a set. We have:

(1) If $A \in R$, then by the definition of R we have $A \notin A$.
(2) If $A \notin R$, then by the definition of R we have not $A \notin A$; that is $A \in A$.

Applying this to the set $A = R$, we obtain:

(1) If $R \in R$, then $R \notin R$.
(2) If $R \notin R$, then $R \in R$.

Thus, we have proved

$$R \in R \Leftrightarrow R \notin R. \tag{5.8}$$

In any case, we have a contradiction. So we conclude that the Russell set, despite its name, is not a set. We discussed the oxymoronic Russell set as a cautionary tale, the moral being that rigorous set theory must find a way to disallow such self-contradictory, fantastic beasts. However, we will not go into the nuances of set theory, as all of the objects we will consider will be well-defined.

5.1.7. Unions, intersections, and complements. Now we consider basic operations on sets. Let A and B be sets. We define their **union** $A \cup B$ to be the set whose elements are in A **or** B. In other words,

$$A \cup B = \{x \;:\; x \in A \text{ or } x \in B\}.$$

That is, an object x is an element of $A \cup B$ if and only if x is an element of A or x is an element of B.

We define the **intersection** $A \cap B$ of A and B to be the set whose elements are in A **and** B. In other words,

$$A \cap B = \{x \;:\; x \in A \text{ and } x \in B\}.$$

That is, an object x is an element of $A \cap B$ if and only if x is an element of A and x is an element of B. Figure 5.1.3 visualizes the union and intersection of two sets.

Example 5.7. Let E be the even integers and let O be the odd integers. We then have

$$E \cup O = \mathbb{Z} \quad \text{and} \quad E \cap O = \emptyset. \tag{5.9}$$

Example 5.8. Given an integer m, let

$$m\mathbb{Z} := \{mx \;:\; x \in \mathbb{Z}\} \tag{5.10}$$

denote the set of multiples of m. If a and b are non-zero integers, then

$$a\mathbb{Z} \cap b\mathbb{Z} = \operatorname{lcm}[a, b]\mathbb{Z}, \tag{5.11}$$

where lcm denotes the least common multiple. We leave this as an exercise for you to prove.

For example,

$$51\mathbb{Z} \cap 68\mathbb{Z} = 204\mathbb{Z}.$$

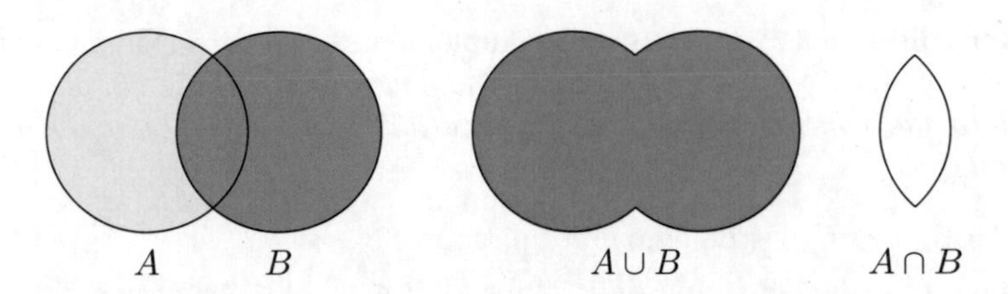

Figure 5.1.3. Two sets, their union, and their intersection.

When we fix a set U such that all sets in our discussion are subsets of it, we call U the **universal set** or **ambient set** or **domain of discourse**.

Caveat: By universal set, we do not mean the standard notion of universal set in set theory, where in that case the universal set is the set of all objects, including itself. This actually leads to Russell's Paradox.

If we fix a universal set U, the **complement** of a set A (A is assumed to be a subset of U) is the set

$$A^c = \{x \in U : x \notin A\}.$$

Figure 5.1.4 visualizes a set and its complement.

Example 5.9. Let $\mathbb{Z}$ be the domain of discourse. Then

$$E^c = O \quad \text{and} \quad O^c = E. \tag{5.12}$$

Example 5.10. Let $\mathbb{R}$ be the domain of discourse. Then $\mathbb{Q}^c$ is the set of irrational numbers.

Example 5.11. Let $\mathbb{Z}$ be the domain of discourse, and let m be a positive integer. Then $(m\mathbb{Z})^c$ is the set of integers that are not multiples of m.

Exercise 5.5 (Double negative). *Show the following:*

(1) $x \in A \Leftrightarrow x \notin A^c$.
(2) $(A^c)^c = A$.

Exercise 5.6 (Contrapositive). *For subsets of a universal set U, $A \subset B$ if and only if $B^c \subset A^c$.*

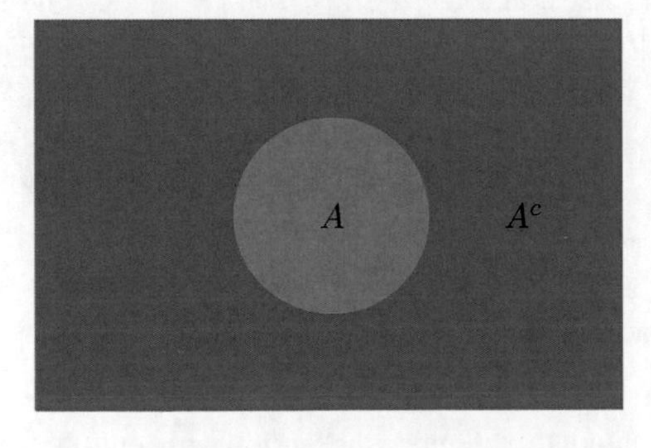

Figure 5.1.4. A set A and its complement A^c.

Solved Problem 5.12 (Subsets and complements). *In Exercise* 5.6 *you showed that for every set A and B, $A \subset B$ if and only if $B^c \subset A^c$. Explain how this is related to the contrapositive of an implication being logically equivalent to the implication.*

Solution. Let U be the domain of discourse.

We have that $A \subset B$ if and only if: for all $x \in U$, $x \in A$ implies $x \in B$.

We have that $B^c \subset A^c$ if and only if: for all $x \in U$, $x \in B^c$ implies $x \in A^c$. That is, for all $x \in U$, not $(x \in B)$ implies not $(x \in A)$. This implication is the contrapositive of the implication in the previous paragraph.

So, from the logical equivalence of an implication with its contrapositive, we see that $A \subset B$ if and only if $B^c \subset A^c$. □

Exercise 5.7 (Returning the "complement"). *Let U be the domain of discourse. For this exercise, do not use Lemma* 5.13 *below. But you may use that $(C^c)^c = C$ for every set C.*

(1) *Prove that a subset A satisfies $A^c = \emptyset$ if and only if $A = U$.*
(2) *Prove that a subset B satisfies $B^c = U$ if and only if $B = \emptyset$.*

Recall from Examples 5.7 and 5.9 that $E \cup O = \mathbb{Z}$, $E \cap O = \emptyset$, $E^c = O$, and $O^c = E$. This is an example of two sets **partitioning** a larger set. We can think of the universal set as a house and each of the sets as a room. The whole house consists only of the two rooms, and the two rooms have nothing in common. More generally, we have the following.

Lemma 5.13. *Let A be a subset of a universal set U. Then:*

(1) $A \cup A^c = U$.
(2) $A \cap A^c = \emptyset$.

Hence, A and A^c partition U.

Exercise 5.8. *Prove Lemma* 5.13. *Do not use the De Morgan laws, which will be discussed below.*

5.1.8. The set difference. So far, we have introduced the operations on sets of unions, intersections, and complements. We now consider a derivative operation. Let U be the universal set, and let A and B be two sets. The **set difference** of A minus B is defined by

$$A - B := A \cap B^c. \tag{5.13}$$

That is, $A - B$ is the set of objects in A that are not also in B. Observe that

$$A - B \subset A \quad \text{and} \quad A - B \subset B^c. \tag{5.14}$$

We may visualize the set difference by Figure 5.1.5.

Example 5.14. For any set A, we have

$$A - A^c = A.$$

The properties discussed in the next subsection will enable you to prove this fact easily. For now, we give a direct proof. Let $x \in A - A^c$. Then $x \in A$ and $x \notin A^c$.

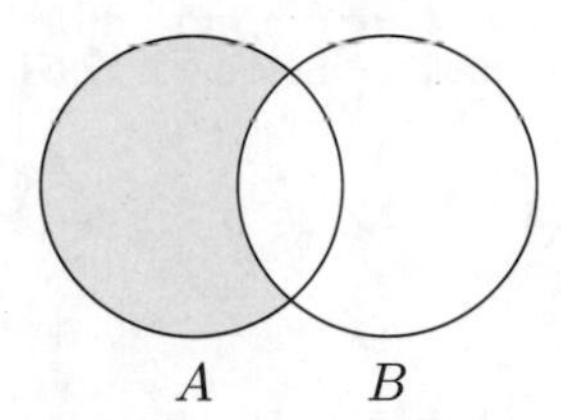

Figure 5.1.5. The set difference $A - B$.

Thus $x \in A$. This proves that $A - A^c \subset A$. Conversely, let $x \in A$. Then $x \in (A^c)^c$, which implies $x \notin A^c$. Hence $x \in A - A^c$. The desired set equality follows.

Exercise 5.9. *Let U be the universal set, and let A and B be sets. Prove the following set equalities:*

(1) $U - B = B^c$.

(2) $A - \emptyset = A$.

(3) $A - A = \emptyset$.

(4) $\emptyset - B = \emptyset$.

(5) $A - U = \emptyset$.

5.1.9. Set operation properties. Given the set operations we have introduced, we can easily prove some basic properties.

Lemma 5.15. *For any sets A and B,*

$$A \cap B \subset A \subset A \cup B. \tag{5.15}$$

Proof. Let $x \in A \cap B$. Then $x \in A$ and $x \in B$. This implies that $x \in A$. Thus we have proved that $A \cap B \subset A$.

Let $x \in A$. Then $x \in A$ or $x \in B$. This says that $x \in A \cup B$. Thus we have proved that $A \subset A \cup B$. □

Figure 5.1.6 visualizes (as Venn diagrams) the set inclusions in (5.15).

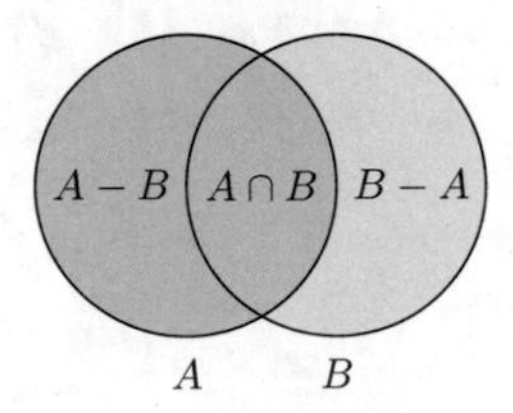

Figure 5.1.6. From this Venn diagram we can see that $A \cap B \subset A \subset A \cup B$, $A - B \subset A$, and $A - B \subset B^c$.

The equality cases of the subset inclusions of the lemma above are characterized by:

Lemma 5.16.

(1) $A \cap B = A \Leftrightarrow A \subset B$.

(2) $A = A \cup B \Leftrightarrow B \subset A$.

Proof. (1) We prove the $\Rightarrow$ direction of the biconditional, and we leave the $\Leftarrow$ direction as an exercise. Suppose $A \cap B = A$. This implies that $A \subset A \cap B \subset B$.

(2) Again, we prove the $\Rightarrow$ direction of the biconditional, and we leave the $\Leftarrow$ direction as an exercise. Suppose $A = A \cup B$. This implies that $B \subset A \cup B \subset A$. □

Note that Lemma 5.16(1) is characterized by the pink region in Figure 5.1.6 having no elements, whereas Lemma 5.16(2) is characterized by the light blue region in Figure 5.1.6 having no elements.

Exercise 5.10. *Let A and B be sets. Prove:*

(1) $A \subset B \Rightarrow A \cap B = A$.

(2) $B \subset A \Rightarrow A = A \cup B$.

This completes the proof of Lemma 5.16.

Note that, as special cases, we have

$$A \cap A = A, \qquad A = A \cup A. \tag{5.16}$$

Exercise 5.11. *Let A and B be sets. Prove that*

$$A \cup B = A \cap B \quad \Leftrightarrow \quad A = B.$$

Regarding the interaction between complements and intersections and unions, the **De Morgan laws** are:

Lemma 5.17. *Let A and B be sets in a universal set U. Then*

$$(A \cap B)^c = A^c \cup B^c, \tag{5.17}$$

$$(A \cup B)^c = A^c \cap B^c. \tag{5.18}$$

Proof. We prove (5.17), and we leave (5.18) as an exercise. We have the following string of biconditionals:

$$\begin{aligned} x \in (A \cap B)^c &\Leftrightarrow x \notin A \cap B \\ &\Leftrightarrow \text{not } (x \in A \cap B) \\ &\Leftrightarrow \text{not } (x \in A \text{ and } x \in B) \\ &\Leftrightarrow \text{not } (x \in A) \text{ or } \text{ not } (x \in B) \\ &\Leftrightarrow x \in A^c \text{ or } x \in B^c \\ &\Leftrightarrow x \in A^c \cup B^c. \end{aligned}$$

□

Figure 3.6.3 visualizes De Morgan's laws.

Exercise 5.12. *Prove $(A \cup B)^c = A^c \cap B^c$. This proves* (5.18) *and hence completes the proof of Lemma* 5.17.

Solved Problem 5.18 (De Morgan laws for logic and sets). *Explain how the De Morgan laws in Lemma* 5.17 *are related to the De Morgan laws in Solved Problem* 3.32.

Solution. We see the relationship by defining

$$P(x) \text{ to be the statement } x \in A \quad \text{and} \quad Q(x) \text{ to be the statement } x \in B.$$

The law $(A \cap B)^c = A^c \cup B^c$ says

$$x \in (A \cap B)^c \quad \text{if and only if} \quad x \in A^c \cup B^c;$$

that is,

$$\text{not } (x \in A \text{ and } x \in B) \quad \text{if and only if} \quad (\text{not } x \in A) \text{ or } (\text{not } x \in B)\ ,$$

or in other words (think about this more abstractly),

$$\text{not } \big(P(x) \text{ and } Q(x)\big) \quad \text{if and only if} \quad \big(\text{not } P(x)\big) \text{ or } \big(\text{not } Q(x)\big)\ .$$

This last statement is one of the two De Morgan laws in Solved Problem 3.32.

Similarly, we see that the law $(A \cup B)^c = A^c \cap B^c$ is equivalent to the biconditional

$$\text{not } \big(P(x) \text{ or } Q(x)\big) \quad \text{if and only if} \quad \big(\text{not } P(x)\big) \text{ and } \big(\text{not } Q(x)\big)$$

is the other De Morgan law in Solved Problem 3.32. □

5.1.10. How unions and intersections interact. We now consider properties restricted to only unions and intersections. It is rather easy to see that both unions and intersections are **commutative**; that is,

$$A \cup B = B \cup A, \qquad A \cap B = B \cap A. \tag{5.19}$$

They are also **associative**:

$$(A \cup B) \cup C = A \cup (B \cup C), \qquad (A \cap B) \cap C = A \cap (B \cap C). \tag{5.20}$$

For this reason we often write the sets above just as

$$A \cup B \cup C, \qquad A \cap B \cap C. \tag{5.21}$$

We also have the following **distributive properties**:

$$A \cup (B \cap C) = (A \cup B) \cap (A \cup C), \tag{5.22a}$$

$$A \cap (B \cup C) = (A \cap B) \cup (A \cap C). \tag{5.22b}$$

We don't bother to prove the properties above but rather leave the details to the interested reader. Don't worry! We allow you to use these properties even though we haven't proved them! You can use Figure 5.1.7 to visualize why these properties are true.

Lemma 5.19. *If A_1, A_2, B_1, B_2 are sets, then*

$$\big(A_1 \cup A_2\big) \cap \big(B_1 \cup B_2\big) = \big(A_1 \cap B_1\big) \cup \big(A_1 \cap B_2\big) \cup \big(A_2 \cap B_1\big) \cup \big(A_2 \cap B_2\big). \tag{5.23}$$

Exercise 5.13. *Prove Lemma* 5.19.

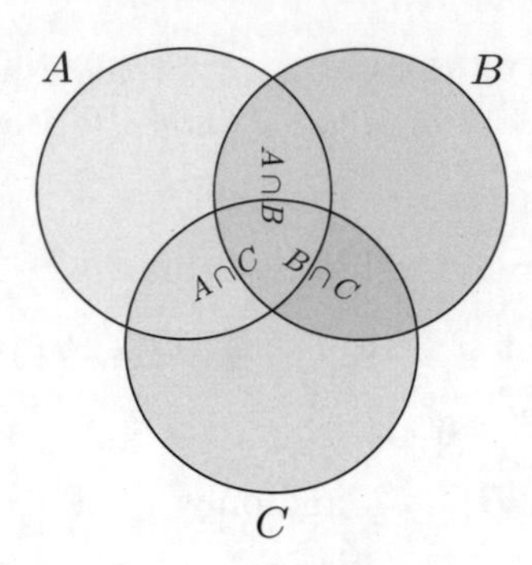

Figure 5.1.7. Venn diagram for the union of sets A, B, C.

5.2. Cartesian products of sets

We now discuss a set operation that impacts how we fundamentally view the universe, classically, as Euclidean space.

Let X and Y be sets. Their **cartesian product** is the set defined by

$$X \times Y := \{(x, y) : x \in X,\ y \in Y\}. \tag{5.24}$$

Here (x, y) denotes an **ordered pair**, so that $(x, y) = (z, w)$ if and only if $x = z$ and $y = w$. We denote

$$X^2 := X \times X. \tag{5.25}$$

Example 5.20. Let $X = \{a, b, c, d\}$ and $Y = \{1, 2, 3\}$. Then their cartesian product is given by

$$X \times Y = \{(a,1),(a,2),(a,3),(b,1),(b,2),(b,3),(c,1),(c,2),(c,3),(d,1),(d,2),(d,3)\}.$$

We can picture the cartesian product as in Figure 5.2.1.

$(a,3)$	$(b,3)$	$(c,3)$	$(d,3)$
$(a,2)$	$(b,2)$	$(c,2)$	$(d,2)$
$(a,1)$	$(b,1)$	$(c,1)$	$(d,1)$

$X \times Y$

Figure 5.2.1. Visualization of the cartesian product of $X = \{a, b, c, d\}$ and $Y = \{1, 2, 3\}$.

We can take the cartesian product of more than two sets. Given sets X, Y, Z, observe that there is a natural bijection

$$b : (X \times Y) \times Z \to X \times (Y \times Z) \tag{5.26}$$

defined by

$$b\big((x, y), z\big) := \big(x, (y, z)\big) \quad \text{for } x \in X,\ y \in Y,\ z \in Z. \tag{5.27}$$

So, instead of distinguishing between the two versions of the triple cartesian product, we simply write either as $X \times Y \times Z$.

Actually, instead of taking cartesian products two sets at a time and appealing to the "associativity" of the cartesian product, we do the following. Let $X_1, X_2, \ldots, X_n$ be sets, where n is an integer greater than 1. Their n-**fold cartesian product** is defined by

$$\prod_{i=1}^{n} X_i = X_1 \times X_2 \times \cdots \times X_n := \{(x_1, x_2, \ldots, x_n) : x_i \in X_i \text{ for } 1 \leq i \leq n\}. \tag{5.28}$$

Here, $(x_1, x_2, \ldots, x_n)$ is called an **ordered** n-**tuple**. We define two n-tuples $(x_1, x_2, \ldots, x_n)$ and $(y_1, y_2, \ldots, y_n)$ to be equal if and only if $x_i = y_i$ for all $1 \leq i \leq n$. Given a set X, we denote

$$X^n := \underbrace{X \times X \times \cdots \times X}_{n \text{ times}}. \tag{5.29}$$

We can visualize triple cartesian products since we live in a 3-dimensional world, but it is a stretch to visualize cartesian products of 4 or more sets.

Remark 5.21 (Thinking decimally). We can even consider cartesian products of an infinite number of sets. For example, recall from (4.11) that

$$R_{10} = \{0, 1, 2, 3, 4, 5, 6, 7, 8, 9\}$$

is the set of remainders for division by 10. Let $X_i = R_{10}$ for $i \in \mathbb{Z}^+$. Then we can take the countably infinite cartesian product of R_{10} with itself; we define:

$$(R_{10})^{\mathbb{Z}^+} := \prod_{i \in \mathbb{Z}^+} X_i := \left\{ f : \mathbb{Z}^+ \to \bigcup_{i \in \mathbb{Z}^+} X_i : f(i) \in X_i \text{ for all } i \in \mathbb{Z}^+ \right\}. \tag{5.30}$$

In other words, an element of $(R_{10})^{\mathbb{Z}^+}$ is an infinite sequence of digits

$$f(1), f(2), f(3), \ldots.$$

How is this set similar to the unit interval $[0, 1]$ of real numbers? How is this different from $[0, 1]$? Hint: A real number may have more than one decimal expansion. For example, $0.1 = 0.0\bar{9}$.

5.2.1. Euclidean space. In calculus, we consider 1-, 2-, and 3-dimensional Euclidean spaces. The n-dimensional Euclidean space can be concretely defined as

$$\mathbb{R}^n = \{(x_1, x_2, \ldots, x_n) : x_i \in \mathbb{R},\ 1 \leq i \leq n\}. \tag{5.31}$$

This is the n-fold cartesian product of $\mathbb{R}$ with itself. We also call $\mathbb{R}^n$ **Euclidean** n-**space**.

Sets may be endowed with additional structures, such as algebraic or geometric structures for example. Euclidean space naturally has a vector space structure (we assume that you are familiar with vector spaces) where **vector addition** is defined by: For $\mathbf{x} = (x_1, x_2, \ldots, x_n)$ and $\mathbf{y} = (y_1, y_2, \ldots, y_n)$,

$$\mathbf{x} + \mathbf{y} := (x_1 + y_1, x_2 + y_2, \ldots, x_n + y_n), \tag{5.32}$$

and **scalar multiplication** is defined by: For $c \in \mathbb{R}$,

$$c\mathbf{x} := (cx_1, cx_2, \ldots, cx_n). \tag{5.33}$$

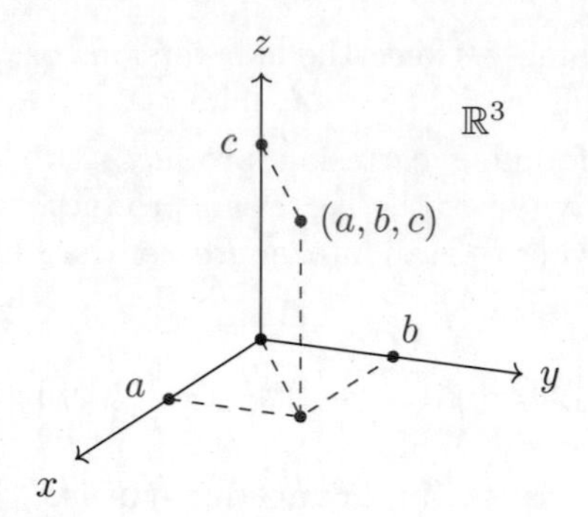

Figure 5.2.2. A visualization of $\mathbb{R}^3 = \mathbb{R} \times \mathbb{R} \times \mathbb{R}$. The dot at the intersection of the axes is the origin $(0, 0, 0)$. Each axis represents a ray emanating from the origin. For example, the z-axis is the set $\{(0, 0, z) : z \geq 0\}$. An element (a, b, c) is visualized.

See the "Vector space" Wikipedia link for visualizations of vector addition and scalar multiplication.

It is easy to see that Euclidean space, with this vector addition and scalar multiplication, satisfies the vector space axioms. Observe that we can consider vector addition as a function from $\mathbb{R}^n \times \mathbb{R}^n$ to $\mathbb{R}^n$, and scalar multiplication as a function from $\mathbb{R} \times \mathbb{R}^n$ to $\mathbb{R}^n$. We discuss functions in more generality in §5.3 below.

5.2.2. Euclidean dot product and norm. Euclidean space is usually endowed with its naturally defined Euclidean dot product, given by

$$\mathbf{x} \cdot \mathbf{y} := x_1 y_1 + x_2 y_2 + \cdots + x_n y_n. \tag{5.34}$$

As we will see below, this is a geometric structure in the guise of an algebraic structure.

Lemma 5.22. *The Euclidean dot product satisfies the following properties:*

(1) $\mathbf{x} \cdot \mathbf{y} = \mathbf{y} \cdot \mathbf{x}$.

(2) $(c\mathbf{x}) \cdot \mathbf{y} = c\,(\mathbf{y} \cdot \mathbf{x})$.

(3) $(\mathbf{x} + \mathbf{y}) \cdot \mathbf{z} = \mathbf{x} \cdot \mathbf{z} + \mathbf{y} \cdot \mathbf{z}$.

Because of these properties, the Euclidean dot product is called a symmetric, bilinear form. Namely, property (1) *is called* ***symmetry****, and in the presence of symmetry, properties* (2) *and* (3) *combined are called* ***bilinearity****.*

Exercise 5.14. *Prove Lemma* 5.22.

The **norm** (or **absolute value**) is defined by

$$|\mathbf{x}| := \sqrt{\mathbf{x} \cdot \mathbf{x}} = \sqrt{x_1^2 + x_2^2 + \cdots + x_n^2}. \tag{5.35}$$

With the norm, we can define the distance between two points as

$$\operatorname{dist}(\mathbf{x}, \mathbf{y}) := |\mathbf{x} - \mathbf{y}|. \tag{5.36}$$

We leave it to you, dear reader, to check that when $n = 2$, this definition of distance follows from the Pythagorean Theorem 3.4.

A more abstract way to define Euclidean n-space is as an n-dimensional vector space with a dot product. We will not go down this route:

> Trinity: "Because you have been down there, Neo. You know that road. You know exactly where it ends. And I know that's not where you want to be."

If, on the other hand, you wish go down this interesting and important route, we note that there are many excellent books on linear algebra.

Exercise 5.15. *Let r be a positive real number, and let $A = [-r, r] \times [-r, r]$ be a square (with side lengths $2r$) in $\mathbb{R}^2$. Let $B = \{(x, y) \in \mathbb{R}^2 : x^2 + y^2 \leq 2r^2\}$. Prove that*

$$A \subset B. \tag{5.37}$$

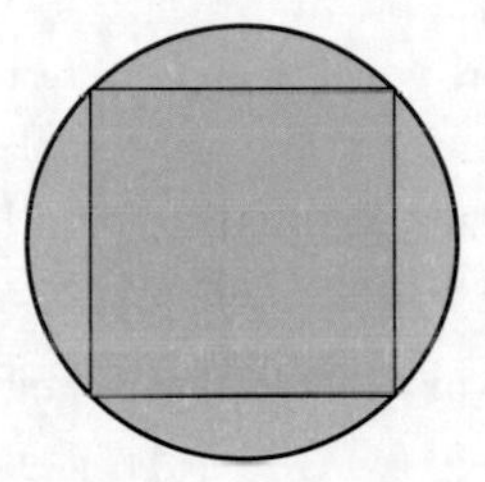

Figure 5.2.3. A square peg in a round hole.

5.2.3. Euclidean vector projection.

Solved Problem 5.23 (Euclidean vector projection)**.** *Let $\mathbf{w}$ be a non-zero vector in $\mathbb{R}^n$ and let $\mathbf{v}$ be a vector in $\mathbb{R}^n$. Derive the formula for the vector projection of $\mathbf{v}$ onto $\mathbf{w}$.*

Solution. The vector projection $\mathrm{proj}_{\mathbf{w}}(\mathbf{v})$ of $\mathbf{v}$ onto $\mathbf{w}$ is

$$c\mathbf{w}, \tag{5.38}$$

where $c \in \mathbb{R}$ is determined by the equation

$$(\mathbf{v} - c\mathbf{w}) \cdot \mathbf{w} = 0, \tag{5.39}$$

which says that the vector $\mathbf{v} - \mathrm{proj}_{\mathbf{w}}(\mathbf{v})$ is orthogonal (perpendicular) to the vector $\mathbf{w}$ (suggestion: draw a picture!). That is,

$$c = \frac{\mathbf{v} \cdot \mathbf{w}}{|\mathbf{w}|^2}. \tag{5.40}$$

Thus the vector projection is

$$\mathrm{proj}_{\mathbf{w}}(\mathbf{v}) = \frac{\mathbf{v} \cdot \mathbf{w}}{|\mathbf{w}|^2}\mathbf{w}. \tag{5.41}$$

See Figure 5.2.4 for a visualization of the vector projection.

Observe that in terms of the unit vector $\mathbf{u} := \frac{\mathbf{w}}{|\mathbf{w}|}$, this says that

$$\mathrm{proj}_{\mathbf{w}}(\mathbf{v}) = \mathrm{proj}_{\mathbf{u}}(\mathbf{v}) = (\mathbf{v} \cdot \mathbf{u})\mathbf{u}. \qquad \square \tag{5.42}$$

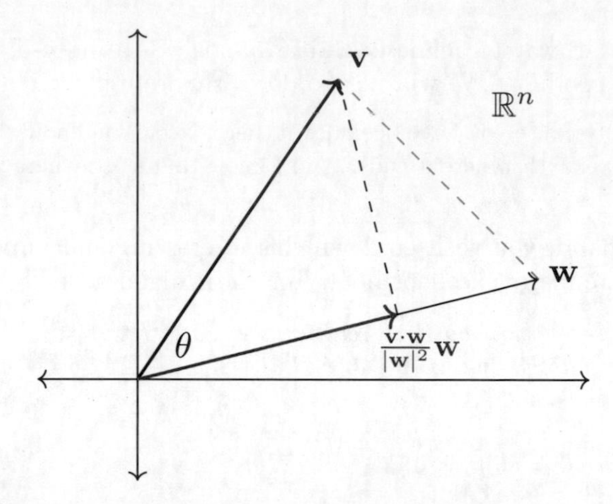

Figure 5.2.4. The vectors $\mathbf{v}$ and $\mathbf{w}$ and the vector projection of $\mathbf{v}$ onto $\mathbf{w}$.

5.2.4. The Cauchy–Schwarz inequality. We have the following fundamental inequality for vectors in $\mathbb{R}^n$ regarding their lengths and dot products.

Theorem 5.24 (Cauchy–Schwarz inequality). *For any* $\mathbf{x}, \mathbf{y} \in \mathbb{R}^n$, *we have*

$$|\mathbf{x} \cdot \mathbf{y}| \le |\mathbf{x}|\,|\mathbf{y}|. \tag{5.43}$$

Equality holds in (5.43) *if and only if one of the vectors* $\mathbf{x}$ *or* $\mathbf{y}$ *is a multiple of the other, i.e., if and only if they are linearly dependent.*

Proof. If $\mathbf{y} = \mathbf{0}$, then the theorem is easy to prove. So, without loss of generality, we may assume that $\mathbf{y} \neq \mathbf{0}$. Let $\lambda \in \mathbb{R}$ be a real number to be chosen below. Whatever λ is, we have

$$\begin{aligned} 0 &\le |\mathbf{x} - \lambda\mathbf{y}|^2 \\ &= (\mathbf{x} - \lambda\mathbf{y}) \cdot (\mathbf{x} - \lambda\mathbf{y}) \\ &= \mathbf{x}\cdot\mathbf{x} + \lambda\mathbf{y}\cdot\lambda\mathbf{y} - \mathbf{x}\cdot\lambda\mathbf{y} - \lambda\mathbf{y}\cdot\mathbf{x} \\ &= |\mathbf{x}|^2 + \lambda^2|\mathbf{y}|^2 - 2\lambda\,\mathbf{x}\cdot\mathbf{y}. \end{aligned} \tag{5.44}$$

The vector projection of $\mathbf{x}$ onto $\mathbf{y}$ is equal to $\frac{\mathbf{x}\cdot\mathbf{y}}{|\mathbf{y}|^2}\mathbf{y}$. Motivated by this, we choose

$$\lambda := \frac{\mathbf{x}\cdot\mathbf{y}}{|\mathbf{y}|^2}, \tag{5.45}$$

which is well-defined since $|\mathbf{y}| \neq 0$. With this choice of λ, the quantity $\mathbf{x} - \lambda\mathbf{y}$ in (5.44) is equal to the vector $\mathbf{x}$ minus its orthogonal projection onto the vector $\mathbf{y}$. Furthermore, inequality (5.44) becomes

$$0 \le |\mathbf{x}|^2 + \left(\frac{\mathbf{x}\cdot\mathbf{y}}{|\mathbf{y}|^2}\right)^2 |\mathbf{y}|^2 - 2\frac{\mathbf{x}\cdot\mathbf{y}}{|\mathbf{y}|^2}\,\mathbf{x}\cdot\mathbf{y} = |\mathbf{x}|^2 - \frac{(\mathbf{x}\cdot\mathbf{y})^2}{|\mathbf{y}|^2}.$$

Therefore

$$(\mathbf{x}\cdot\mathbf{y})^2 \le |\mathbf{x}|^2|\mathbf{y}|^2. \tag{5.46}$$

Inequality (5.43) follows from taking square roots of both non-negative sides.

Finally, we characterize the equality case. By the discussion above, equality holds in (5.44) if and only if we have equality in (5.46). This, in turn, is equivalent to equality in (5.43). Now, equality in (5.44) holds if and only if either $\mathbf{y} = 0$ or $\mathbf{x}$ is a scalar multiple of $\mathbf{y}$. This completes the proof of the theorem. □

In what sense have we been an architect (in the sense of §3.1.4) for the proof above?

Let θ be the angle between non-zero vectors $\mathbf{x}$ and $\mathbf{y}$. Since the vector projection of $\mathbf{x}$ onto $\mathbf{y}$ has length $\frac{\mathbf{x}\cdot\mathbf{y}}{|\mathbf{y}|}$, we have

$$\cos(\theta) = \frac{\frac{\mathbf{x}\cdot\mathbf{y}}{|\mathbf{y}|}}{|\mathbf{x}|} = \frac{\mathbf{x}\cdot\mathbf{y}}{|\mathbf{x}||\mathbf{y}|}. \tag{5.47}$$

That is,

$$\theta = \cos^{-1}\left(\frac{\mathbf{x}\cdot\mathbf{y}}{|\mathbf{x}||\mathbf{y}|}\right). \tag{5.48}$$

Alternatively, we can use the law of cosines to derive (5.47). We have a triangle in Figure 5.2.4 with side lengths $|\mathbf{x}|$, $|\mathbf{y}|$, $|\mathbf{x}-\mathbf{y}|$ and angle θ. By the law of cosines for this triangle, we have

$$|\mathbf{x}-\mathbf{y}|^2 = |\mathbf{x}|^2 + |\mathbf{y}|^2 - 2|\mathbf{x}||\mathbf{y}|\cos(\theta). \tag{5.49}$$

Exercise 5.16. *Rederive* (5.47) *from equality* (5.49).

A fundamentally important consequence of the Cauchy–Schwarz inequality is:

Corollary 5.25 (Triangle inequality). *For any* $\mathbf{x},\mathbf{y} \in \mathbb{R}^n$, *we have*

$$|\mathbf{x}+\mathbf{y}| \le |\mathbf{x}| + |\mathbf{y}|. \tag{5.50}$$

Exercise 5.17. *Prove Corollary* 5.25. *Hint: Square both sides and expand.*

The reason why (5.50) is called the triangle inequality is that it implies for every $\mathbf{a},\mathbf{b},\mathbf{c} \in \mathbb{R}^n$,

$$\mathrm{dist}(\mathbf{a},\mathbf{c}) \le \mathrm{dist}(\mathbf{a},\mathbf{b}) + \mathrm{dist}(\mathbf{b},\mathbf{c}), \tag{5.51}$$

where distance is defined by (5.36). We leave it to you, dear reader, to prove this. In other words, going straight from point $\mathbf{a}$ to point $\mathbf{c}$ is at least as short as going straight from $\mathbf{a}$ to $\mathbf{b}$ and then going straight from $\mathbf{b}$ to $\mathbf{c}$.

5.2.5. Planes. With respect to the vector space structure of $\mathbb{R}^n$, we can define special subsets. We say that a subset L of $\mathbb{R}^n$ is a linear subspace (a.k.a. **vector subspace**) provided it has the property that if $\mathbf{x},\mathbf{y} \in L$ and $a, b \in \mathbb{R}$, then

$$a\mathbf{x} + b\mathbf{y} \in L. \tag{5.52}$$

Given $\mathbf{x} \in \mathbb{R}^n$, the line passing through $\mathbf{x}$ and the origin $\mathbf{0}$ is the set

$$\ell_{\mathbf{x}} := \{t\mathbf{x} : t \in \mathbb{R}\}. \tag{5.53}$$

Observe that if L is a linear subspace of $\mathbb{R}^n$ and if $\mathbf{x} \in L$, then $\ell_{\mathbf{x}} \subset L$.

It turns out that if L is a linear subspace of $\mathbb{R}^n$, then L with the induced vector addition and scalar multiplication is itself a vector space of dimension between 0

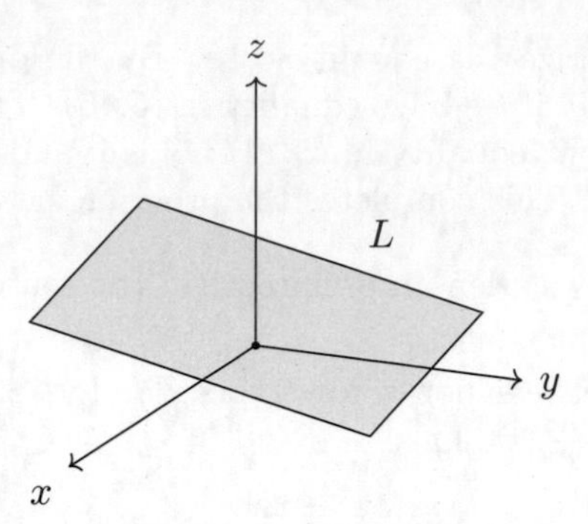

Figure 5.2.5. A linear subspace is a "plane" passing through the origin.

and n, inclusive. The only subspace of dimension 0 is $\{\mathbf{0}\}$, and the only subspace of dimension n is $\mathbb{R}^n$.

Exercise 5.18. *Let $\ell : \mathbb{R}^n \to \mathbb{R}$ be a linear function. Prove that*

$$\ker(\ell) := \{\mathbf{x} \in \mathbb{R}^n : \ell(\mathbf{x}) = 0\} \tag{5.54}$$

is a linear subspace.

We say that a subset A of $\mathbb{R}^n$ is an affine subspace if *there exists* $\mathbf{a} \in A$ such that the subset

$$A_{\mathbf{a}} := \{\mathbf{x} - \mathbf{a} : \mathbf{x} \in A\} \tag{5.55}$$

is a linear subspace of $\mathbb{R}^n$.

We see that a linear subspace is necessarily an affine subspace since if L is a linear subspace, then $L_{\mathbf{0}} = L$, where $\mathbf{0}$ is the origin. We won't go into detail about the linear algebra of affine subspaces, but an affine subspace A of $\mathbb{R}^n$ has a dimension d, where $0 \leq d \leq n$ is an integer. We may think of A as a "copy" of $\mathbb{R}^d$ in $\mathbb{R}^n$. In other words, A is a d-dimensional plane in $\mathbb{R}^n$. Now, linear subspaces must contain the origin (see Figure 5.2.5), whereas affine subspaces may not necessarily contain the origin. See Figure 5.2.6.

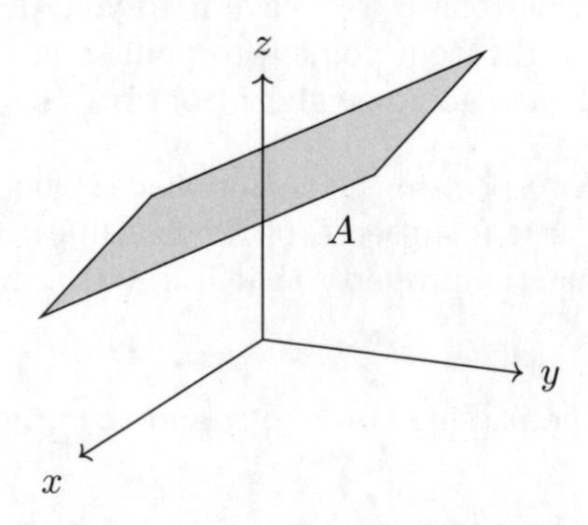

Figure 5.2.6. An affine space is a "plane" which may or may not pass through the origin.

Exercise 5.19. *Show that, given a non-zero vector* $\mathbf{v} \in \mathbb{R}^n$ *and* $c \in \mathbb{R}$,

$$\{\mathbf{x} \in \mathbb{R}^n : \mathbf{x} \cdot \mathbf{v} = c\} \tag{5.56}$$

is an affine subspace.

Explain why it is a linear subspace if and only if $c = 0$.

Lemma 5.26. *A subset* A *of* $\mathbb{R}^n$ *is an affine subspace if and only if* for all $\mathbf{a} \in A$, *the subset* $A_{\mathbf{a}} = \{\mathbf{x} - \mathbf{a} : \mathbf{x} \in A\}$ *is a linear subspace of* $\mathbb{R}^n$.

Exercise 5.20. *Prove Lemma* 5.26. *Note that the difference is the replacement of the existential quantifier "there exists" with the universal quantifier "for all".*

Lemma 5.27. *If* L *is a linear subspace of* $\mathbb{R}^n$ *and if* $\mathbf{x}_0 \in \mathbb{R}^n$, *then*

$$L + \mathbf{x}_0 := \{\mathbf{x} + \mathbf{x}_0 : \mathbf{x} \in L\} \tag{5.57}$$

is an affine subspace of $\mathbb{R}^n$.

Exercise 5.21. *Prove Lemma* 5.27.

5.2.6. Spheres. In addition to special linear subsets of Euclidean space, we have special non-linear subsets.

The $(n-1)$-dimensional **sphere** of radius $r \in \mathbb{R}^+$ centered at the origin $\mathbf{0}$ in $\mathbb{R}^n$ is defined by

$$S^{n-1}(r) := \{\mathbf{x} \in \mathbb{R}^n : |\mathbf{x}| = r\}. \tag{5.58}$$

We also call $S^{n-1}(r)$ the $(n-1)$-sphere of radius r.

Writing a point in terms of its Euclidean coordinates as

$$\mathbf{x} =: (x_1, x_2, \ldots, x_n), \tag{5.59}$$

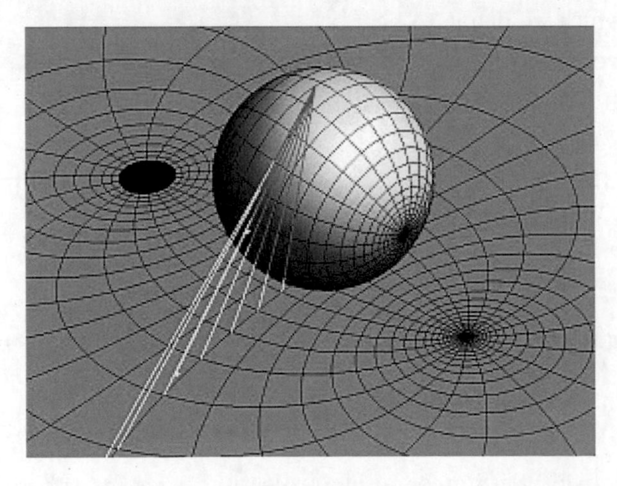

Figure 5.2.7. A 2-sphere. The curves in the plane are the images under stereographic projection (see §5.2.7 below) of the curves in the sphere. From Virtual Mathematics Museum by the 3D-XplorMath Consortium. Public Domain.

we have that

$$S^{n-1}(r) = \{(x_1, x_2, \ldots, x_n) \in \mathbb{R}^n : x_1^2 + x_2^2 + \cdots + x_n^2 = r^2\}, \tag{5.60}$$

where we denote $x_i^2 := (x_i)^2$.

In low dimensions, we often use the indexless coordinates $x, y, z, \ldots$. So the 1-sphere, a.k.a. the **circle**, of radius r is given by

$$S^1(r) = \{(x, y) \in \mathbb{R}^2 : x^2 + y^2 = r^2\}, \tag{5.61}$$

and the 2-sphere, a.k.a. the **sphere**, of radius r is given by

$$S^2(r) = \{(x, y, z) \in \mathbb{R}^3 : x^2 + y^2 + z^2 = r^2\}. \tag{5.62}$$

The **unit** $(n-1)$-sphere is defined to be the $(n-1)$-sphere of radius 1. We use the notation

$$S^{n-1} := S^{n-1}(1). \tag{5.63}$$

5.2.7. Stereographic projection. An interesting and useful function defined on spheres is stereographic projection, which we now describe. Functions, in more generality, will be discussed in the next section.

Let S^{n-1} be the unit $(n-1)$-sphere as defined in (5.63). Let $\mathbf{n} = (0, \ldots, 0, 1) \in S^{n-1}$ be the "north pole". Let L be the hyperplane ($(n-1)$-dimensional subspace of $\mathbb{R}^n$) defined by

$$L = \{\mathbf{y} \in \mathbb{R}^n : \mathbf{y} \cdot \mathbf{n} = -1\} = \{(\mathbf{z}, -1) : \mathbf{z} \in \mathbb{R}^{n-1}\} = \mathbb{R}^{n-1} \times \{-1\}. \tag{5.64}$$

Then L is tangent to S^{n-1} at the south pole $\mathbf{s} := -\mathbf{n} = (0, \ldots, 0, -1)$. Note also that another way to express L is as the set of points $\mathbf{x} = (x_1, x_2, \ldots, x_n)$ in $\mathbb{R}^n$ satisfying $x_n = -1$.

We will define stereographic projection f from the sphere minus a point $S^{n-1} - \{\mathbf{n}\}$ to the hyperplane L. The idea is that given a point $\mathbf{x}$ on the sphere minus the point $\mathbf{n}$, the unique line passing through $\mathbf{n}$ and $\mathbf{x}$ intersects the hyperplane L at a unique point, which we define to be $f(\mathbf{x})$.

Define **stereographic projection** $f : S^{n-1} - \{\mathbf{n}\} \to L$ by

$$f(\mathbf{x}) = \frac{2\mathbf{x} - (1 + \mathbf{x} \cdot \mathbf{n})\,\mathbf{n}}{1 - \mathbf{x} \cdot \mathbf{n}} \tag{5.65}$$

for $\mathbf{x} \in S^{n-1} - \{\mathbf{n}\}$. See Figure 5.2.8 for a visualization of stereographic projection. In this picture, we have $n = 3$ and

$$f(x, y, z) = \left(\frac{2x}{1-z}, \frac{2y}{1-z}, -1 \right). \tag{5.66}$$

Observe that $f(\mathbf{s}) = \mathbf{s}$ since $\mathbf{s} \cdot \mathbf{n} = -1$. The equator of S^{n-1} consists of those points $\mathbf{x} \in S^{n-1}$ satisfying $\mathbf{x} \cdot \mathbf{n} = 0$. For such points, $f(\mathbf{x}) = 2\mathbf{x} - \mathbf{n}$. So f maps the equator (which is a circle of radius 1) to a circle of radius 2 in the hyperplane L.

Solved Problem 5.28 (Stereographic projection from the sphere to the plane). *Derive formula* (5.65) *for stereographic projection. That is, show for any* $\mathbf{x} \in S^{n-1} - \{\mathbf{n}\}$ *that* $f(\mathbf{x})$ *defined by* (5.65) *is the unique point in the intersection of* L *and the line passing through* $\mathbf{n}$ *and* $\mathbf{x}$.

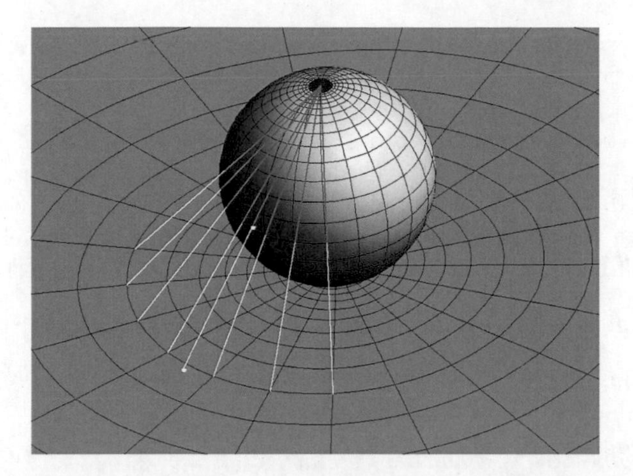

Figure 5.2.8. Visualizing stereographic projection. Meridians and parallels are mapped to lines and circles, respectively. From the Virtual Mathematics Museum by the 3D-XplorMath Consortium. Public Domain.

Solution. Let us derive the formula for the stereographic projection $f(\mathbf{x})$ of a point $\mathbf{x}$. This is equal to the point $\mathbf{y}$ in the hyperplane L which also lies on the unique line determined by $\mathbf{n}$ and $\mathbf{x}$. So

$$\mathbf{y} = (1-t)\mathbf{n} + t\mathbf{x}, \tag{5.67}$$

where t is determined by the equation $\mathbf{y} \in L$; that is,

$$-1 = \mathbf{y} \cdot \mathbf{n} = (1-t) + t\mathbf{x} \cdot \mathbf{n}. \tag{5.68}$$

Hence

$$t = \frac{2}{1 - \mathbf{x} \cdot \mathbf{n}}, \quad \text{so that} \quad 1 - t = -\frac{1 + \mathbf{x} \cdot \mathbf{n}}{1 - \mathbf{x} \cdot \mathbf{n}}. \tag{5.69}$$

We then conclude that

$$f(\mathbf{x}) = \mathbf{y} = -\frac{1 + \mathbf{x} \cdot \mathbf{n}}{1 - \mathbf{x} \cdot \mathbf{n}}\mathbf{n} + \frac{2}{1 - \mathbf{x} \cdot \mathbf{n}}\mathbf{x}. \tag{5.70}$$

Equation (5.65) follows easily from this. □

Exercise 5.22 (Stereographic projection). *Let $f : S^{n-1} - \{\mathbf{n}\} \to L$ be stereographic projection.*

(i) *Prove that $f(\mathbf{x}) \cdot \mathbf{n} = -1$ for all $\mathbf{x} \in S^{n-1} - \{\mathbf{n}\}$. This proves that f maps into L as advertised; i.e., f is well-defined.*

(ii) *Prove that the function $g : L \to S^{n-1} - \{\mathbf{n}\}$ defined by*

$$g(\mathbf{y}) = \frac{4\mathbf{y} + (|\mathbf{y}|^2 - 1)\,\mathbf{n}}{|\mathbf{y}|^2 + 3} \tag{5.71}$$

is the inverse of the function f.

For simplicity, let $n = 3$. Then

$$g(u, v, -1) = \frac{(4u, 4v, u^2 + v^2 - 4)}{u^2 + v^2 + 4}. \tag{5.72}$$

By the exercise, we also have that f is a bijecton from $S^2 - \{\mathbf{n}\}$ to $\mathbb{R}^2 \times \{-1\}$, which in turn is bijective with $\mathbb{R}^2$. So, in a sense, we can think of S^2 as being "equal to" $\mathbb{R}^2$ with a point added, which we may call "infinity", ∞. We write $S^2 \cong \mathbb{R}^2 \cup \{\infty\}$. So $\infty \in \mathbb{R}^2 \cup \{\infty\}$ corresponds to $\mathbf{n} \in S^2$. On the 2-sphere S^2, if we start at the south pole $\mathbf{s} = (0,0,-1)$ and travel along any meridian, then we end up at the north pole $(0,0,1)$ after travelling distance π. Correspondingly, via stereographic projection, if in the plane $\mathbb{R}^2$ we start at the origin $(0,0)$ and travel along any ray, we can go an infinite distance, and in the limit we reach the added point ∞.[2]

5.2.8. Real projective space. Related to spheres are projective spaces.

Exercise 5.23. *A line through the origin in $\mathbb{R}^{n+1}$ may be described as*

$$\{t\mathbf{v} : t \in \mathbb{R}\}, \tag{5.73}$$

where $\mathbf{v} \in \mathbb{R}^{n+1} - \{\mathbf{0}\}$ is a non-zero vector.

Prove that any line L passing through the origin in $\mathbb{R}^{n+1}$ intersects the unit sphere S^n at exactly two points, which are antipodes of each other.

Exercise 5.24. *Define $\mathbb{R}P^n$ to be the* set of lines *passing through the origin in $\mathbb{R}^{n+1}$. Define the function $f : S^n \to \mathbb{R}P^n$ by $f(\mathbf{x})$ is the unique line passing through the origin in $\mathbb{R}^{n+1}$ containing $\mathbf{x}$. Prove that such a function indeed exists and for each $L \in \mathbb{R}P^n$ we have $|f^{-1}(L)| = 2$.*

$\mathbb{R}P^n$ is called the (n-dimensional) ***real projective space****. We call $\mathbb{R}P^2$ the* ***real projective plane****.*

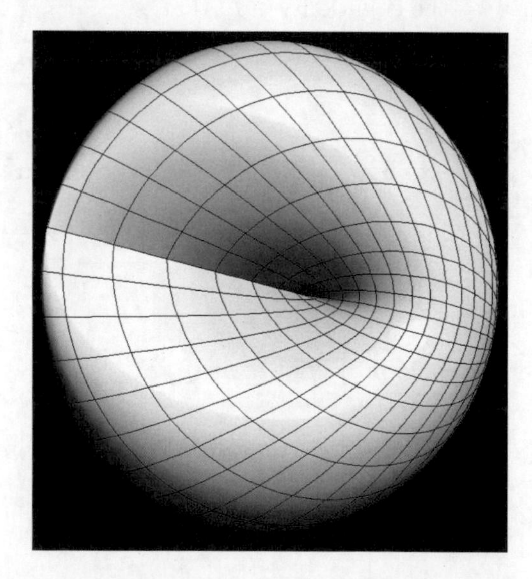

Figure 5.2.9. A cross-cap. This is a model for the real projective plane $\mathbb{R}P^2$. Fun fact: $\mathbb{R}P^2$ can be **immersed** in $\mathbb{R}^3$ but cannot be **embedded** in $\mathbb{R}^3$. From the Virtual Mathematics Museum by the 3D-XplorMath Consortium. Public Domain.

[2] So we say that stereographic projection does not preserve distances. But, amazingly, it preserves angles. Alas, we do not prove this here.

5.3. Functions and their properties

Sets without functions are like goalposts without balls. There is not much point playing the game without them!

We have so far been discussing functions from a working point of view. Now we will go a bit into the rigorous foundations for the notion of function.

5.3.1. What is a function?

When you think about a function, probably something like $y = \sin(x)$ comes to mind. Other functions are polynomials, and you've probably spent some time considering quadratic functions such as $f(x) = ax^2+bx+c$, where a, b, c are real numbers. But what is the general definition of a function?

Definition 5.29. Given two sets X and Y, a **function** f from X to Y assigns to each element in X a unique element in Y. So, to each $x \in X$, the function f assigns a unique element of Y, which we denote by $f(x)$.

Example 5.30. Let $X = \{a, b, c\}$ and let $Y = \{1, 2, 3\}$. Defining a function f from X to Y is equivalent to specifying elements $f(a)$, $f(b)$, and $f(c)$ in Y. For example, we can define f by $f(a) = 3$, $f(b) = 1$, and $f(c) = 3$. See Figure 5.3.1 for a visualization of this function.

Historical note. The song "ABC" by The Jackson 5 (not to be confused with Maroon 5) was a #1 hit in the year 1970. Part of the lyrics is: "A B C, one, two, three, ...".

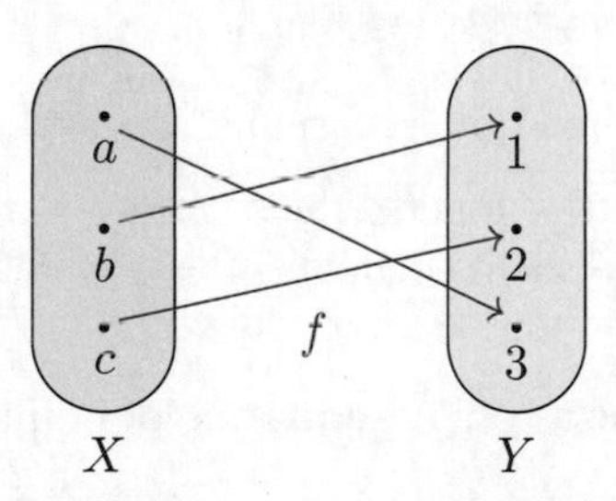

Figure 5.3.1. Visualizing the function f in Example 5.30.

The set X is called the **domain** of f and the set Y is called the **codomain** of f. We denote the function usually by

$$f : X \to Y.$$

We also use the following notation:

$$x \mapsto f(x).$$

Here we can think of x as the **input** and $f(x)$ as the **output**.

Exercise 5.25. *Again let $X = \{a, b, c\}$ and let $Y = \{1, 2, 3\}$.*

(1) *If we define $f(a) = 1$ and $f(b) = 2$, does this define a function $f : X \to Y$?*

(2) *If we define $g(a) = 1$, $g(b) = 2$, and $g(c) = 4$, does this define a function $g : X \to Y$?*

(3) *If we define $h(a) = 1$, $h(b) = 2$, $h(c) = 3$, and $h(d) = 1$, does this define a function $h : X \to Y$?*

5.3.2. Examples of functions. An important, albeit somewhat trivial looking, function is the identity function. Given a set X, the **identity function** of X is the function $I_X : X \to X$ defined by $I_X(x) = x$ for all $x \in X$.

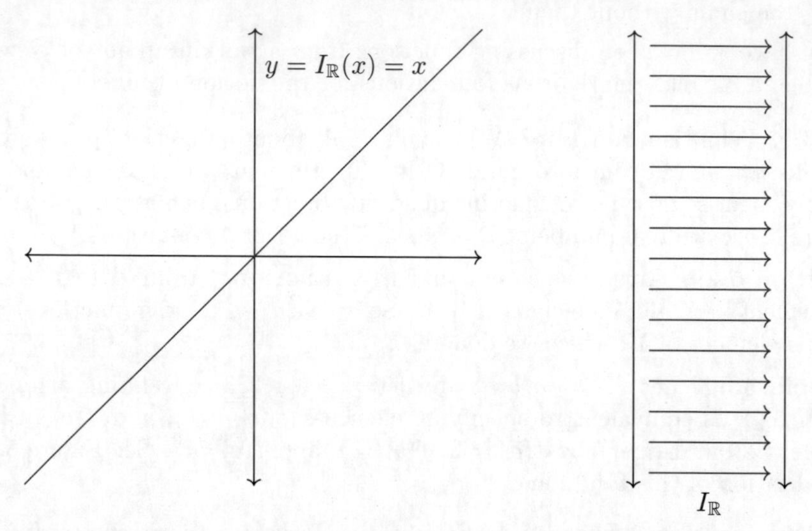

Figure 5.3.2. The identity function $I_{\mathbb{R}}$ of the set $\mathbb{R}$ viewed in two ways. Left: The graph of $I_{\mathbb{R}}$ is a diagonal line in $\mathbb{R}^2 = \mathbb{R} \times \mathbb{R}$. Right: The arrows show that $I_{\mathbb{R}}$ maps each real number to itself.

Many functions are defined by a "formula".

Example 5.31. Consider the function $y = x^2$. What we really mean by this is the function f from $\mathbb{R}$ to $\mathbb{R}$ defined by $f(x) = x^2$. We will call f the **squaring function**.

Definition 5.32. Let X be a set. A **sequence** of points in the set X is a function $f : \mathbb{Z}^+ \to X$.

For example, a sequence of real numbers is a function $f : \mathbb{Z}^+ \to \mathbb{R}$. However, we usually write a sequence using the notation $a_1, a_2, a_3, \ldots$ or $\{a_n\}_{n=1}^{\infty}$. The correspondence between the definition and this notation is given by the equality

$$a_n = f(n) \quad \text{for } n \in \mathbb{Z}^+.$$

For example, consider the **harmonic sequence** $\left\{\frac{1}{n}\right\}_{n=1}^{\infty}$. This is equivalent to the function $f : \mathbb{Z}^+ \to \mathbb{R}$ defined by $f(n) = \frac{1}{n}$.

Remark 5.33. We should not confuse a sequence $\{a_n\}_{n=1}^{\infty}$ with the set $\{a_n : n \in \mathbb{Z}^+\}$. Indeed, the latter is the *image* of the sequence as a function; see §5.3.5.1 below. For example, $\{(-1)^n\}_{n=1}^{\infty}$ is an alternating sequence $-1, 1, -1, 1, -1, 1, \ldots$, whereas $\{(-1)^n : n \in \mathbb{Z}^+\} = \{-1, 1\}$ is a set with two elements.

Example 5.34. Let S be an infinite subset of $\mathbb{Z}^+$. Associated to S, we can define a strictly increasing sequence $f : \mathbb{Z}^+ \to \mathbb{Z}^+$ by induction as follows.

(1) $f(1)$ is the minimum element of S, which exists by the well-ordering principle.

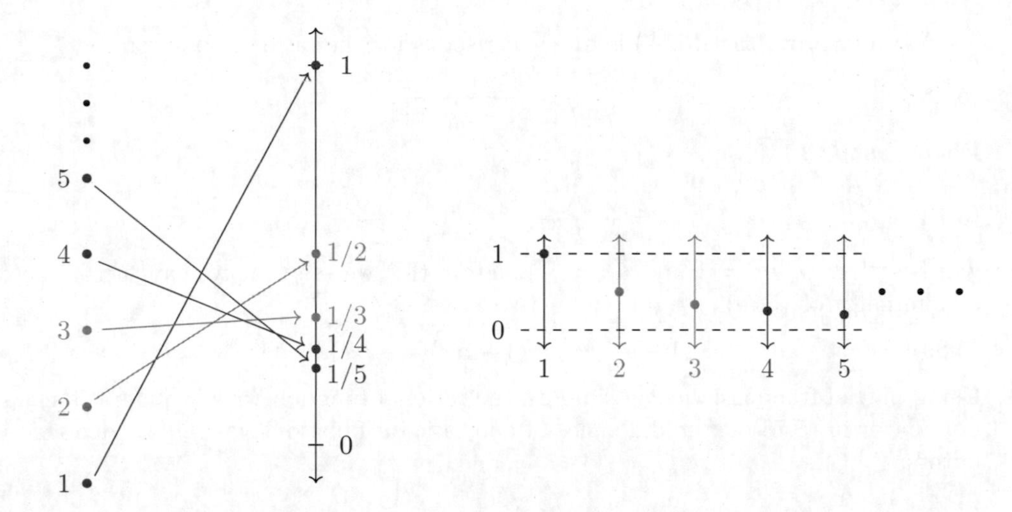

Figure 5.3.3. The harmonic sequence $\{\frac{1}{n}\}_{n=1}^{\infty}$ viewed in two ways. Left: As a function mapping inputs (positive integers) to their reciprocals. Right: The colored dots represent the graph of the function $n \mapsto 1/n$.

(2) Given a positive integer k for which $f(1), f(2), \ldots, f(k)$ are defined, we define $f(k+1)$ to be the minimum element of the set $S - \{f(1), f(2), \ldots, f(k)\}$. Note that this minimum element exists because S is infinite.

As an easy example of all of this, the function associated to the subset $E \cap \mathbb{Z}^+$ of even positive integers is the function defined by $f(n) = 2n$.

5.3.3*. Generating functions. Given a sequence $\{a_n\}_{n=0}^{\infty}$, its **generating function** is defined to be the power series

$$\sum_{n=0}^{\infty} a_n x^n. \tag{5.74}$$

So the generating function of a sequence is not a function per se, for the radius of convergence of the series may be finite or even zero. But generating functions are useful ways to encode the information, and hence properties, of a sequence.

Example 5.35. Consider the constant sequence $\{1\}_{n=0}^{\infty}$. Its generating function is the power series

$$\sum_{n=0}^{\infty} x^n. \tag{5.75}$$

If $|x| < 1$, then this series converges to $\frac{1}{1-x}$.

Let $\{f_n\}_{n=0}^{\infty}$ be the Fibonacci sequence (defined in §2.5), where $f_0 := 0$:

$$0, 1, 1, 2, 3, 5, 8, 13, 21, 34, 55, 89, \ldots. \tag{5.76}$$

Its generating function is

$$\sum_{n=0}^{\infty} f_n x^n. \tag{5.77}$$

We now prove that (5.77) is the power series for the rational function

$$f(x) := \frac{x}{1 - x - x^2}. \tag{5.78}$$

Define, as in (2.116),

$$\alpha = \frac{1+\sqrt{5}}{2}, \qquad \beta = \frac{1-\sqrt{5}}{2}, \tag{5.79}$$

which satisfy $\alpha + \beta = 1$ and $\alpha\beta = -1$. From this we see that we can factor the denominator of (5.78) as

$$1 - x - x^2 = (1 - \alpha x)(1 - \beta x). \tag{5.80}$$

Using partial fractions, we can simplify the rational function with a quadratic denominator in (5.78) as the difference of two rational functions with linear denominators:

$$\frac{x}{1 - x - x^2} = \frac{1}{\sqrt{5}}\left(\frac{1}{1-\alpha x} - \frac{1}{1-\beta x}\right). \tag{5.81}$$

This leads to the geometric series

$$\begin{aligned}\frac{1}{1-\alpha x} &= \sum_{n=0}^{\infty} \alpha^n x^n \qquad \text{for } |x| < \frac{1}{|\alpha|},\\ \frac{1}{1-\beta x} &= \sum_{n=0}^{\infty} \beta^n x^n \qquad \text{for } |x| < \frac{1}{|\beta|}.\end{aligned}$$

We have

$$\frac{x}{1 - x - x^2} = \sum_{n=0}^{\infty} \frac{\alpha^n - \beta^n}{\sqrt{5}} x^n = \sum_{n=0}^{\infty} f_n x^n \tag{5.82}$$

for $|x| < \frac{1}{|\beta|}$, where we used Binet's formula (2.117).

In the argument above, we used Binet's formula to prove that $\frac{x}{1-x-x^2} = \sum_{n=0}^{\infty} f_n x^n$. Conversely, we can prove this last formula to give a new *derivation* of Binet's formula. To see this, let

$$F(x) := \sum_{n=0}^{\infty} f_n x^n = \sum_{n=1}^{\infty} f_n x^n = x + \sum_{n=2}^{\infty} f_n x^n, \tag{5.83}$$

where the second two equalities follow from $f_0 = 0$ and $f_1 = 1$. We then compute using the Fibonacci number recursive definition (2.99) that

$$\begin{aligned}F(x) &= x + \sum_{n=2}^{\infty} (f_{n-1} + f_{n-2}) x^n\\ &= x + \sum_{n=2}^{\infty} f_{n-1} x^{n-1} x + \sum_{n=2}^{\infty} f_{n-2} x^{n-2} x^2\\ &= x + \sum_{n=1}^{\infty} f_n x^n x + \sum_{n=0}^{\infty} f_n x^n x^2\\ &= x + xF(x) + x^2 F(x).\end{aligned}$$

From this it follows that

$$F(x) = \frac{x}{1 - x - x^2}. \tag{5.84}$$

From seeing this new derivation of Binet's formula using a generating function, we can imagine the general usefulness of generating functions.

5.3.4. Equality of functions. We say that two functions $f : X \to Y$ and $g : Z \to W$ are **equal** provided their domains are equal, their codomains are equal, and all of their values are equal, that is, provided:

(1) $X = Z$ and $Y = W$.

(2) For all $x \in X = Z$, $f(x) = g(x) \in Y = W$.

Example 5.36. Let X be a set and let $f : X \to \mathbb{R}$ be a function. Define $g : X \to \mathbb{R}$ by

$$g(x) = (-1)^2 f(x) \quad \text{for } x \in X. \tag{5.85}$$

Then $f = g$ (as functions).

5.3.5. New functions out of old functions. Given a function $f : X \to Y$, there are few simple things we can do to the function to create a new function. In other words, just as we have operations on sets to create new sets, we have operations on functions to create new functions.

5.3.5.1. *Images.* The **image** of a function $f : X \to Y$ is defined by

$$\mathrm{Im}(f) = \{f(x) : x \in X\} \subset Y.$$

The terminology **range** is also used for image.

When we think about real-valued functions of a single variable, we often picture their graph. For example, consider the sine function $f : \mathbb{R} \to \mathbb{R}$ defined by $f(x) = \sin x$. Its graph looks like a wave. Indeed, the periodicity of the sine function, that is, $f(x + 2\pi) = f(x)$ for all $x \in \mathbb{R}$, means that the graph of sine repeats itself with period 2π.

Let sin denote the sine function and let cos denote the cosine function. Then

$$\mathrm{Im}(\sin) = [-1, 1] = \mathrm{Im}(\cos). \tag{5.86}$$

If $\mathrm{sq} : \mathbb{R} \to \mathbb{R}$ is the squaring function defined by $\mathrm{sq}(x) = x^2$, then

$$\mathrm{Im}(\mathrm{sq}) = \mathbb{R}^{\geq}. \tag{5.87}$$

Indeed, for every $y \in \mathbb{R}^{\geq}$, we have $\mathrm{sq}(\sqrt{y}) = (\sqrt{y})^2 = y$. Moreover, for every $x \in \mathbb{R}$, we have that $\mathrm{sq}(x) \geq 0$.

5.3.5.2. *Restrictions and extensions.* Let $A \subset X$. We can define the **restriction** of f to A by

$$f|_A : A \to X, \tag{5.88}$$

where

$$f|_A(x) = f(x) \quad \text{for } x \in A. \tag{5.89}$$

For example, we can restrict the sine function $\sin : \mathbb{R} \to \mathbb{R}$ to the subset $[-\pi/2, \pi/2]$. On this subset, sin is (strictly) increasing since its derivative $\sin'(x) = \cos(x)$ is positive for $x \in (-\pi/2, \pi/2)$. See Figure 5.3.4.

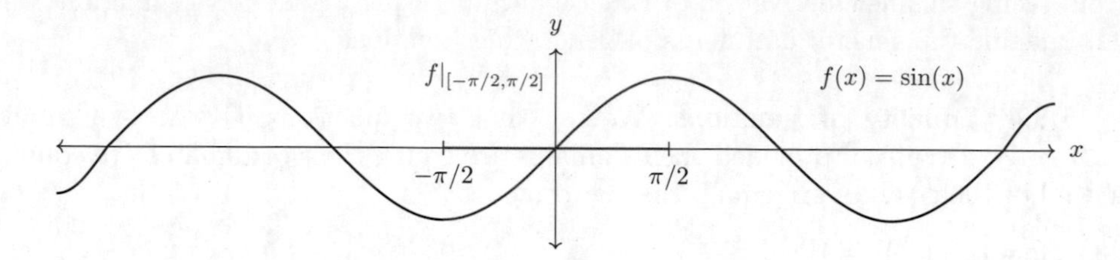

Figure 5.3.4. The graph of the restriction of the sine function to the closed interval $[-\pi/2, \pi/2]$ is in red.

Suppose, on the other hand, $X \subset B$. Suppose $g : B \to Y$ satisfies

$$g(x) = f(x) \quad \text{for } x \in X. \tag{5.90}$$

Then we say that g is an **extension** of f. Note that in this case $f = g|_X$.

Besides restricting and extending the domain, we can restrict and extend the codomain: Suppose that $Y \subset C$. Then we can define $h : X \to C$ by $h(x) = f(x)$ for $x \in X$. Note that h is well-defined since $f(x) \in Y$ implies that $f(x) \in C$.

If D is a set satisfying $\mathrm{Im}(f) \subset D \subset Y$, then we can define $k : X \to D$ by $k(x) = f(x)$ for $x \in X$. Here, k is well-defined since for each $x \in X$ we have $f(x) \in \mathrm{Im}(f) \subset D$. In particular, we have the restriction $k : X \to \mathrm{Im}(f)$ defined by $k(x) = f(x)$ for $x \in X$.

So, to summarize, without any essential change to how the function is defined, we can shrink the domain or enlarge the codomain. We can shrink the codomain provided it is to a set that contains the image. If we enlarge the domain, we have to specify how we define the function outside its original domain.

5.3.5.3. *Compositions.* Let

$$f : X \to Y$$

be a function. Then, given $x \in X$, we have that $f(x) \in Y$. So if we have another function

$$g : Y \to Z,$$

we can then take $g(f(x))$ to obtain an element of Z. This combined process is the **composition** of the functions f and g. That is, we define

$$g \circ f : X \to Z$$

by

$$g \circ f(x) = g(f(x)) \quad \text{for } x \in X.$$

Remark 5.37. For the composition to be well-defined, we just need that the image of f is contained in the domain of g. Note that in this case, by either enlarging the codomain of f or restricting the domain of g, if necessary, we can make them equal as sets.

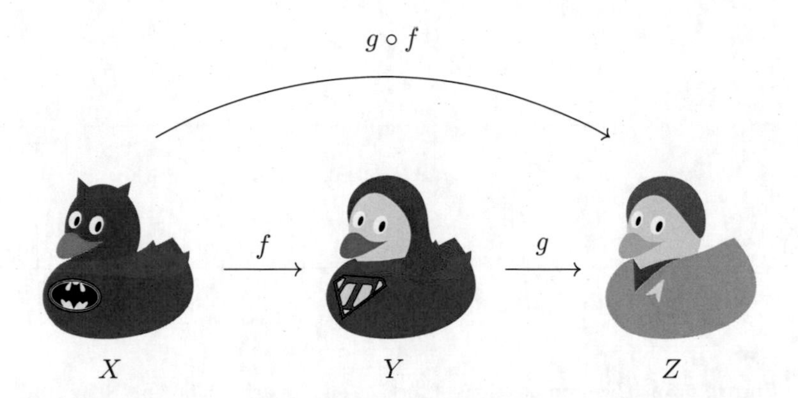

Figure 5.3.5. The function f maps from Batduck to Superduck, and the function g maps from Superduck to Starduck. Their composition $g \circ f$ maps from Batduck to Starduck. From "'The duck pond': showcase of TikZ-drawn animals/ducks" Stack Exchange answer by samcarter_is_at_topanswers.xyz. Licensed under Creative Commons Attribution-Share Alike 4.0 International License (https://creativecommons.org/licenses/by-sa/4.0/deed.en).

Example 5.88. Define $\alpha : \mathbb{R}^n \to \mathbb{R}^n$ by

$$\alpha(\mathbf{x}) = -\mathbf{x} \quad \text{for } \mathbf{x} \in \mathbb{R}^n. \tag{5.91}$$

Then

$$\alpha \circ \alpha = I_{\mathbb{R}^n}, \tag{5.92}$$

where $I_{\mathbb{R}^n}$ is the identity function of $\mathbb{R}^n$. Indeed, we calculate that

$$(\alpha \circ \alpha)(\mathbf{x}) = \alpha\big(\alpha(\mathbf{x})\big) = \alpha(-\mathbf{x}) = -(-\mathbf{x}) = \mathbf{x}. \tag{5.93}$$

The function α is called the antipodal map. It is an example of an involution; that is, α is its own inverse; see §5.4.4 below.

Exercise 5.26 (Mirror, mirror on the wall). *Let* $\mathbf{u}$ *be a unit vector in* $\mathbb{R}^n$. *Prove that the function* $f : \mathbb{R}^n \to \mathbb{R}^n$ *defined by*

$$f : \mathbf{x} \to \mathbf{x} - 2(\mathbf{x} \cdot \mathbf{u})\,\mathbf{u} \tag{5.94}$$

is an involution.

Can you figure out (geometrically) what kind of function this is? Hint: Firstly, what are the fixed points of the function? By definition, a point $x \in X$ *is a* ***fixed point*** *of a function* $f : X \to X$ *if* $f(x) = x$.

An example from calculus class is: $f(x) = x^2$ and $g(y) = \sin y$. Both of these functions have domain and codomain $\mathbb{R}$. Then

$$g \circ f(x) = \sin(x^2), \qquad f \circ g(x) = \sin^2(x).$$

Caveat: The sine function has often been confused with Lombard Street. ☺

Here is another example. Define $g : \mathbb{Z} \to \{0, 1\}$ by

$$g(n) = \begin{cases} 0 & \text{if } n \text{ is even}, \\ 1 & \text{if } n \text{ is odd} \end{cases} \tag{5.95}$$

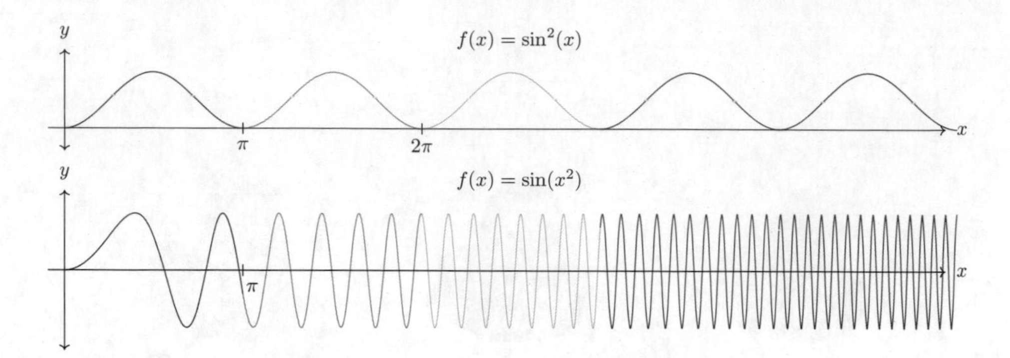

Figure 5.3.6. The compositions of the sine and squaring functions. The order of composition makes a difference!

(g is the characteristic function of the subset of odd integers in the set of integers), and for each positive integer k define $f_k : \mathbb{Z} \to \mathbb{Z}$ by

$$f_k(n) = n^k. \tag{5.96}$$

We leave it to the reader to show that

$$g \circ f_k = g \quad \text{for all } k \in \mathbb{Z}^+. \tag{5.97}$$

Exercise 5.27. *Prove* $g \circ f_k = g$ *in* (5.97).

Some basic properties of composite functions are given by:

Theorem 5.39. *Suppose that* $f : X \to Y$, $g : Y \to Z$, *and* $h : Z \to W$ *are functions. Then:*

(1) *(Associativity)* $(h \circ g) \circ f = h \circ (g \circ f)$.
(2) *(Identity function)* $f \circ I_X = f = I_Y \circ f$.

Exercise 5.28. *Prove Theorem* 5.39.

5.3.5.4. *Graphs.* We are accustomed to visualizing functions by their graphs, at least for real-valued functions of one or two variables.

Formally, the **graph** of a function $f : X \to Y$ is defined by

$$G_f = \{(x, f(x)) : x \in X\} \subset X \times Y. \tag{5.98}$$

Equivalently,

$$G_f = \{(x, y) : x \in X, \, y = f(x)\}.$$

Example 5.40. Define the function

$$f(x, y) = \ln\left(\frac{\cos(x)}{\cos(y)}\right) \tag{5.99}$$

on the largest subset of elements (x, y) in $\mathbb{R}^2$ for which the expression on the right-hand side is well-defined. Call D the domain of f. The graph of f is the (graphical) surface in $\mathbb{R}^3 = \mathbb{R}^2 \times \mathbb{R}$ defined by

$$G_f = \left\{\left(x, y, \ln\left(\frac{\cos(x)}{\cos(y)}\right)\right) : (x, y) \in D\right\}. \tag{5.100}$$

In what way does the domain D of f remind you of an (infinite) chessboard? Hint: A ratio is well-defined if and only if the denominator is non-zero. The logarithm of a number is well-defined if and only if the number is positive.

Figure 5.3.7. Scherk's surface. Wikimedia Commons, author: Erminia Naccarato (Miriane), licensed under GNU Free Documentation License, Version 1.2 or any later version.

The surface G_f is known as Scherk's first surface. This is an example of a minimal surface, a concept in differential geometry related to soap bubbles.

Exercise 5.29. *Find the domain of the function defined in Example* 5.40. *That is, find the largest subset of* $\mathbb{R}^2$ *on which the expression* $f(x,y)$ *given by* (5.99) *is well-defined.*

5.3.6. The floor and ceiling functions. The floor function

$$x \mapsto \lfloor x \rfloor \tag{5.101}$$

of a real number input x is defined to be the greatest integer less than or equal to x. That is,

$$\lfloor x \rfloor := \sup\{n \in \mathbb{Z} : n \leq x\} = \max\{n \in \mathbb{Z} : n \leq x\}. \tag{5.102}$$

The **ceiling function**

$$x \mapsto \lceil x \rceil \tag{5.103}$$

of a real number input x is defined to be the least integer greater than or equal to x. That is,

$$\lceil x \rceil := \inf\{n \in \mathbb{Z} : n \geq x\} = \min\{n \in \mathbb{Z} : n \geq x\}. \tag{5.104}$$

We leave it as an exercise to graph the ceiling function.

The fractional part (or **decimal part**) of a non-negative real number is defined by

$$\text{frac}(x) := x - \lfloor x \rfloor; \tag{5.105}$$

that is,

$$\lfloor x \rfloor + \text{frac}(x) = x. \tag{5.106}$$

The functions above are nicely understood by drawing their graphs.

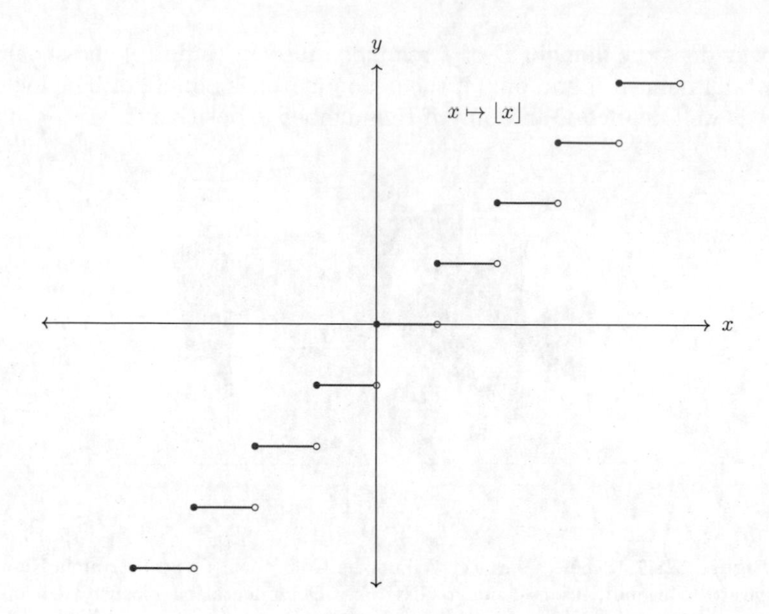

Figure 5.3.8. The graph of the floor function.

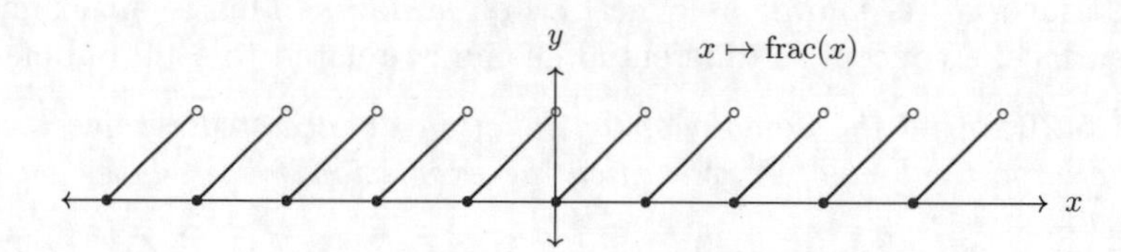

Figure 5.3.9. The graph of the fractional part function.

5.4. Types of functions: Injections, surjections, and bijections

For any function $f : X \to Y$, simply because it is a function, for any elements x_1 and x_2 of X,

$$\text{if } f(x_1) \neq f(x_2), \quad \text{then } x_1 \neq x_2. \tag{5.107}$$

This is simply because: If $x_1 = x_2$, then by the definition of function we have that $f(x_1) = f(x_2)$. The preceding statement is the contrapositive of this statement.

5.4.1. Injections, surjections, and bijections.

5.4.1.1. *Surjections.* For a function, is every element of the codomain the image of some element of the domain? Not necessarily, but if it is, then we have the following.

Definition 5.41. We say that $f : X \to Y$ is **surjective** if for every $y \in Y$, there exists $x \in X$ such that $f(x) = y$. We also call f a **surjection**.

In other words, a function f is surjective if and only if f takes *all* values in Y. Equivalently, the image of f is Y:

$$\mathrm{Im}(f) = Y. \tag{5.108}$$

Examples. (1) The (linear) function $f : \mathbb{R} \to \mathbb{R}$ defined by $f(x) = 2x - 3$ is surjective. Indeed, for every $y \in \mathbb{R}$, there exists $x \in \mathbb{R}$ such that $y = 2x - 3$; namely $x = \frac{y+3}{2}$.

(2) The function $g : \mathbb{R} \to \mathbb{R}$ defined by $g(x) = x^2$ is not surjective. For example, for $-1 \in \mathbb{R}$ there does not exist $x \in \mathbb{R}$ such that $x^2 = -1$.

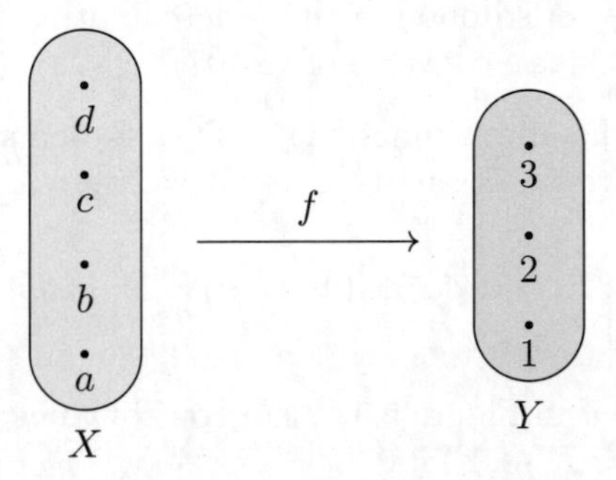

Figure 5.4.1. For example, the function $f : X \to Y$ defined by $f(a) = 2$, $f(b) = 1$, $f(c) = 2$, $f(d) = 3$ is a surjection.

Solved Problem 5.42. *Define* $f : \mathbb{Z}^+ \to \mathbb{Z}^+$ *by*

$$f(n) = \begin{cases} \frac{n}{13} & \text{if } n \text{ is divisible by 13,} \\ n^2 & \text{if } n \text{ is not divisible by 13.} \end{cases}$$

Prove that f *is surjective.*

Solution. Let $k \in \mathbb{Z}^+$. Then $n = 13k$ is divisible by 13. Therefore

$$f(n) = \frac{n}{13} = k. \tag{5.109}$$

This proves that f is a surjection. □

Exercise 5.30. *Show that any cubic function* $f : \mathbb{R} \to \mathbb{R}$*, defined by* $f(x) = ax^3 + bx^2 + cx + d$*, where* $a, b, c, d \in \mathbb{R}$ *with* $a \neq 0$*, is surjective. Hint: Take limits as* $x \to \pm\infty$ *and use the Intermediate Value Theorem from calculus.*

Exercise 5.31. *Let* Y *be a set and let* $f : \mathbb{Z} \to Y$ *be a function satisfying the following properties:*

(1) f *is a surjection,*

(2) $f(2n) = f(2n+1)$ *for all* $n \in \mathbb{Z}$.

Let $E \subset \mathbb{Z}$ *denote the set of even integers. Define the function*

$$g : E \to Y, \quad \text{where } g(k) = f(k) \text{ for } k \in E.$$

Prove that g *is a surjection. Hint: You may wish to consider two cases. Can you draw a picture to visualize* f *and* g*?*

5.4.1.2. *Injections.* For a function, can two different elements in the domain have the same image? Possibly, but if we require not, then we have the following.

Definition 5.43. We say that $f : X \to Y$ is **injective** if $x_1, x_2 \in X$ with $x_1 \neq x_2$ implies $f(x_1) \neq f(x_2)$. We also call f an **injection**.

Note that, equivalently, $f : X \to Y$ is injective if $x_1, x_2 \in X$ with $f(x_1) = f(x_2)$ implies $x_1 = x_2$. Also observe that the condition in Definition 5.43 for a function to be injective is the converse of (5.107).

Example 5.44. (1) The above function $f : \mathbb{R} \to \mathbb{R}$ defined by $f(x) = 2x - 3$ is injective. Indeed, if $x_1, x_2 \in \mathbb{R}$ satisfy $2x_1 - 3 = 2x_2 - 3$, then $2x_1 = 2x_2$, which implies $x_1 = x_2$.

(2) The function $g : \mathbb{R} \to \mathbb{R}$ defined by $g(x) = x^2$ is not injective. For example, for $-1 \neq 1$ we have $(-1)^2 = 1^2$.

For the sets X and Y in Figure 5.4.1, can you convince yourself that there does not exist an injection $g : X \to Y$?

On the other hand, for the sets X and Y in Figure 5.4.1, there exists an injection $h : Y \to X$. For example, we can define h by $h(1) = b$, $h(2) = a$, and $h(3) = d$.

Exercise 5.32. *Define $g : \mathbb{Z}^+ \to \mathbb{Z}$ by $g(n) = (-1)^n n$. Prove that g is injective, but not surjective.*

Exercise 5.33. *Define $f : \mathbb{Z}^+ \to \mathbb{Z}^+$ by*

$$f(n) = \begin{cases} n & \text{if } n \text{ is even,} \\ \frac{n+1}{2} & \text{if } n \text{ is odd.} \end{cases} \tag{5.110}$$

Is f injective? Is f surjective? For each $k \in \mathbb{Z}^+$, what integers does the set $f^{-1}(k)$ consist of?

Exercise 5.34. *Let X and Y be disjoint sets. Suppose that $f : X \to Y$ and $g : Y \to X$ are injections. Prove that $h : X \cup Y \to X \cup Y$, defined by*

$$h(x) = f(x) \text{ if } x \in X \quad \text{and} \quad h(x) = g(x) \text{ if } x \in Y, \tag{5.111}$$

is an injection.

Exercise 5.35. *Let $m \in \mathbb{Z}^+$ and let $a \in R_m - \{0\} = \mathbb{N}_{m-1}$. Define $f_a : R_m \to R_m$ by*

$$f_a(x) = \mathbf{r}(ax), \tag{5.112}$$

where $\mathbf{r} : \mathbb{Z} \to R_m$ is the remainder function.

(1) For which $a \in R_3 - \{0\}$ is $f_a : R_3 \to R_3$ an injection?

(2) For which $a \in R_4 - \{0\}$ is $f_a : R_4 \to R_4$ an injection?

(3) Suppose that $a \in R_m - \{0\}$ satisfies $\gcd(a, m) > 1$. Is $f_a : R_m \to R_m$ an injection?

(4) Suppose that $a \in R_m - \{0\}$ satisfies $\gcd(a, m) = 1$. Is $f_a : R_m \to R_m$ an injection?

Exercise 5.36. *Answer the same questions as in the previous exercise except with "injection" replaced by "surjection". (Is it easier to answer these questions now that you have done the previous exercise?)*

5.4.1.3. *Bijections.* When does a function provide a one-to-one correspondence between the elements of the domain and the elements of the codomain? In the following case.

Definition 5.45. $f : X \to Y$ is **bijective** if f is surjective and injective. We also call f a **bijection**.

Example 5.46. (1) The above function $f : \mathbb{R} \to \mathbb{R}$ defined by $f(x) = 2x - 3$ is bijective since it is both surjective and injective.

(2) Of course, $g : \mathbb{R} \to \mathbb{R}$ defined by $g(x) = x^2$ is not bijective since it is neither surjective nor injective (to be not bijective it **only needs** to be not surjective **or** not injective). On the other hand, $h : \mathbb{R}^{\geq} \to \mathbb{R}^{\geq}$ defined by $h(x) = x^2$ is bijective.

Exercise 5.37. *Define $f_k : \mathbb{R} \to \mathbb{R}$ by $f_k(x) = x^k$. Determine whether each of the following functions is a surjection, injection, and/or bijection:*

(1) *f_k, where k is an odd positive integer,*

(2) *f_k, where k is an even positive integer.*

Now let $f : X \to Y$ be a function. Then we can restrict the codomain Y to the subset $\operatorname{Im}(f)$ and define the function

$$\bar{f} : X \to \operatorname{Im}(f) \tag{5.113}$$

by $\bar{f}(x) := f(x)$ for all $x \in X$. The reason why this function is well-defined is simply because for all $x \in X$ we have $\bar{f}(x) = f(x) \in \operatorname{Im}(f)$. We leave it to you, dear reader, to show that $\bar{f}$ is a surjection. In particular, if f is an injection, then $\bar{f}$ is a bijection.

5.4.2. The pre-image.

> Question: What do you call it when you fire pre-images at your enemy?
> Answer: A preimage attack.

Let $f : X \to Y$ be a function. Let B be a subset of Y. The **pre-image** of B is the set of elements in X whose values are in B. That is,

$$f^{-1}(B) = \{x \in X : f(x) \in B\}. \tag{5.114}$$

For example, $f^{-1}(Y) = X$ since $f(x) \in Y$ for all $x \in X$. If $y \in Y$, then we denote

$$f^{-1}(y) = f^{-1}(\{y\}) = \{x \in X : f(x) = y\}. \tag{5.115}$$

So the pre-image of a subset of the codomain is the set of elements of the domain whose images are elements of the subset of the codomain.

Another way to say that f is a surjection is: For all $y \in Y$, the pre-image $f^{-1}(y)$ is non-empty (i.e., consists of at least one element).

Another way to say that f is an injection is: For all $y \in Y$, the pre-image $f^{-1}(y)$ consists of at most one element.

Another way to say that f is a bijection is: For all $y \in Y$, the pre-image $f^{-1}(y)$ consists of exactly one element.

Example 5.47. Define the function $g : \mathbb{Z} \to \{0,1\}$ by

$$g(n) = \begin{cases} 0 & \text{if } n \text{ is even,} \\ 1 & \text{if } n \text{ is odd.} \end{cases}$$

We then have that $g^{-1}(0)$ is the set $\mathbb{E}$ of even integers and that $g^{-1}(1)$ is the set $\mathbb{O}$ of odd integers.

Example 5.48. Define $h : \mathbb{Z} \to \mathbb{Z}$ by $h(n) = 2n$. If k is odd, then $h^{-1}(k) = \emptyset$. If k is even, then $h^{-1}(k) = k/2$.

Example 5.49. For the function $f : X \to Y$ in Figure 5.4.1, the pre-image of 2 is

$$f^{-1}(2) = \{a, c\}.$$

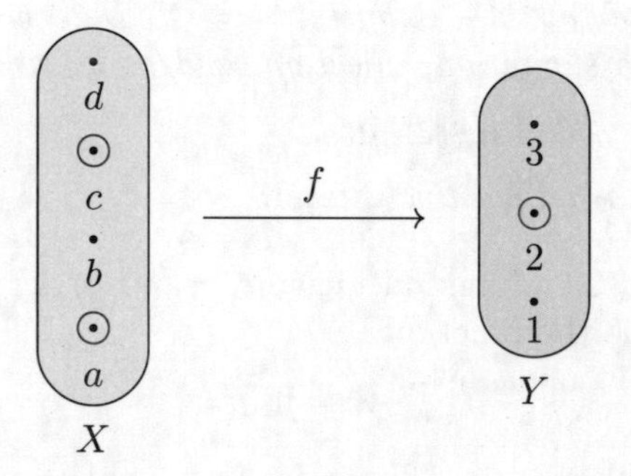

Figure 5.4.2. Since a and c are the only elements in X that map to 2, we have $f^{-1}(2) = \{a, c\}$.

Solved Problem 5.50. *Let $f : X \to Y$ be a function, and let y_1 and y_2 be distinct elements of Y. Prove that $f^{-1}(y_1) \cap f^{-1}(y_2) = \emptyset$.*

Solution. Suppose for a contradiction that there exists $x \in f^{-1}(y_1) \cap f^{-1}(y_2)$. Then $x \in f^{-1}(y_1)$ and $x \in f^{-1}(y_2)$. This implies that $f(x) = y_1$ and $f(x) = y_2$. Therefore, $y_1 = y_2$, which is a contradiction. □

Solved Problem 5.51 (Pre-images and the subset relation)**.** *Let $f : X \to Y$ be a function, and let $B_1 \subset B_2 \subset Y$. Prove that $f^{-1}(B_1) \subset f^{-1}(B_2)$.*

Solution. Let $x \in f^{-1}(B_1)$. Then $f(x) \in B_1$. Since $B_1 \subset B_2$, this implies that $f(x) \in B_2$. Therefore $x \in f^{-1}(B_2)$. □

Exercise 5.38. *Define $f_k : \mathbb{R} \to \mathbb{R}$ by $f_k(x) = x^k$ for $k \in \mathbb{Z}^+$. Explain how the pre-image set $f_k^{-1}(y)$ (e.g., its cardinality $|f_k^{-1}(y)|$) depends on $y \in \mathbb{R}$ and whether k is even or odd.*

Note that this exercise is more about understanding what the pre-image of a real number for a power function is in various cases rather than about giving rigorous proofs.

Exercise 5.39. *Let $f : X \to Y$ be a function. Suppose that B and C are subsets of Y such that $B \cap C = \emptyset$. Prove that*

$$f^{-1}(B) \cap f^{-1}(C) = \emptyset. \tag{5.116}$$

Exercise 5.40. *Let f_n be the n-th Fibonacci number. Define $g : \mathbb{Z}^+ \to \{-1, 1\}$ by*

$$g(n) = (-1)^{f_n}. \tag{5.117}$$

Describe $g^{-1}(1)$ and $g^{-1}(-1)$.

Exercise 5.41. *Given $m \in \mathbb{Z}^+$, let $\mathbf{r} : \mathbb{Z} \to R_m$ be the remainder function defined by (4.14). Describe the pre-image*

$$\mathbf{r}^{-1}(a) \tag{5.118}$$

for $a \in \mathbb{Z}$.

Exercise 5.42. *Let $f : X \to Y$ be a function. Now suppose that B and C are any subsets of Y. Can you find and prove an "identity" for $f^{-1}(B) \cap f^{-1}(C)$? Namely, is there a different expression that this is equal to?*

Exercise 5.43. *Let $f : X \to Y$ be a function. Suppose that B is a subset of Y.*

(1) *Show that*

$$f(f^{-1}(B)) \subset B. \tag{5.119}$$

(2) *Describe under what circumstance*

$$f(f^{-1}(B)) = B. \tag{5.120}$$

(3) *Complementarily, when is $f(f^{-1}(B))$ a proper subset of B?*

Exercise 5.44. *Let $f : X \to Y$ be a function. Suppose that A is a subset of X.*

(1) *Show that*

$$A \subset f^{-1}(f(A)). \tag{5.121}$$

(2) *Describe under what circumstance*

$$A = f^{-1}(f(A)). \tag{5.122}$$

5.4.3. Compositions of injections, surjections, bijections.

Injections, surjections, and bijections have the following nice properties regarding their compositions.

Theorem 5.52.

(1) *The composition of two injections is an injection.*
(2) *The composition of two surjections is a surjection.*
(3) *The composition of two bijections is a bijection.*

Proof. Let $f : X \to Y$ and $g : Y \to Z$ be two functions.

(1) Let f and g be injections. Suppose that $x_1, x_2 \in X$ are such that $g \circ f(x_1) = g \circ f(x_2)$. This says that $g(f(x_1)) = g(f(x_2))$. Since g is an injection, this implies that $f(x_1) = f(x_2)$. Now, f being an injection implies that $x_1 = x_2$. We have proved that $g \circ f$ is an injection.

(2) Let f and g be surjections. Let $z \in Z$. Since g is a surjection, there exists $y \in Y$ such that $g(y) = z$. Now, since f is a surjection, there exists $x \in X$ such that $f(x) = y$. We conclude that $g \circ f(x) = g(f(x)) = g(y) = z$. We have proved that $g \circ f$ is a surjection.

Part (3) follows from parts (1) and (2). □

Exercise 5.45. *Let $f : X \to Y$ be an injection such that $f(X) \subset B$, where $B \subset Y$. Let $g : Y \to Z$ be a function with the property that if $y_1, y_2 \in B$ satisfy $g(y_1) = g(y_2)$, then $y_1 = y_2$. Prove that $g \circ f : X \to Z$ is an injection. (Again, drawing a picture may be helpful!)*

5.4.4. Bijections and inverses. We say that a function $f : X \to Y$ is **invertible** if there exists a function $g : Y \to X$ which "undoes what f does"; that is:

$$\text{For all } x \in X,\ g(f(x)) = x \text{ and for all } y \in Y,\ f(g(y)) = y.$$

We call g the **inverse** of f.

For example, let $a \in \mathbb{R} - \{0\}$ and let $b \in \mathbb{R}$ and define the function $f : \mathbb{R} \to \mathbb{R}$ by $f(x) = ax + b$. Then we can find the inverse of f as follows. Write the equation $y = ax + b$; that is, y is the value of f at x. Then we solve for x: $x = \frac{y-b}{a}$, which is well-defined since $x \neq 0$. That is, if an input x has value y under f, then $x = \frac{y-b}{a}$. So we define $g : \mathbb{R} \to \mathbb{R}$ by $g(y) = \frac{y-b}{a}$. It is easy to check that then g is the inverse of f.

The following says that being a *bijection* and being *invertible* are equivalent.

Theorem 5.53. *Let $f : X \to Y$. Then f is invertible if and only if it is a bijection. Furthermore, if it is invertible, then its inverse function is unique.*

Proof. *Invertible implies bijection.* Suppose that $f : X \to Y$ is invertible, so that it has an inverse $g : Y \to X$.

(1) f is injective. Suppose that $x_1, x_2 \in X$ are such that $f(x_1) = f(x_2)$. Then $x_1 = g(f(x_1)) = g(f(x_2)) = x_2$. This proves that f is injective.

(2) f is surjective. Let $y \in Y$. Let $x = g(y) \in X$. Then $f(x) = f(g(y)) = y$. This proves that f is surjective.

Bijection implies invertible. Suppose that $f : X \to Y$ is bijective.

Let $y \in Y$. Since f is surjective, there exists $x \in X$ such that $f(x) = y$. Since f is injective, x is unique (exercise: understand why). We then define $g(y) = x$. By definition,

$$x = g(y) = g(f(x)).$$

This equation is true for all $x \in X$. (Subtle explanation for the last sentence: This is true for all $y \in Y$, where $x = g(y)$. So given $x \in X$, we let $y = f(x)$, so that $x = g(y)$.)

On the other hand, given $x \in X$, let $y = f(x)$. Since x is the unique element of X with $f(x) = y$, we have that $g(y) = x$. Thus

$$y = f(x) = f(g(y))$$

for all $y \in Y$. □

Example 5.54 (Inverse trigonometric functions)**.** An important use of inverse functions is to define inverse trigonometric functions. Consider the sine function

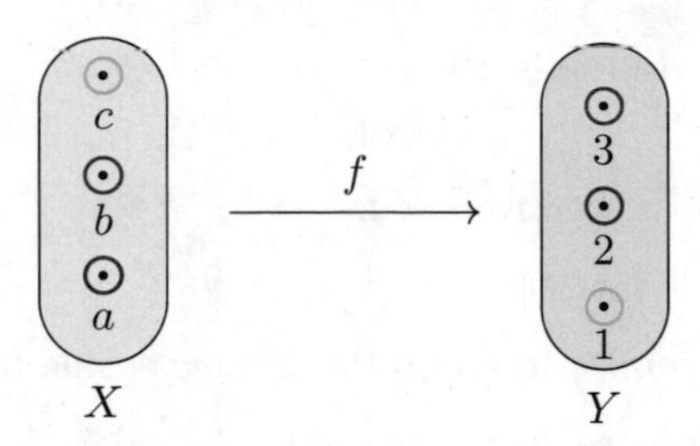

Figure 5.4.3. This illustrates a one-to-one correspondence between two 3-element sets X and Y. Correspondingly, we have bijections (that are inverses of each other) $f : X \to Y$ and $g : Y \to X$ defined by $f(a) = 2$, $f(b) = 3$, $f(c) = 1$, and $g(1) = c$, $g(2) = a$, $g(3) = b$.

$\sin : \mathbb{R} \to \mathbb{R}$. This function is neither injective nor surjective. We can easily make it surjective by restricting the codomain to the image and defining

$$\sin : \mathbb{R} \to [-1, 1].$$

(Prove that this function is surjective by using the Intermediate Value Theorem.) To make the function injective, as in §5.3.5.2 we can restrict it to a suitable interval; namely, we consider the restriction

$$\sin : [-\pi/2, \pi/2] \to [-1, 1].$$

Indeed, since $\sin'(x) > 0$ for all $x \in (-\pi/2, \pi/2)$, we have that sin is a strictly increasing function on $[-\pi/2, \pi/2]$ (the Mean Value Theorem gives a rigorous proof of this). For simplicity, we have used the same notation for the restricted functions as for the original function. So, $\sin : [-\pi/2, \pi/2] \to [-1, 1]$ is a bijective function. Thus, by Theorem 5.53, sine has an inverse:

$$\sin^{-1} : [-1, 1] \to [-\pi/2, \pi/2].$$

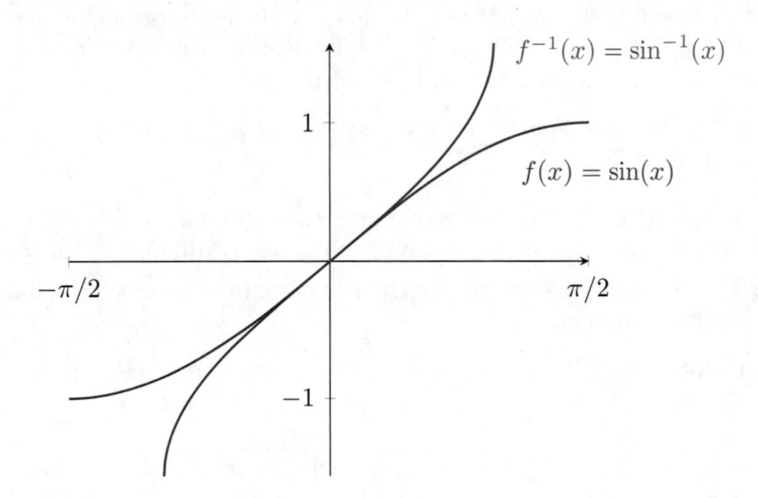

Figure 5.4.4. The graph of the inverse sine function.

Similarly, the restriction of cosine,

$$\cos : [0, \pi] \to [-1, 1],$$

is bijective. Thus we have the inverse function

$$\cos^{-1} : [-1, 1] \to [0, \pi].$$

We leave it as an exercise to graph the inverse cosine function.

Example 5.55 (The square root function). Consider the bijective function $f : \mathbb{R}^{\geq} \to \mathbb{R}^{\geq}$ defined by $f(x) = x^2$. This function is bijective and its inverse is the function $g : \mathbb{R}^{\geq} \to \mathbb{R}^{\geq}$ defined by $g(x) = \sqrt{x}$.

Exercise 5.46. *Let*

$$\frac{\pi}{2} + \pi\mathbb{Z} = \left\{\frac{\pi}{2} + k\pi : k \in \mathbb{Z}\right\}. \tag{5.123}$$

Explain why the tangent function $\tan : \mathbb{R} - \left(\frac{\pi}{2} + \pi\mathbb{Z}\right) \to \mathbb{R}$, *when restricted to the open interval* $(-\frac{\pi}{2}, \frac{\pi}{2})$, *has an inverse function.*

What other open intervals yield similar inverse functions?

How do the graphs of the various inverse tangent functions compare?

5.4.5. Injections, surjections, and bijections between finite sets.

In this section we consider functions between finite sets from the point of view of counting. We will take a less rigorous approach, usually omitting proofs, since the results are especially intuitive.

5.4.5.1. *The cardinality of a finite set.*

Definition 5.56. Given a set X, if there is a bijection from $\mathbb{N}_n = \{1, 2, 3, \ldots, n\}$ to X, then we say that X has **cardinality** n and we write $|X| = n$. If $|X| = n$ for some $n \geq 0$, then we say that X is **finite**. Otherwise, we say X is **infinite**.

Simply put, the cardinality of a finite set is the number of (distinct) elements it has. In the definition, we have made this notion more formal by choosing the "model set" $\mathbb{N}_n$ of n elements. A set X is "equivalent" to this set from the viewpoint of cardinality if there exists a bijection between X and $\mathbb{N}_n$.

Lemma 5.57. *Suppose that Y is a finite set and B is a subset of Y. Then B is a finite set and $|B| \leq |Y|$.*

That is, the number of elements of a subset of a finite set is less than or equal to that finite set. For example, any subset of $\mathbb{N}_n$ has cardinality at most n.

5.4.5.2. *Injections between finite sets.*

The domain of an injection has no more elements than the codomain.

Theorem 5.58. *If $f : X \to Y$ is an injection between finite sets, then*

$$|X| \leq |Y|. \tag{5.124}$$

In particular, for any injection from $\mathbb{N}_n$, the codomain has cardinality at least n. And for any injection to $\mathbb{N}_n$, the domain has cardinality at most n. The basic reason the theorem above is true is that the image set $f(X)$, which is a subset of the codomain Y, has the same cardinality as X.

By taking the contrapositive of the statement of this theorem, we obtain the following, which is referred to as the **pigeonhole principle**.

Corollary 5.59. *Let X and Y be finite sets. If $|X| > |Y|$ and $f : X \to Y$ is a function, then f is not an injection.*

Intuitively, if the domain has more elements than the codomain, then the domain has more elements than the image, which means that there must exist two distinct elements which map to the same element.

5.4.5.3. *Surjections between finite sets.* The codomain of a surjection has no more elements than the domain.

Theorem 5.60. *Let X and Y be finite sets. If there is a surjection $f : X \to Y$, then $|X| \geq |Y|$.*

In particular, for any surjection from $\mathbb{N}_n$, the codomain has cardinality at most n. And for any surjection to $\mathbb{N}_n$, the domain has cardinality at least n.

The contrapositive of this is:

Corollary 5.61. *Let X and Y be finite sets. If $|X| < |Y|$ and $f : X \to Y$, then f is not a surjection.*

The basic reason for the theorem above is: For any function between finite sets, the cardinality of the image is at most the cardinality of the domain, and in the case of a surjection, the image is equal to the codomain.

5.4.5.4. *Bijections between finite sets.* Since bijections are one-to-one correspondences, we have the following.

Theorem 5.62. *Let X and Y be finite sets. If $f : X \to Y$ is a bijection, then $|X| = |Y|$.*

Proof. This is a direct consequence of Theorems 5.58 and 5.60. $\square$

In particular, for any bijection from $\mathbb{N}_n$ to $\mathbb{N}_m$, we must have that $n = m$.

If the domain and codomain of a function have the same number of elements, then being injective is the same as being surjective.

Theorem 5.63. *Let X and Y be finite sets with $|X| = |Y|$. Then $f : X \to Y$ is an injection if and only if $f : X \to Y$ is a surjection.*

Intuitively, we can think of things as follows. Let $f : X \to Y$ and define $f(X) = \{f(x) : x \in X\}$ (another notation for $f(X)$ is $\operatorname{Im} f$), the image of f. Then

$$|f(X)| \leq |X|, \tag{5.125}$$

and since $f(X) \subseteq Y$ we also have

$$|f(X)| \leq |Y|. \tag{5.126}$$

If f is an injection, then $|f(X)| = |X|$, so that $|X| \leq |Y|$ by (5.126).

If f is a surjection, then $f(X) = Y$, so that $|f(X)| = |Y|$. Hence $|Y| \leq |X|$ by (5.125).

5.5. Arbitrary unions, intersections, and cartesian products

Let X be a set (called the ambient set), and let I be a set (called the index set). A family of subsets of X indexed by I is a function

$$A : I \to \mathcal{P}(X). \tag{5.127}$$

We consider an indexed family of subsets as a collection of sets and we usually denote it by

$$\{A_i\}_{i \in I}. \tag{5.128}$$

The **union** is defined by

$$\bigcup_{i \in I} A_i := \{x \in X : \exists i \in I \text{ such that } x \in A_i\}. \tag{5.129}$$

For example, if X is any set, then

$$\bigcup_{x \in X} \{x\} = X.$$

The **intersection** is defined by

$$\bigcap_{i \in I} A_i := \{x \in X : \forall i \in I,\ x \in A_i\}. \tag{5.130}$$

Example 5.64. Let the ambient set be $\mathbb{R}$. Define $A_i := [-i, i]$, and define $B_i = [-1/i, 1/i]$, for $i \in \mathbb{Z}^+$. Then

$$\bigcup_{i \in I} A_i = \mathbb{R}, \qquad \bigcap_{i \in I} B_i = \{0\}. \tag{5.131}$$

Exercise 5.47. *Prove the facts in Example* 5.64.

Exercise 5.48. *Give an example of a collection* $\{A_i\}_{i \in \mathbb{Z}^+}$ *of non-empty subsets of* $\mathbb{R}$ *satisfying the following properties:*

(i) $A_{i+1} \subset A_i$ *for all* $i \in \mathbb{Z}^+$.

(ii) $\bigcap_{i \in \mathbb{Z}^+} A_i = \emptyset$.

Example 5.65. Let $f : X \to Y$ be a function. Consider the family of subsets of X indexed by Y defined by

$$A : Y \to \mathcal{P}(X), \tag{5.132}$$

where

$$A(y) = f^{-1}(y). \tag{5.133}$$

Then

$$\bigcup_{y \in Y} f^{-1}(y) = X \tag{5.134}$$

is a disjoint union.

Let $\{X_i\}_{i\in I}$ be an indexed family of sets. We define its cartesian product

$$\prod_{i\in I} X_i \tag{5.135}$$

to be the set of functions

$$f : I \to \bigcup_{i\in I} X_i \tag{5.136}$$

satisfying

$$f(i) \in X_i. \tag{5.137}$$

In the case where the index set I is infinite, we will assume the:

Axiom of Choice. Let $\{X_i\}_{i\in I}$ be an indexed family of non-empty sets. Then there exists a choice function

$$\text{choice} : I \to \bigcup_{i\in I} X_i \tag{5.138}$$

satisfying

$$\text{choice}(i) \in X_i \quad \text{for all } i \in I. \tag{5.139}$$

With this assumption, we see that if all the sets X_i, $i \in I$, are non-empty, then their cartesian product is also non-empty.

Let us see how this definition compares with the n-fold cartesian product definition in the case where I is a finite set. Without loss of generality, assume that $I = \mathbb{N}_n$, where $n \in \mathbb{Z}^+$. Define the function

$$F : \prod_{i\in \mathbb{N}_n} X_i \to X_1 \times X_2 \times \cdots \times X_n \tag{5.140}$$

by

$$F(f) = (f(1), f(2), \ldots, f(n)). \tag{5.141}$$

We leave it as an exercise to show that the inverse of F is given by the function

$$G : X_1 \times X_2 \times \cdots \times X_n \to \prod_{i\in \mathbb{N}_n} X_i \tag{5.142}$$

defined by

$$G(x_1, x_2, \ldots, x_n) = f_{\mathbf{x}}, \tag{5.143}$$

where $f_{\mathbf{x}} : \mathbb{N}_n \to \bigcup_{i\in\mathbb{N}_n} X_i$ is given by

$$f_{\mathbf{x}}(i) = x_i. \tag{5.144}$$

Thus F and G are bijective. For this reason we consider the two definitions of the cartesian product of n sets to be equivalent.

5.6*. Universal properties of surjections and injections

In this section, we consider what may be considered as a more advanced topic, "universal properties". This perspective is useful in abstract algebra and its related fields.

We have the following elegant, but abstract, characterization of injective functions.

Theorem 5.66. *Let $f : X \to Y$ be a function. Then*

$$f \text{ is an injection} \quad \Leftrightarrow \quad \text{there exists a function } g : Y \to X \text{ such that } g \circ f = I_X.$$

The function g is called a ***left inverse*** *of f.*

Proof. We assume that $X \neq \emptyset$; otherwise it is tautological.

($\Leftarrow$) Suppose that there exists a function $g : Y \to X$ such that $g \circ f = I_X$. Let $x_1, x_2 \in X$ be such that $f(x_1) = f(x_2)$. Then $g(f(x_1)) = g(f(x_2))$, just because g is a function. Since $g \circ f = I_X$, this says that $x_1 = x_2$. We have proved that f is an injection.

($\Rightarrow$) Suppose that f is an injection. We need to define the left inverse g. For this purpose, we choose an element $x_0 \in X$. Let $y \in Y$.

Case 1. $y \in \operatorname{Im} f$. Since f is an injection, there *exists* a *unique* $x \in X$ such that $f(x) = y$.[3] Define $g(y) = x$.

Case 2. $y \notin \operatorname{Im} f$. Define $g(y) = x_0$. (It doesn't matter how we define g in this case.)

Now that we have defined g, we prove that $g \circ f = I_X$. Let $x \in X$. Then since $f(x) \in \operatorname{Im} f$, we have $g(f(x)) = x$ by Case 1. This proves $g \circ f = I_X$. □

Theorem 5.66 may be summarized by the following **commutative diagram**:

(5.145)
$$\begin{array}{ccc} X & & \\ \downarrow f & \searrow I_X & \\ Y & \overset{g}{\dashrightarrow} & X \end{array}$$

The arrow $\rightarrowtail$ denotes that f is injective. The arrow $\dashrightarrow$ denotes that g is the function posited to exist. By the diagram being commutative, we mean that both ways of getting from X to X are equal; namely, $g \circ f = I_X$.

The surjection universal property is:

Theorem 5.67. *Let $f : X \to Y$ be a function. Then*

$$f \text{ is a surjection} \quad \Leftrightarrow \quad \text{there exists a function } h : Y \to X \text{ such that } f \circ h = I_Y.$$

The function h is called a ***right inverse*** *of f.*

Exercise 5.49. *Prove Theorem* 5.67.

[3]Details: **(i)** *Existence.* Such an x exists by the definition of $y \in \operatorname{Im} f$.
(ii) *Uniqueness.* Suppose $x' \in X$ is also such that $f(x') = y$. Then $f(x') = f(x)$. Since f is an injection, $x' = x$.

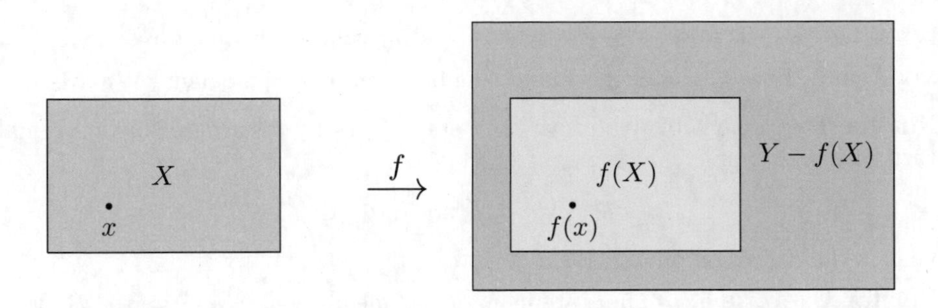

Figure 5.6.1. Schematic picture of an injective function $f : X \to Y$.

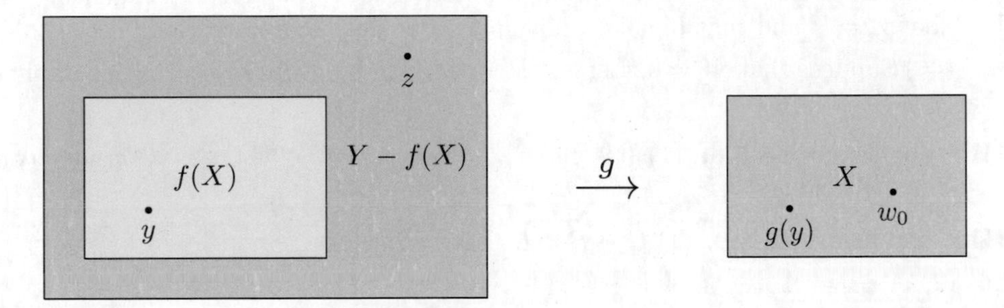

Figure 5.6.2. Picturing a left inverse $g : Y \to X$ of f. On $f(X)$, g undoes what f does. On $Y - f(X)$, g can be defined arbitrarily; e.g., given $w_0 \in X$, we can define $g(z) = w_0$ for all $z \in Y - f(X)$.

Theorem 5.67 may be summarized by the following commutative diagram:

$$\begin{array}{ccc} Y & & \\ \downarrow h & \searrow I_Y & \\ X & \overset{f}{\twoheadrightarrow} & Y \end{array} \tag{5.146}$$

The arrow $\twoheadrightarrow$ denotes that f is surjective.

5.7. Hints and partial solutions for the exercises

Hint for Exercise 5.1. What are the four elements of $\mathcal{S}$? Is $Q\heartsuit$ one of these elements?

Hint for Exercise 5.2. What can you say about the implication $x \in \emptyset \Rightarrow x \in A$ for every object x?

Hint for Exercise 5.3. Let A be a subset of $\emptyset$. Let $x \in A$. Use that $A \subset \emptyset$ and continue to get a contradiction.

Hint for Exercise 5.4.

(i) We actually have equality.

(ii) $\{\emptyset\} \in \{\emptyset\}$. What is the sole element of $\{\emptyset\}$? Is $\{\emptyset\}$ equal to this element?

(iii) What is the sole element of $\{\{\emptyset\}\}$?

(iv) $\{\{\emptyset\}\}$ has exactly two subsets. Is $\{\emptyset\}$ one of them?

(v) What property does the empty set have regarding the subset relation?

(vi) Again, the set on the right-hand side has exactly one element. What is it?

Hint for Exercise 5.5. We prove both statements by a string of biconditionals. Start with

$$x \in A \Leftrightarrow \text{not(not } (x \in A)).$$

What is the definition of $x \in A^c$?

Hint for Exercise 5.6. The statement $A \subset B$ means that $x \in A \Rightarrow x \in B$. What is the contrapositive of this implication?

Hint for Exercise 5.7. (1) ($\Leftarrow$) Suppose that $A = U$. Let $x \in A^c$. Then $x \notin A$. Use that $A = U$ and that U is the universal set to get a contradiction.

($\Rightarrow$) Suppose that $A^c = \emptyset$. Let $x \in U$. Clearly $x \notin \emptyset$. Since $A^c = \emptyset$, this implies that $x \notin A^c$. Continue.

Hint for Exercise 5.8. (1) It suffices to prove that $U \subset A \cup A^c$. Consider two cases: $x \in A$ and $x \notin A$.

Hint for Exercise 5.9. (1) $U - B = U \cap B^c = B^c$.

(2) $A - \emptyset = A \cap \emptyset^c = A \cap U = A$.

Hint for Exercise 5.10. (1) Suppose that $A \subset B$. Prove that $A \cap B \subset A$ and $A \subset A \cap B$.

Hint for Exercise 5.11. Prove ($\Leftarrow$) and ($\Rightarrow$), where in the latter case you may wish to use Lemma 5.15.

Hint for Exercise 5.12. Take the proof of (5.17) and interchange $\cap$ and $\cup$ and "and" and "or".

Hint for Exercise 5.13. Use the distributivity property for the union and intersection of sets (5.22).

Hint for Exercise 5.14. (1) We start the calculation by

$$\mathbf{x} \cdot \mathbf{y} = x_1 y_1 + x_2 y_2 + \cdots + x_n y_n.$$

Continue.

Hint for Exercise 5.15. Let $(x, y) \in A = [-r, r] \times [-r, r]$. Then $-r \le x \le r$ and $-r \le y \le r$, so that $|x| \le r$ and $|y| \le r$. Continue.

Hint for Exercise 5.16. Use that

$$|\mathbf{x} - \mathbf{y}|^2 = |\mathbf{x}|^2 + |\mathbf{y}|^2 - 2\mathbf{x} \cdot \mathbf{y}. \tag{5.147}$$

Hint for Exercise 5.17. Show that

$$|\mathbf{x} + \mathbf{y}|^2 \le |\mathbf{x}| + |\mathbf{y}|^2. \tag{5.148}$$

Hint for Exercise 5.18. Let $\mathbf{x}, \mathbf{y} \in \ker(\ell)$, and let $a, b \in \mathbb{R}$. Using that ℓ is a linear function, compute

$$\ell(a\mathbf{x} + b\mathbf{y}). \tag{5.149}$$

Hint for Exercise 5.19. Use that $\ell(\mathbf{x}) := \mathbf{x}\cdot\mathbf{v}$ is a linear function. Also observe that

$$\{\mathbf{x}\in\mathbb{R}^n : \mathbf{x}\cdot\mathbf{v}=c\} = \left\{\mathbf{x}\in\mathbb{R}^n : \left(\mathbf{x}-\frac{c\mathbf{v}}{|\mathbf{v}|^2}\right)\cdot\mathbf{v}=0\right\}$$
$$= \left\{\mathbf{y}+\frac{c\mathbf{v}}{|\mathbf{v}|^2}\in\mathbb{R}^n : \mathbf{y}\cdot\mathbf{v}=0\right\}.$$

Hint for Exercise 5.20. Since the universal quantifier "for all" is stronger than the existential quantifier "there exists", we have the $\Leftarrow$ direction of the biconditional being true.

Proof of the $\Rightarrow$ direction: By definition, we have that there exists $\mathbf{b}\in A$ such that

$$A_{\mathbf{b}} = \{\mathbf{x}-\mathbf{b} : \mathbf{x}\in A\}$$

is a linear subspace of $\mathbb{R}^n$.

Now let $\mathbf{a}\in A$. Show that

$$A_{\mathbf{a}} = A_{\mathbf{b}}. \tag{5.150}$$

Hint for Exercise 5.21. Let

$$A := L+\mathbf{x}_0 = \{\mathbf{x}+\mathbf{x}_0 : \mathbf{x}\in L\}, \tag{5.151}$$

where L is a linear subspace of $\mathbb{R}^n$ and $\mathbf{x}_0\in\mathbb{R}^n$.

Show that

$$A_{-\mathbf{x}_0} = \{\mathbf{y}-\mathbf{x}_0 : \mathbf{y}\in A\} = L.$$

Hint for Exercise 5.22. (i) We start the calculation by

$$f(\mathbf{x})\cdot\mathbf{n} = \frac{\mathbf{x}-(\mathbf{x}\cdot\mathbf{n})\,\mathbf{n}}{1-\mathbf{x}\cdot\mathbf{n}}\cdot\mathbf{n}.$$

Continue, while using that $|\mathbf{n}|=1$.

(ii) One way to answer this part is for $\mathbf{y}\in L$ to show that one can uniquely solve for s in the equation $|\mathbf{x}|^2=1$, where $\mathbf{x}=(1-s)\mathbf{n}+s\mathbf{y}$. This will derive the formula for the inverse function g of f.

Another way to see that g is the inverse of f is to make two calculations. Firstly, letting

$$\mathbf{x} := g(\mathbf{y}) = \frac{4\mathbf{y}+(|\mathbf{y}|^2-1)\,\mathbf{n}}{|\mathbf{y}|^2+3}, \tag{5.152}$$

show that

$$\mathbf{x}\cdot\mathbf{n} = \frac{|\mathbf{y}|^2-5}{|\mathbf{y}|^2+3}, \quad \text{so that} \quad 1-\mathbf{x}\cdot\mathbf{n} = \frac{8}{|\mathbf{y}|^2+3}, \quad 1+\mathbf{x}\cdot\mathbf{n} = \frac{2(|\mathbf{y}|^2-1)}{|\mathbf{y}|^2+3}.$$

Using this, show that

$$f(g(\mathbf{y})) = \mathbf{y}.$$

This will show that g is a right inverse of f.

Secondly, show that g is a left inverse of f.

Hint for Exercise 5.23. The intersection of the line $\{t\mathbf{v} : t \in \mathbb{R}\}$ with the unit sphere S^n is the set of all $t\mathbf{v}$, where $t \in \mathbb{R}$ is such that $|t\mathbf{v}| = 1$. Using that $\mathbf{v}$ is assumed to be non-zero, we see that $|t| = 1/|\mathbf{v}|$. Continue.

Hint for Exercise 5.24. It is easy to show that the function exists since $\mathbf{x} \in S^n$ implies that it is distinct from the origin, and for every two distinct points there exists a unique line containing the two points. Continue the explanation.

Hint for Exercise 5.25. None of these define functions from X to Y, for different reasons.

Hint for Exercise 5.26. The calculation that

$$f(f(\mathbf{x})) = f(\mathbf{x} - 2(\mathbf{x} \cdot \mathbf{u})\,\mathbf{u}) = \mathbf{x}$$

is rather straightforward. This function is a reflection about the hyperplane

$$H := \{\mathbf{x} \in \mathbb{R}^n : \mathbf{x} \cdot \mathbf{u} = 0\}. \tag{5.153}$$

Observe that if $\mathbf{x} \in H$, then

$$f(\mathbf{x}) = \mathbf{x} - 2(\mathbf{x} \cdot \mathbf{u})\,\mathbf{u} = \mathbf{x}. \tag{5.154}$$

So the function f fixes points on the hyperplane H. Now, let $\mathbf{x} \in \mathbb{R}^n$. Then the projection π of $\mathbf{x}$ onto H is given by

$$\pi(\mathbf{x}) := \mathbf{x} - (\mathbf{x} \cdot \mathbf{u})\,\mathbf{u}. \tag{5.155}$$

To show that the function f is a mirror image, we add to $\pi(\mathbf{x})$ the vector $\pi(\mathbf{x}) - \mathbf{x}$ to obtain

$$\pi(\mathbf{x}) + \pi(\mathbf{x}) - \mathbf{x} = \mathbf{x} - 2(\mathbf{x} \cdot \mathbf{u})\,\mathbf{u}.$$

Draw some pictures to explain what this shows about the function f.

Hint for Exercise 5.27. We have

$$(g \circ f_k)(n) = g(f_k(n)) = g(n^k). \tag{5.156}$$

Consider the cases where n is even and where n is odd.

Hint for Exercise 5.28. (1) We compute that

$$((h \circ g) \circ f)(x) = (h \circ g)(f(x)) = h(g(f(x))).$$

On the other hand,

$$(h \circ (g \circ f))(x) = h((g \circ f)(x)) = h(g(f(x))).$$

Therefore $(h \circ g) \circ f = h \circ (g \circ f)$.

Hint for Exercise 5.29. Since the domain of the natural logarithm function is the set of positive real numbers, the domain D of f is the set

$$D = \left\{(x, y) \in \mathbb{R}^2 : \frac{\cos(x)}{\cos(y)} > 0\right\}. \tag{5.157}$$

Show that the domain D is the union of open squares (that is, the interior of squares not including their boundaries) of side lengths π centered at the points

$$(2m\pi, 2n\pi) \quad \text{and} \quad ((2m+1)\pi, (2n+1)\pi), \quad \text{where } m, n \in \mathbb{Z}. \tag{5.158}$$

Hint for Exercise 5.30. Since $a \neq 0$, we have $a > 0$ or $a < 0$. We assume $a > 0$, and leave it to you to work out the similar case where $a < 0$. We have

$$\lim_{x\to\infty} f(x) = +\infty, \tag{5.159}$$

whereas

$$\lim_{x\to-\infty} f(x) = -\infty. \tag{5.160}$$

Let y_0 be any real number. By (5.159), there exists $b \in \mathbb{R}$ such that $f(b) > y_0$. On the other hand, by (5.160), there exists $a \in \mathbb{R}$ such that $f(a) < y_0$. Apply the Intermediate Value Theorem.

Hint for Exercise 5.31. Let $y \in Y$. Since f is a surjection, there exists $m \in \mathbb{Z}$ such that $f(m) = y$.

Case 1: m is even. That is, $m \in E$. So, in this case,

$$g(m) = f(m) = y.$$

Case 2: m is odd. Then $m = 2\ell + 1$ for some integer ℓ. So

$$g(2\ell) = f(2\ell) = f(2\ell + 1) = f(m) = y.$$

We have proved that g is a surjection.

Hint for Exercise 5.32. Suppose that $m, n \in \mathbb{Z}^+$ satisfy $g(m) = g(n)$. Then $|g(m)| = |g(n)|$. Continue to prove that g is injective.

Suppose that $g(n) = 1$. Derive a contradiction.

Hint for Exercise 5.33. This function is surjective, but it is not injective.

Hint for Exercise 5.34. Suppose that $z_1, z_2 \in X \cup Y$ are such that $h(z_1) = h(z_2)$.

Case 1: $z_1 \in X$. Then $h(z_2) = h(z_1) = f(z_1) \in Y$. Since X and Y are disjoint, this implies that $z_2 \in X$ (if $z_2 \in Y$, then $h(z_2) \in X$, which is a contradiction since X and Y are disjoint). Thus we have that

$$f(z_2) = h(z_2) = f(z_1).$$

Since f is injective, this implies that $z_1 = z_2$.

Case 2: $z_1 \in Y$. Then $h(z_2) = h(z_1) = g(z_1) \in X$. Since X and Y are disjoint, this implies that $z_2 \in Y$. Thus we have that

$$g(z_2) = h(z_2) = g(z_1).$$

Since g is injective, this implies that $z_1 = z_2$.

We have proved that h is an injection.

Remark. A variant on the proof above is to split the two cases instead as to whether $h(z_1) = h(z_2)$ is (1) an element of X or (2) an element of Y. In the first case one shows that $z_1, z_2 \in Y$ and uses that g is an injection, and in the second case one shows that $z_1, z_2 \in X$ and uses that f is an injection. We leave it to the reader to fill in the details.

Hint for Exercise 5.35. (1) $m = 3$. The functions f_1 and f_2 are injections.

(2) $m = 4$. The functions f_1 and f_3 are injections. The function f_2 is not an injection.

(3) No.

(4) Yes.

Hint for Exercise 5.36. As we will prove in a later chapter, for functions between finite sets of the same cardinality, they are injections if and only if they are surjections. So the answers for this exercise are the same as the answers for the previous exercise.

Hint for Exercise 5.37.

(1) If k is an odd positive integer, then f_k is a bijection (and hence an injection and a surjection).

(2) If k is an even positive integer, then f_k is neither an injection nor a surjection, and hence not a bijection.

Hint for Exercise 5.38. Firstly, if k is a positive odd integer, then $f_k : \mathbb{R} \to \mathbb{R}$ is a bijection with inverse function given by

$$(f_k)^{-1}(y) = y^{1/k}. \tag{5.161}$$

What does this say about $f_k^{-1}(y)$ for each $y \in \mathbb{R}$?

Secondly, suppose that k is a positive even integer. Consider the three cases $y < 0$, $y = 0$, and $y > 0$.

Hint for Exercise 5.39. Suppose that $x \in f^{-1}(B) \cap f^{-1}(C)$. Show that $f(x) \in B \cap C$. Derive a contradiction.

Hint for Exercise 5.40. The Fibonacci sequence has the pattern "odd, odd, even".

Hint for Exercise 5.41. For $a \in \mathbb{Z}$, the pre-image

$$\mathbf{r}^{-1}(a) = \{a + qm : q \in \mathbb{Z}\}. \tag{5.162}$$

Hint for Exercise 5.42. Based on our solution to the previous exercise, we propose the identity

$$f^{-1}(B) \cap f^{-1}(C) = f^{-1}(B \cap C). \tag{5.163}$$

Show that

$$f^{-1}(B) \cap f^{-1}(C) \subset f^{-1}(B \cap C) \tag{5.164}$$

and

$$f^{-1}(B \cap C) \subset f^{-1}(B) \cap f^{-1}(C). \tag{5.165}$$

Hint for Exercise 5.43. (1) Let $y \in f(f^{-1}(B))$. Show that $y \in B$.

(2) Suppose that $f(f^{-1}(B)) = B$. Let $y \in B$. Show that $y \in \text{Im}(f)$.

Conversely, suppose that $B \subset \text{Im}(f)$. Let $y \in B$. Show that $y \in f(f^{-1}(B))$.

(3) We have $f(f^{-1}(B))$ a proper subset of B if and only if $f(f^{-1}(B)) \neq B$, which by part (2) is equivalent to $B \not\subset \text{Im}(f)$.

Hint for Exercise 5.44. (1) Let $x \in A$. Then $f(x) \in f(A)$. Thus $x \in f^{-1}(f(A))$.

(2) We have $x \in f^{-1}(f(A))$ if and only if $f(x) \in f(A)$. So $A = f^{-1}(f(A))$ means the same thing as

$$x \in A \text{ if and only if } f(x) \in f(A).$$

In particular, if $x \notin A$, then $f(x) \notin f(A)$.

Hint for Exercise 5.45. Let $x_1, x_2 \in X$ be such that $(g \circ f)(x_1) = (g \circ f)(x_2)$. Then $g(f(x_1)) = g(f(x_2))$. Since $f(X) \subset B$, we have

$$f(x_1), f(x_2) \in B.$$

By our hypothesis on g, this implies that $f(x_1) = f(x_2)$. Now, since f is an injection, we have that $x_1 = x_2$. We conclude that $g \circ f$ is an injection.

Hint for Exercise 5.46. The tangent function has the property that it is strictly increasing on the open interval $(-\frac{\pi}{2}, \frac{\pi}{2})$ (open means that the interval does not include its endpoints). In fact,

$$\lim_{x \to \pi/2} \tan(x) = +\infty, \qquad \lim_{x \to -\pi/2} \tan(x) = -\infty. \tag{5.166}$$

Apply the Intermediate Value Theorem.

We also have that the derivative

$$\frac{d}{dx} \tan(x) = \sec^2(x) > 0 \tag{5.167}$$

for $x \in (-\frac{\pi}{2}, \frac{\pi}{2})$. What can you conclude from this?

The analogous facts hold on any open interval of the form

$$\left(k\pi - \frac{\pi}{2}, k\pi + \frac{\pi}{2}\right), \tag{5.168}$$

where $k \in \mathbb{Z}$.

The (infinite number of) graphs of the various inverse tangent functions are stacked on top of each other, where any given inverse is a "distance" π above the inverse right below it (and so of course a "distance" π below the inverse right above it).

Hint for Exercise 5.47. (1) Let $x \in \mathbb{R}$. Then there exists a positive integer i such that $i \geq |x|$.

(2) Let x be a non-zero real number. Then there exists a positive integer i such that $1/i < |x|$.

Hint for Exercise 5.48. There are many examples!

For instance, we may define

$$A_i = \{x \in \mathbf{F} : x > i\}, \tag{5.169}$$

where $\mathbf{F}$ is $\mathbb{R}$, $\mathbb{Q}$, or $\mathbb{Z}$.

Another example is

$$A_i = \{x \in \mathbf{F} : 0 < x < 1/i\}, \tag{5.170}$$

where $\mathbf{F}$ is $\mathbb{R}$ or $\mathbb{Q}$.

Hint for Exercise 5.49. ($\Leftarrow$) Let $y \in Y$. By assumption, there exists a function $h : Y \to X$ such that $f \circ h = I_Y$. We have

$$y = I_Y(y) = (f \circ h)(y) = f(h(y)). \tag{5.171}$$

Thus, there exists $x \in X$, namely $x = h(y)$, such that $f(x) = y$. This proves that f is a surjection.

($\Rightarrow$) Now suppose that f is a surjection. Let $y \in Y$. Since f is a surjection, the pre-image set $f^{-1}(y)$ is non-empty. Choose $x \in f^{-1}(y)$ and define $h(y) = x$. This defines a function $h : Y \to X$. We compute that for every $y \in Y$,

$$(f \circ h)(y) = f(h(y)) = f(x), \quad \text{where } x \in f^{-1}(y). \tag{5.172}$$

Since $x \in f^{-1}(y)$, $f(x) = y$. Therefore

$$(f \circ h)(y) = f(x) = y = I_Y(y). \tag{5.173}$$

We have proved that $f \circ h = I_Y$; i.e., h is a right inverse of f.

Chapter 6

Modular Arithmetic

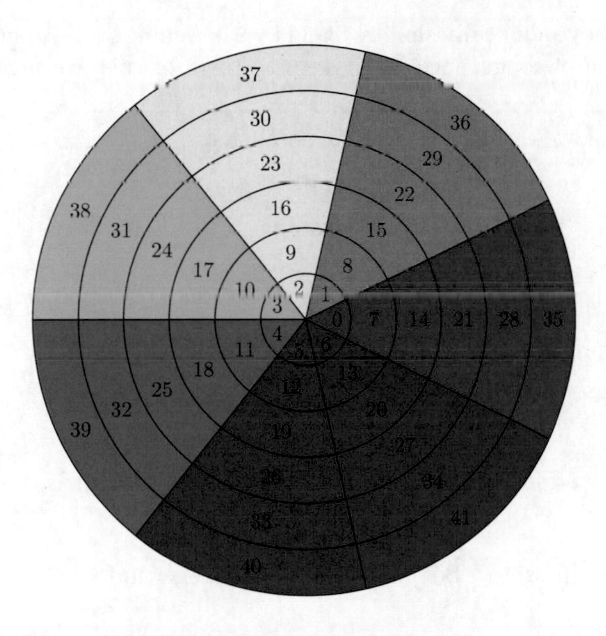

Figure 6.0.1. A "modulo 7" number wheel. Can you explain what this wheel represents? (It's not a dartboard. ☺)

Goals of this chapter: To understand a powerful concept in number theory: congruence, which is based on division. To solve linear congruence equations. To prove an important result in number theory: Fermat's Little Theorem. To see its application to cryptography.

6.1. Multiples of 3 and 9 and the digits of a number in base 10

Modular arithmetic is actually a topic whose basics you are familiar with. We start with a couple of division facts which you may have learned in secondary school.

There is a nice way to see if a (positive) integer is divisible by 3. For example, consider the integer 123456789. We may add up all the digits to get

$$1+2+3+4+5+6+7+8+9=45. \tag{6.1}$$

If the result (in this case the number 45) is divisible by 3, then the original number is divisible by 3. Since 45 is indeed divisible by 3, we conclude that 123456789 is divisible by 3. Note that we can even check that 45 is divisible by 3 by adding up its digits to get 9, which is divisible by 3.

Moreover, if the result is divisible by 9, then the original number is divisible by 9. So, since 45 is actually divisible by 9, we obtain the stronger result that 123456789 is divisible by 9. Indeed, we check that

$$123456789 = 9 \cdot 13717421. \tag{6.2}$$

The facts above about division by 3 and by 9 are some of the many consequences of the arithmetic of congruence. To give us a taste of this, we now give proofs of these facts.

Observe that we can write a 2-digit number as

$$a_1 a_0 = a_1 10^1 + a_0 10^0, \tag{6.3}$$

where $0 \leq a_0, a_1 \leq 9$ and $a_1 > 0$. For example, $45 = 4 \cdot 10^1 + 5 \cdot 10^0$. In general, we can write an $(n+1)$-digit number as

$$a_n a_{n-1} \cdots a_2 a_1 a_0 = a_n 10^n + a_{n-1} 10^{n-1} + \cdots + a_2 10^2 + a_1 10^1 + a_0 10^0, \tag{6.4}$$

where $0 \leq a_0, a_1, a_2, \ldots, a_{n-1}, a_n \leq 9$ and $a_n > 0$.

Next, we observe that each of the numbers

$$9,\ 99,\ 999,\ 9999,\ \ldots \tag{6.5}$$

is divisible by 9 and in particular divisible by 3. In other words,

$$10^k - 1 \text{ is divisible by } 9$$

for every positive integer k. So, if we divide our integer in (6.4) by 9, we can obtain a remainder of

$$a_n + a_{n-1} + \cdots + a_2 + a_1 + a_0. \tag{6.6}$$

Another way of saying this is that the integers $a_n a_{n-1} \cdots a_2 a_1 a_0$ and $a_n + a_{n-1} + \cdots + a_2 + a_1 + a_0$ differ by a multiple of 9. For example, $345 = 3+4+5+3 \cdot 99+4 \cdot 9$.

Therefore, we have proved:

Proposition 6.1. The $(n+1)$-digit integer $a_n a_{n-1} \cdots a_2 a_1 a_0$ is divisible by 9 if and only if the integer $a_n + a_{n-1} + \cdots + a_2 + a_1 + a_0$ formed by summing its digits is divisible by 9.

Since 3 divides 9, we have that $10^k - 1$ is divisible by 3. So, as a bonus we get the additional fact that $a_n a_{n-1} \cdots a_2 a_1 a_0$ is divisible by 3 if and only if the integer $a_n + a_{n-1} + \cdots + a_2 + a_1 + a_0$ is divisible by 3.

For example, $345 = 3+4+5+3\cdot 33\cdot 3+4\cdot 3\cdot 3$. So 345 is divisible by 3 if and only if $3+4+5=12$ is divisible by 3. Since 12 is divisible by 3, we conclude that 345 is divisible by 3. Note that we may also conclude that 345 is not divisible by 9.

In the arguments above, we actually used *modular arithmetic*, a.k.a. the arithmetic of congruence. After you feel that you have read enough of this chapter, we invite you to revisit the discussion above and prove the facts by modular arithmetic.

6.2. Congruence modulo m

Consider the days of a non-leap year, numbered from 1 to 365. If January 1 is a Tuesday, then we know that March 27 is a Wednesday. This is because March 27 is the 86th day of the year and $86 = 7\cdot 12+2$, so March 27 is the same day of the week as the second day of the year, which is a Wednesday. Here, we doing arithmetic "modulo 7".

Recall the modulo 7 number wheel in Figure 6.0.1 at the beginning of this chapter. If, for example, we look to the right of its center, then we see the numbers

$$2, 9, 16, 23, 30.$$

Another way of saying this is that we see

$$7k+2, \quad \text{where } k = 0, 1, 2, 3, 4.$$

So each of these integers, after being divided by 7, has a remainder of 2.

The integers in a wedge of a given color in Figure 6.0.1 have the same remainder modulo 7. When integers have the same remainder (modulo a fixed number such as 7), we can think of the integers as being the same or equivalent from a perspective. In the next subsection we will make this precise.

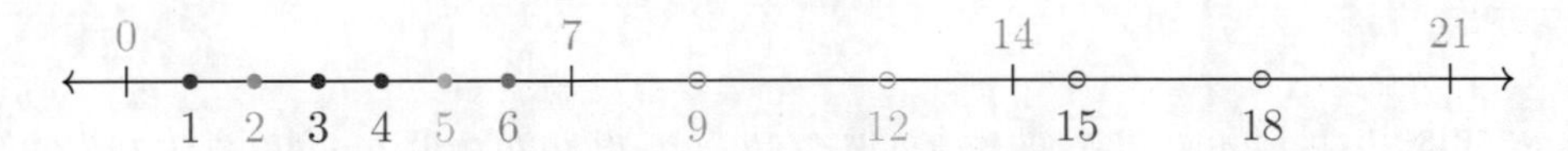

Figure 6.2.1. The first six positive multiples of 3 and their remainders after dividing by 7. The remainder of each multiple is congruent to the multiple, and hence they are colored the same.

6.2.1. Definition of congruence modulo m.

Now we dig into the mathematics of modular arithmetic.

Let m be a positive integer.

Definition 6.2. We say that integers a and b are **congruent modulo** m if and only if $a-b$ is divisible by m; that is, $a-b$ is a multiple of m.

For example:

(1) The integers 55 and 55 itself are congruent modulo 3.
(2) The integers 277 and 2 are congruent modulo 5.
(3) The integers -3, 17, 29, and 37 are all congruent to each other modulo 4.
(4) The integers 2, 9, 16, 23, 30 are all congruent to each other modulo 7.

(5) In Figure 6.2.1, numbers of the same color are congruent to each other modulo 7. E.g., $18 \equiv 4 \bmod 7$.

Non-examples are:

(1) The integers -2 and 2 are not congruent modulo 5.

(2) No two of the integers $16, 25, 49, 122, 253$ are congruent modulo 5.

(3) The integers 123456789 and 9876543210 are not congruent modulo m if m is even.

(4) A fun fact is that if m is odd, then $-a$ and a are not congruent modulo m if and only if a and 0 are not congruent modulo m. At some point soon during your reading of this chapter, you should be able to prove this.

We often write the condition that a and b are congruent modulo m as

$$m \,|\, (a - b) \tag{6.7}$$

(i.e., m divides $a - b$) and we often use the fact that this means that there *exists* $q \in \mathbb{Z}$ such that

$$a - b = m\,q.$$

For example,

$$(-23) \,|\, 92, \qquad 34 \,|\, 204. \tag{6.8}$$

One writes

$$a \equiv b \mod m$$

for a and b being congruent modulo m. For example,

$$277 \equiv 2 \bmod 5 \quad \text{and} \quad -3 \equiv 17 \bmod 4.$$

Exercise 6.1. *Show that two integers a and b are congruent modulo 2 if and only if either a and b are both even or a and b are both odd.*

Idea: The difference of two even integers is even. The difference of two odd integers is even. Etc.

Exercise 6.2. (1) *Prove by induction that for every non-negative integer k,*

$$10^k \equiv 1 \bmod 9. \tag{6.9}$$

(2) *Use part* (1) *to show that for every non-negative integer k,*

$$10^k \equiv 1 \bmod 3. \tag{6.10}$$

6.2.2. Properties of congruence modulo m. The relation of congruence has the following general properties:

(1) Reflexive. Any integer a is congruent to itself modulo m; that is,

$$a \equiv a \bmod m.$$

This is because $a - a = 0$ is divisible by m since $0 = m \cdot 0$.

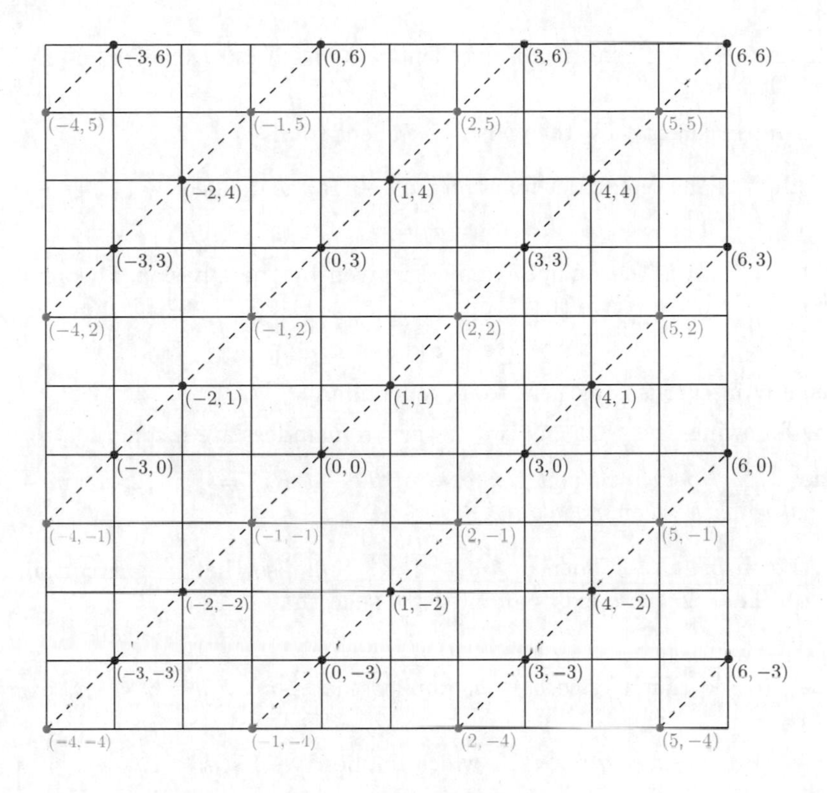

Figure 6.2.2. Pairs of integers (a, b), where $a \equiv b \bmod 3$.

(2) Symmetric. If a is congruent to b modulo m, then b is congruent to a modulo m; i.e.,

$$a \equiv b \bmod m \quad \text{implies} \quad b \equiv a \bmod m.$$

Indeed, $m \mid (a - b)$ implies $m \mid (b - a)$.

(3) Transitive. If a is congruent to b modulo m and if b is congruent to c modulo m, then a is congruent to c modulo m; i.e.,

$$a \equiv b \bmod m \quad \text{and} \quad b \equiv c \bmod m \quad \text{implies} \quad a \equiv c \bmod m.$$

Indeed, $m | (a - b)$ and $m | (b - c)$ implies

$$m | ((a - b) + (b - c)); \quad \text{i.e.,} \quad m | (a - c). \tag{6.11}$$

Remark 6.3. A relation that satisfies the properties above of reflexivity, symmetry, and transitivity is called an **equivalence relation** (see §8.4 below for a general discussion).

If we think of being congruent to as being excellent to, the equivalence relation properties of congruence say the following:

(1) (Self-love) An integer is always excellent to itself.

(2) (Mutual love) If a is excellent to b, then b is excellent to a.

(3) (Transference of love) If a is excellent to b and if b is excellent to c, then a is excellent to c.

So, if a isn't excellent to b, then b isn't excellent to a!

Recall that the remainder function $\mathbf{r} : \mathbb{Z} \to R_m$ is defined by (4.14) as

$$\mathbf{r}(a) = r, \tag{6.12}$$

where $0 \leq r < m$ is the unique remainder given by the Division Theorem. Since $a = m\,\mathbf{q}(a) + \mathbf{r}(a)$, we have that

$$a \equiv \mathbf{r}(a) \bmod m. \tag{6.13}$$

That is, any integer is congruent to its remainder.

The following says that any two distinct remainders are incongruent.

Lemma 6.4. *No two distinct elements of the set* $R_m = \{0, 1, 2, \ldots, m-1\}$ *are congruent to each other modulo* m.

Proof. Let $a, b \in R_m$. Then $0 \leq a, b < m$. Suppose that $a \equiv b \bmod m$. Then $m|(b-a)$. That is, there exists an integer k such that

$$mk = b - a. \tag{6.14}$$

It is easy to see that $0 \leq a, b < m$ implies that $-m < b - a < m$ (exercise!). Therefore

$$-m < mk < m, \quad \text{which implies} \quad -1 < k < 1. \tag{6.15}$$

Since k is an integer, we conclude that $k = 0$, which implies that $a = b$. □

Exercise 6.3. *Let* a, b, m *be real numbers. Prove that if* $0 \leq a, b < m$, *then* $-m < b - a < m$.

Hint: You may use the fact that if $x \leq y$ *and* $z < w$, *then* $x + z < y + w$.

The following says that if two integers are congruent modulo an integer, then they are congruent modulo any positive divisor of that integer.

Lemma 6.5. *If* $x \equiv y \bmod m$ *and if* b *is a positive divisor of* m, *then* $x \equiv y \bmod b$.

Exercise 6.4. *Prove Lemma* 6.5.

Example 6.6. By Lemma 6.5: If $x \equiv y \bmod 140$, then $x \equiv y \bmod 35$.

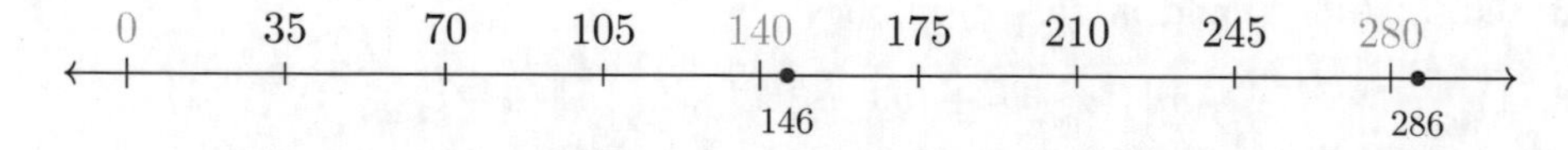

Figure 6.2.3. $286 \equiv 146 \bmod 140$ implies that $286 \equiv 146 \bmod 35$. Indeed, $286 - 146 = 1 \cdot 140 = 4 \cdot 35$.

Solved Problem 6.7. *The prime factorization of* 123456789 *is*

$$123456789 = 3^2 \cdot 3607 \cdot 3803. \tag{6.16}$$

Using this fact, find the smallest integer greater than 123460000 *which is congruent to* 1 *modulo* 3607.

Solution. Firstly, we observe that 123460000 is greater than 123456789 and that their difference

$$123460000 - 123456789 = 3211 \tag{6.17}$$

is less than 3607. By (6.16), $123456789 = 3607q$ for some integer q (note that $q = 3^2 \cdot 3803$, but we won't use this). Thus,

$$123456789 + 1 \equiv 1 \bmod 3607; \tag{6.18}$$

however, $123456789 + 1 < 123460000$. Thus, the smallest integer greater than 123460000 which is congruent to 1 modulo 3607 is

$$123456789 + 3607 + 1 = 123460397. \quad \square \tag{6.19}$$

Exercise 6.5. *Note that* $7 \cdot 11 \cdot 13 = 1001$.

(1) *List the smallest* 5 *integers greater than* 1000 *that are congruent to* 1 *modulo* 13.

(2) *List the largest* 5 *integers less than* 1000 *that are congruent to* 12 *modulo* 7.

Exercise 6.6. (1) *Find the largest integer less than* 456456 *that is congruent to* -1002 *modulo* 1001.

(2) *Find the largest multiple of* 13 *less than* 567567.

(3) *Let* $\mathbf{r} : \mathbb{Z} \to R_{13}$ *be the remainder function. Find* $\mathbf{r}(579583)$ *by "calculating by hand".*

6.2.3. Congruence properties of addition and multiplication. Congruence is compatible with addition, subtraction, and multiplication in the following sense.

Example 6.8 (Congruence and addition and multiplication). Since $46 \equiv 3 \bmod 43$ and $51 \equiv 8 \bmod 43$, we have

$$46 + 51 \equiv 3 + 8 \equiv 11 \bmod 43$$

and

$$46 \cdot 51 \equiv 3 \cdot 8 \equiv 24 \bmod 43.$$

The congruences in the example above are special cases of the following result. This result says that the sums, differences, and products of congruent integers are congruent.

Theorem 6.9 (Congruence and addition, subtraction, and multiplication). *If* $a_1 \equiv a_2 \bmod m$ *and* $b_1 \equiv b_2 \bmod m$, *then*

(1) $a_1 + b_1 \equiv a_2 + b_2 \bmod m$,

(2) $a_1 - b_1 \equiv a_2 - b_2 \bmod m$,

(3) $a_1 b_1 \equiv a_2 b_2 \bmod m$.

Proof. **(1)** By hypothesis, there exist integers k and ℓ such that $a_1 - a_2 = mk$ and $b_1 - b_2 = m\ell$. Hence

$$\begin{aligned}(a_1 + b_1) - (a_2 + b_2) &= (a_1 - a_2) + (b_1 - b_2) \\ &= mk + m\ell \\ &= m\,(k + \ell)\,.\end{aligned}$$

Since $k+\ell$ is an integer, we conclude from the definition of congruence that $a_1+b_1 \equiv a_2+b_2 \bmod m$.

(2) We leave this as an exercise.

(3) By hypothesis, there exist integers k and ℓ such that $a_1 - a_2 = mk$ and $b_1 - b_2 = m\ell$. Hence

$$\begin{aligned}(a_1b_1) - (a_2b_2) &= a_1b_1 + (-a_2b_1 + a_2b_1) - a_2b_2 \\ &= (a_1b_1 - a_2b_1) + (a_2b_1 - a_2b_2) \\ &= (a_1 - a_2)\, b_1 + (b_1 - b_2)\, a_2 \\ &= mkb_1 + m\ell a_2 \\ &= m\,(kb_1 + \ell a_2)\,.\end{aligned}$$

Since $kb_1 + \ell a_2$ is an integer, we conclude from the definition of congruence that $a_1b_1 \equiv a_2b_2 \bmod m$. □

Exercise 6.7. *Prove part* (2) *of Theorem* 6.9.

By using induction, we can extend the sum property of congruence.

Exercise 6.8 (Extending Theorem 6.9(1))**.** *Let m and k be positive integers. Prove that if we have integers a_i and b_i, where $1 \le i \le k$, satisfying*

$$a_i \equiv b_i \bmod m \quad \text{for } 1 \le i \le k, \tag{6.20}$$

then their sums satisfy

$$a_1 + a_2 + \cdots + a_k \equiv b_1 + b_2 + \cdots + b_k \bmod m. \tag{6.21}$$

Revisiting the subject of multiples of 3 and 9 discussed at the beginning of this chapter, we have:

Exercise 6.9. *Prove that*

$$a_n a_{n-1} \cdots a_2 a_1 a_0 \equiv a_n + a_{n-1} + \cdots + a_2 + a_1 + a_0 \bmod 9. \tag{6.22}$$

Explain why in the congruence above, one can replace "mod 9" by "mod 3".

Hint:

$$a_n a_{n-1} \cdots a_3 a_2 a_1 = \sum_{k=1}^{n} a_k 10^{k-1}. \tag{6.23}$$

Exercise 6.10. *Consider an integer $a_n a_{n-1} \cdots a_2 a_1 a_0$, where the a_i's are the digits of this $(n+1)$-digit integer. Suppose that*

$$a_n + a_{n-1} + \cdots + a_2 + a_1 + a_0 \equiv b_k b_{k-1} \cdots b_2 b_1 b_0 \bmod 9, \tag{6.24}$$

where the b_j's are the digits of the $(k+1)$-digit integer $b_k b_{k-1} \cdots b_2 b_1 b_0$. Explain why

$$a_n a_{n-1} \cdots a_2 a_1 a_0 \equiv b_k + b_{k-1} + \cdots + b_2 + b_1 + b_0 \bmod 9. \tag{6.25}$$

Finally, at the risk of being too pedantic (if you liked the proofs above, no need to read further!), here are slightly different proofs of parts (1) and (3) of Theorem 6.9: By hypothesis, there exist $k, \ell \in \mathbb{Z}$ such that

$$a_1 = a_2 + km, \qquad b_1 = b_2 + \ell m.$$

Then mod m,

$$\begin{aligned} a_1 + b_1 &\equiv (a_2 + km) + (b_2 + \ell m) \\ &\equiv a_2 + b_2 + (k + \ell)m \\ &\equiv a_2 + b_2, \end{aligned}$$

and also mod m,

$$\begin{aligned} a_1 b_1 &\equiv (a_2 + km)\,(b_2 + \ell m) \\ &\equiv a_2 b_2 + (a_2 \ell + k b_2 + k\ell m)\, m \\ &\equiv a_2 b_2. \end{aligned}$$

6.2.4. Congruence multiplication tables.

> Question: Why did the mathematician do multiplication problems on the floor?
> Answer: Because they were taught not to use tables.

We can define multiplication on the set of remainders $R_m = \{0, 1, 2, \ldots, m-1\}$ as follows. (The set R_m is also called the least residue system modulo m.) Define

$$\times_m : R_m \times R_m \to R_m \tag{6.26}$$

by

$$a \times_m b := \times_m(a, b) := \mathbf{r}(ab), \tag{6.27}$$

where $\mathbf{r} : \mathbb{Z} \to R_m$ is the remainder function.

Example 6.10. Modulo 5, we have $2 \times_5 3 = 1$, and the multiplication table for $(R_5, \times_5)$ is given by

$\times_5$	0	1	2	3	4
0	0	0	0	0	0
1	0	1	2	3	4
2	0	2	4	1	3
3	0	3	1	4	2
4	0	4	3	2	1

Exercise 6.11. *Give the multiplication table for* $(R_7, \times_7)$.

Solved Problem 6.11. *Find all* $1 \le a < 6$ *for which the function* $f_a : R_6 \to R_6$ *defined by* (5.112) *is a bijection.*

Solution. Recall that $f_a : R_6 \to R_6$ defined by

$$f_a(x) = \mathbf{r}(ax),$$

where $\mathbf{r} : \mathbb{Z} \to R_6$ is the remainder function. Using this, we express the values of the functions f_a, $1 \le a < 6$, in the table in Figure 6.2.4. From this table we see that the functions f_1 and f_5 are both bijections. On the other hand, we also see that each of the functions f_2, f_3, f_4 is neither an injection nor a surjection. Observe that $\gcd(1,6) = \gcd(5,6) = 1$, whereas $\gcd(2,6), \gcd(3,6), \gcd(4,6) > 1$.

x	$f_1(x)$	$f_2(x)$	$f_3(x)$	$f_4(x)$	$f_5(x)$
0	0	0	0	0	0
1	1	2	3	4	5
2	2	4	0	2	4
3	3	0	3	0	3
4	4	2	0	4	2
5	5	4	3	2	1

Figure 6.2.4. The green shaded functions are bijections, while the red shaded functions are neither injections nor surjections.

6.2.5. Congruence properties of perfect squares, products, and powers. Let us first consider perfect squares modulo 4. Let n be a perfect square. Then $n = a^2$, where a is a non-negative integer.

Case 1: a is even. Then there exists an integer k such that $a = 2k$. We calculate that

$$n = a^2 = (2k)^2 = 4 \cdot k^2. \tag{6.28}$$

Hence $n \equiv 0 \bmod 4$.

Case 2: a is odd. Then there exists an integer k such that $a = 2k + 1$. We calculate that

$$n = a^2 = (2k+1)^2 = 4 \cdot (k^2 + k) + 1. \tag{6.29}$$

Hence $n \equiv 1 \bmod 4$.

We conclude:

Lemma 6.12. *If n is a perfect square, then $n \equiv 0$ or $1 \bmod 4$.*

For example, we can immediately deduce that $10^{101} + 2$ and $10^{101} + 3$ are not perfect squares since they are congruent to 2 and 3 mod 4, respectively.

Now we make a few remarks about more general moduli m. Recall from (6.13) that $a \equiv \mathbf{r}(a) \bmod m$. By Theorem 6.9(3), we have

$$a^2 \equiv \mathbf{r}(a)^2 \bmod m. \tag{6.30}$$

Since $\mathbf{r}(a)$ is contained in the remainder set R_m for every integer a, we have that any perfect square is congruent modulo m to r^2 for some remainder r.

For example, taking $m = 3$, we see that modulo 3 we have $0^2 \equiv 0$, $1^2 \equiv 1$, and $2^2 \equiv 1$. Hence:

Lemma 6.13. *For any integer a, $a^2 \equiv 0$ or $1 \bmod 3$.*

Exercise 6.12. *Can an integer of the form $3k + 2$, where $k \in \mathbb{Z}$, be a perfect square?*

Exercise 6.13. *Let n be a perfect square. Prove the following:*

(1) $n \equiv 0, 1,$ *or* $4 \bmod 5$.

(2) $n \equiv 0, 1, 3,$ *or* $4 \bmod 6$.

(3) $n \equiv 0, 1, 2,$ *or* $4 \bmod 7$.

(4) $n \equiv 0, 1,$ *or* $4 \bmod 8$.

By using induction, we can extend the product property of congruence.

Lemma 6.14 (Congruence and multiple multiplications!)**.** *Let k be a positive integer. If we have integers satisfying*

$$a_i \equiv b_i \bmod m \quad \text{for } 1 \leq i \leq k, \tag{6.31}$$

then their products satisfy

$$a_1 a_2 \cdots a_k \equiv b_1 b_2 \cdots b_k \bmod m. \tag{6.32}$$

In particular, if $a \equiv b \bmod m$, then

$$a^k \equiv b^k \bmod m. \tag{6.33}$$

Proof. We prove this by induction.

The base case ($k = 1$) is tautologically true.

Now suppose that k is a positive integer with the property that for all sets of integers $a_1, \ldots, a_k$ and $b_1, \ldots, b_k$ satisfying $a_i \equiv b_i \bmod m$ for $1 \leq i \leq k$, we have

$$a_1 a_2 \cdots a_k \equiv b_1 b_2 \cdots b_k \bmod m. \tag{6.34}$$

Now suppose that $a_1, \ldots, a_{k+1}$ and $b_1, \ldots, b_{k+1}$ are integers satisfying $a_i \equiv b_i \bmod m$ for $1 \leq i \leq k + 1$. Then, by (6.34) and $a_{k+1} \equiv b_{k+1} \bmod m$, we have

$$\begin{aligned} a_1 a_2 \cdots a_{k+1} &\equiv (a_1 a_2 \cdots a_k) \cdot a_{k+1} \\ &\equiv (b_1 b_2 \cdots b_k) \cdot b_{k+1} \\ &\equiv b_1 b_2 \cdots b_{k+1} \bmod m. \end{aligned} \tag{6.35}$$

By induction, the theorem has been proven! □

Exercise 6.14. *Using Lemma* 6.14, *give a shorter proof that $10^k \equiv 1 \bmod 9$ for every positive integer k.*

Exercise 6.15. *Using Lemma* 6.14, *give another proof that the product of any finite number of odd integers is odd.*

Exercise 6.16. *Prove that for every positive integer k, $1000000^k \equiv 1 \bmod 13$. Hint:* $1000000 = 1000^2$.

6.3. Inverses, coprimeness, and congruence

In the set of integers, only 1 and -1 have multiplicative inverses. On the other hand, in the set of rational numbers, any non-zero number has a multiplicative inverse. We now consider inverses modulo m.

Definition 6.15. Let m be a positive integer. Let a be an integer. We say that an integer b is a (multiplicative) **inverse** of a modulo m if

$$ab \equiv 1 \bmod m. \tag{6.36}$$

Theorem 6.16. *Let m be a positive integer, and let a be an integer. Then:*

$$a \text{ has an inverse modulo } m \Leftrightarrow \gcd(a, m) = 1. \tag{6.37}$$

Proof. ($\Leftarrow$) Suppose that a and m are coprime. By Theorem 4.20, there exist integers b and n such that

$$ab + mn = 1. \tag{6.38}$$

This implies that b is an inverse of a modulo m.

($\Rightarrow$) Suppose that an integer a has an inverse b modulo m. Then $ab \equiv 1 \bmod m$. This implies that there exists an integer q such that

$$ab - 1 = qm. \tag{6.39}$$

By Theorem 4.20, this implies that a and m are coprime. □

In particular, if m is a prime, then a has an inverse modulo m if and only if a is not a multiple of m.

Can you find the inverses modulo 7 of each of the integers $1, 2, 3, 4, 5, 6$ by just looking at the colorful multiplication table in Figure 8.1.4 in Chapetr 8? Check your answers by calculating remainders of products!

Solved Problem 6.17. *Find all pairs of inverses modulo* 30; *namely, find all* (x, y) *with* $x \le y$ *satisfying* $xy \equiv 1 \bmod 30$.

Solution. By straightforward calculations, one sees that the pairs of all inverses are

$$(1,1),\ (7,13),\ (11,11),\ (17,23),\ (19,19),\ (29,29). \tag{6.40}$$

The integers between 1 and 29 without inverses are those not coprime with 30, which are

$$2,\ 3,\ 4,\ 5,\ 6,\ 8,\ 9,\ 10,\ 12,\ 14,\ 15,\ 16,\ 18,\ 20,\ 21,\ 22,\ 24,\ 25,\ 26,\ 27,\ 28. \tag{6.41}$$

Observe that for each integer a in this list, there exists an integer $0 < b < 30$ satisfying $ab \equiv 0 \bmod 30$.

Inverses, if they exist, are unique with respect to any modulus.

Exercise 6.17. *Suppose that a has an inverse b modulo m. Prove that any inverse of a is congruent to b modulo m. Hint: Use Theorem* 4.23.

Thus, if $0 < a < m$ and $\gcd(a, m) = 1$, then a has a unique inverse b modulo m satisfying $0 < b < m$.

Solving linear congruence equations of a *single variable* is generally easy. For now, we just content ourselves with a few small modulus examples.

Exercise 6.18. *For integers* $2 < a < m < 8$ *satisfying* $\gcd(a, m) = 1$*, find the unique integer* $0 < x < m$ *satisfying*

$$ax \equiv 1 \bmod m. \tag{6.42}$$

Exercise 6.19. *Find the smallest* 5 *positive integers* x *satisfying the equation*

$$2x \equiv 1 \bmod 3. \tag{6.43}$$

Do you see a pattern?

In view of how we take negatives modulo m, we have:

Lemma 6.18. *Let* m *be a positive integer, and let* $0 < a, b < m$*. If* $ab \equiv 1 \bmod m$*, then*

$$(m - a)(m - b) \equiv 1 \bmod m. \tag{6.44}$$

Exercise 6.20. *Prove Lemma* 6.18.

As an application of Lemma 6.18, consider for example the inverse pairs in Solved Problem 6.17. Since $(7, 13)$ is an inverse pair, we can deduce that $(30 - 7, 30 - 13)$ is an inverse pair; that is, $(17, 23)$ is an inverse pair.

6.4. Congruence and multiplicative cancellation

In this section we discuss a way to simplify congruence equations. We suggest that the readers practice with lots of examples until they get the hang of it.

Here is the general result about dividing congruence equations. In other words, this is the multiplicative cancellation property. This is a very useful property for simplifying congruence equations.

Theorem 6.19 (Congruence and division). *Let* $m \in \mathbb{Z}^+$ *and* $a \in \mathbb{Z}$ *and let* $g = \gcd(a, m)$*. We have*

$$ab_1 \equiv ab_2 \bmod m \quad \Leftrightarrow \quad b_1 \equiv b_2 \bmod \frac{m}{g}. \tag{6.45}$$

Proof. Suppose that $ab_1 \equiv ab_2 \bmod m$. This is equivalent to m dividing $a\,(b_1 - b_2)$. By Theorem 4.32, this in turn is equivalent to $\frac{m}{g}$ dividing $b_1 - b_2$, which finally is equivalent to $b_1 \equiv b_2 \bmod \frac{m}{g}$. □

As special cases of the multiplicative cancellation property, we have the following two corollaries. They represent the extreme cases for $\gcd(a, m)$.

Corollary 6.20 (Congruence and division by a divisor of the modulus). *Suppose that* a *is an integer which divides* m *(i.e.,* m/a *is an integer). Then*

$$ab_1 \equiv ab_2 \bmod m \quad \Leftrightarrow \quad b_1 \equiv b_2 \bmod m/a.$$

Proof. This is the theorem in the special case where a divides m, which implies that $\gcd(a, m) = a$. □

Corollary 6.21 (Congruence and coprime division). *Suppose that a is an integer such that a and m are coprime (i.e.,* $\gcd(a, m) = 1$*). Then*

$$ab_1 \equiv ab_2 \bmod m \quad \Leftrightarrow \quad b_1 \equiv b_2 \bmod m.$$

Proof. This is the theorem in the special case where $\gcd(a, m) = 1$. □

Example 6.22. We have

(1) By Corollary 6.20,

$$6b_1 \equiv 6b_2 \bmod 18 \quad \Leftrightarrow \quad b_1 \equiv b_2 \bmod 3.$$

(2) By Corollary 6.21,

$$35b_1 \equiv 35b_2 \bmod 18 \quad \Leftrightarrow \quad b_1 \equiv b_2 \bmod 18.$$

(3) By Theorem 6.19,

$$15b_1 \equiv 15b_2 \bmod 18 \quad \Leftrightarrow \quad b_1 \equiv b_2 \bmod 6.$$

Observe that these results follow from:

(1) $6b \equiv 0 \bmod 18$ if and only if $b \equiv 0 \bmod 3$. That is, $6b = 18k$ for some $\ell \in \mathbb{Z}$ if and only if $b = 3\ell$ for some $k \in \mathbb{Z}$.

(2) $35b \equiv 0 \bmod 18$ if and only if $b \equiv 0 \bmod 18$. That is, $35b = 18k$ for some $k \in \mathbb{Z}$ if and only if $b = 18\ell$ for some $\ell \in \mathbb{Z}$.

(3) $15b \equiv 0 \bmod 18$ if and only if $b \equiv 0 \bmod 6$. That is, $15b = 18k$ for some $k \in \mathbb{Z}$ if and only if $b = 6\ell$ for some $\ell \in \mathbb{Z}$.

Exercise 6.21. *For each of the following congruences, write an equivalent congruence of the form* $b_1 \equiv b_2 \bmod m$*:*

(1) $14b_1 \equiv 14b_2 \bmod 35$.
(2) $9b_1 \equiv 9b_2 \bmod 21$.
(3) $28b_1 \equiv 28b_2 \bmod 45$.

We take this opportunity to observe the following.

Lemma 6.23. *Let m be a positive integer. If* $a \equiv b \bmod m$*, then*

$$\gcd(a, m) = \gcd(b, m). \tag{6.46}$$

Proof. Suppose that $a \equiv b \bmod m$. Then there exists an integer q such that

$$a = mq + b. \tag{6.47}$$

By Theorem 4.16, we conclude that $\gcd(a, m) = \gcd(b, m)$. □

Exercise 6.22. *Let p be a prime, and let $a, b \in \mathbb{Z}$. Prove that* $ab \equiv 0 \bmod p$ *if and only if* $a \equiv 0 \bmod p$ *or* $b \equiv 0 \bmod p$*. Hint: Rephrase the conditions in terms of division, and use properties of primes and division.*

6.5*. Fun congruence facts

6.5.1. Squares of primes. Squares of primes have the following interesting congruence property.

Theorem 6.24. *If $p > 3$ is a prime, then $p^2 \equiv 1 \bmod 24$.*

Proof. Since p is odd, both $p-1$ and $p+1$ are even. Hence, by Exercise 4.7, 4 divides one of them (while the other of course is divisible by 2). Thus 8 divides $(p-1)p(p+1) = p(p^2-1)$. Since we also have that 3 divides $(p-1)p(p+1)$ and since 3 and 8 are coprime, we have that 24 divides $p(p^2-1)$. On the other hand, since $p > 3$ is a prime, we have that 24 and p are coprime. Therefore 24 divides $p^2 - 1$. □

Of course, the conclusion of the theorem is not true if $p = 2$ or $p = 3$.

Example 6.25 (Squares of primes $p < 20$)**.** The squares of the primes between 5 and 19, inclusive, are $25 = 24 + 1$, $49 = 24 \cdot 2 + 1$, $121 = 24 \cdot 5 + 1$, $169 = 24 \cdot 7 + 1$, $289 = 24 \cdot 12 + 1$, $361 = 24 \cdot 15 + 1$.

Example 6.26. Consider primes that are congruent to 3 modulo 4. The list of the first 12 primes $p_1, p_2, \ldots, p_{12}$ satisfying $p_i \equiv 3 \bmod 4$ is:

$$3,\ 7,\ 11,\ 19,\ 23,\ 31,\ 43,\ 47,\ 59,\ 67,\ 71,\ 79,\ 83. \tag{6.48}$$

Observe that for each $1 \leq k \leq 7$, the prime factorization of $2p_1 \cdots p_k + 1$ generates a prime q which is congruent to 3 modulo 4 but is different than all of $p_1, \ldots, p_k$. For example, when $k = 5$, the new such prime is 271. See Table 6.5.1. This example illustrates the key idea of the proof of Theorem 6.27 below.

Table 6.5.1. The new primes congruent to 3 modulo 4 are colored in red. Observe that regarding the prime factorization of $2p_1p_2p_3p_4p_5 + 1$, although $5 \equiv 1 \bmod 4$ and $149 \equiv 1 \bmod 4$, we have $271 \equiv 3 \bmod 4$.

k	$p_1, \ldots, p_k$	$2p_1 \cdots p_k + 1$	prime factorization
1	3	7	7
2	$3, 7$	43	43
3	$3, 7, 11$	463	463
4	$3, 7, 11, 19$	8779	8779
5	$3, 7, 11, 19, 23$	201895	$5 \cdot 149 \cdot 271$
6	$3, 7, 11, 19, 23, 31$	6258715	$5 \cdot 1251743$
7	$3, 7, 11, 19, 23, 31, 43$	269124703	$6151 \cdot 43753$

6.5.2. Infinitude of primes congruent to 3 mod 4. A result stronger than Theorem 4.12 on the infinitude of primes is the following.

Theorem 6.27 (Bountiful primes congruent to 3 mod 4)**.** *There are an infinite number of primes p satisfying $p \equiv 3 \bmod 4$.*

Proof. Suppose that there are only a finite number of primes p satisfying $p \equiv 3 \bmod 4$. Let

$$\{p_1, p_2, \ldots, p_k\} \tag{6.49}$$

be the complete list of such primes. Note that since $p_i \equiv 3 \bmod 4$, we have that p_i is odd for each $1 \le i \le k$. Define

$$q := 2p_1p_2\cdots p_k + 1. \tag{6.50}$$

Since $p_1p_2\cdots p_k$ is odd, we have that $2p_1p_2\cdots p_k \equiv 2 \bmod 4$, and hence

$$q \equiv 3 \bmod 4. \tag{6.51}$$

Of course, the integer q is odd.

Note that for each $1 \le i \le k$, p_i divides $2p_1p_2\cdots p_k$, and hence p_i does not divide q.

Since $q > p_i$ for all $1 \le i \le k$, we have that q is not a prime. Let

$$q = q_1^{\ell_1} q_2^{\ell_2} \cdots q_s^{\ell_s} \tag{6.52}$$

be the prime factorization of q. Since q is odd, so is each q_j, $1 \le j \le s$. Thus, for each j, q_j is either congruent to 1 or 3 modulo 4.

Since each p_i does not divide q and since the p_i are all the primes congruent to 3 mod 4, we must have that each q_j is congruent to 1 modulo 4. However, this then implies that q is congruent to 1 modulo 4, which is a contradiction to (6.51).

Since the sole assumption we made to obtain the contradiction is that there are only a finite number of primes p satisfying $p \equiv 3 \bmod 4$, we conclude that there must exist an infinite number of primes p satisfying $p \equiv 3 \bmod 4$. □

The theorem begs the question: Are there an infinite number of primes p satisfying $p \equiv 1 \bmod 4$? With the help of Fermat's Little Theorem 6.63 below, we can answer this in the affirmative! See Exercise 6.38 below.

6.6. Solving linear congruence equations

Having discussed the basic properties of congruence, we may now solve linear congruence equations.

6.6.1. How to solve linear congruences via examples.

A **linear congruence equation** of a single variable is an equation of the form

$$ax \equiv b \bmod m. \tag{6.53}$$

Here, we wish to find all integers x which satisfy this linear congruence equation.

Remark 6.28. If $a = 1$ (or more generally if $a \equiv 1 \bmod m$), then the answer is easy: The solutions are the integers satisfying $x \equiv b \bmod m$, which are the integers

$$b, b+m, b-m, b+2m, b-2m, b+3m, b-3m, \ldots. \tag{6.54}$$

Example 6.29. Consider the linear congruence equation

$$6x \equiv 15 \bmod 21.$$

By inspection, we see that $x = 6$ is a solution, because $6 \cdot 6 = 36$ is congruent to 15 modulo 21. But what are the other solutions? We may then observe that $x = 6 + 7 = 13$ is also a solution. This is because then $6x = 6(6+7) = 36 + 42$ and as we just saw, 36 is congruent to 15 modulo 21, whereas 42 is a multiple of 21. We may guess that the general solution to our example linear congruence equation is

$$x = 6 + 7k, \quad k \in \mathbb{Z}.$$

This is indeed the case. We now discuss how to solve this and other congruence equations.

To be more concrete, we approach solving linear congruence equations by working some examples. Recall from Theorem 6.19 the following: Let $m \in \mathbb{Z}^+$ and $a \in \mathbb{Z}$ and let $g = \gcd(a, m)$. Then

$$ab_1 \equiv ab_2 \bmod m \quad \Leftrightarrow \quad b_1 \equiv b_2 \bmod \frac{m}{g}. \tag{6.55}$$

Example 6.30. We use this result to see how we can simplify linear congruence equations. Consider

$$4x \equiv 12 \bmod 14.$$

By (6.55), we have that

$$4x \equiv 12 \bmod 14 \quad \Leftrightarrow \quad 4x \equiv 4 \cdot 3 \bmod 14 \quad \Leftrightarrow \quad x \equiv 3 \bmod 7$$

since $\gcd(4, 14) = 2$ and $\dfrac{14}{\gcd(4, 14)} = 7$. That is, x is a solution to the linear congruence equation if and only if $x = 3 + 7k$, where $k \in \mathbb{Z}$.

Recall from Corollary 6.20: Suppose a divides m (so that $\gcd(a, m) = a$). Then

$$ab_1 \equiv ab_2 \bmod m \quad \Leftrightarrow \quad b_1 \equiv b_2 \bmod \frac{m}{a}.$$

Example 6.31. Let us look again at the first example of this section:

$$6x \equiv 15 \bmod 21 \quad \Leftrightarrow \quad 3 \cdot 2x \equiv 3 \cdot 5 \bmod 21 \quad \Leftrightarrow \quad 2x \equiv 5 \bmod 7$$

since $\frac{21}{3} = 7$. At this point, we still haven't derived the solution to this linear congruence equation. We'll do that later.

Recall from Corollary 6.21: Suppose that a and m are coprime; i.e., $\gcd(a, m) = 1$. Then

$$ab_1 \equiv ab_2 \bmod m \quad \Leftrightarrow \quad b_1 \equiv b_2 \bmod m.$$

Example 6.32. Consider the linear congruence equation

$$2x \equiv 12 \bmod 7 \quad \Leftrightarrow \quad 2x \equiv 2 \cdot 6 \bmod 7 \quad \Leftrightarrow \quad x \equiv 6 \bmod 7$$

since $\gcd(2, 7) = 1$. This solves the linear congruence equation; i.e., the solutions are the integers of the form $x = 6 + 7k$, $k \in \mathbb{Z}$.

The next result helps us when we are stuck in trying to solve a linear congruence equation.

Lemma 6.33. *$ax \equiv b \bmod m$ is equivalent to $ax \equiv b + cm \bmod m$ for every integer c.*

Example 6.34.

$$2x \equiv 5 \bmod 7 \quad \Leftrightarrow \quad 2x \equiv 12 \bmod 7$$

since $12 = 5 + 7$. The motivation for considering the equivalence above is that 5 is not divisible by 2, but the integer 12 congruent to it (mod 7) is.

Can you solve $2x \equiv 5 \bmod 7$ by looking at the colorful multiplication table in Figure 8.1.4 in Chapter 8? Which row or column of the multiplication table did you look at?

Putting things together, we solve the first example of this subsection.

Example 6.35. *Solve*

$$6x \equiv 15 \bmod 21.$$

All of the following steps we have actually carried out above!

Step 1. This linear congruence equation is equivalent to

$$2x \equiv 5 \bmod 7.$$

Step 2. The equation $2x \equiv 5 \bmod 7$ is equivalent to

$$2x \equiv 12 \bmod 7.$$

Step 3. The equation $2x \equiv 12 \bmod 7$ is equivalent to

$$x \equiv 6 \bmod 7.$$

The key is that the coefficient in front of x is 1 so we can just read off the answer as: The general solution to the linear congruence equation is $x = 6 + 7k$, where $k \in \mathbb{Z}$. This proves the claimed solution to Example 6.29.

We can also choose to describe the solutions using the "original" modulus 21 instead of the "new" modulus 7. We see that x is a solution to the linear congruence equation if and only if

$$x \equiv 6,\ 13,\ \text{or}\ 20 \bmod 21.$$

That is, we add 7's to 6 until we get back to an integer congruent to 6 modulo 21. Observe that if we add 7 to 20, then we get 27, which is congruent to 6 modulo 21.

Observe also that the number of solutions modulo 21 to the linear congruence equation is equal to 3, which is the gcd of 6 and 15 coming from $6x \equiv 15 \bmod 21$. We shall see later that this true in general.

Exercise 6.23. *For each of* $a = 1, 2, 3, 4, 5, 6$*, find the unique* $0 < x < 7$ *satisfying*

$$5x \equiv a \bmod 7. \tag{6.56}$$

6.6.2. How to solve a general linear congruence. Let $a, b \in \mathbb{Z}$ be non-zero integers. Consider the general linear congruence equation for $x \in \mathbb{Z}$:

$$ax \equiv b \bmod m. \tag{6.57}$$

Remark 6.36. Equation (6.57) is indeed the *general form* of a linear congruence equation. Consider the apparently more general linear congruence equation

$$a_1 x + a_0 \equiv b \bmod m. \tag{6.58}$$

All we have to do is move the a_0 to the other side to get an equation of the form (6.57)!

Lemma 6.37 (Criterion for non-existence). *There does not exist a solution $x \in \mathbb{Z}$ of the linear congruence equation $ax \equiv b \mod m$ if and only if $g = \gcd(a, m)$ does not divide b.*

Equivalently:

Lemma 6.38 (Criterion for existence). *There exists a solution $x \in \mathbb{Z}$ of the linear congruence equation $ax \equiv b \mod m$ if and only if $g = \gcd(a, m)$ divides b.*

Proof. We have the following equivalences of statements:

$$\begin{aligned} & \text{There exists } x \in \mathbb{Z},\ ax \equiv b \bmod m \\ \Leftrightarrow\ & \text{there exists } x \in \mathbb{Z} \text{ such that there exists } y \in \mathbb{Z},\ ax - b = -ym \\ \Leftrightarrow\ & \text{there exists } x, y \in \mathbb{Z},\ ax + ym = b. \end{aligned}$$

The last equation is a linear Diophantine equation and we have already proved that it has a solution if and only if $g = \gcd(a, m)$ divides b. □

Example 6.39 (Non-existence). The linear congruence equation

$$21x \equiv 5 \bmod 33$$

does not have a solution since $\gcd(21, 33) = 3$ does not divide 5.

Example 6.40 (Existence). The linear congruence equation

$$21x \equiv 6 \bmod 33$$

has a solution since $\gcd(21, 33) = 3$ divides 6.

A particular solution is $x = 5$ since

$$21 \cdot 5 = 105 = 33 \cdot 3 + 6.$$

(This particular solution was found "by inspection". A more systematic way would be to solve the linear Diophantine equation $21x+33y = 6$ as we have done in earlier chapters.)

Since $21 \cdot 5 \equiv 6 \bmod 33$, x is a solution to

$$\begin{aligned} & 21x \equiv 6 \bmod 33 \\ \Leftrightarrow\ & 21\,(x - 5) \equiv 0 \bmod 33 \\ \Leftrightarrow\ & 33 \text{ divides } 21\,(x - 5) \\ \Leftrightarrow\ & 11 \text{ divides } 7\,(x - 5) \\ \Leftrightarrow\ & 11 \text{ divides } x - 5\ , \end{aligned}$$

where the last equivalence is true since $\gcd(7, 11) = 1$. So a complete set of solutions x modulo 33 is $x = 5, 16, 27$. Again, we see that the number of solutions modulo the "original modulus" 33 is equal to 3, which is the gcd of 21 and 33.

Exercise 6.24. *List all the integers $0 < b < 15$ such that the linear congruence equation*

$$6x \equiv b \bmod 15 \tag{6.59}$$

has solution(s).

For each such integer b, find all integer solutions $0 < x < 15$.

By arguing more generally along the lines of Example 6.40, we obtain:

Theorem 6.41 (General solution from a particular solution). *Let* $\tilde{m} = \frac{m}{g}$, *where* $g = \gcd(a, m)$. *If* x_0 *is a solution to*

$$ax \equiv b \bmod m, \tag{6.60}$$

then a complete set of solutions x *modulo* m *is given by*

$$x_0, x_0 + \tilde{m}, x_0 + 2\tilde{m}, \ldots, x_0 + (g-1)\tilde{m}. \tag{6.61}$$

In particular, there are g *solutions modulo* m.

Example 6.42. Consider the linear congruence equation

$$6x \equiv 3 \bmod 15. \tag{6.62}$$

By inspection we see that $x_0 = 3$ is a solution. We have that $\gcd(6, 15) = 3$. Thus $\tilde{m} = \frac{15}{3} = 5$. So, by Theorem 6.41, a complete set of solutions x modulo m is given by

$$3,\ 8,\ 13. \tag{6.63}$$

Observe that this result agrees with Corollary 6.20 telling us that the congruence equation (6.62) is equivalent to

$$2x \equiv 1 \bmod 5. \tag{6.64}$$

We have the following important consequence of Lemma 6.38.

Corollary 6.43. *Let* p *be a prime. There exists a solution* $x \in \mathbb{Z}$ *to the linear congruence equation*

$$ax \equiv 1 \bmod p \tag{6.65}$$

if and only if a *is not a multiple of* p.

Proof. The ($\Leftarrow$) direction is easy to see by considering its contrapositive.

The ($\Rightarrow$) direction follows from Lemma 6.38 and the fact that if a is not a multiple of p, then $\gcd(a, p) = 1$ (see Proposition 2.17). $\square$

Again, let p be a prime. Suppose that a is not a multiple of p. Observe that if integers x and y satisfy (exercise!)

$$ax \equiv 1 \bmod p, \qquad ay \equiv 1 \bmod p, \tag{6.66}$$

then

$$x \equiv y \bmod p. \tag{6.67}$$

So the solution x in Corollary 6.43 is unique. A solution x is called an **inverse** of a modulo p.

An interesting question is: When is an integer its own inverse modulo a prime p?

Lemma 6.44. *Let* p *be a prime. An integer* a *is its own inverse modulo* p *if and only if* $a \equiv 1 \bmod p$ *or* $a \equiv -1 \bmod p$.

Proof. Suppose that $a \in \mathbb{Z}$ satisfies

$$a^2 \equiv 1 \bmod p. \tag{6.68}$$

Then

$$(a-1)(a+1) \equiv a^2 - 1 \equiv 0 \bmod p. \tag{6.69}$$

So, by Exercise 6.22, we have that $a - 1 \equiv 0 \bmod p$ or $a + 1 \equiv 0 \bmod p$. □

Example 6.45 (Pairs of inverses modulo a prime). (1) For $p = 7$, modulo 7 the pairs of inverses (x, y) with $xy \equiv 1$ and $x \leq y$ are

$$(1,1),\ (2,4),\ (3,5),\ (6,6). \tag{6.70}$$

(2) For $p = 11$, modulo 11 the pairs of inverses are

$$(1,1),\ (2,6),\ (3,4),\ (5,9),\ (7,8), (10,10). \tag{6.71}$$

Solved Problem 6.46. *Find all pairs of inverses (x, y) modulo the prime 23 with $xy \equiv 1 \bmod 23$ and $x \leq y$.*

Solution. By inspection, where we provide the relevant calculations below, we see that these pairs are

$$\begin{gathered}(1,1),\ (2,12),\ (3,8),\ (4,6),\ (5,14),\ (7,10),\\(9,18),\ (11,21),\ (13,16),\ (15,20),\ (17,19),\ (22,22).\end{gathered}$$

For the pairs of inverses in the first row, the calculations we made to obtain these answers are

$$\begin{aligned}2 \cdot 12 &= 24 = 23 + 1,\\3 \cdot 8 &= 24 = 23 + 1,\\4 \cdot 6 &= 24 = 23 + 1,\\5 \cdot 14 &= 70 = 3 \cdot 23 + 1,\\7 \cdot 10 &= 70 = 3 \cdot 23 + 1,\end{aligned}$$

and for the pairs in the second row, we applied Lemma 6.18 to the first row with the calculations

$$\begin{aligned}&23 - 1 = 22,\\&23 - 2 = 21, \quad 23 - 12 = 11,\\&23 - 3 = 20, \quad 23 - 8 = 15,\\&23 - 4 = 19, \quad 23 - 6 = 17,\\&23 - 5 = 18, \quad 23 - 14 = 9,\\&23 - 7 = 16, \quad 23 - 10 = 13.\end{aligned}$$

Exercise 6.25. *Similarly to Example 6.45, write out the pairs of inverses (x, y) with $xy \equiv 1 \bmod p$ and $x \leq y$, for $p = 13$ and for $p = 17$.*

Non-Example 6.47. *If m is not a prime, then we have more possibilities for integers to be their own inverse modulo m.*

(1) *Let $m = 8$. Then 1, 3, 5, and 7 are each their own inverse.*
(2) *Let $m = 15$. Then 1, 4, 11, and 14 are each their own inverse.*
(3) *Let $m = 20$. Then 1, 9, 11, and 19 are each their own inverse.*

We end this subsection with the following generalization of Corollary 6.43.

Theorem 6.48. *Let m be a positive integer. There exists a solution $x \in \mathbb{Z}$ to the linear congruence equation*

$$ax \equiv 1 \bmod m \tag{6.72}$$

if and only if a is coprime to m.

Exercise 6.26. *Prove Theorem* 6.48.

Example 6.49. For $m = 15$ the remainders that are coprime to 15 are

$$1, 2, 4, 7, 8, 11, 13, 14.$$

The set of all pairs of inverses (x, y) with $xy \equiv 1 \bmod 15$ and $x \leq y$ are

$$(1,1),\ (2,8),\ (4,4),\ (7,13),\ (11,11),\ (14,14). \tag{6.73}$$

Exercise 6.27. *For each $1 \leq a \leq 17$ find $1 \leq b \leq 17$ such that $ab \equiv 1 \bmod 18$ if such an integer b exists. If you like, you may write your answer in a format similar to Example* 6.49.

6.6.3. Solving congruences revisited and examples. Consider

$$ax \equiv b \bmod m$$

and let $g = \gcd(a, m)$. If g does not divide b, then there are no solutions. So we may assume that g divides b. We then get that our linear congruence equation is equivalent to

$$\frac{a}{g}x \equiv \frac{b}{g} \bmod \frac{m}{g},$$

where $\gcd\left(\frac{a}{g}, \frac{m}{g}\right) = 1$ as we know from a previous result.

We will take this as saying that we only need to consider

$$ax \equiv b \bmod m,$$

where $\gcd(a, m) = 1$, which we now assume.

6.6.3.1. *Solving linear congruences by solving linear Diophantine equations.* One way of proceeding from here is to solve the linear Diophantine equation

$$a\bar{x} + m\bar{y} = 1,$$

which we can by reversing the Euclidean algorithm since $\gcd(a, m) = 1$. We then let $x = b\bar{x}$ and $y = b\bar{y}$. Then

$$ax + my = b.$$

Then x is a solution to

$$ax \equiv b \bmod m.$$

We then add to x the multiples of $\frac{m}{g}$ from 1 to $g - 1$ to get all solutions modulo m.

Example 6.50.

$$6x \equiv 18 \bmod 14.$$

We have that $g = \gcd(6, 14) = 2$. So our equation is equivalent to

$$3x \equiv 9 \bmod 7,$$

where we observe that $\gcd(3, 7) = 1$. By reversing the Euclidean algorithm, we obtain that

$$3 \cdot (-2) + 7 \cdot 1 = 1.$$

(In the notation above, $\bar{x} = -2$ and $\bar{y} = 1$. Thus $x = 2 \cdot (-2) = -4$ is a solution to our linear congruence equation. We then add to this solution $\frac{14}{2} = 7$ to get the only other solution modulo 14, which is $-4 + 7 = 3$. That is, our two solutions modulo 14 are

$$x = -4 \quad \text{and} \quad x = 3.$$

Another way we could have done this is simply to apply Theorem 6.19 to get that our linear congruence equation is equivalent to

$$x \equiv 3 \bmod 7$$

since $\gcd(6, 14) = 2$. Thus, modulo 14, the solutions are $x = 3$ and $x = 10$ (note that $10 \equiv -4 \bmod 14$).

6.6.3.2. *Solving linear congruences by adding a good multiple of m.* We recall the observation that the congruence equation $ax \equiv b \bmod m$ is equivalent to there existing y such that $ax - b = my$; that is, $ax = b + my$. In fact, for every $y \in \mathbb{Z}$, we have

$$ax \equiv b \bmod m \quad \Leftrightarrow \quad ax \equiv b + my \bmod m. \tag{6.74}$$

So if we add the right multiple of m to b, the result will be divisible by a.

Example 6.51 ($6x \equiv 15 \bmod 21$ revisited)**.** Of course 15 is not divisible by 6. So we keep adding 21 to it until it is:

$$15 + 21 = 36,$$

which is divisible by 6 (we were lucky on the first try). Thus our linear congruence equation is equivalent to

$$6x \equiv 36 \bmod 21,$$

which on the other hand is equivalent to

$$x \equiv 6 \bmod 7$$

since $\gcd(6, 21) = 3$. Thus we obtained the answer rather easily.

Solved Problem 6.52. *Solve the linear congruence equation*

$$52x \equiv 24 \bmod 18. \tag{6.75}$$

Solution. Equation (6.75) is equivalent to

$$13x \equiv 6 \bmod 9$$

since $\gcd(4,18) = 2$ and $18/2 = 9$. The quickest way to solve this equation is to add multiples of 9 to 6 until we get a multiple of 13. In this vein we consider the equivalent equation

$$13x \equiv 6 + 8 \cdot 9 = 78 \bmod 9.$$

Since $78 = 13 \cdot 6$ and $\gcd(13,9) = 1$, we see that this equation is in turn equivalent to

$$x \equiv 6 \bmod 9.$$

Thus, a set of incongruent solutions modulo the original modulus 18 is $\{6, 15\}$. The set of all integer solutions is $\{6 + 9k : k \in \mathbb{Z}\}$.

6.7*. The Chinese Remainder Theorem

Can we solve for an integer being congruent to various integers modulo various moduli? Equivalently, can we solve *systems* of linear congruence equations? For example, does there exist an integer x such that

$$x \equiv 8 \bmod 24, \tag{6.76a}$$

$$x \equiv 45 \bmod 23? \tag{6.76b}$$

Yes. It turns out that the general solution to this system of two congruence equations is given by

$$x \equiv 344 \bmod 552. \tag{6.77}$$

Observe that $552 = 24 \cdot 23$.

On the other hand, the system of equations

$$x \equiv 8 \bmod 24, \tag{6.78a}$$

$$x \equiv 13 \bmod 45 \tag{6.78b}$$

has no solutions! This is because the first equation implies that $x \equiv 2 \bmod 3$, whereas the second equation implies that $x \equiv 1 \bmod 3$.

Now let's analyze the first system (6.76) of congruence equations. Firstly, $x_0 = 344$ is a solution, as well as is any x satisfying $x \equiv x_0 \bmod 552$. Conversely, suppose that x is a solution to (6.76). Then $x \equiv x_0 \bmod 24$ and $x \equiv x_0 \bmod 23$. Since 23 and 24 are coprime, this implies that $x \equiv x_0 \bmod 24 \cdot 23 = 552$. Thus, the solution 344 is the unique solution modulo 552.

Secondly, how can one find the solution 344? The general solution to the first equation is $x = 8 + 24k$, where $k \in \mathbb{Z}$. We can plug this into the second equation to obtain

$$8 + 24k \equiv 45 \bmod 23; \quad \text{that is,} \quad 24k \equiv 37 \bmod 23.$$

Since 24 and 23 are coprime, this congruence equation has a solution. Note that $24 \equiv 1 \bmod 23$, and so $k = 37$ is a solution. We conclude that $x = 8 + 24 \cdot 37 = 896$ is a solution to (6.76). Now $896 - 552 = 344$, so that 344 is also a solution.

The following result appeared in the treatise Sūnzǐ Suànjīng (The Mathematical Classic of Master Sun). This result gives us a sufficient condition for solutions to a system of linear congruence equations to exist, but it doesn't tell us how to solve them! The reader interested in learning how to solve such systems of congruences may consult the "Chinese Remainder Theorem" Wikipedia link.

Theorem 6.53 (Chinese Remainder). *Let $m_1, m_2, \ldots, m_k$ be pairwise coprime integers greater than 1. That is, for each $i \neq j$, $m_i > 1$ and $\gcd(m_i, m_j) = 1$. Then for any integers $b_1, b_2, \ldots, b_k$, the system of linear congruence equations*

$$x \equiv b_1 \bmod m_1, \tag{6.79a}$$

$$x \equiv b_2 \bmod m_2, \tag{6.79b}$$

$$\vdots \tag{6.79c}$$

$$x \equiv b_k \bmod m_k \tag{6.79d}$$

has a solution $x_0 \in \mathbb{Z}$.

Moreover, if x_0 is a solution, then $x \in \mathbb{Z}$ is a solution to (6.79) *if and only if*

$$x \equiv x_0 \bmod m_1 m_2 \cdots m_k. \tag{6.80}$$

In other words, (6.79) *has a unique solution modulo $m_1 m_2 \cdots m_k$.*

Proof. (1) *Uniqueness of the solution.* Suppose that x and x_0 are integer solutions to (6.79). Then we have

$$x - x_0 \equiv 0 \bmod m_1,$$
$$x - x_0 \equiv 0 \bmod m_2,$$
$$\vdots$$
$$x - x_0 \equiv 0 \bmod m_k.$$

Thus, each of $m_1, m_2, \ldots, m_k$ divides $x - x_0$. Since the m_i's are pairwise coprime by hypothesis, from Corollary 4.34 we have that the product $m_1 m_2 \cdots m_k$ divides $x - x_0$. We conclude that

$$x \equiv x_0 \bmod m_1 m_2 \cdots m_k.$$

Conversely, it is easy to see that if x_0 is a solution to (6.79) and if x satisfies (6.80), then x is a solution to (6.79). This completes the uniqueness proof.

(2) *Existence of a solution.* Recall that $\mathbf{r}_m : \mathbb{Z} \to R_m$ denotes the remainder function. Define the function

$$f : R_{m_1 m_2 \cdots m_k} \to R_{m_1} \times R_{m_2} \times \cdots \times R_{m_k} \tag{6.81}$$

by

$$f(x) = (\mathbf{r}_{m_1}(x), \mathbf{r}_{m_2}(x) \ldots, \mathbf{r}_{m_k}(x)). \tag{6.82}$$

Observe for the function f that both its domain $R_{m_1 m_2 \cdots m_k}$ and its codomain $R_{m_1} \times R_{m_2} \times \cdots \times R_{m_k}$ have the same cardinality $m_1 m_2 \cdots m_k$ (the latter by the multiplication principle for sets (Corollary 7.6 below)).

By part (1) of the present proof, the function f is an injection. Indeed, the argument of part (1) says that if $x, x_0 \in R_{m_1 m_2 \cdots m_k}$ are such that $f(x) = f(x_0)$, that is, $\mathbf{r}_{m_i}(x) = \mathbf{r}_{m_i}(x_0)$ for all $1 \leq i \leq k$, then $\mathbf{r}_{m_1 m_2 \cdots m_k}(x) = \mathbf{r}_{m_1 m_2 \cdots m_k}(x_0)$. Since $x, x_0 \in R_{m_1 m_2 \cdots m_k}$, this implies $x = x_0$, so that f is an injection. Thus, by Theorem 5.63, f is a surjection (and hence a bijection)!

Finally, given any integers $b_1, b_2, \ldots, b_k$, since f is a surjection, there exists $x \in R_{m_1 m_2 \cdots m_k}$ such that

$$f(x) = \big(\mathbf{r}_{m_1}(b_1), \mathbf{r}_{m_2}(b_2) \ldots, \mathbf{r}_{m_k}(b_k)\big). \tag{6.83}$$

This implies that $\mathbf{r}_{m_i}(x) = \mathbf{r}_{m_i}(b_i)$ for all $1 \le i \le k$, so that $x \equiv b_i \bmod m_i$; i.e., x is a solution to (6.79). This completes the existence proof. □

Exercise 6.28. *Let $p_1, p_2, \ldots, p_k$ be distinct primes. Prove that for every integer $b_1, b_2, \ldots, b_k$, the system of linear congruence equations*

$$x \equiv b_1 \bmod p_1, \tag{6.84a}$$

$$x \equiv b_2 \bmod p_2, \tag{6.84b}$$

$$\vdots \tag{6.84c}$$

$$x \equiv b_k \bmod p_k \tag{6.84d}$$

has a unique solution modulo $p_1 p_2 \cdots p_k$.

6.8. Quadratic residues

This section is an introductory discussion of quadratic equations on the finite set R_m, which are important in number theory. We will return to this topic later in the book.

6.8.1. Definition of quadratic residue and examples.

Let $m \ge 3$ be an integer. Define the function $f : R_m \to R_m$ by

$$f(x) = \mathbf{r}(x^2), \tag{6.85}$$

where $\mathbf{r} : \mathbb{Z} \to R_m$ is the remainder function. In other words, we are considering the squaring function modulo m.

Definition 6.54. Let $m \ge 2$ be an integer. We say that $z \in R_m$ is a **quadratic residue** if z is an element of the image of f, that is, if there exists $x \in R_m$ such that $x^2 \equiv z \bmod m$.

Let

$$QR_m = \operatorname{Im}(f) - \{0\} \tag{6.86}$$

denote the **set of non-zero quadratic residues.**

In other words, a quadratic residue is a perfect square modulo m.

Example 6.55. Our discussion of quadratic residues actually began in §4.1.7 on division and squares. The results of Exercises 4.8, 4.10, and 4.11 and Solved Problem 4.11 are summarized by Table 4.1.1, which says

$$QR_2 = \{1\},$$
$$QR_3 = \{1\},$$
$$QR_4 = \{1\},$$
$$QR_5 = \{1, 4\},$$
$$QR_7 = \{1, 2, 4\}.$$

Exercise 6.29. *Show that*

$$QR_6 = \{1, 3, 4\},$$
$$QR_8 = \{1, 4\},$$
$$QR_9 = \{1, 4, 7\},$$
$$QR_{10} = \{1, 4, 5, 6, 9\}.$$

Explain why this exercise, together with Example 6.55, *show that for each* $3 \leq m \leq 10$, $f : R_m \to R_m$ *is neither injective nor surjective.*

Solved Problem 6.56. *Find all positive integers* x *such that* $x^2 \equiv 4 \bmod 6$.

Solution. Firstly, $x^2 \equiv 4 \bmod 6$ if and only if there exists an integer q such that $x^2 = 6q + 4$. Let x be a solution. Since x^2 is even, we have that x is even. This implies that x is congruent modulo 6 to 0, 2, or 4. If $x \equiv 0 \bmod 6$, then $x^2 \equiv 0 \bmod 6$. If $x \equiv 2 \bmod 6$, then $x^2 \equiv 4 \bmod 6$. If $x \equiv 4 \bmod 6$, then $x^2 \equiv 16 \equiv 4 \bmod 6$. Thus x is a positive integer satisfying $x^2 \equiv 4 \bmod 6$ if and only if $x = 6k + r$, where $k \in \mathbb{Z}^{\geq}$ and $r = 2$ or $r = 4$. Another way to characterize these integers is that they are the even integers that are not multiples of 6. □

Exercise 6.30. *Find all positive integers* x *such that* $x^2 \equiv 36 \bmod 5$.

6.8.2. Elementary properties of quadratic residues. We have the following property regarding products of quadratic residues.

Exercise 6.31. *Show that if* $x, y \in QR_m$, *then* $\mathbf{r}(xy) \in QR_m$ *provided* xy *is not a multiple of* m.

In particular, if $x \in QR_m$, *then* $\mathbf{r}(x^2) \in QR_m$ *provided* x^2 *is not a multiple of* m.

Example 6.57. For $m = 9$, we have

$$\mathbf{r}(4^2) = 7, \quad \mathbf{r}(7^2) = 4, \quad \mathbf{r}(4 \cdot 7) = 1. \tag{6.87}$$

Quadratic residues have the following nice properties modulo primes p at least 3. As a special case, you may consider $p = 23$ as in Solved Problem 6.46.

Solved Problem 6.58 (Quadratic residues modulo an odd prime). *Let* p *be an odd prime. Define the* mod p *squaring function* $f : R_p - \{0\} \to R_p - \{0\}$ *by*

$$f(a) = \mathbf{r}(a^2). \tag{6.88}$$

Prove the following:

(1) f *is well-defined.*

(2) (f is 2-to-1) *For each* $r \in \text{Im}(f)$, *there exist exactly two elements of* $R_p - \{0\}$ *whose images under* f *are equal to* r.

(3) $\text{Im}(f) = f(R_p - \{0\})$ *has cardinality (number of elements) equal to* $(p-1)/2$.

Solution. (1) To see that f is well-defined, we just need to show that for every $a \in R_p - \{0\}$, $f(a) \neq 0$. Suppose, for a contradiction, that there exists $a \in R_p - \{0\}$ such that $\mathbf{r}(a^2) = f(a) = 0$. Then p divides a^2. Since p is a prime and by Corollary

4.24, this implies that p divides a. Since $a \in R_p$, we conclude that $a = 0$, which is a contradiction to $a \in R_p - \{0\}$.

(2) Suppose that $a, b \in R_p - \{0\}$ satisfy $f(a) = f(b)$. Then $\mathbf{r}(a^2) = \mathbf{r}(b^2)$; that is, p divides $a^2 - b^2 = (a - b)(a + b)$.

Case 1: p divides $a - b$. Since $a, b \in R_p - \{0\} \subset R_p$, this implies that $a = b$.

Case 2: p divides $a + b$. Since $a, b \in R_p - \{0\}$, this implies that $a + b = p$; that is, $b = p - a$. Observe that since p is odd, we have $a \neq b$.

Conversely, if $a = b$ or if $a + b = p$, then one easily checks that $f(a) = f(b)$.

Hence, we have proved that $f(a) = f(b)$ if and only if $a = b$ or $a + b = p$. This proves that the function $f : R_p - \{0\} \to R_p - \{0\}$ is two-to-one; that is, for each $r \in \text{Im}(f)$, there exist exactly two elements of $R_p - \{0\}$ whose values under f equal r.

(3) Consequently, by a counting argument, the cardinality of $\text{Im}(f)$ is equal to one-half of the cardinality of $R_p - \{0\}$; i.e., $|\,\text{Im}(f)| = (p - 1)/2$. □

Example 6.59 (Perfect squares modulo 11). For $m = 11$, we have

$$QR_{11} = \{1, 3, 4, 5, 9\},$$

which has cardinality 5. We have $f(1) = f(10) = 1$, $f(2) = f(9) = 4$, $f(3) = f(8) = 9$, $f(4) = f(7) = 5$, and $f(5) = f(6) = 3$.

Exercise 6.32. *Suppose that x is a quadratic residue modulo ab, where a and b are positive integers. Let $\mathbf{r}_a : \mathbb{Z} \to R_a$ and $\mathbf{r}_b : \mathbb{Z} \to R_b$ denote the remainder functions. Show that $\mathbf{r}_a(x)$ is a quadratic residue modulo a and $\mathbf{r}_b(x)$ is a quadratic residue modulo b. In other words, a quadratic residue modulo an integer is a quadratic residue modulo a divisor of that integer.*

Now suppose, for example, that m is a prime or a product of two distinct primes. Define, as before,

$$f : QR_m \to QR_m$$

by $f(x) = \mathbf{r}(x^2)$ for $x \in QR_m$.

Exercise 6.33. *Using that m is a prime or a product of two distinct primes, show that f is well-defined.*

Exercise 6.34. *Continuing with the previous exercise, investigate under the additional restriction that $3 \leq m \leq 10$ whether for each m the function $f : QR_m \to QR_m$ is a bijection.*

6.8.3. Quadratic residues modulo a product of distinct primes. Consider, as above, the squaring function on the set of quadratic residues.

Example 6.60. The function $f : QR_{21} \to QR_{21}$ defined by $f(x) = \mathbf{r}(x^2)$ satisfies $QR_{21} = \{1, 4, 7, 9, 15, 16, 18\}$ and

$$f(1) = 1,\ f(4) = 16,\ f(7) = 7,\ f(9) = 18,\ f(15) = 15,\ f(16) = 4,\ f(18) = 9.$$

Thus, f is a bijection.

Observe that $21 = 3 \cdot 7$. Now suppose more generally that $m = pq$, where p and q are distinct primes. Suppose that $x, y \in QR_{pq}$ satisfy $f(x) = f(y)$. By definition, we have that

$$x^2 \equiv y^2 \bmod pq. \tag{6.89}$$

Let us try to show that $x = y$; that is, show that f is injective. We will have to impose a condition on the primes p and q to accomplish this (see Theorem 6.62 below, generalizing Example 6.60, which this is leading to).

Suppose for a hypothetical contradiction that $x \neq y$. Since $x, y \in QR_{pq}$, this is equivalent to the assumption that

$$x \not\equiv y \bmod pq. \tag{6.90}$$

Exercise 6.35. *Show that*

$$x \not\equiv y \bmod p \quad \textit{or} \quad x \not\equiv y \bmod q. \tag{6.91}$$

By Exercise 6.35, without loss of generality, we may assume that

$$x \not\equiv y \bmod p. \tag{6.92}$$

On the other hand, by (6.89),

$$x^2 \equiv y^2 \bmod p. \tag{6.93}$$

Exercise 6.36. *Show that y has an inverse z modulo p. Hint: Can you derive a contradiction from assuming that p divides y?*

Show that $z \in QR_{pq}$.

We have, for the inverse z of y modulo p from Exercise 6.36, that

$$(xz)^2 \equiv 1 \bmod p. \tag{6.94}$$

Exercise 6.37. (1) *Explain why* (6.94) *implies that*

$$xz \equiv 1 \bmod p \quad \textit{or} \quad xz \equiv -1 \bmod p. \tag{6.95}$$

(2) *Explain why we must have that $xz \equiv -1 \bmod p$. Hint: Appeal to* (6.92) *and the uniqueness of the inverse modulo p.*

Now, since x and z are non-zero quadratic residues, we have that xz is a non-zero quadratic residue, i.e., $xz \in QR_{pq}$, which implies that xz is a quadratic residue modulo p. Since $xz \equiv -1 \bmod p$, we obtain that -1 is a non-zero quadratic residue.

We now need the following result, which we do not prove.

Proposition 6.61. If p is congruent to 3 modulo 4, then -1 is not a quadratic residue modulo p.

By this result and the work above, we have proved the following.

Theorem 6.62. *Suppose that p and q are distinct primes that are both congruent to* 3 *modulo* 4. *Then the function $f : QR_{pq} \to QR_{pq}$ defined by $f(x) = \mathbf{r}(x^2)$ is a bijection.*

6.9. Fermat's Little Theorem

6.9.1. The statement of Fermat's Little Theorem. We have considered examples of the remainders of perfect squares after dividing by a positive integer m. One may ask is there are "special" exponents and moduli for which the remainders are nice. Here is a beautiful and fundamental result in this vein.

Theorem 6.63 (Fermat's Little Theorem). *If p is a prime and if a is a positive integer which is not a multiple of p, then*

$$a^{p-1} \equiv 1 \bmod p. \tag{6.96}$$

6.9.2. Examples of Fermat's Little Theorem. Let us look at some examples. Clearly, for every prime p, $1^{p-1} = 1 \equiv 1 \bmod p$.

(1) The prime $p = 5$:
   (a) We have $2^4 = 16 = 3 \cdot 5 + 1 \equiv 1 \bmod 5$.
   (b) We have $3^4 = 81 = 16 \cdot 5 + 1 \equiv 1 \bmod 5$.
   (c) We have $4^4 = 256 = 51 \cdot 5 + 1 \equiv 1 \bmod 5$.

(2) The prime $p = 7$:
   (a) We have $2^6 = 64 = 9 \cdot 7 + 1 \equiv 1 \bmod 7$.
   (b) We have $3^6 = 729 = 104 \cdot 7 + 1 \equiv 1 \bmod 7$.
   (c) We have $4^6 = 4096 = 585 \cdot 7 + 1 \equiv 1 \bmod 7$.
   (d) We have $5^6 = 15625 = 2232 \cdot 7 + 1 \equiv 1 \bmod 7$.
   (e) We have $6^6 \equiv (-1)^6 \equiv 1 \bmod 7$.

6.9.3. The idea of the proof of Fermat's Little Theorem. We consider the effect of multiplication of integers by a given integer a modulo p. At first glance, this topic has nothing to do with Fermat's Little Theorem!

Figure 6.9.1. Pierre de Fermat (1607–1665). Wikimedia Commons, Public Domain.

For example, suppose that $p = 5$ and $a = 2$. Consider the set of non-zero remainders modulo 5, which are the integers $\{1, 2, 3, 4\} = \mathbb{N}_4$. Their multiples by 2 are, in order,

$$\{2, 4, 6, 8\}. \tag{6.97}$$

Modulo 5, these are, in order,

$$\{2, 4, 1, 3\}. \tag{6.98}$$

Observe that, as a set this is also equal to $\mathbb{N}_4$. Likewise, multiplication by 3 yields

$$\{3, 6, 9, 12\} \equiv \{3, 1, 4, 2\}, \tag{6.99}$$

and multiplication by 4 yields

$$\{4, 8, 12, 16\} \equiv \{4, 3, 2, 1\}. \tag{6.100}$$

In each of the cases above the set of multiples, modulo 5, is equal to the set $\mathbb{N}_4$!

More generally, suppose that p is a prime and a is an integer that is not a multiple of p. Consider each of the integers $1, 2, 3, \ldots, p-1$ as representing both the number of a chair and the number of the person sitting on that chair. So we have $p-1$ chairs and $p-1$ people. Now change each person's number by multiplying it by a and assigning the remainder to be the person's new number.

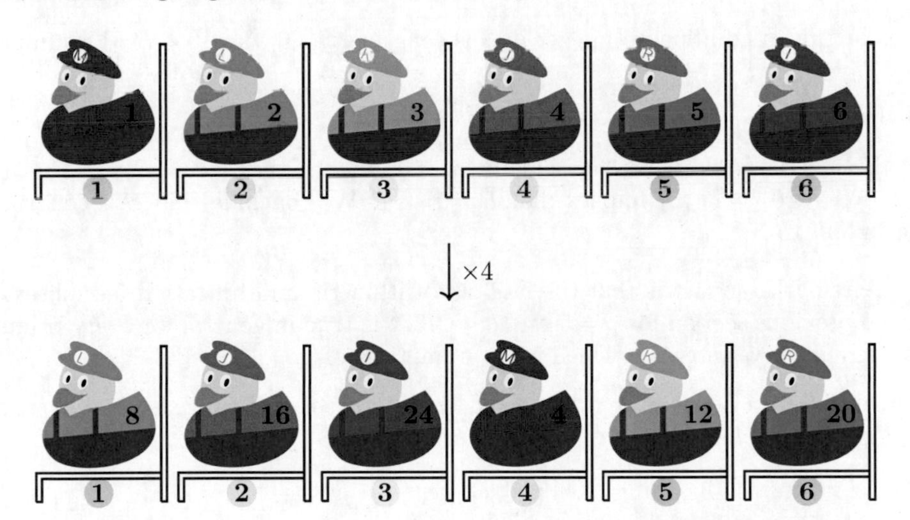

Figure 6.9.2. Multiplication by 4 rearranges the six brothers in the six chairs. Tikz LaTeX code by Sigur answering the question "Drawing Super Mario Bros ... in LaTeX" in StackExchange. Licensed under Creative Commons Attribution-Share Alike 4.0 International License (https://creativecommons.org/licenses/by-sa/4.0/deed.en).

Figure 6.9.2 shows the anatine example of musical chairs (but with no chairs removed!) with 6 Super Mario Bros. brothers (Mario, Luigi, and their heretofore unknown four brothers Kevin, Jovani, Romeo, and Ignazio) and 6 chairs. Multiplying each brother by 4 and taking the remainder modulo 7 assigns a new number to each brother and hence a new chair (six chairs for six brothers instead of Seven Brides for Seven Brothers!).

We wish to prove this property in general. We start with a little help from our friends.[1]

6.9.4. A proof of Fermat's Little Theorem. The general musical chairs fact we use is the following.

Lemma 6.64. *Let m be a positive integer. If S is a (finite) set of integers such that no two integers in S are congruent to each other modulo m, then S has at most m elements.*

If, in addition, no integer in S is congruent to 0, then S has at most $m-1$ elements. In this case, if S has exactly $m-1$ elements, then S is congruent modulo m to $\mathbb{N}_{m-1}$.

Proof. Recall from (6.13) that for every integer a, it is congruent to its remainder: $a \equiv \mathbf{r}(a) \bmod m$. So, up to congruence, we can replace each element a of S by $\mathbf{r}(a)$, and we call the new set R. Since no two integers in S are congruent to each other modulo m, we have that no two integers in R are congruent to each other modulo m. Since $R \subset R_m$, by Lemma 6.4 this implies that the elements of R are distinct. By Lemma 5.57 we conclude that

$$|S| = |R| \leq m. \tag{6.101}$$

Next, if in addition no integer in S is congruent to 0, then $0 \notin R$. This implies that

$$|S| = |R| \leq m - 1. \tag{6.102}$$

Finally, if in addition S has exactly $m-1$ elements, then $|R| = m-1$. This and $R \subset R_m - \{0\} = \mathbb{N}_{m-1}$ implies that $R = \mathbb{N}_{m-1}$. We conclude that S is congruent modulo m to $\mathbb{N}_{m-1}$. □

A remarkable fact is that the modular arithmetic multiplication fact observed in the previous section for $p = 5$ and $a = 2, 3, 4$ is true in general for every prime p and any positive integer a which is not a multiple of p.

Lemma 6.65. *Let p be a prime, and let a be a positive integer which is not a multiple of p. Then the set of multiples of a,*

$$S := \{a, 2a, 3a, \ldots, (p-1)a\}, \tag{6.103}$$

is, up to congruence modulo p, the same as the set

$$\mathbb{N}_{p-1} = \{1, 2, 3, \ldots, p-1\}. \tag{6.104}$$

Proof. We first show that no two elements of the set S in (6.103) are congruent to each other modulo p. Suppose that integers $1 \leq i, j \leq p-1$ satisfy

$$ia \equiv ja \bmod p.$$

Since a is not a multiple of p, it follows that $\gcd(a, p) = 1$. Hence we can factor out the a in the congruence equation to get

$$i \equiv j \bmod p.$$

[1] "With a Little Help From My Friends" is the title of a Beatles' song.

By Lemma 6.4 we conclude that $i = j$. Thus, any two integers in S cannot be congruent to each other.

Moreover, if $ia \equiv 0 \bmod p$, we obtain $i \equiv 0 \bmod p$. Assuming $i \not\equiv 0 \bmod p$, we obtain $ia \not\equiv 0 \bmod p$. Hence, any integer in S cannot be congruent to 0 modulo p.

Since S has $p-1$ elements, by Lemma 6.64 we conclude that S is, up to congruence modulo p, the same as $\mathbb{N}_{p-1}$. □

Example 6.66. We have seen above with the six Super Mario Brothers that corresponding to the prime $p = 7$ and the positive integer $a = 4$, the set of multiples of 4,

$$S = \{4, 8, 12, 16, 20, 24\},$$

is, up to congruence modulo 7, the same as the set

$$\mathbb{N}_6 = \{1, 2, 3, 4, 5, 6\}.$$

We are now ready to prove:

Theorem 6.67 (Fermat's Little Theorem)**.** *If p is a prime and if a is a positive integer which is not a multiple of p, then*

$$a^{p-1} \equiv 1 \bmod p. \tag{6.105}$$

Proof. Since p is a prime and p does not divide a, by Lemma 6.65 we have that "modulo p" the set of integers

$$S = \{1a, 2a, \ldots, (p-1)a\}$$

is "the same" as the set

$$\mathbb{N}_{p-1} = \{1, 2, \ldots, p-1\}.$$

That is, each integer in S is congruent to exactly one integer in $\mathbb{N}_{p-1}$ (so different integers in S are congruent to different integers in $\mathbb{N}_{p-1}$). The KEY observation is that this implies that

$$1a \cdot 2a \cdots (p-1)a \equiv 1 \cdot 2 \cdots (p-1) \mod p.$$

That is,

$$(p-1)!\, a^{p-1} \equiv (p-1)! \mod p.$$

Since p is a prime, by Corollary 2.18, we have

$$\gcd(p, (p-1)!) = 1. \tag{6.106}$$

Hence, by Corollary 6.21, we conclude that

$$a^{p-1} \equiv 1 \mod p.$$

□

Example 6.68. Taking $p = 7$ and $a = 4$ as in Example 6.66, we have that

$$6! \cdot 4^6 \equiv 4 \cdot 8 \cdot 12 \cdot 16 \cdot 20 \cdot 24 \equiv 1 \cdot 2 \cdot 3 \cdot 4 \cdot 5 \cdot 6 \equiv 6! \bmod 7.$$

Since $\gcd(6!, 7) = 1$, this implies that

$$4^6 \equiv 1 \bmod 7.$$

We have the following characterization of primes using modular arithmetic and factorials.

Theorem 6.69 (Wilson's Theorem). *Let $n \geq 2$ be an integer. Then*

$$(n-1)! \equiv -1 \bmod n \quad \Leftrightarrow \quad n \text{ is a prime.} \tag{6.107}$$

Proof. (1): $\Rightarrow$ We prove the contrapositive of this implication. Suppose that $n = ab$, where $a > 1$ and $b > 1$, is a composite number. Suppose for a contradiction that $(n-1)! \equiv -1 \bmod n$. Then, by Lemma 6.5 we have that

$$(n-1)! \equiv -1 \bmod a. \tag{6.108}$$

But since $1 \leq a \leq n-1$, we have that a divides $(n-1)!$, so that $(n-1)! \equiv 0 \bmod a$. Since $a \geq 2$, we have a contradiction. We conclude that $(n-1)! \not\equiv -1 \bmod n$.

(2): $\Leftarrow$ Let n be a prime. If $n = 2$, it is easy to see that the implication is true. So we may suppose that n is an odd prime. Since n is a prime, by Corollary 6.43 each of the (even number of) integers $1, 2, \ldots, n-1$ has an inverse modulo n. Furthermore, by Lemma 6.44, the only integers of these which are their own inverse are 1 and $n-1$. Thus, the (even number of) integers $2, \ldots, n-2$ may be paired up with their inverses, where the product of each pair a, a^{-1} satisfies $a^{-1}a \equiv 1 \bmod n$. We conclude that

$$2 \cdot 3 \cdots (n-2) \equiv 1 \bmod n, \tag{6.109}$$

and hence

$$(n-1)! \equiv 1 \cdot (n-1) \equiv -1 \bmod n. \tag{6.110}$$

□

Example 6.70. We have for the prime $p = 7$

$$6! \equiv 720 \equiv -1 \bmod 7 \tag{6.111}$$

since $721 = 7 \cdot 103$. Here, the pairs of inverses are $(2, 4)$ and $(3, 5)$.

And also, for the prime $p = 11$,

$$10! \equiv 3628800 \equiv -1 \bmod 11 \tag{6.112}$$

since $3628801 = 11 \cdot 329891$. Here, the pairs of inverses are $(2, 6)$, $(3, 4)$, $(5, 9)$, and $(7, 8)$.

And another(!), for the prime $p = 13$,

$$12! \equiv 479001600 \equiv -1 \bmod 13 \tag{6.113}$$

since $479001601 = 13 \cdot 36846277$. We leave it to you, dear reader, to list the pairs of inverses modulo the prime 13.

Exercise 6.38. *Let $n \geq 2$ be an integer. Observe that $(n!)^2 + 1$ is an odd integer. Let p be a prime divisor of $(n!)^2 + 1$.*

(1) *Prove that p is odd and $p > n$.*
(2) *Prove that $(n!)^{p-1} \equiv (-1)^{\frac{p-1}{2}} \bmod p$. Hint: What is $(n!)^2 \bmod p$?*
(3) *Prove that $(n!)^{p-1} \equiv 1 \bmod p$. Hint: Fermat's Little Theorem 6.67.*
(4) *Prove that $p \equiv 1 \bmod 4$.*
(5) *Explain how we can use the above to prove that there are an infinite number of primes p satisfying $p \equiv 1 \bmod 4$, thus proving Theorem 6.27.*

6.10*. Euler's totient function and Euler's Theorem

Definition 6.71. For any integer $n \geq 2$, let $\varphi(n)$ be the number of positive integers k less than n satisfying $\gcd(k, n) = 1$. The function

$$\varphi : \mathbb{Z}^+ \to \mathbb{Z}^+ \tag{6.114}$$

thus defined is called Euler's totient function.

Observe that $1 \leq \varphi(n) \leq n - 1$. Define the set

$$R_n^* := \{k \in \mathbb{Z} : 1 \leq k \leq n - 1 \text{ and } k \text{ and } n \text{ are coprime}\}. \tag{6.115}$$

We have that $\varphi(n)$ is equal to the number of elements (cardinality) of the set R_n^*. Note that R_n^* is a subset of the remainder set $R_n = \{0, 1, 2, \ldots, n - 1\}$. We may write

$$R_n^* := \{k \in R_n : \gcd(k, n) = 1\}. \tag{6.116}$$

By Theorem 6.48, R_n^* is the set of remainders that have inverses modulo n.

Example 6.72. If p is a prime, then

$$\varphi(p) = p - 1. \tag{6.117}$$

So

$$\varphi(2) = 1,\ \varphi(3) = 2,\ \varphi(5) = 4,\ \varphi(7) = 6,\ \varphi(11) = 10, \text{ etc.} \tag{6.118}$$

Example 6.73. For the integer $n = 15$, the integers between 1 and 14 inclusive that are coprime with 15 are (see Example 6.49)

$$1, 2, 4, 7, 8, 11, 13, 14. \tag{6.119}$$

Hence, we count that $\varphi(15) = 8$.

Figure 6.10.1. Leonhard Euler (1707–1783). Wikimedia Commons, Public Domain.

Wikipedia gives a table of values of $\varphi(n)$ for $1 \leq n \leq 100$. They are

(6.120)

	+1	+2	+3	+4	+5	+6	+7	+8	+9	+10
0	1	1	2	2	4	2	6	4	6	4
10	10	4	12	6	8	8	16	6	18	8
20	12	10	22	8	20	12	18	12	28	8
30	30	16	20	16	24	12	36	18	24	16
40	40	12	42	20	24	22	46	16	42	20
50	32	24	52	18	40	24	36	28	58	16
60	60	30	36	32	48	20	66	32	44	24
70	70	24	72	36	40	36	60	24	78	32
80	54	40	82	24	64	42	56	40	88	24
90	72	44	60	46	72	32	96	42	60	40

We observe formula (6.117) for the primes, such as

$$\varphi(67) = 66.$$

We will use right below the following result.

Exercise 6.39. *Let p be a prime and let k be a positive integer. Prove that if m is a positive divisor of p^k, then m is equal to one of the integers $1, p, p^2, \ldots, p^k$.*

In view of the Prime Factorization Theorem, we first consider Euler's totient function for inputs that are powers of primes.

Lemma 6.74 (Euler totient function of prime powers)**.** *If p is a prime and if k is a positive integer, then*

$$\varphi(p^k) = p^k \left(1 - \frac{1}{p}\right). \tag{6.121}$$

Proof. Let m be any integer. Since $\gcd(m, p^k)$ is a positive integer dividing p^k, by Exercise 6.39 we have that $\gcd(m, p^k)$ is equal to one of the following integers:

$$1,\ p,\ p^2, \ldots, p^k. \tag{6.122}$$

In particular, it is easy to see from this that

$$\gcd(m, p^k) > 1 \text{ if and only if } m \text{ is a multiple of } p. \tag{6.123}$$

Now, the multiples of p between 1 and p^k inclusive are

$$1p,\ 2p,\ 3p, \ldots,\ p^{k-1}p. \tag{6.124}$$

We see that there are exactly p^{k-1} such multiples (corresponding to $1, 2, 3, \ldots, p^{k-1}$). That is, there are exactly p^{k-1} integers m between 1 and p^k inclusive satisfying $\gcd(m, p^k) > 1$. We conclude by the definition of Euler's totient function that

$$\varphi(p^k) = p^k - p^{k-1} = p^k \left(1 - \frac{1}{p}\right). \tag{6.125}$$

□

Example 6.75. Consider $3^3 = 27$. The multiples of 3 between 1 and 27 inclusive are

$$3, 6, 9, 12, 15, 18, 21, 24, 27,$$

which number nine in all. We have

$$\begin{aligned}\varphi(27) &= \big|\{1, 2, 4, 5, 7, 8, 10, 11, 13, 14, 16, 17, 19, 20, 22, 23, 25, 26\}\big| \\ &= 18 = 27 \cdot \left(1 - \frac{1}{3}\right) = 27 - 9.\end{aligned}$$

Exercise 6.40. *Find all (unordered) pairs of inverses modulo* 27. *You may refer to Example* 6.75.

Secondly, we consider Euler's totient function of products of coprime integers.

Theorem 6.76 (Euler totient function for coprime products). *If $m, n \in \mathbb{Z}^+$ satisfy* $\gcd(m, n) = 1$, *then*

$$\varphi(mn) = \varphi(m)\varphi(n). \tag{6.126}$$

Proof. Recall that $\varphi(n)$ is the number of elements of the set R_n^* defined by (6.115). Formula (6.126) follows from showing that the two sets $R_m^* \times R_n^*$ and R_{mn}^* have the same number of elements. By Theorem 5.62, it suffices to find a *bijection* between these two sets. In turn, by Theorem 5.53 it suffices to find a function between the two sets which has an *inverse*.

Define the function $f : R_m^* \times R_n^* \to R_{mn}^*$ as follows. Let $(a, b) \in R_m^* \times R_n^*$. Since $\gcd(m, n) = 1$, by the Chinese Remainder Theorem 6.53, there exists a unique integer $x_0 \in R_{mn}$ satisfying

$$x_0 \equiv a \bmod m, \tag{6.127a}$$

$$x_0 \equiv b \bmod n. \tag{6.127b}$$

We define

$$f(a, b) = x_0. \tag{6.128}$$

Since $(a, b) \in R_m^* \times R_n^*$, we have $\gcd(a, m) = 1$ and $\gcd(b, n) = 1$. Thus, by (6.127) and Lemma 6.23, we have

$$\gcd(x_0, m) = \gcd(a, m) = 1, \qquad \gcd(x_0, n) = \gcd(b, n) = 1. \tag{6.129}$$

By Lemma 4.29, this implies that $\gcd(x_0, mn) = 1$. Hence $f(a, b) = x_0$ is indeed in the specified codomain R_{mn}^*, so that the function f is well-defined.

Next, we define the function $g : R_{mn}^* \to R_m^* \times R_n^*$ by

$$g(x) := (\mathbf{r}_m(x), \mathbf{r}_n(x)), \tag{6.130}$$

where $\mathbf{r}_m : \mathbb{Z} \to R_m$ and $\mathbf{r}_n : \mathbb{Z} \to R_n$ are the remainder functions. Let $x \in R_{mn}^*$. Then $\gcd(x, mn) = 1$, which implies that $\gcd(x, m) = 1$ and $\gcd(x, n) = 1$. Hence, by Lemma 6.23 again, we have that $\gcd(\mathbf{r}_m(x), m) = 1$ and $\gcd(\mathbf{r}_n(x), n) = 1$, so that $\mathbf{r}_m(x) \in R_m^*$ and $\mathbf{r}_n(x) \in R_n^*$. Hence, the function g is well-defined.

Now we check that f and g are inverses of each other. Let $(a, b) \in R_m^* \times R_n^*$. As above, we use the notation $x_0 := f(a, b)$. Then, by (6.127),

$$x_0 \equiv a \bmod m, \qquad x_0 \equiv b \bmod n. \tag{6.131}$$

Hence,

(6.132) $$\mathbf{r}_m(x_0) = a, \qquad \mathbf{r}_n(x_0) = b.$$

This proves that $g(f(a,b)) = (a,b)$.

Now let $x \in R^*_{mn}$. Then

$$f(g(x)) = f(\mathbf{r}_m(x), \mathbf{r}_n(x)) = x$$

since

$$x \equiv \mathbf{r}_m(x) \bmod m, \qquad x \equiv \mathbf{r}_n(x) \bmod n.$$

We have proved that f and g are inverses of each other. The theorem follows. □

For example, the theorem implies that

(6.133) $$\varphi(15) = \varphi(3)\varphi(5) = 2 \cdot 4 = 8, \quad \varphi(30) = \varphi(5)\varphi(6) = 4 \cdot 2 = 8.$$

Thirdly, and finally, we can extend Lemma 6.74 and Theorem 6.76 to the following beautiful general formula.

Theorem 6.77 (Euler's product formula). *For any integer $n \geq 2$,*

(6.134) $$\varphi(n) = n \prod_{p|n} \left(1 - \frac{1}{p}\right),$$

where the product on the right-hand side is over all distinct primes p dividing n.

Proof. By the Prime Factorization Theorem, there exist primes $p_1 < p_2 < \cdots < p_r$ and positive integers $k_1, k_2, \ldots, k_r$ such that

(6.135) $$n = p_1^{k_1} \, p_2^{k_2} \cdots p_r^{k_r}.$$

By Theorem 6.76, since $\gcd(p_i^{k_i}, p_j^{k_j}) = 1$ for $i \neq j$, we have

(6.136) $$\varphi(n) = \varphi(p_1^{k_1}) \, \varphi(p_2^{k_2}) \cdots \varphi(p_r^{k_r}).$$

Actually, we should be a bit more detailed in our justification: Firstly, since $\gcd(p_1^{k_1}, p_2^{k_2} \cdots p_r^{k_r}) = 1$, we have

(6.137) $$\varphi(n) = \varphi(p_1^{k_1}) \, \varphi(p_2^{k_2} \cdots p_r^{k_r}).$$

Starting with this, we can prove (6.136) by induction on r. We leave this as an exercise.

Now we are in a perfect position to apply Lemma 6.74. Indeed, by applying this lemma to (6.136), we obtain

$$\varphi(n) = p_1^{k_1} \left(1 - \frac{1}{p_1}\right) p_2^{k_2} \left(1 - \frac{1}{p_2}\right) \cdots \, p_r^{k_r} \left(1 - \frac{1}{p_r}\right).$$

The lemma now follows from (6.135) and the fact that $p_1, p_2, \ldots, p_r$ are the primes dividing n. □

Example 6.78. Since the only prime divisors of 18 are 2 and 3, by Theorem 6.77 we have

$$\varphi(18) = 18\left(1 - \frac{1}{2}\right)\left(1 - \frac{1}{3}\right) = 18 \cdot \frac{1}{2} \cdot \frac{2}{3} = 6. \tag{6.138}$$

Similarly, $36 = 2^2 3^2$ satisfies

$$\varphi(36) = 36 \cdot \frac{1}{2} \cdot \frac{2}{3} = 12. \tag{6.139}$$

Exercise 6.41. *Use Theorem* 6.77 *to compute Euler's totient function* $\varphi(n)$ *for all composite numbers that are between* 21 *and* 40 *inclusive. You can also use Theorem* 6.76 *if you like.*

Recall that if m is prime, then $\varphi(m) = m - 1$. A generalization of Fermat's Little Theorem is Euler's Theorem (we will prove this theorem in §8.5.12 below ☺), which says:

Theorem 6.79. *If positive integers* a *and* m *are coprime, then*

$$a^{\varphi(m)} \equiv 1 \bmod m. \tag{6.140}$$

6.10.1. Reduced residue systems.

Definition 6.80. A (finite) set of integers R is called a reduced residue system modulo m if the following are true:

(1) For each $r \in R$, $\gcd(r, m) = 1$.
(2) The cardinality $|R| = \varphi(m)$.
(3) Distinct elements of R are incongruent modulo m.

Example 6.81. The set

$$R = \{1, 3, 7, 9\} \tag{6.141}$$

is a reduced residue system modulo 10. We also have that

$$R' = \{101, 213, 747, 909\} \tag{6.142}$$

is a reduced residue system modulo 10.

Exercise 6.42. *Let* $R = \{r_1, r_2, \ldots, r_{\varphi(m)}\}$ *be a reduced residue system modulo* m. *Prove that if* $m > 2$, *then*

$$\sum_{i=1}^{\varphi(m)} r_i \equiv 0 \bmod m. \tag{6.143}$$

6.10.2. Schneider's formula. Let $\alpha = \frac{1+\sqrt{5}}{2}$ be the golden ratio (as in (2.116)). Then a formula of Schneider is

$$\alpha = -\sum_{n=1}^{\infty} \frac{\varphi(n)}{n}\left(1 - \frac{1}{\alpha^n}\right). \tag{6.144}$$

We do not prove this formula. ☹

6.11*. An application of Fermat's Little Theorem: The RSA algorithm

In this section we discuss an application of Fermat's Little Theorem to cryptography, namely, the RSA algorithm, named after Rivest, Shamir, and Adleman. Our discussion is only a tiny glimpse of the application of number theory to cryptography.

Let

(6.145) $$\mathcal{A} = \{A, B, C, \ldots, Z\}, \qquad \mathcal{B} = \{65, 66, 67, \ldots, 90\}.$$

Define function $f : \mathcal{A} \to \mathcal{B}$ to be the unique bijection which takes the alphabetical ordering of $\mathcal{A}$ to the numerical ordering of $\mathcal{B}$, so that $f(A) = 65$, $f(B) = 66$, etc.[2]

For example, consider the word[3]

B L A D E R U N N E R ,

which is a string of 11 letters. Our friend Leon Kowalski would like to secretly send this word to us via the internet, in a way that anyone intercepting his message will not be able to decipher it.

By using the function f, the word BLADERUNNER corresponds to a string of 11 2-digit integers:

(6.146) $$66\,76\,65\,68\,69\,82\,85\,78\,78\,69\,82.$$

This forms a 22-digit integer

(6.147) $$a := 6676656869828578786982$$

from which we can recover the word BLADERUNNER by applying the inverse function $f^{-1} : \mathcal{B} \to \mathcal{A}$ to each of the 11 2-digit integers comprising a. So the question becomes: How do we enable our friend Leon to encrypt the integer a so that only we can decrypt it?

Let p and q be large and distinct primes. For example, you could take (but not advisable!) p and q to be the twin primes in (1.90). Let

$$m = pq.$$

By Theorem 6.76,

(6.148) $$\varphi(m) = (p-1)(q-1).$$

On the other hand, without knowing p and q, it would be very hard to compute $\varphi(m)$.

Next, we choose a positive integer e that is coprime with $\varphi(m)$; i.e.,

$$\gcd(e, \varphi(m)) = 1.$$

By (6.36) there exists a positive integer d such that

(6.149) $$de \equiv 1 \bmod \varphi(m).$$

Choose such an integer d.

[2] This is the ASCII code.

[3] Of course, we could choose another word, such as MACGUFFIN.

Definition 6.82. The **public key** is the pair of integers

$$(m, e). \tag{6.150}$$

The **private key** is the triplet of integers

$$(p, q, d). \tag{6.151}$$

The public key is used to **encrypt** an integer a and the private key is used to **decrypt** the resulting encryption. This is how it is done. We choose m so large that

$$a < m$$

for every positive integer a we wish to encrypt and decrypt. The encryption is given by defining

$$b := \mathbf{r}(a^e), \tag{6.152}$$

where $\mathbf{r} : \mathbb{Z} \to R_m$ is the remainder function. The key is that we can recover a from b by the formula

$$a = \mathbf{r}(b^d). \tag{6.153}$$

To see this, we observe that (6.152) implies that

$$b \equiv a^e \bmod m.$$

Thus

$$b^d \equiv a^{de} \bmod m.$$

By (6.149), there exists an integer k such that

$$de - 1 = k\varphi(m). \tag{6.154}$$

On the other hand, by Euler's Theorem 6.79,

$$a^{k\varphi(m)} \equiv 1^k \equiv 1 \bmod m. \tag{6.155}$$

Therefore,

$$b^d \equiv a^1 \equiv a \bmod m. \tag{6.156}$$

Since $0 < a < m$, this implies (6.153).

6.12*. The Euclid–Euler Theorem characterizing even perfect numbers

As we have just seen, Euler's totient function φ is very useful. In this section, using the arithmetic function σ defined in the next paragraph, we give a well-known characterization of even perfect numbers.

6.12.1. Sum of positive divisors function. The sum of positive divisors function is defined for $m \in \mathbb{Z}^+$ by

$$\sigma(m) := \sum_{d|m} d, \tag{6.157}$$

where the sum is over all positive divisors d of m. In general, a divisor function is an arithmetic function which depends on the divisors of the positive integer input.

Firstly, the following result is immediate from the definition of σ.

Lemma 6.83. *If p is a prime, since its only positive divisors are 1 and p, we then have*

$$\sigma(p) = p + 1. \tag{6.158}$$

A table of values of the sum of positive divisors function σ for the first 50 positive integers is

(6.159)

	+1	+2	+3	+4	+5	+6	+7	+8	+9	+10
0	1	3	4	7	6	12	8	15	13	18
10	12	28	14	24	24	31	18	39	20	42
20	32	36	24	60	31	42	40	56	30	72
30	32	63	48	54	48	91	38	60	56	90
40	42	96	44	84	78	72	48	124	57	93

For example, this table says that

$$\sigma(35) = 1 + 5 + 7 + 35 = 48$$

and that

$$\sigma(36) = 1 + 2 + 3 + 4 + 6 + 9 + 12 + 18 + 36 = 91.$$

Secondly, we have the following reduction formula for σ of the product of coprime integers.

Lemma 6.84. *If a and b are coprime positive integers, then*

$$\sigma(ab) = \sigma(a)\,\sigma(b). \tag{6.160}$$

Proof. Let $\{c_i\}_{i=1}^k = \{c_1, c_2, \ldots, c_k\}$ be the set of positive divisors of a, and let $\{d_j\}_{j=1}^\ell = \{d_1, d_2, \ldots, d_\ell\}$ be the set of positive divisors of b. Then (exercise!), since a and b are coprime, the set of positive divisors of ab is equal to the set

$$\{c_i d_j\}_{1 \le i \le k,\, 1 \le j \le \ell}. \tag{6.161}$$

From this we conclude that

$$\sigma(ab) = \sum_{i=1}^k \sum_{j=1}^\ell c_i d_j = \sum_{i=1}^k c_i \sum_{j=1}^\ell d_j = \sigma(a)\,\sigma(b). \tag{6.162}$$ □

For example, $\sigma(42) = \sigma(6)\,\sigma(7) = 12 \cdot 8 = 96$.

By Lemmas 6.83 and 6.84, we have the following.

Corollary 6.85. *If $m = pq$ is a* ***semiprime****, that is, p and q are distinct primes, then*

$$\sigma(m) = (p+1)(q+1) = m+1+p+q. \tag{6.163}$$

For example, $\sigma(35) = \sigma(5 \cdot 7) = 35 + 1 + 5 + 7 = 48$.

Thirdly, generalizing Lemma 6.83, we have:

Lemma 6.86. *If q is a prime and if k is a positive integer, then*

$$\sigma(q^k) = \frac{q^{k+1}-1}{q-1}. \tag{6.164}$$

Proof. The positive divisors of q^k are

$$1,\ q,\ q^2, \ldots,\ q^k. \tag{6.165}$$

Thus,

$$\sigma(q^k) = \sum_{i=0}^{k} q^i = \frac{q^{k+1}-1}{q-1}. \tag{6.166}$$

□

For example, $\sigma(27) = \sigma(3^3) = \frac{3^4-1}{3-1} = 40$.

Fourthly and finally, from Lemmas 6.84 and 6.86, we have the following general formula for σ.

Corollary 6.87. *If $n = p_1^{k_1} p_2^{k_2} \cdots p_r^{k_r}$ is the prime factorization of an integer $n \geq 2$, then*

$$\sigma(n) = \prod_{i=1}^{r} \frac{p_i^{k_i+1}-1}{p_i - 1}. \tag{6.167}$$

Example 6.88. Using Corollary 6.87, we compute that

$$\begin{aligned}
\sigma(30) &= \sigma(2 \cdot 3 \cdot 5) = 3 \cdot 4 \cdot 6 = 72, \\
\sigma(35) &= \sigma(5 \cdot 7) = 6 \cdot 8 = 48, \\
\sigma(36) &= \sigma(2^2 3^2) = \frac{2^3-1}{2-1} \cdot \frac{3^3-1}{3-1} = 7 \cdot 13 = 91.
\end{aligned}$$

6.12.2. Statement and proof of the Euclid–Euler Theorem. Recall from §1.7* that a positive integer n is a **perfect number** if n is equal to the sum of its positive divisors less than n. In terms of the sum of divisors function, this says that

$$n = \sigma(n) - n; \tag{6.168}$$

that is, we have:

Lemma 6.89. *A positive integer n is a perfect number if and only if*

$$\sigma(n) = 2n. \tag{6.169}$$

We have the following characterization of even perfect numbers.

Theorem 6.90 (Euclid–Euler Theorem). *An even positive integer n is a perfect number if and only if*

$$n = 2^{p-1}(2^p - 1) = \frac{M_p(M_p + 1)}{2}, \tag{6.170}$$

where $M_p := 2^p - 1$ is a Mersenne prime!

Proof. ($\Leftarrow$) Suppose that $2^p - 1$ is a Mersenne prime. We then have that

$$\gcd(2^{p-1}, 2^p - 1) = 1 \tag{6.171}$$

since $1 < 2^{p-1} < 2^p - 1$. Thus, by Lemma 6.84, we have the sum of positive divisors satisfies

$$\sigma(2^{p-1}(2^p - 1)) = \sigma(2^{p-1})\,\sigma(2^p - 1). \tag{6.172}$$

By Lemma 6.86 we have

$$\sigma(2^{p-1}) = 2^p - 1. \tag{6.173}$$

Since $2^p - 1$ is assumed to be prime, we have

$$\sigma(2^p - 1) = 2^p. \tag{6.174}$$

Therefore, by plugging (6.172) and (6.173) into (6.171), we obtain

$$\sigma(2^{p-1}(2^p - 1)) = (2^p - 1)2^p = 2 \cdot 2^{p-1}(2^p - 1). \tag{6.175}$$

Thus, by the characterization (6.169), $2^{p-1}(2^p - 1)$ is a perfect number.

($\Rightarrow$) Suppose that n is an even perfect number. Then we may write n as

$$n = 2^k \cdot q, \tag{6.176}$$

where $k \geq 1$ and q is an odd integer (i.e., $\gcd(q, 2) = 1$). Since $2^k \cdot q$ is a perfect number, by (6.169) and by Lemmas 6.84 and 6.86, we have

$$2^{k+1} \cdot q = 2 \cdot 2^k \cdot q = \sigma(2^k \cdot q) = (2^{k+1} - 1)\,\sigma(q), \tag{6.177}$$

where we also used that $\gcd(q, 2^k) = 1$. Observe that since $k \geq 1$, we have $2^{k+1} - 1 \geq 3$. Moreover, since $2^{k+1} - 1$ is odd and divides $2^{k+1} \cdot q$ from (6.177), by Theorem 4.23 we have that $2^{k+1} - 1$ divides q.

Now, since $2^{k+1} - 1 > 1$ is a positive divisor of q, we have that

$$r := \frac{q}{2^{k+1} - 1} = \frac{\sigma(q)}{2^{k+1}} \tag{6.178}$$

(where we used (6.177) for the second equality) is a divisor of q which is less than q. Hence, r and q are distinct positive divisors of q. This implies that

$$\sigma(q) \geq r + q, \tag{6.179}$$

with equality if and only if r and q are the only divisors of q. Note that in this equality case we have $r = 1$ and q is prime.

Now, (6.178) and (6.179) imply

$$2^{k+1}r = \sigma(q) \geq r + q = r + (2^{k+1} - 1)r = 2^{k+1}r. \tag{6.180}$$

This implies that $\sigma(q) = r + q$, which in turn implies that

$$q = (2^{k+1} - 1)r = 2^{k+1} - 1 \tag{6.181}$$

is a prime. We are done since we now have by (6.176) that

$$n = 2^k(2^{k+1} - 1). \tag{6.182}$$

Finally, we let $p = k + 1$ to get $n = 2^{p-1}(2^p - 1)$, where $2^p - 1$ is prime. □

Remark 6.91. Euclid proved the "if" implication direction of the theorem, whereas Euler proved the "only if" implication direction of the theorem. This is even though just about every mathematician between Euclid and Euler believed the "only if" implication to be true!

6.13*. Twin prime pairs

Recall that we encountered twin primes in §1.9*. More importantly, the twin prime conjecture that there are an infinite number of twin primes is still unsolved! We have the following result which we will not prove. ☹

Theorem 6.92. *A pair $(m, m + 2)$, where $m \geq 2$, is a twin prime pair if and only if*

$$4((m - 1)! + 1) \equiv -m \bmod m(m + 2). \tag{6.183}$$

For example, for $m = 3$ we have

$$4((3 - 1)! + 1) = 12 \equiv -3 \bmod 3(3 + 2) = 15, \tag{6.184}$$

corresponding to $(3, 5)$ being a twin prime pair. For $m = 7$ we have

$$4((7 - 1)! + 1) = 2884 \not\equiv -7 \bmod 7(7 + 2) = 63 \tag{6.185}$$

since $2891 = 63 \cdot 45 + 56$, corresponding to $(7, 9)$ not being a twin prime pair. For $m = 11$ we have

$$4((11 - 1)! + 1) = 14515204 \equiv -11 \bmod 11(11 + 2) = 143 \tag{6.186}$$

since $14515215 = 143 \cdot 101505$, corresponding to $(11, 13)$ being a twin prime pair.

Factorials quickly get large. For example, the next smallest twin prime pair is $(17, 19)$. We compute that

$$4(16! + 1) = 83691159552004 \equiv -17 \bmod 323 \tag{6.187}$$

since $83691159552021 = 323 \cdot 259105757127$.

6.14. Chameleons roaming around in a zoo

We conclude this chapter with a fun puzzle regarding the interactions of social chameleons, which can be solved using modular arithmetic.

Solved Problem 6.93. *There are several chameleons roaming around in the San Diego Zoo. Of these chameleons, R_0 are red, B_0 are blue, and G_0 are green, where all of R_0, B_0, G_0 are positive integers. Whenever two chameleons meet, if they are the same color, then their colors remain unchanged after the meeting. On the other hand, if they are different colors, then both of their colors change* *to the third color. Are there any restrictions on R_0, G_0, and B_0 for it to be possible that all of the chameleons are eventually the same color? (Assume that the Fountain of Youth is in the San Diego Zoo and so the chameleons never die; that is, they are immortal!)*

Figure 6.14.1. Indian chameleon From Kanakpura, Karnataka. Photo credit: Wikimedia Commons, author: Girish Gowda, licensed under Creative Commons Attribution-Share Alike 4.0 International (https://creativecommons.org/licenses/by-sa/4.0/deed.en) license.

Solution. If two same colored chameleons meet, then there is no change to the colors due to this meeting. So we just need to consider the case where two different colored chameleons meet.

Suppose that a red chameleon and a blue chameleon meet. Then both of these chameleons become green. So, if R, B, G is the number of red, blue, and green chameleons right before the meeting, respectively, then

$$\tilde{R} := R - 1, \quad \tilde{B} := B - 1, \quad \tilde{G} := G + 2 \tag{6.188}$$

is the number of red, blue, and green chameleons right after the meeting, respectively. The key observation is that

$$\tilde{R} - \tilde{B} \equiv R - B \bmod 3, \ \tilde{B} - \tilde{G} \equiv B - G \bmod 3, \ \tilde{G} - \tilde{R} \equiv G - R \bmod 3. \tag{6.189}$$

Indeed, this is a simple calculation using (6.188). Note that a (somewhat) trivial consequence of this is

$$\tilde{B} - \tilde{R} \equiv B - R \bmod 3, \ \tilde{G} - \tilde{B} \equiv G - B \bmod 3, \ \tilde{R} - \tilde{G} \equiv R - G \bmod 3. \tag{6.190}$$

The displayed congruence equations above are also true: (1) when a blue chameleon and a green chameleon meet and (2) when a green chameleon and a red chameleon meet. That is, they are true whenever two different colored chameleons meet. On the other hand, when two same colored chameleons meet, they are true trivially.

Let R_0, B_0, G_0 denote the initial number of red, blue, and green chameleons, and let R_i, B_i, G_i denote the number of red, blue, and green chameleons after the i-th meeting, where $1 \leq i \leq n$ and n is the total number of chameleon meetings (which can be any positive integer). By (6.189), we have that modulo 3, $R_i - B_i$,

$B_i - G_i$, and $G_i - R_i$ are independent of i. In particular, the initial differences and the end differences are congruent modulo 3:
(6.191)
$R_n - B_n \equiv R_0 - B_0 \bmod 3,\ B_n - G_n \equiv B_0 - G_0 \bmod 3,\ G_n - R_n \equiv G_0 - R_0 \bmod 3.$

Now suppose that after the n-th meeting, all of the chameleons are one color. Firstly, assume that this color is red. Then $B_n = G_n = 0$, and so (6.191) says that

$$R_n \equiv R_0 - B_0 \bmod 3,\ 0 \equiv B_0 - G_0 \bmod 3,\ -R_n \equiv G_0 - R_0 \bmod 3. \tag{6.192}$$

Notice that any one of these equations follows from the other two. The only restrictive equation we have is the middle equation, which says that

$$B_0 \equiv G_0 \bmod 3. \tag{6.193}$$

Similarly, if all of the chameleons are blue after the n-th meeting, then we must have that

$$G_0 \equiv R_0 \bmod 3. \tag{6.194}$$

And, if all of the chameleons are green after the n-th meeting, then we must have that

$$R_0 \equiv B_0 \bmod 3. \tag{6.195}$$

Thus, if R_0, B_0, G_0 are all incongruent modulo 3, then it is impossible for all of the chameleons to ever be the same color. Moreover, if the numbers of chameleons for two different colors are not congruent modulo 3, then the chameleons can never all be the third color.

Conversely, suppose that (at least) two of R_0, B_0, G_0 are congruent modulo 3. Then, without loss of generality, we may assume that $G_0 \equiv R_0 \bmod 3$. In addition, suppose that $G_0 \geq R_0$. We will show in this case that it is possible to arrange a sequence of meetings of the chameleons so that eventually all of the chameleons are blue. By symmetry in green and red, it follows that we can remove the assumption that $G_0 \geq R_0$.

We first arrange for R_0 pairs of green and red chameleons to meet. After these R_0 arranged meetings, we have that

$$R_1 = 0, \quad G_1 = G_0 - R_0, \quad B_1 = B_0 + 2R_0. \tag{6.196}$$

Case 1: $k := \frac{1}{3}G_1 \leq B_1$; i.e., $G_1 \leq 3B_1$. Note that since $G_0 \equiv R_0 \bmod 3$, we have that k is a non-negative integer. Since $k \leq B_1$, we can arrange for k meetings between green and blue chameleons. After these k arranged meetings, we have that

$$R_2 = 2k, \quad G_2 = 2k, \quad B_2 = B_1 - k. \tag{6.197}$$

Then all we have to do is arrange for $2k$ pairs of red and green chameleons to meet to obtain

$$R_3 = 0, \quad G_3 = 0, \quad B_3 = B_1 + 3k = R_0 + B_0 + G_0, \tag{6.198}$$

at which point we have all blue chameleons!

Case 2: $k := \frac{1}{3}G_1 > B_1$; i.e., $G_1 > 3B_1$. In this case we can arrange for B_1 green and blue chameleons to meet. After these B_1 arranged meetings, we have that

$$R_2 = 2B_1, \quad G_2 = G_1 - B_1, \quad B_2 = 0. \tag{6.199}$$

Since $G_1 > 3B_1$, we have $G_2 > R_2$. So we can arrange for R_2 of the red and green chameleons to meet. After these R_2 arranged meetings, we have that

$$R_3 = 0, \quad G_3 = G_2 - R_2 = G_1 - 3B_1, \quad B_3 = B_2 + 2R_2 = 4B_1. \tag{6.200}$$

Now, let $\ell := \frac{1}{3}G_3$, which is an integer. If $\ell = \frac{1}{3}G_3 \leq B_3$, we can use the argument of Case 1 to obtain (in two steps) all blue chameleons! This condition is equivalent to $G_1 \leq 15B_1$. So we have reduced the problem to the case $G_1 > 15B_1$, as compared to the previous reduction to the case $G_1 > 3B_1$. Continuing in this way, we have a sequence of reductions of the problem to the case $G_1 > M_i B_1$, $i \geq 1$, where if the sequence of reductions is not finite, then $M_i \to \infty$ as $i \to \infty$ (we do not attempt a rigorous justification of this). Since the ratio of G_1 to B_1 is bounded, eventually these reductions must halt, and so we must have all blue chameleons after a finite number of reductions! □

6.15. Hints and partial solutions for the exercises

Hint for Exercise 6.1. Two integers a and b are congruent modulo 2 if and only if $\mathbf{r}(a) = \mathbf{r}(b)$, where $\mathbf{r} : \mathbb{Z} \to R_2$ is the remainder function. Consider the following two cases: (1) $\mathbf{r}(a) = \mathbf{r}(b) = 0$ and (2) $\mathbf{r}(a) = \mathbf{r}(b) = 1$.

Hint for Exercise 6.2. (1) One way to see this is to observe that

$$10^k = \underbrace{99\cdots9}_{k\text{ 9's}} + 1 = 9 \cdot \underbrace{11\cdots1}_{k\text{ 1's}} + 1. \tag{6.201}$$

(2) We have

$$10^k = 3 \cdot \underbrace{33\cdots3}_{k\text{ 3's}} + 1. \tag{6.202}$$

Hint for Exercise 6.3. Since $a \geq 0$ and $b < m$, so that $-b > -m$, we have

$$a - b = a + (-b) > 0 + (-m) = -m. \tag{6.203}$$

Continue.

Hint for Exercise 6.4. Suppose that $x \equiv y \bmod m$ and that a is a positive divisor of m. Then m divides $x - y$. Use that b divides m.

Hint for Exercise 6.5. (1)

$$1002, 1015, 1028, 1041, 1054. \tag{6.204}$$

Hint for Exercise 6.6. (1) Use that

$$456456 = 456 \cdot 1001 \equiv 0 \bmod 1001. \tag{6.205}$$

(2) Use that 567567 is a multiple of 1001, which in turn is a multiple of 13.

(3) Use that $579583 = 579579 + 4$.

Hint for Exercise 6.7. Suppose that $a_1 \equiv a_2 \bmod m$ and $b_1 \equiv b_2 \bmod m$. Then m divides $a_1 - a_2$, and m divides $b_1 - b_2$. Use that

$$(a_1 - b_1) - (a_2 - b_2) = (a_1 - a_2) - (b_1 - b_2). \tag{6.206}$$

Hint for Exercise 6.8. A key implication in proving the inductive step is that if $a_1 \equiv b_2 \bmod m$ and if

$$a_1 + a_2 + \cdots + a_k \equiv b_1 + b_2 + \cdots + b_k \bmod m, \tag{6.207}$$

then

$$(a_1 + a_2 + \cdots + a_k) + a_{k+1} \equiv (b_1 + b_2 + \cdots + b_k) + b_{k+1} \bmod m. \tag{6.208}$$

Hint for Exercise 6.9. Use the hint, Exercise 6.2(1), Theorem 6.9(3), and Exercise 6.8.

Hint for Exercise 6.10. Use the transitivity of congruence.

Hint for Exercise 6.11. The answer is given by

$\times_7$	0	1	2	3	4	5	6
0	0	0	0	0	0	0	0
1	0	1	2	3	4	5	6
2	0	2	4	6	1	3	5
3	0	3	6	2	5	1	4
4	0	4	1	5	2	6	3
5	0	5	3	1	6	4	2
6	0	6	5	4	3	2	1

Hint for Exercise 6.12. No. Explain why.

Hint for Exercise 6.13. See Table 4.1.1.

Hint for Exercise 6.14. $10^k \equiv 1^k \bmod 9$.

Hint for Exercise 6.15. An integer is odd if and only if it is congruent to 1 mod 2.

Hint for Exercise 6.16. $1000 \equiv -1 \bmod 13$.

Hint for Exercise 6.17. Suppose $ab \equiv 1 \bmod m$ and $ac \equiv 1 \bmod m$. Using $\gcd(a, m) = 1$ (why is this true?), show that $b \equiv c \bmod m$.

Hint for Exercise 6.18. The solutions, except for the $m = 7$ case (which contributes 4 rows), is given by the following table:

m	a	x
4	3	3
5	3	2
5	4	4
6	5	5

Hint for Exercise 6.19. The smallest such solution is 2 and the largest such solution is 14.

Hint for Exercise 6.20. We have

$$(m - a)(m - b) = ab + m(m - a - b). \tag{6.209}$$

Hint for Exercise 6.21. The answers are (explain why):

(1) $b_1 \equiv b_2 \bmod 5$.

(2) $b_1 \equiv b_2 \bmod 7$.

(3) $b_1 \equiv b_2 \bmod 45$.

Hint for Exercise 6.22. If a prime p divides ab, then p divides a or p divides b.

Hint for Exercise 6.23. We present the solutions in a table:

a	x
1	3
2	6
3	2
4	5
5	1
6	4

Hint for Exercise 6.24. The integers b are $3, 6, 9, 12$. We have the following table of solutions:

b	x
3	3, 8, 13
6	1, 6, 11
9	4, 9, 14
12	2, 7, 12

Hint for Exercise 6.25. The pairs of inverses modulo 13 are

$$(1,1),\ (2,7),\ (3,9),\ (4,10),\ (5,8),\ (6,11),\ (12,12). \tag{6.210}$$

Hint for Exercise 6.26.

($\Rightarrow$) Suppose that a has the property that there exists an integer b such that $ab \equiv 1 \bmod m$. Then there exists an integer q such that

$$ab - 1 = qm. \tag{6.211}$$

Show from this that a and m are coprime.

($\Leftarrow$) Suppose that a and m are coprime. By Theorem 4.20 there exist integers b and c such that

$$(b)a + (c)m = 1. \tag{6.212}$$

Continue.

Hint for Exercise 6.27. The pairs (a, b) such that $ab \equiv 1 \bmod 18$ are

$$(1,1),\ (5,11),\ (7,13),\ (11,5),\ (13,7),\ (17,17).$$

Hint for Exercise 6.28. If $p_1, p_2, \ldots, p_k$ are distinct primes, then $\gcd(p_i, p_j) = 1$ for $i \neq j$.

Hint for Exercise 6.29. For $m = 6$, we have modulo 6 the following congruences:

$$\begin{aligned} 1^2 &\equiv 1, \\ 2^2 &\equiv 4, \\ 3^2 &\equiv 3, \\ 4^2 &\equiv 4, \\ 5^2 &\equiv 1. \end{aligned}$$

Hint for Exercise 6.30. Since $36 \equiv 1 \bmod 5$, the congruence equation in this exercise is equivalent to $x^2 \equiv 1 \bmod 5$. The set of all positive solutions is

$$\{1 + 5k : k \in \mathbb{Z}^{\geq}\}. \tag{6.213}$$

Hint for Exercise 6.31. For the first part, we just need to show that $\mathbf{r}(xy) \neq 0$, which is true if xy is not a multiple of m.

Hint for Exercise 6.32. Suppose that x is a quadratic residue modulo ab. Then $x = \mathbf{r}_{ab}(y^2)$ for some $y \in \mathbb{Z}$. Now use that $\mathbf{r}_a \circ \mathbf{r}_{ab} = \mathbf{r}_a$.

Hint for Exercise 6.33. f is well-defined if and only if $\mathbf{r}(x^2) \neq 0$ for $x \in QR_m$. That is, m does not divide x^2. Consider the two types of m's, and for each type use properties of division and primes.

Hint for Exercise 6.34. Example 6.55 and Exercise 6.29 calculate for each $3 \leq m \leq 10$ the set QR_m. All of these integers m are primes or products of two distinct primes except for $m = 4, 8, 9$. By inspection, we see that f is not a bijection for $m = 5, 10$ and that f is a bijection for $m = 3, 6, 7$.

We also observe that f is well-defined and a bijection for $m = 4, 9$ even though they are not the product of distinct primes.

Hint for Exercise 6.35. The contrapositive of the statement we want to prove is: $(x \equiv y \bmod p)$ and $(x \equiv y \bmod q)$ imply $(x \equiv y \bmod pq)$. One can prove this implication using standard properties of division and the fact that p and q are distinct primes.

Hint for Exercise 6.36. Suppose p divides y. Then $y \equiv 0 \bmod p$ and $y^2 \equiv 0 \bmod p$. Derive a contradiction from this and from the equations that x and y satisfy.

Hint for Exercise 6.37. (1) Equation (6.94) implies that p divides the product $(xz - 1)(xz + 1)$. Continue.

(2) If $xz \equiv 1 \bmod p$, then z is an inverse of $x \bmod p$. Why is this a contradiction?

Hint for Exercise 6.38. (1) It is easy to show that p is odd. Obtain a contradiction if $p \leq n$.

(2) Show that $(n!)^2 \equiv -1 \bmod p$ and hence $(n!)^{p-1} \equiv (-1)^{\frac{p-1}{2}} \bmod p$.

(3) Show that $n!$ is not a multiple of p, and then use Fermat's Little Theorem.

(4) Show that $(-1)^{\frac{p-1}{2}} = 1$ and that this implies that $\frac{p-1}{2}$ is even.

(5) Suppose for a contradiction that there are only a finite number of primes that are congruent to 1 modulo 4. Let p_0 be the largest such prime. Let p be a prime divisor of $(p_0!)^2+1$. Explain why $p > p_0$ and $p \equiv 1 \bmod 4$ and why this is a contradiction.

Hint for Exercise 6.39. By the Fundamental Theorem of Arithmetic, m is a product of primes. Let q be a prime in this product. Then q divides m. Since m divides p^k, we have that q divides p^k. This implies that $q = p$. Since m is positive, this implies that m is a non-negative power of p, so we write $m = p^\ell$, where $\ell \in \mathbb{Z}^{\geq}$. Now, since p^ℓ divides p^k, we have $\ell \leq k$.

Hint for Exercise 6.40. The set of integers between 1 and 27 that are coprime with 27 is

$$\{1,2,4,5,7,8,10,11,13,14,16,17,19,20,22,23,25,26\}. \tag{6.214}$$

Hint for Exercise 6.41. By (6.134), we have

$$\varphi(n) = n\prod_{p|n}\left(1-\frac{1}{p}\right). \tag{6.215}$$

Theorem 6.76 says that if $m,n \in \mathbb{Z}^+$ satisfy $\gcd(m,n)=1$, then

$$\varphi(mn) = \varphi(m)\varphi(n). \tag{6.216}$$

By (6.121), we have

$$\varphi(p^k) = p^k\left(1-\frac{1}{p}\right). \tag{6.217}$$

We also recall from (1.75) that the prime factorizations of the integers between 21 and 40 are given by

$$\begin{matrix} 3\cdot 7 & 2\cdot 11 & 23 & 2^3 3 & 5^2 & 2\cdot 13 & 3^3 & 2^2 7 & 29 & 2\cdot 3\cdot 5 \\ 31 & 2^5 & 3\cdot 11 & 2\cdot 17 & 5\cdot 7 & 2^2 3^2 & 37 & 2\cdot 19 & 3\cdot 13 & 2^3 5 \end{matrix} \tag{6.218}$$

We compute that

$$\begin{aligned}
\varphi(21) &= \varphi(3)\varphi(7) = 2\cdot 6 = 12,\\
\varphi(22) &= \varphi(2)\varphi(11) = 10,\\
\varphi(24) &= \varphi(2^3 3) = 2^3\frac{1}{2}2 = 8,\\
\varphi(25) &= 25\frac{4}{5} = 20,\\
\varphi(26) &= \varphi(13) = 12,\\
\varphi(27) &= 27\frac{2}{3} = 18,\\
\varphi(28) &= \varphi(2^2 7) = 2\cdot 6 = 12,\\
\varphi(30) &= 30\frac{1}{2}\frac{2}{3}\frac{4}{5} = 8.
\end{aligned}$$

Hint for Exercise 6.42. Let $R = \{r_1, r_2, \ldots, r_{\varphi(m)}\}$ be a reduced residue system modulo m. Consider the restriction of the remainder map:

$$\mathbf{r} : R \to R_m. \tag{6.219}$$

By hypothesis, $|R| = |R_m| = \varphi(m)$. Moreover, since distinct elements of R are incongruent modulo m, we have that $\mathbf{r}$ is an injection. Therefore $\mathbf{r}$ is a bijection.

Now, for each $1 \le i \le \varphi(m)$, there exists an integer q_i such that

$$r_i = \mathbf{r}(r_i) + q_i m. \tag{6.220}$$

Therefore,

$$\begin{aligned}\sum_{i=1}^{\varphi(m)} r_i &= \sum_{i=1}^{\varphi(m)} (\mathbf{r}(r_i) + q_i m) \\ &= \sum_{i=1}^{\varphi(m)} \mathbf{r}(r_i) + m \sum_{i=1}^{\varphi(m)} q_i \\ &= \sum_{r \in R_m,\, \gcd(r,m)=1} r + m \sum_{i=1}^{\varphi(m)} q_i.\end{aligned}$$

The result now follows from the claim that $\sum_{r \in R_m,\, \gcd(r,m)=1} r$ is divisible by m.

To see this, observe that for $r \in R_m$,

$$\gcd(r, m) = 1 \quad \Leftrightarrow \quad \gcd(-r, m) = 1 \quad \Leftrightarrow \quad \gcd(m - r, m) = 1. \tag{6.221}$$

We also have that if $m > 2$ is even, then $\frac{m}{2}$ is not coprime with m. Thus, the elements $r \in R_m$ satisfying $\gcd(r, m) = 1$ may be paired as $(r, m - r)$ for $r < m/2$. Since each such r satisfies $r + (m - r) = m$, we conclude that $\sum_{r \in R_m,\, \gcd(r,m)=1} r$ is divisible by m.

Chapter 7

Counting Finite Sets

$$
\begin{array}{ccccccccccccc}
 & & & & & & 1 & & & & & & \\
 & & & & & a & + & b & & & & & \\
 & & & & 1a^2 & + & 2ab & + & 1b^2 & & & & \\
 & & & 1a^3 & + & 3a^2b & + & 3ab^2 & + & 1b^3 & & & \\
 & & 1a^4 & + & 4a^3b & + & 6a^2b^2 & + & 4ab^3 & + & 1b^4 & & \\
 & 1a^5 & + & 5a^4b & + & 10a^3b^2 & + & 10a^2b^3 & + & 5ab^4 & + & 1b^5 & \\
1a^6 & + & 6a^5b & + & 15a^4b^2 & + & 20a^3b^3 & + & 15a^2b^4 & + & 6ab^5 & + & 1b^6 \\
 & & & & & & \text{etc.} & & & & & &
\end{array}
$$

Figure 7.0.1. The Binomial Theorem visualized as Pascal's triangle!

Goals of this chapter: Discuss the basics of counting finite sets. Prove the inclusion-exclusion principle for counting finite unions of sets. To understand binomial coefficients and to prove the Binomial Theorem. To count functions.

Question: Are monsters good at math?
Answer: Not unless you Count Dracula.

Recall from Definition 5.56 that a set X is *finite* if there exists a non-negative integer n and a bijection from $\mathbb{N}_n := \{1, 2, 3, \ldots, n\}$ to X (where $\mathbb{N}_0 := \emptyset$). Furthermore, we say that X has **cardinality** n and we write this as $|X| = n$.

7.1. The addition principle

A key question is: How does cardinality relate to set operations? The addition principle answers this for the unions of disjoint sets.

7.1.1. The addition principle for two sets. Suppose that we have two baskets X and Y, each containing mysterious objects. How do we count the total number in the baskets altogether? Simple! We add the number of objects in each of the two baskets together.

Theorem 7.1 (The addition principle for two disjoint sets). *Suppose that X and Y are disjoint finite sets. Then $X \cup Y$ is finite and*

$$|X \cup Y| = |X| + |Y|. \tag{7.1}$$

That is, if X and Y are disjoint finite sets, then the cardinality of the union is the sum of the cardinality of the sets.

Proof. See Figure 7.1.1 for a visualization of this proof.

Let $n = |X|$ and let $k = |Y|$. By the definition of cardinality, there exist bijections $f : \mathbb{N}_n \to X$ and $g : \mathbb{N}_k \to Y$. To prove the addition principle, we need to define a function

$$h : \mathbb{N}_{n+k} \to X \cup Y$$

and show that it is a bijection. We "stack" the function g on top of the function f to define h. Namely, we define h by

$$h(i) = \begin{cases} f(i) & \text{if } 1 \leq i \leq n, \\ g(i-n) & \text{if } n+1 \leq i \leq n+k. \end{cases} \tag{7.2}$$

It is easy to see that h is well-defined. We now show that h is a bijection, from which the theorem follows.

(1) Proof that h is a surjection. Let $z \in X \cup Y$. Then $z \in X$ or $z \in Y$ (exclusive "or" since X and Y are disjoint). So we consider two cases:

(1a) If $z \in X$, then there exists $i \in \mathbb{N}_n$ such that $f(i) = z$ since f is a surjection. Since $1 \leq i \leq n$, we have

$$h(i) = f(i) = z.$$

(1b) If $z \in Y$, then there exists $j \in \mathbb{N}_k$ such that $g(j) = z$ since g is a surjection. Since $1 \leq j \leq k$, we have $n + 1 \leq j + n \leq n + k$ and

$$h(j+n) = g((j+n) - n) = g(j) = z.$$

This proves h is a surjection.

(2) Proof that h is an injection. Suppose that $h(i_1) = h(i_2) \in X \cup Y$. Then $h(i_1) = h(i_2) \in X$ or $h(i_1) = h(i_2) \in Y$. So again we consider two cases:

(2a) Suppose $h(i_1) = h(i_2) \in X$. Then $1 \leq i_1, i_2 \leq n$. Hence $f(i_1) = f(i_2)$. Since f is an injection,

$$i_1 = i_2.$$

(2b) Suppose $h(i_1) = h(i_2) \in Y$. Then $n + 1 \leq i_1, i_2 \leq n + k$. Hence $g(i_1 - n) = g(i_2 - n)$. Since g is an injection, $i_1 - n = i_2 - n$, so that

$$i_1 = i_2.$$

This proves h is an injection. □

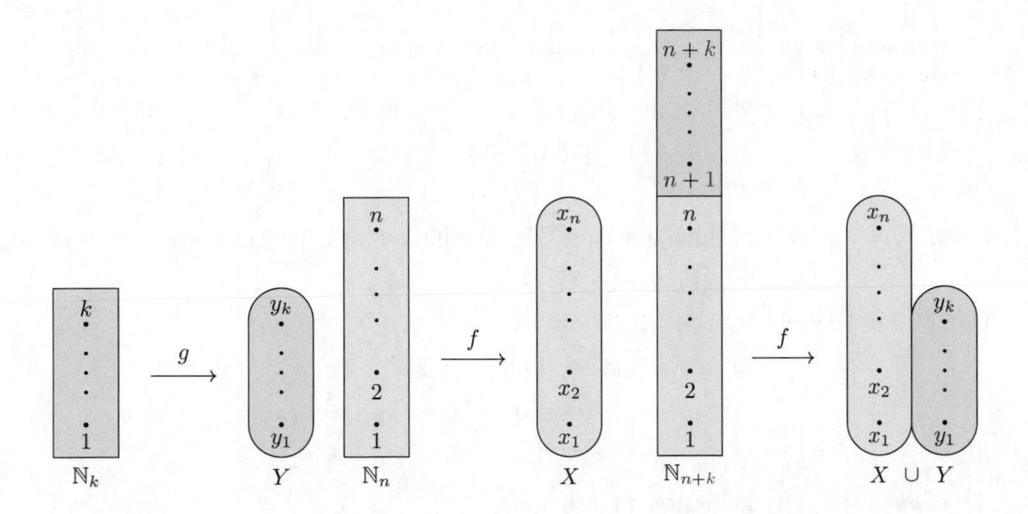

Figure 7.1.1. The addition principle visualized.

Example 7.2. If X is a set of n numbers and Y is a set of k letters, then $X \cup Y$ has cardinality $n + k$.

Exercise 7.1. *Let m be a positive integer, and let $\mathbb{N}_m = \{1, 2, \ldots, m\}$. Prove that if X and Y are subsets of $\mathbb{N}_m$ satisfying $|X| + |Y| > m$, then*

$$X \cap Y \neq \emptyset. \tag{7.3}$$

7.1.2. Thinking about the addition principle for two sets. Can you convince yourself that the idea of the statement and proof of the addition principle (Theorem 7.1) is encoded in Figure 7.1.2?

$$\begin{array}{ccccccccccc}
x_1 & x_2 & x_3 & & x_n & y_1 & y_2 & y_3 & & y_k \\
\bullet & \bullet & \bullet & \cdots & \bullet & \bullet & \bullet & \bullet & \cdots & \bullet \\
1 & 2 & 3 & & n & n+1 & n+2 & n+3 & & n+k
\end{array}$$

Figure 7.1.2. Another visualization of the proof of the addition principle for sets.

7.1.3. The addition principle for n sets. From the addition principle for two sets and by mathematical induction, one can prove:

Corollary 7.3 (Addition principle for n disjoint sets). *If $X_1, X_2, \ldots, X_n$ are disjoint finite sets, i.e., $X_i \cap X_j = \emptyset$ for all $1 \leq i < j \leq n$, then $X_1 \cup X_2 \cup \cdots \cup X_n$ is a finite set and*

$$|X_1 \cup X_2 \cup \cdots \cup X_n| = |X_1| + |X_2| + \cdots + |X_n| \,. \tag{7.4}$$

Proof. *Base case.* The addition principle for two disjoint sets says that this is true for $n = 2$.

Inductive step. Suppose $k \geq 2$ is such that for every disjoint finite set $X_1, \ldots, X_k$ we have $|X_1 \cup \cdots \cup X_k| = |X_1| + \cdots + |X_k|$. Let $Y_1, \ldots, Y_{k+1}$ be disjoint finite sets. Since $Y_1 \cup \cdots \cup Y_k$ and Y_{k+1} are disjoint finite sets, by the base $n = 2$ case, we have

$$\begin{aligned} |Y_1 \cup \cdots \cup Y_{k+1}| &= |(Y_1 \cup \cdots \cup Y_k) \cup Y_{k+1}| \\ &= |Y_1 \cup \cdots \cup Y_k| + |Y_{k+1}| . \end{aligned}$$

Since $Y_1, \ldots, Y_k$ are disjoint finite sets, by the inductive hypothesis, we have

$$|Y_1 \cup \cdots \cup Y_k| = |Y_1| + \cdots + |Y_k| .$$

Combining the two displays above, we have

$$\begin{aligned} |Y_1 \cup \cdots \cup Y_{k+1}| &= |Y_1 \cup \cdots \cup Y_k| + |Y_{k+1}| \\ &= (|Y_1| + \cdots + |Y_k|) + |Y_{k+1}| \\ &= |Y_1| + \cdots + |Y_{k+1}| . \end{aligned}$$

Conclusion. By induction we are done. □

Example 7.4. Let m be a positive integer, and let X be a finite set of integers. Define $X_i := \{x \in X : x \equiv i \bmod m\}$ for $1 \leq i \leq m$. Then

$$|X| = |X_1| + |X_2| + \cdots + |X_m| .$$

7.2. Cartesian products and the multiplication principle

Let $n \geq 2$ be an integer and let $X_1, X_2, \ldots, X_n$ be sets. Recall that their n-fold cartesian product is defined by (5.28) as

$$X_1 \times X_2 \times \cdots \times X_n = \{(x_1, x_2, \ldots, x_n) : x_i \in X_i \text{ for } i = 1, 2, \ldots, n\} .$$

Recall also, as in (5.29), that X^n is the n-fold cartesian product where $X_i = X$ for each $i = 1, 2, \ldots, n$. For example, as in (5.31) Euclidean n-space $\mathbb{R}^n$ is the n-fold cartesian product of $\mathbb{R}$.

Theorem 7.5 (Multiplication principle for two sets)**.** *If X and Y are finite sets, then*

$$|X \times Y| = |X| \cdot |Y| . \tag{7.5}$$

Proof. See Figure 7.2.1 for a visualization of the proof.

Let $n = |X|$ and $m = |Y|$. Then there exist bijections $f : \mathbb{N}_n \to X$ and $g : \mathbb{N}_m \to Y$. Define $h : \mathbb{N}_n \times \mathbb{N}_m \to X \times Y$ by

$$h(j,k) = (f(j), g(k)) \quad \text{for } j \in \mathbb{N}_n \text{ and } k \in \mathbb{N}_m.$$

Then h is a bijection (exercise). Define $b : \mathbb{N}_n \times \mathbb{N}_m \to \mathbb{N}_{nm}$ by

$$b(j,k) = j + (k-1)n \quad \text{for } j \in \mathbb{N}_n \text{ and } k \in \mathbb{N}_m. \tag{7.6}$$

Then b is a bijection (exercise; observe that $b(j,1) = j$, $b(j,2) = j + n$, $b(j,3) = j + 2n$, etc.). Hence b has an inverse $c : \mathbb{N}_{nm} \to \mathbb{N}_n \times \mathbb{N}_m$ which is a bijection. Therefore the composition $h \circ c : \mathbb{N}_{nm} \to X \times Y$ is a bijection. □

In the proof above, we counted the elements of a cartesian product using the invariance of cardinality under bijections. We can also use the addition principle for sets, as in the following exercise.

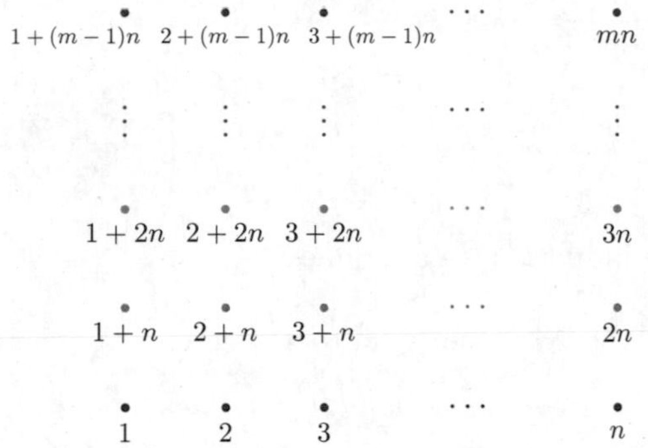

$$\begin{array}{ccccc}
\bullet & \bullet & \bullet & \cdots & \bullet \\
1+(m-1)n & 2+(m-1)n & 3+(m-1)n & & mn \\
\vdots & \vdots & \vdots & \cdots & \vdots \\
\bullet & \bullet & \bullet & \cdots & \bullet \\
1+2n & 2+2n & 3+2n & & 3n \\
\bullet & \bullet & \bullet & \cdots & \bullet \\
1+n & 2+n & 3+n & & 2n \\
\bullet & \bullet & \bullet & \cdots & \bullet \\
1 & 2 & 3 & & n
\end{array}$$

Figure 7.2.1. A visual proof of the multiplication principle (Theorem 7.5). We can think of this figure as representing the bijection b defined by (7.6).

Exercise 7.2. *Give another proof of Theorem 7.5: Let $|X| = n$ and let*

$$X = \{x_1, x_2, \ldots, x_n\}. \tag{7.7}$$

Show that

$$f : X \times Y \to \bigcup_{i=1}^{n} \{x_i\} \times Y \tag{7.8}$$

defined by

$$f(x, y) = (x, y) \tag{7.9}$$

is a bijection. Apply the addition principle for counting finite sets.

By induction, we can soup up the previous theorem to the following.

Corollary 7.6 (Multiplication principle for n sets). *If $X_1, X_2, \ldots, X_n$ are finite sets, where $n \geq 2$, then*

$$|X_1 \times X_2 \times \cdots \times X_n| = |X_1| \cdot |X_2| \cdot \cdots \cdot |X_n| . \tag{7.10}$$

Proof. *Base case.* Theorem 7.5 says this is true for $n = 2$.

Inductive step. Suppose $k \geq 2$ is such that for any collection of finite sets $X_1, \ldots, X_k$ we have $|X_1 \times X_2 \times \cdots \times X_k| = |X_1| \cdot |X_2| \cdot \cdots \cdot |X_k|$. Let $Y_1, \ldots, Y_{k+1}$ be finite sets. Since $Y_1 \times \cdots \times Y_k$ and Y_{k+1} are finite sets, by the base $n = 2$ case, we have

$$\begin{aligned} |Y_1 \times \cdots \times Y_{k+1}| &= |(Y_1 \times \cdots \times Y_k) \times Y_{k+1}| \\ &= |Y_1 \times \cdots \times Y_k| \times |Y_{k+1}| . \end{aligned}$$

Since $Y_1, \ldots, Y_k$ are finite sets, by the inductive hypothesis, we have

$$|Y_1 \times \cdots \times Y_k| = |Y_1| \times \cdots \times |Y_k| .$$

Combining the two displays above, we have

$$\begin{aligned} |Y_1 \times \cdots \times Y_{k+1}| &= |Y_1 \times \cdots \times Y_k| \times |Y_{k+1}| \\ &= (|Y_1| \times \cdots \times |Y_k|) \times |Y_{k+1}| \\ &= |Y_1| \times \cdots \times |Y_{k+1}| . \end{aligned}$$

Conclusion. By induction we are done. □

If we take all of the X_i to be equal to a single set X, then we obtain:

Corollary 7.7. *If X is a finite set and $n \geq 1$, then*

$$|X^n| = |X|^n. \tag{7.11}$$

Example 7.8. Let n be a positive integer. Then

$$\left|\{0,1\}^n\right| = 2^n.$$

This formula is related to *power sets*, which we will consider in §7.4 below.

7.3. The inclusion-exclusion principle

In this section we discuss the formula for the cardinality of the union of finitely many finite sets, called the *inclusion-exclusion principle.* This generalizes the addition principle in §7.1 to the case where the sets may intersect.

Recall that, given two sets X and Y, their *set difference* is defined to be

$$X - Y = \{x \in X \mid x \notin Y\}. \tag{7.12}$$

Theorem 7.9. *For any sets X and Y, we have*

$$X \cup Y = (X - Y) \cup (Y - X) \cup (X \cap Y), \tag{7.13}$$

where the right side is a disjoint union of sets

Exercise 7.3. *Without using Venn diagrams, prove Theorem 7.9.*

Example 7.10. Let k and n be positive integers. Let $X = \{0, 1, 2, \ldots, n-1, n\}$, and let $Y = \{-k, -k+1, \ldots, -1, 0\}$. Then

$$\begin{aligned} X - Y &= \{1, 2, \ldots, n-1, n\}, \\ Y - X &= \{-k, -k+1, \ldots, -1\}, \\ X \cap Y &= \{0\}, \end{aligned}$$

and the disjoint union of these three sets is equal to $X \cup Y$.

7.3.1. Inclusion-exclusion principle for two finite sets.

Using the theorem above, the following result about the cardinality of the union of two finite sets is easy to prove from the additional principle for sets. It says that if you count the things in a set X and then add the count of things in a set Y, you will have counted everything. But, you will have counted the things in both sets twice. So just subtract the number of things in the intersection. See Figure 7.3.2 for the case of 3 sets. Can you draw the analogous picture for 2 sets X and Y?

Theorem 7.11 (Inclusion-exclusion principle for two sets). *If X and Y are finite sets, then*

$$|X \cup Y| = |X| + |Y| - |X \cap Y|. \tag{7.14}$$

Proof. *You may wish to draw Venn diagrams to picture the proof; cf. Figure* 5.1.3. By Theorem 7.9 and Corollary 7.3,

$$|X \cup Y| = |X - Y| + |Y - X| + |X \cap Y|.$$

We wish to express the right side more in terms of $|X|$ and $|Y|$. To this end, since $X = (X - Y) \cup (X \cap Y)$ is a disjoint union, we have

$$|X| = |X - Y| + |X \cap Y|.$$

Similarly,

$$|Y| = |Y - X| + |X \cap Y|.$$

Hence $|X - Y| = |X| - |X \cap Y|$ and $|Y - X| = |Y| - |X \cap Y|$. We conclude that

$$\begin{aligned} |X \cup Y| &= |X - Y| + |Y - X| + |X \cap Y| \\ &= |X| + |Y| - |X \cap Y|. \end{aligned}$$

□

7.3.2. Inclusion-exclusion principle for three finite sets.

Using the same essential idea as for two sets, we can obtain a formula for the cardinality of the union of three sets.

Theorem 7.12 (Inclusion-exclusion principle for three sets). *If X, Y, and Z are finite sets, then*

$$|X \cup Y \cup Z| = |X| + |Y| + |Z| - |X \cap Y| - |X \cap Z| - |Y \cap Z| + |X \cap Y \cap Z|. \tag{7.15}$$

We can use **Venn diagrams** to visualize the inclusion-exclusion principle. A Venn diagram for counting the union of three finite sets is given in Figure 7.3.1.

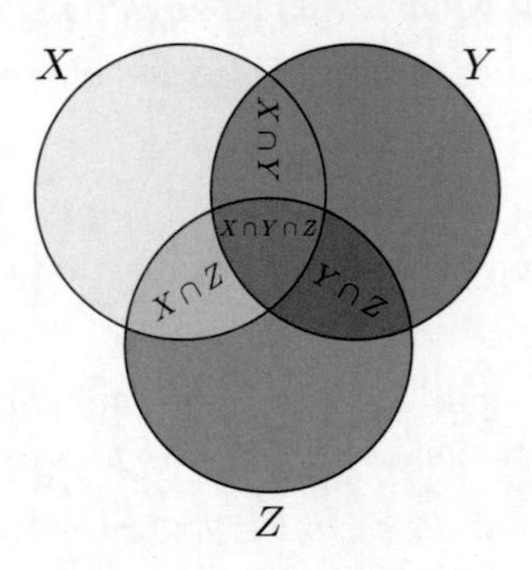

Figure 7.3.1. Venn diagram for the union of sets X, Y, Z.

A schematic diagram for using the inclusion-exclusion principle for counting the union of three sets is given in Figure 7.3.2: $|X| + |Y| + |Z|$ overcounts. A correction is subtract: $|X \cap Y| + |X \cap Z| + |Y \cap Z|$. But then this undercounts by: $|X \cap Y \cap Z|$.

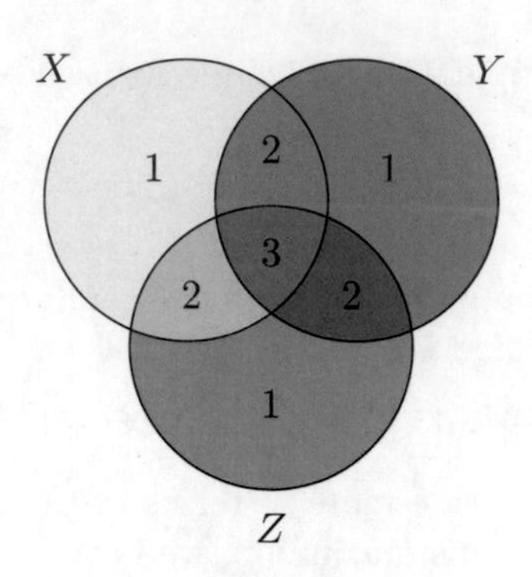

Figure 7.3.2. Counting overlaps in the Venn diagram for the union $X \cup Y \cup Z$.

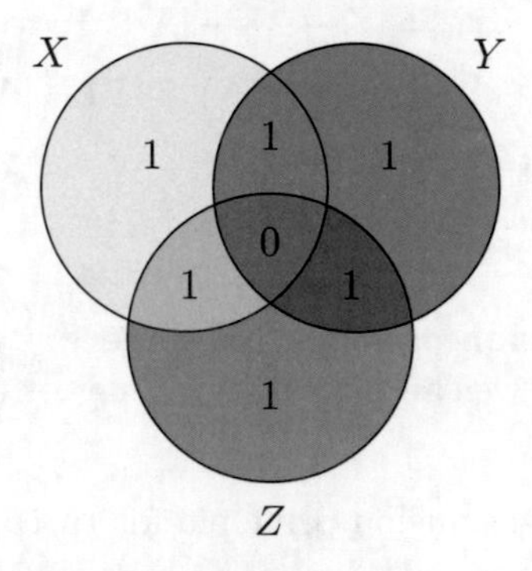

Figure 7.3.3. After subtracting the cardinalities of the double intersections $X \cap Y$, $X \cap Z$, and $Y \cap Z$, we undercount by the triple intersection $X \cap Y \cap Z$.

Exercise 7.4. *Of the* 202 *dinosaurs at Jurassic Park,*

(1) 125 *eat apples,* 131 *eat bananas, and* 136 *eat cherries,*

(2) 81 *eat apples and bananas,* 84 *eat apples and cherries, and* 88 *eat bananas and cherries,*

(3) 41 *eat all three fruits.*

Using the inclusion-exclusion principle, find how many dinosaurs at Jurassic Park eat none of the three fruits.

Exercise 7.5. *For the previous exercise, write out the formula for how many dinosaurs at Jurassic Park eat none of the three fruits assuming the following:*

(1) *There are d dinosaurs at Jurassic Park.*

(2) *a dinosaurs eat apples, b dinosaurs eat bananas, and c dinosaurs eat cherries.*

(3) *ab dinosaurs eat apples and bananas, ac dinosaurs eat apples and cherries, and bc dinosaurs eat bananas and cherries.*

(4) *abc dinosaurs eat all three fruits.*

Here, ab, ac, bc, and abc are variable names and not the products of a, b, c!

The readers may wish to prove for themselves the inclusion-exclusion principle for three sets in a similar spirit as the case for two sets. We will prove the general case of n sets below.

7.3.3. Inclusion-exclusion principle for a finite collection of finite sets. Here, we will find it convenient to introduce a notation to describe intersections of subcollections of the collection of finite sets. Let $X_1, \ldots, X_k$ be a collection of finite sets, where k is a positive integer. Given a set of j distinct indices

$$I = \{i_1, i_2, \ldots, i_j\} \subset \mathbb{N}_k = \{1, 2, \ldots, k\}, \tag{7.16}$$

where $1 \leq j \leq k$, define

$$X_I = \bigcap_{i \in I} X_i = X_{i_1} \cap X_{i_2} \cap \cdots \cap X_{i_j}. \tag{7.17}$$

That is, the set $\mathbb{N}_k$ is the set of indices of our collection of finite sets. A subset I of $\mathbb{N}_k$ with $|I| = j$ corresponds to a choice of j sets of this collection. The set X_I denotes the intersection of these j sets. To see this more concretely, we consider the case $k = 3$:

Example 7.13. For each non-empty subset $I \subset \mathbb{N}_3$, we write out the set X_I. Note that the number of such sets X_I is equal to $2^3 - 1 = 7$.

(1) $I = \{1\}$, $X_I = X_1$.
(2) $I = \{2\}$, $X_I = X_2$.
(3) $I = \{3\}$, $X_I = X_3$.
(4) $I = \{1, 2\}$, $X_I = X_1 \cap X_2$.
(5) $I = \{1, 3\}$, $X_I = X_1 \cap X_3$.
(6) $I = \{2, 3\}$, $X_I = X_2 \cap X_3$.
(7) $I = \{1, 2, 3\}$, $X_I = X_1 \cap X_2 \cap X_3$.

Theorem 7.14 (Inclusion-exclusion principle for n sets). *Let $X_1, X_2, \ldots, X_n$ be a collection of n finite sets. Then*

$$|X_1 \cup X_2 \cup \cdots \cup X_n| = \sum_{\emptyset \neq I \subset \mathbb{N}_n} (-1)^{|I|-1} |X_I|. \tag{7.18}$$

Example 7.15. For $n = 3$, we write out the sum on the right-hand side of (7.18) as

$$\begin{aligned}\sum_{\emptyset \neq I \subset \mathbb{N}_3} (-1)^{|I|-1} |X_I| = |X_1| &+ |X_2| + |X_3| \\ &- |X_1 \cap X_2| - |X_1 \cap X_3| - |X_2 \cap X_3| \\ &+ |X_1 \cap X_2 \cap X_3|.\end{aligned}$$

Observe that the first line has the terms with $|I| = 1$, the second line has the terms with $|I| = 2$, and the third line has the terms with $|I| = 3$. From this, we see that when $n = 3$, Theorem 7.14 reduces to Theorem 7.12.

Exercise 7.6. *Count the overlaps (numbering 1 through 4) for the Venn diagram of 4 sets in Figure 7.3.4. Your picture should be analogous to the Venn diagram in Figure 7.3.2. If you prefer a more symmetric picture, see Figure 7.3.5.*

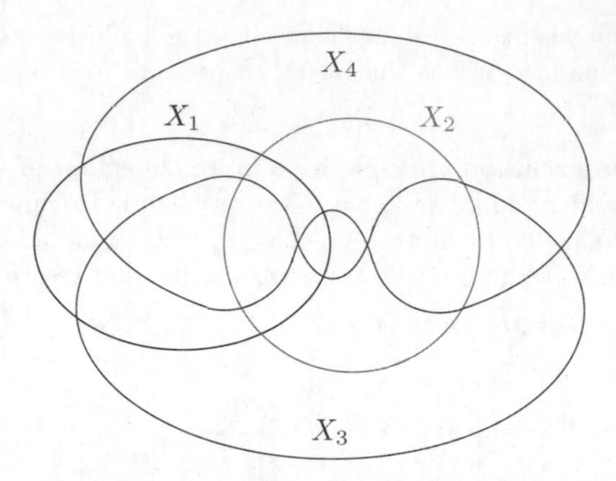

Figure 7.3.4. A Venn diagram for 4 sets! Can you count overlaps as in Figure 7.3.2?

Figure 7.3.5. A more symmetric Venn diagram for 4 sets X_1, X_2, X_3, and X_4, each represented by an isosceles triangle. See "Venn Symmetry and Prime Numbers: A Seductive Proof Revisited" by Stan Wagon and Peter Webb, in *American Mathematical Monthly*, August 2008, pp. 645–648.

7.3.4. A conceptual proof of the inclusion-exclusion principle. Why the inclusion-exclusion principle is true. Here is the intuition for why Theorem 7.14 is true. Consider the sum

$$|X_1| + |X_2| + \cdots + |X_n|. \tag{7.19}$$

This sum multiply counts certain elements of the union $|X_1 \cup X_2 \cup \cdots \cup X_n|$. On the other hand, all of the elements outside of the union of the double intersections are singly counted correctly. So we add to the sum in (7.19) the non-positive correction term

$$-\sum_{1 \leq i < j \leq n} |X_i \cap X_j|. \tag{7.20}$$

We claim that with this correction term we are correctly counting the elements outside of the union of the triple intersections. Indeed, for elements in the union of the double intersections, but outside the union of the triple intersections, we are correctly removing exactly 1 from the double counting to get the correct single counting of these elements. On the other hand, for the elements in the union of the triple intersections, we have overcorrected and have hence undercounted these elements. We thus next add the non-negative correction term

$$(7.21)\qquad + \sum_{1\le i<j<k\le n} |X_i \cap X_j \cap X_k|.$$

Continuing this way, we see that the inclusion-exclusion formula should be an alternating sum exactly as it is written on the right-hand side of (7.18).

Did you glimpse at what lies ahead? For the impatient reader who just wants the verbose author to get to the point, you may proceed directly from here to §7.3.6 below for a quick and dirty proof before reading the following detailed proof!

7.3.5. The full proof of the inclusion-exclusion principle.

Proof of Theorem 7.14. Inductive step and base case. *Inductive hypothesis*: Assume that $k \ge 2$ is an integer with the following property (the *inclusion-exclusion principle for k sets*): For any finite sets $Y_1, \dots, Y_k$, we have

$$(7.22)\qquad |Y_1 \cup \cdots \cup Y_k| = \sum_{\emptyset \ne I \subset \mathbb{N}_k} (-1)^{|I|-1} |Y_I|.$$

Note that by Theorem 7.11, $k = 2$ has this property and hence the base case is true.

Proof of the inductive conclusion: Let $X_1, \dots, X_{k+1}$ be finite sets. Of course, we have

$$X_1 \cup \cdots \cup X_{k+1} = (X_1 \cup \cdots \cup X_k) \cup X_{k+1}.$$

By the inclusion-exclusion principle for two sets and by the inductive hypothesis, we have that

(7.23)

$$\begin{aligned} |X_1 \cup \cdots \cup X_{k+1}| &= |(X_1 \cup \cdots \cup X_k) \cup X_{k+1}| \\ &= |X_1 \cup \cdots \cup X_k| + |X_{k+1}| - |(X_1 \cup \cdots \cup X_k) \cap X_{k+1}| \\ &= \sum_{\emptyset \ne I \subset \mathbb{N}_k} (-1)^{|I|-1} |X_I| + |X_{k+1}| - |(X_1 \cup \cdots \cup X_k) \cap X_{k+1}|. \end{aligned}$$

To prove the inductive conclusion, that is, the inclusion-exclusion principle for $k+1$ sets, we need to rewrite the right-hand side of the display above. To this end:

Claim 1.

$$(7.24)\qquad (X_1 \cup \cdots \cup X_k) \cap X_{k+1} = (X_1 \cap X_{k+1}) \cup \cdots \cup (X_k \cap X_{k+1}),$$

and hence

$$(7.25)\qquad |(X_1 \cup \cdots \cup X_k) \cap X_{k+1}| = |Y_1 \cup \cdots \cup Y_k|,$$

where

$$(7.26)\qquad Y_i := X_i \cap X_{k+1}.$$

Proof of Claim 1. Let $P(k)$ be the statement that for every $k+1$ sets $X_0, X_1, \dots, X_k$,

$$(7.27)\qquad (X_1 \cup \cdots \cup X_k) \cap X_0 = (X_1 \cap X_0) \cup \cdots \cup (X_k \cap X_0).$$

It is easy to see that by calling X_{k+1} to be X_0, Claim 1 is equivalent to the assertion that $P(k)$ is true for all $k \in \mathbb{Z}^+$.

Base case for Claim 1: $k = 1$. In this case the statement is that $(X_1) \cap X_0 = (X_1 \cap X_0)$, which is obvious.

Inductive step for Claim 1. Suppose that $k \geq 1$ is such that for every $k+1$ sets $Y_0, Y_1, \ldots, Y_k$,

$$(Y_1 \cup \cdots \cup Y_k) \cap Y_0 = (Y_1 \cap Y_0) \cup \cdots \cup (Y_k \cap Y_0).$$

We want to prove the next case. So let $X_0, X_1, \ldots, X_k, X_{k+1}$ be $(k+1)+1 = k+2$ sets. Then, by the distributive law and by the inductive hypothesis, we have

$$\begin{aligned}
(X_1 \cup \cdots \cup X_k \cup X_{k+1}) \cap X_0 &= ((X_1 \cup \cdots \cup X_k) \cup X_{k+1}) \cap X_0 \\
&= ((X_1 \cup \cdots \cup X_k) \cap X_0) \cup (X_{k+1} \cap X_0) \\
&= ((X_1 \cap X_0) \cup \cdots \cup (X_k \cap X_0)) \cup (X_{k+1} \cap X_0) \\
&= (X_1 \cap X_0) \cup \cdots \cup (X_k \cap X_0) \cup (X_{k+1} \cap X_0).
\end{aligned}$$

This proves the desired inductive conclusion. Therefore this completes both the inductive step and the proof of Claim 1 by induction.

Now, by applying Claim 1 to (7.23), we have

$$|X_1 \cup \cdots \cup X_{k+1}| = \sum_{\emptyset \neq I \subset \mathbb{N}_k} (-1)^{|I|-1} |X_I| + |X_{k+1}| - |Y_1 \cup \cdots \cup Y_k|. \tag{7.28}$$

As for the intersections of X's, define for the intersections of Y's the notation

$$Y_J = \bigcap_{j \in J} Y_j = Y_{j_1} \cap Y_{j_2} \cap \cdots \cap Y_{j_\ell} \tag{7.29}$$

for subsets $J = \{j_1, \ldots, j_\ell\} \subset \mathbb{N}_k$.

Claim 2. *If* $J \subset \mathbb{N}_k$, *then*

$$Y_J = X_J \cap X_{k+1} = X_{J \cup \{k+1\}}. \tag{7.30}$$

Proof of Claim 2. Let $J = \{j_1, \ldots, j_\ell\} \subset \mathbb{N}_k$. Then

$$\begin{aligned}
Y_J &= Y_{j_1} \cap \cdots \cap Y_{j_\ell} \\
&= (X_{j_1} \cap X_{k+1}) \cap \cdots \cap (X_{j_\ell} \cap X_{k+1}) \\
&= (X_{j_1} \cap \cdots \cap X_{j_\ell}) \cap X_{k+1} \\
&= X_J \cap X_{k+1} \\
&= X_{J \cup \{k+1\}},
\end{aligned}$$

where the last equality is true since J and $\{k+1\}$ are disjoint.

Now, by (7.28), the inductive hypothesis, and Claim 1, we obtain

(7.31)

$$\begin{aligned}|X_1 \cup \cdots \cup X_{k+1}| &= \sum_{\emptyset \neq I \subset \mathbb{N}_k} (-1)^{|I|-1}|X_I| + |X_{k+1}| - \sum_{\emptyset \neq J \subset \mathbb{N}_k} (-1)^{|J|-1}|Y_J| \\ &= \sum_{\emptyset \neq I \subset \mathbb{N}_k} (-1)^{|I|-1}|X_I| + |X_{k+1}| + \sum_{\emptyset \neq J \subset \mathbb{N}_k} (-1)^{|J|}|X_{J\cup\{k+1\}}| \\ &= \sum_{\emptyset \neq I \subset \mathbb{N}_k} (-1)^{|I|-1}|X_I| + \sum_{J \subset \mathbb{N}_k} (-1)^{|J|}|X_{J\cup\{k+1\}}|,\end{aligned}$$

where for the last equality we used that $X_{\emptyset\cup\{k+1\}} = X_{k+1}$. Regarding one of the exponents of (-1), notice that $|J| = |J \cup \{k+1\}| - 1$. We are getting close to completing the proof of the inductive step.

Claim 3. *If $\emptyset \neq I \subset \mathbb{N}_{k+1}$, then exactly one of the following holds:*

Case (i): $\emptyset \neq I \subset \mathbb{N}_k$.

Case (ii): $I = J \cup \{k+1\}$, *where* $J \subset \mathbb{N}_k$.

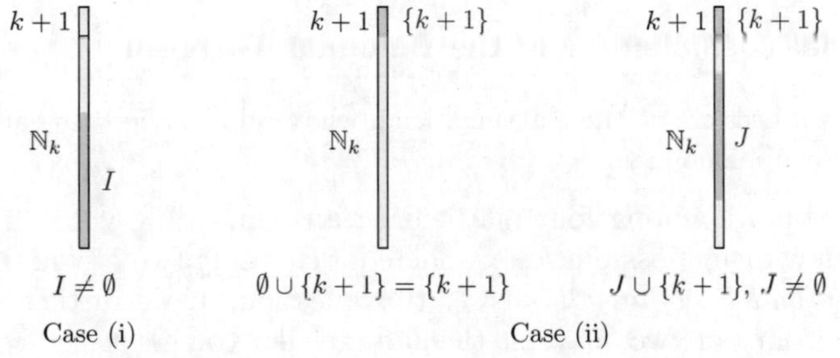

Figure 7.3.6. Picturing schematically the two types of non-empty subsets of $\mathbb{N}_{k+1}$.

Proof of Claim 3. Let $\emptyset \neq I \subset \mathbb{N}_{k+1}$. Then Case (i) corresponds to $k+1 \notin I$, and Case (ii) corresponds to $k+1 \in I$. This proves Claim 3.

Now, by applying Claim 3 to (7.31), we have

$$\sum_{\emptyset \neq I \subset \mathbb{N}_k} (-1)^{|I|-1}|X_I| + |X_{k+1}| + \sum_{\emptyset \neq J \subset \mathbb{N}_k} (-1)^{|J|}|X_{J\cup\{k+1\}}| = \sum_{\emptyset \neq I \subset \mathbb{N}_{k+1}} (-1)^{|I|-1}|X_I|.$$

The inductive conclusion follows.

By induction, we conclude that Theorem 7.14 is true. □

7.3.6. A quick and dirty proof of the inclusion-exclusion principle. *Less is more!* To understand proofs better, it is convenient (but sometimes frustrating for the reader!) to skip steps. Having or not having read the detailed proof above, can you fill in the details of the following inductive step in the proof of the general inclusion-exclusion principle?

By the inclusion-exclusion principle for two sets and by Claim 1, we have

$$|X_1 \cup \cdots \cup X_{k+1}| - |X_1 \cup \cdots \cup X_k| = |X_{k+1}| - |Y_1 \cup \cdots \cup Y_k|, \tag{7.32}$$

where $Y_i = X_i \cap X_{k+1}$. So the inductive step is true provided

$$\sum_{I \subset \mathbb{N}_{k+1},\, I \not\subset \mathbb{N}_k} (-1)^{|I|-1} |X_I| = |X_{k+1}| - |Y_1 \cup \cdots \cup Y_k|. \tag{7.33}$$

This formula is equivalent to

$$\sum_{\emptyset \neq J \subset \mathbb{N}_k} (-1)^{|J|} |X_{J \cup \{k+1\}}| = -|Y_1 \cup \cdots \cup Y_k|. \tag{7.34}$$

Since $X_{J \cup \{k+1\}} = Y_J$, this in turn is equivalent to

$$\sum_{\emptyset \neq J \subset \mathbb{N}_k} (-1)^{|J|-1} |Y_J| = |Y_1 \cup \cdots \cup Y_k|, \tag{7.35}$$

which is true by the inductive hypothesis!

Exercise 7.7. *Now that you know the idea of the proof of the inclusion-exclusion principle:*

(1) *Prove the $n = 3$ case of the inclusion-exclusion principle directly.*
(2) *Prove the $n = 4$ case of the inclusion-exclusion principle directly.*

7.4. Binomial coefficients and the Binomial Theorem

> Did you hear about the mathematician who couldn't afford to eat out?
> They could binomials.

Suppose you are among four mathematicians (including yourself) eating at a restaurant. How many possibilities are there for choosing two of you to play a game of tennis? *Binomial coefficients* answer this question. If we further ask to list all the possible groups of two players, then we are led to the *power set* of the four mathematicians.

7.4.1. The power set of a set. Let X be a set. The **power set** of X, denoted by $\mathcal{P}(X)$, is the set of all subsets of X. An *element* A of $\mathcal{P}(X)$, denoted by $A \in \mathcal{P}(X)$, is the same as a *subset* A of X, denoted by $A \subset X$. To wit,

$$A \in \mathcal{P}(X) \quad \Leftrightarrow \quad A \subset X.$$

We are most interested in the case where X is the set

$$\mathbb{N}_n = \{1, 2, \ldots, n\} = \{k \in \mathbb{Z}^+ : 1 \leq k \leq n\}.$$

This is because $\mathbb{N}_n$ is the prototypical (model case) finite set. Of course, $|\mathbb{N}_n| = n$.

Example 7.16 (Power sets of $\mathbb{N}_k$ for $k \leq 4$)**.**

(1) The power set of $\mathbb{N}_1$ is
$$\mathcal{P}(\mathbb{N}_1) = \{\emptyset, \{1\}\}.$$
(2) The power set of $\mathbb{N}_2$ is
$$\mathcal{P}(\mathbb{N}_2) = \{\emptyset, \{1\}, \{2\}, \{1, 2\}\}.$$
(3) The power set of $\mathbb{N}_3$ is (for a visualization, see Figure 7.4.1)
$$\mathcal{P}(\mathbb{N}_3) = \{\emptyset, \{1\}, \{2\}, \{3\}, \{2, 3\}, \{1, 3\}, \{1, 2\}, \{1, 2, 3\}\}.$$

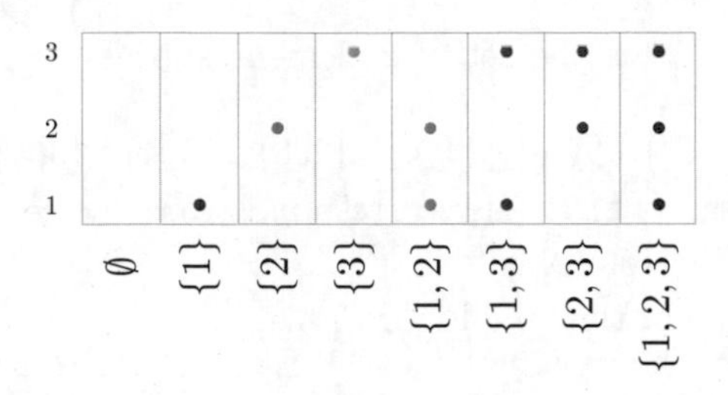

Figure 7.4.1. Visualizing the power set $\mathcal{P}(\mathbb{N}_3)$.

(4) The power set of $\mathbb{N}_4$ is the set whose elements are

$$
\begin{aligned}
&\emptyset, \\
&\{1\},\ \{2\},\ \{3\},\ \{4\}, \\
&\{1,2\},\ \{1,3\},\ \{1,4\},\ \{2,3\},\ \{2,4\},\ \{3,4\}, \\
&\{1,2,3\},\ \{1,2,4\},\ \{1,3,4\},\ \{2,3,4\}, \\
&\{1,2,3,4\}.
\end{aligned}
$$

Exercise 7.8 (Properties of power sets).

(1) *Explain why* $\emptyset \in \mathcal{P}(X)$ *for every set* X*, where* $\mathcal{P}(X)$ *is the power set of* X.

(2) *Suppose that* $A \subset B$*. Prove that* $\mathcal{P}(A) \subset \mathcal{P}(B)$.

(3) *Let* U *be the universal set. Suppose that* A *and* B *are sets. Explain why* $\mathcal{P}(A \cap B) \subset \mathcal{P}(A) \cap \mathcal{P}(B)$.

Exercise 7.9. *Define* $f : \mathcal{P}(\mathbb{Z}) \times \mathcal{P}(\mathbb{Z}) \to \mathcal{P}(\mathbb{Z})$ *by*

$$f(A, B) = A \cup B.$$

Let E *be the set of even integers and let* O *be the set of odd integers. Find all* $(A, B) \in \mathcal{P}(\mathbb{Z}) \times \mathcal{P}(\mathbb{Z})$ *satisfying*

$$f(A, B) = \mathbb{Z}, \quad A \subset E, \quad \text{and} \quad B \subset O.$$

7.4.2. Binomial coefficients.

Given an integer $k \geq 0$, let $\mathcal{P}_k(X)$ be the **set of** k**-element subsets** of X; that is,

$$\mathcal{P}_k(X) = \{A \in \mathcal{P}(X) : |A| = k\}.$$

We have

$$\mathcal{P}_k(\mathbb{N}_n) = \{A \in \mathcal{P}(\mathbb{N}_n) : |A| = k\}.$$

In the British television series The Prisoner, the protagonist is Number Six. Suppose that ten residents of the Village, including the prisoner, are numbered 1 through 10.[1] Then $\mathcal{P}_4(\mathbb{N}_{10})$ can be interpreted as the possible groups of four residents chosen to play the card game Bridge.

[1]However, Number One has never been seen, and Number Two, the Village administrator, constantly changes.

Example 7.17. $\mathcal{P}_2(\mathbb{N}_4)$ is the set of 2-element subsets of the set $\{1,2,3,4\}$; its elements are

$$\{1,2\},\ \{1,3\},\ \{1,4\},\ \{2,3\},\ \{2,4\},\ \{3,4\}.$$

We see that each element of $\mathcal{P}_2(\mathbb{N}_4)$ corresponds to a choice of 2 distinct elements out of the 4 elements of $\mathbb{N}_4$.

For a visualization of this, see Figure 7.4.2.

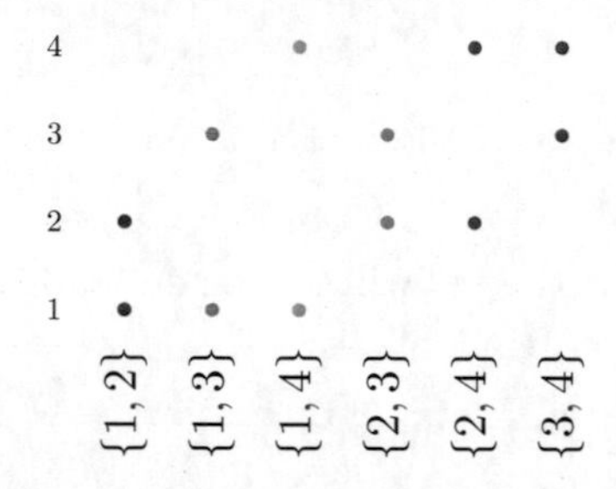

Figure 7.4.2. Visualizing $\mathcal{P}_2(\mathbb{N}_4)$.

Since $\mathcal{P}_k(\mathbb{N}_n)$ is empty if $k < 0$ or $k > n$, we shall assume that $0 \leq k \leq n$. Since every subset of $\mathbb{N}_n$ has between 0 and n elements, inclusive, we have

$$\mathcal{P}(\mathbb{N}_n) = \bigsqcup_{k=0}^{n} \mathcal{P}_k(\mathbb{N}_n),$$

where the "disjoint union" symbol $\bigsqcup$ indicates that the sets $\mathcal{P}_k(\mathbb{N}_n)$ that we are taking the union of are disjoint; i.e.,

$$\mathcal{P}_k(\mathbb{N}_n) \cap \mathcal{P}_\ell(\mathbb{N}_n) = \emptyset \quad \text{for } k \neq \ell. \tag{7.36}$$

Effectively, we have **partitioned** the power set of $\mathbb{N}_n$ into the 0-element subsets, the 1-element subsets, the 2-element subsets, ..., up to the n-element subsets.

Example 7.18. We have

$$\mathcal{P}(\mathbb{N}_4) = \mathcal{P}_0(\mathbb{N}_4) \sqcup \mathcal{P}_1(\mathbb{N}_4) \sqcup \mathcal{P}_2(\mathbb{N}_4) \sqcup \mathcal{P}_3(\mathbb{N}_4) \sqcup \mathcal{P}_4(\mathbb{N}_4),$$

where the right side is a disjoint union.

This all may seem somewhat complicated, but we greatly simplify things by only concerning ourselves with counting, i.e., the cardinality of finite sets. Define the **binomial coefficient**

$$\binom{n}{k} = |\mathcal{P}_k(\mathbb{N}_n)|$$

to be the number of (how many) k-element subsets of $\mathbb{N}_n$. Now, each element of $\mathcal{P}_k(\mathbb{N}_n)$ corresponds to a *choice* of k distinct elements out of the n elements of $\mathbb{N}_n$. For this reason, the binomial coefficient $\binom{n}{k}$ is also called n **choose** k.

Fact: The power set of a finite set is a finite set (we will prove this in the next section). Hence, each $\mathcal{P}_k(\mathbb{N}_n)$ is a finite set and hence each $\binom{n}{k}$ is a (finite) non-negative integer.

Example 7.19. By Example 7.17, we have

$$\binom{4}{2} = |\mathcal{P}_2(\mathbb{N}_4)| = 6.$$

By the addition principle for sets, we have

$$|\mathcal{P}(\mathbb{N}_n)| = \sum_{k=0}^{n} |\mathcal{P}_k(\mathbb{N}_n)| = \sum_{k=0}^{n} \binom{n}{k}.$$

One may think of a particular subset of $\mathbb{N}_n$ as corresponding to a particular set of n choices of whether to put each of the elements $1, 2, \ldots, n$ in the subset. Since for each element of $\mathbb{N}_n$ we have two choices, *in* or *out*, we obtain

$$|\mathcal{P}(\mathbb{N}_n)| = 2^n$$

(see also (7.109) below). Therefore, we have the binomial coefficient summation identity

$$\sum_{k=0}^{n} \binom{n}{k} = 2^n \tag{7.37}$$

Example 7.20. $\sum_{k=0}^{4} \binom{4}{k} = 1 + 4 + 6 + 4 + 1 = 16 = 2^4$.

Binomial coefficients have the following symmetry property.

Lemma 7.21. *If n is a non-negative integer and if k is an integer satisfying $0 \leq k \leq n$, then*

$$\binom{n}{k} = \binom{n}{n-k}. \tag{7.38}$$

Proof. Define the "complement function"

$$c : \mathcal{P}_k(\mathbb{N}_n) \to \mathcal{P}_{n-k}(\mathbb{N}_n) \tag{7.39}$$

by

$$c(A) = A^c. \tag{7.40}$$

Its inverse function is given by (exercise!)

$$c' : \mathcal{P}_{n-k}(\mathbb{N}_n) \to \mathcal{P}_k(\mathbb{N}_n) \tag{7.41}$$

where

$$c'(B) = B^c. \tag{7.42}$$

Hence, c is a bijection. □

{3,4,5}	•	•	•	•	•	{1,2}
{2,4,5}	•	•	•	•	•	{1,3}
{2,3,5}	•	•	•	•	•	{1,4}
{2,3,4}	•	•	•	•	•	{1,5}
{1,4,5}	•	•	•	•	•	{2,3}
{1,3,5}	•	•	•	•	•	{2,4}
{1,3,4}	•	•	•	•	•	{2,5}
{1,2,5}	•	•	•	•	•	{3,4}
{1,2,4}	•	•	•	•	•	{3,5}
{1,2,3}	•	•	•	•	•	{4,5}

Figure 7.4.3. Visualizing the $\binom{5}{3} = \binom{5}{2}$ case of Lemma 7.21. The 3-element sets consist of red dots and the 2-element complementary sets consist of blue dots.

Example 7.22. $\dbinom{15}{13} = \dbinom{15}{2} = 105.$

Thinking about it, this all makes sense. The binomial coefficient $\binom{n}{k}$ can be viewed as the number of ways, given n mice, to choose k of them to be members of The Mickey Mouse Club. Now, the number of ways of forming such k-member mice clubs (consisting of the **hoity-toity cool mice**) is the same as the number of ways of choosing which $n-k$ mice will not be in the club (the **down-and-out nerdy mice**). The point is that choosing which mice are "cool" by exclusion is the same as choosing which mice are "nerdy". For example, suppose we have 3 mice named Moe, Curly, and Larry. Let us look at the number of ways of choosing 2 cool mice to be in the club. This is the same as choosing $3-2=1$ nerdy mouse not to be in the club. There are $\binom{3}{1} = 3$ ways to do this: (1) choose Moe to be nerdy, (2) choose Curly to be nerdy, or (3) choose Larry to be nerdy. These three choices are in one-to-one correspondence with the three choices: (1) choose Curly and Larry to be cool, (2) choose Moe and Larry to be cool, or (3) Choose Moe and Curly to be cool.

Exercise 7.10. *Let m be an odd positive integer. Let*

$$X := \Big\{A \in \mathcal{P}(\mathbb{N}_m) : |A| < \frac{m}{2}\Big\}, \qquad Y := \Big\{A \in \mathcal{P}(\mathbb{N}_m) : |A| > \frac{m}{2}\Big\}. \tag{7.43}$$

Prove that

$$|X| = |Y|. \tag{7.44}$$

7.4.3. Relations between binomial coefficients. We have the following **recursive formula** for binomial coefficients (for the application to Pascal's triangle, see §7.4.4 below).

Theorem 7.23. *For positive integers n and r such that $1 \le r \le n$,*

$$\binom{n}{r} = \binom{n-1}{r-1} + \binom{n-1}{r}. \tag{7.45}$$

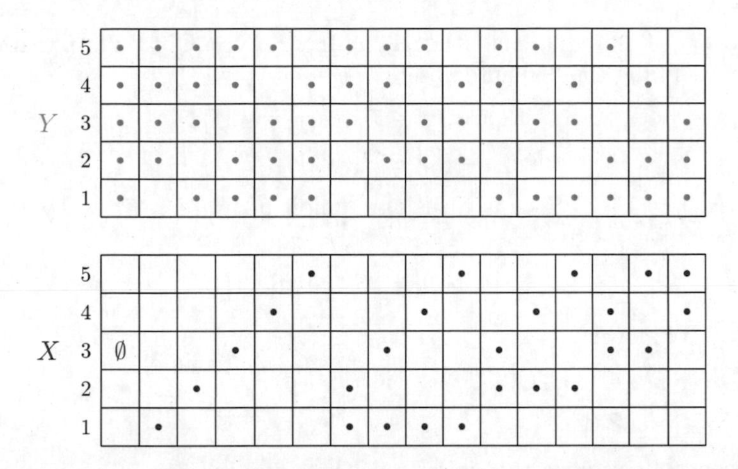

Figure 7.4.4. For Exercise 7.10, the sets X and Y when $m = 5$. Each column of the lower and upper grid represents an element of X and Y, respectively. For example, the lower rightmost column represents the subset $\{4, 5\}$ and upper rightmost column represents the subset $\{1, 2, 3\}$. If a subset A represents a lower column, then its complement $A^c = \mathbb{N}_5 - A$ represents the column directly above it.

Proof. For a visualization of this proof, see Figure 7.4.5.

For an r-element subset A of $\mathbb{N}_n$,

(1) call A **Type-In** if $n \in A$,

(2) call A **Type-Out** if $n \notin A$.

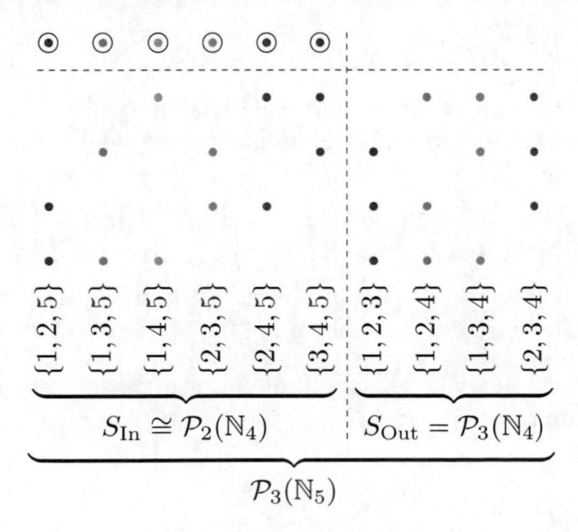

Figure 7.4.5. Visualization of the proof of Theorem 7.23 for $n = 5$ and $r = 3$. From this picture we conclude that $\binom{5}{3} = \binom{4}{2} = \binom{4}{3}$.

Clearly, every $A \in \mathcal{P}_r(\mathbb{N}_n)$ is either Type-In or Type-Out. We partition the set $\mathcal{P}_r(\mathbb{N}_n)$ by defining the disjoint subsets:

$$S_{\text{In}} = \{A \in \mathcal{P}_r(\mathbb{N}_n) \mid n \in A\},$$
$$S_{\text{Out}} = \{A \in \mathcal{P}_r(\mathbb{N}_n) \mid n \notin A\}.$$

Then $\mathcal{P}_r(\mathbb{N}_n) = S_{\text{In}} \cup S_{\text{Out}}$ is a disjoint union of finite sets, so by the addition principle we have

$$|\mathcal{P}_r(\mathbb{N}_n)| = |S_{\text{In}}| + |S_{\text{Out}}|. \tag{7.46}$$

Claim. We have:

(i) $|S_{\text{In}}| = |\mathcal{P}_{r-1}(\mathbb{N}_{n-1})|$.

(ii) $|S_{\text{Out}}| = |\mathcal{P}_r(\mathbb{N}_{n-1})|$.

The theorem follows from the claim since the claim implies

$$\binom{n}{r} = |\mathcal{P}_r(\mathbb{N}_n)| = |\mathcal{P}_{r-1}(\mathbb{N}_{n-1})| + |\mathcal{P}_r(\mathbb{N}_{n-1})| = \binom{n-1}{r-1} + \binom{n-1}{r}.$$

Proof of the claim.

Practical proof of (i). Elements of S_{In} are those r-element subsets of $\mathbb{N}_n$ which contain the element n. By removing the element n from these subsets, this corresponds to arbitrary $(r-1)$-element subsets of $\mathbb{N}_{n-1}$. Thus

$$|S_{\text{In}}| = |\mathcal{P}_{r-1}(\mathbb{N}_{n-1})|. \tag{7.47}$$

Formal proof of (i). We define

$$F : S_{\text{In}} \to \mathcal{P}_{r-1}(\mathbb{N}_{n-1})$$

by $F(A) = A - \{n\}$. Clearly $F(A) \subset \mathbb{N}_{n-1}$. Since $n \in A$, $F(A) \sqcup \{n\} = A$, so by the addition principle $|F(A)| + 1 = r$. Thus $|F(A)| = r - 1$ and hence $F(A) \in \mathcal{P}_{r-1}(\mathbb{N}_{n-1})$. This shows that the function F is well-defined. Conversely, define

$$G : \mathcal{P}_{r-1}(\mathbb{N}_{n-1}) \to S_{\text{In}}$$

by $G(B) = B \cup \{n\}$. Since B is an $(r-1)$-element subset of $\mathbb{N}_{n-1}$, $B \cup \{n\}$ is an r-element subset of $\mathbb{N}_n$; so G is well-defined. We now show that F and G are inverses of each other.

Let $B \in \mathcal{P}_{r-1}(\mathbb{N}_{n-1})$. Then $F(G(B)) = F(B \cup \{n\}) = (B \cup \{n\}) - \{n\} = B$ with the last equality true since $n \notin B$, which is true since $B \subset \mathbb{N}_{n-1}$.

Let $A \in S_{\text{In}}$. Then $n \in A$. We have $G(F(A)) = G(A - \{n\}) = (A - \{n\}) \cup \{n\} = A$ with the last equality true since $n \in A$.

Practical proof of (ii). Elements of S_{In} are those r-element subsets of $\mathbb{N}_n$ which do not contain the element n. This corresponds to arbitrary r-element subsets of $\mathbb{N}_{n-1}$. Thus

$$|S_{\text{Out}}| = |\mathcal{P}_r(\mathbb{N}_{n-1})|. \tag{7.48}$$

We leave the formal proof of (ii) as an exercise. The theorem follows. □

Exercise 7.11. *Give a formal proof of claim* (ii) *in the proof of Theorem* 7.23.

Suppose we have a basket with m objects and another basket with n objects. How can we describe the possible choices of a total of k objects from both baskets? The following gives us an answer.

Exercise 7.12. *Let m and n be positive integers, and let $0 \leq k \leq m+n$ be an integer. Given disjoint finite sets X and Y with $|X| = m$ and $|Y| = n$, define*

$$f : \mathcal{P}_k(X \cup Y) \to \bigcup_{i=0}^{k} \mathcal{P}_i(X) \times \mathcal{P}_{k-i}(Y) \tag{7.49}$$

by

$$f(C) = (C \cap X, C \cap Y). \tag{7.50}$$

(For example, we may take $X = \mathbb{N}_m$ and $Y = \{m+1, m+2, \ldots, m+n\}$.) Also define

$$g : \bigcup_{i=0}^{k} \mathcal{P}_i(X) \times \mathcal{P}_{k-i}(Y) \to \mathcal{P}_k(X \cup Y) \tag{7.51}$$

by

$$g(A, B) = A \cup B. \tag{7.52}$$

(1) *Explain why the functions f and g are well-defined. In particular, why are the function values contained in the codomain as advertised?*

(2) *Prove that the functions f and g are inverses of each other. Did your proof use that X and Y are finite sets? Did your proof use that X and Y are disjoint?*

(3) *Show that one can conclude that*

$$\binom{m+n}{k} = \sum_{i=0}^{k} \binom{m}{i} \binom{n}{k-i}. \tag{7.53}$$

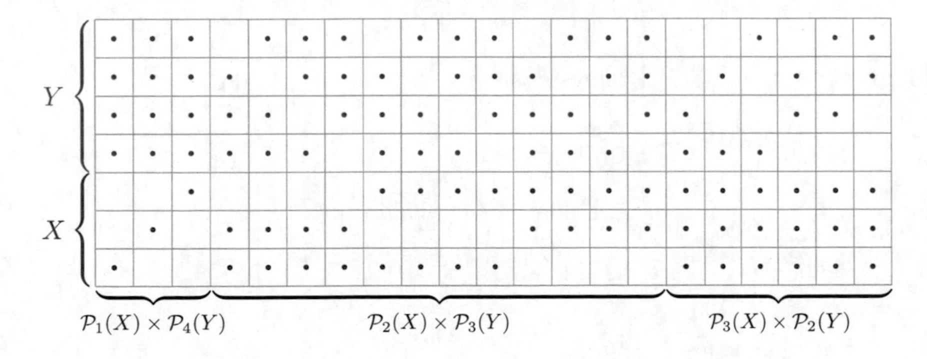

Figure 7.4.6. For Exercise 7.12, suppose that $|X| = 3$ and $|Y| = 4$. We have $\mathcal{P}_5(X \cup Y) \cong \bigcup_{i=1}^{3} \mathcal{P}_i(X) \times \mathcal{P}_{5-i}(Y)$, where $\cong$ means "is bijective to". Note that Figure 7.4.10 below corresponds to the 10th column from the left in this figure. Can you explain how this figure is related to (7.73)?

7.4.4. Pascal's triangle. **Pascal's triangle** is

$$
\begin{array}{ccccccccccccccc}
&&&&&&& 1 &&&&&&& \\
&&&&&& 1 && 1 &&&&&& \\
&&&&& 1 && 2 && 1 &&&&& \\
&&&& 1 && 3 && 3 && 1 &&&& \\
&&& 1 && 4 && 6 && 4 && 1 &&& \\
&& 1 && 5 && 10 && 10 && 5 && 1 && \\
& 1 && 6 && 15 && 20 && 15 && 6 && 1 & \\
1 && 7 && 21 && 35 && 35 && 21 && 7 && 1
\end{array}
$$

etc.

As you may see, an integer in the triangle is the sum of the two integers above it and closest to it. For example $35 = 15 + 20$ occurs in $\begin{array}{ccc} 15 && 20 \\ & 35 & \end{array}$.

Remarkably, Pascal's triangle is made up of binomial coefficients as we see by

$$
\begin{array}{ccccccccccc}
&&&&& \binom{0}{0} &&&&& \\
&&&& \binom{1}{0} && \binom{1}{1} &&&& \\
&&& \binom{2}{0} && \binom{2}{1} && \binom{2}{2} &&& \\
&& \binom{3}{0} && \binom{3}{1} && \binom{3}{2} && \binom{3}{3} && \\
& \binom{4}{0} && \binom{4}{1} && \binom{4}{2} && \binom{4}{3} && \binom{4}{4} & \\
\binom{5}{0} && \binom{5}{1} && \binom{5}{2} && \binom{5}{3} && \binom{5}{4} && \binom{5}{5}
\end{array}
$$

etc.

Note that $\begin{array}{ccc} \binom{4}{1} && \binom{4}{2} \\ & \binom{5}{2} & \end{array}$ represents

$$\binom{5}{2} = \binom{4}{1} + \binom{4}{2}. \tag{7.54}$$

7.4.5. Pascal's triangle and Fibonacci numbers. In Figure 2.6.1 we saw that the first several diagonal sums give Fibonacci numbers $\{f_n\}_{n=1}^{\infty}$. Let us try to formulate this as a mathematical statement and prove this statement in general.

From Figure 2.6.1 we see that

$$\begin{aligned}
\binom{0}{0} &= f_1 = 1, \\
\binom{1}{0} &= f_2 = 1, \\
\binom{2}{0} + \binom{1}{0} &= f_3 = 2, \\
\binom{3}{0} + \binom{2}{1} &= f_4 = 3, \\
\binom{4}{0} + \binom{3}{1} + \binom{2}{2} &= f_5 = 5, \\
\binom{5}{0} + \binom{4}{1} + \binom{3}{2} &= f_6 = 8,
\end{aligned} \tag{7.55}$$

etc.

Exercise 7.13. *Define* $\binom{m}{s} = 0$ *if* $m < 0$, $s < 0$, *or* $s > m$. *Prove that the recursive formula of Theorem 7.23 hold for all integers* r *and* n.

The following generalizes (7.55).

Exercise 7.14. *Let* f_m *denote the* m*-th Fibonacci number. Prove that for* $n \in \mathbb{Z}^{\geq}$,

$$f_{n+1} = \sum_{j+k=n} \binom{k}{j}, \tag{7.56}$$

where in the sum we may assume that $k \geq j \geq 0$ *since otherwise* $\binom{k}{j} = 0$.

7.4.6. Modular Pascal triangles. The first eleven rows of Pascal's triangle are given by Figure 7.4.7.

We can consider the remainders modulo 3 of the binomial coefficients to form the "mod 3 Pascal triangle". See Figure 7.4.8.

We will leave it to the reader to investigate what patterns there are for modular Pascal triangles!

7.4.7. Binomial coefficient factorial formula. We have the following **factorial formula** for the binomial coefficients.

Theorem 7.24. *For non-negative integers* n *and* r *such that* $r \leq n$,

$$\binom{n}{r} = \frac{n(n-1)\cdots(n-r+1)}{r!} = \frac{n!}{r!\,(n-r)!}. \tag{7.57}$$

Proof (less formal, more intuitive). The binomial $\binom{n}{r}$ counts the number of unordered r-element subsets of a set of cardinality n, such as $\mathbb{N}_n$. Such a subset can be obtained by first choosing r elements of $\mathbb{N}_n$. There are n choices for the first element. Given any choice of the first element, there are always $n-1$ choices for

Figure 7.4.7. The first eleven rows of Pascal's triangle.

Figure 7.4.8. Pascal's triangle modulo 3. Can you fill in the last two blank rows and beyond?!

the second element, which has to be distinct from the first element (independent of what the choice was for the first element). Then, given any choices for the first two distinct elements, there are always $n-2$ choices for the third element. We continue in this way until we have chosen the r-th element, for which there are $n - r + 1$ choices. Thus there are

$$n(n-1)(n-2)\cdots(n-r+1) \tag{7.58}$$

possibilities for ordered r-element subsets of $\mathbb{N}_n$. Since we are counting unordered r-element subsets of $\mathbb{N}_n$ and since there are $r!$ permutations of any set of cardinality r, we have that there are

$$\begin{aligned}
&\frac{n(n-1)(n-2)\cdots(n-r+1)}{r!} \\
&= \frac{n(n-1)(n-2)\cdots(n-r+1)\cdot(n-r)(n-r-1)\cdots 2\cdot 1}{r!\cdot(n-r)(n-r-1)\cdots 2\cdot 1} \\
&= \frac{n!}{r!(n-r)!}
\end{aligned}$$

unordered r-element subsets of $\mathbb{N}_n$, and this number by definition is equal to $\binom{n}{r}$. □

Proof. The proof is by induction. Let $n \geq 0$ and let $P(n)$ be the statement that for all $0 \leq r \leq n$

$$\binom{n}{r} = \frac{n!}{r!\,(n-r)!}.$$

The base case is $n = 0$. One assumes $P(k)$ for some $k \geq 0$. To prove $P(k+1)$, one uses the identity from Theorem 7.23 (take $n = k + 1$)

$$\binom{k+1}{r} = \binom{k}{r-1} + \binom{k}{r}$$

and one uses the inductive hypothesis to calculate the right side in terms of factorials. Here one verifies the formula for $0 < r < k + 1$. Separately, one checks that the formula in the theorem is true for $r = 0$ and for $r = k + 1$. □

Observe that Lemma 7.21 follows immediately from Theorem 7.24.

Example 7.25 (Special values of binomial coefficients). For $n \geq 0$,

$$\binom{n}{0} = \binom{n}{n} = 1. \tag{7.59}$$

For $n \geq 1$,

$$\binom{n}{1} = \binom{n}{n-1} = n. \tag{7.60}$$

For $n \geq 2$,

$$\binom{n}{2} = \binom{n}{n-2} = \frac{n(n-1)}{2}. \tag{7.61}$$

Exercise 7.15 (A binomial identity). *Prove that*

$$k\binom{n}{k} = n\binom{n-1}{k-1}. \tag{7.62}$$

Exercise 7.16 (Property of a prime choose k). *Let p be a prime and let $1 \leq k \leq p-1$. Prove that the binomial coefficient $\binom{p}{k}$ is divisible by p.*

For example, the binomial coefficients $\binom{7}{k}$, $1 \leq k \leq 6$, are $7, 21, 35, 35, 21, 7$, where each of these integers is divisible by 7.

Exercise 7.17 (Another binomial identity). *Prove that*

$$\sum_{k=0}^{n} k \binom{n}{k} = n\, 2^{n-1}. \tag{7.63}$$

7.4.8. Proof of the Binomial Theorem. Monomials of one variable are nonnegative integer powers of a single variable, denoted such as a^n, where $\mathbb{Z}^{\geq}$. More interestingly, we can consider (powers of) binomials of the form $(a+b)^n$. The following result explains the terminology "binomial coefficients".

Theorem 7.26 (Binomial Theorem). *For all real numbers a and b and nonnegative integers n, we have the expansion*

$$(a+b)^n = \sum_{i=0}^{n} \binom{n}{i} a^{n-i} b^i \tag{7.64}$$
$$= a^n + \cdots + \binom{n}{i} a^{n-i} b^i + \cdots + b^n.$$

For example,

$$(a+b)^2 = a^2 + 2ab + b^2, \tag{7.65a}$$
$$(a+b)^3 = a^3 + 3a^2b + 3ab^2 + b^3, \tag{7.65b}$$
$$(a+b)^4 = a^4 + 4a^3b + 6a^2b^2 + 4ab^3 + b^4, \tag{7.65c}$$
$$(a+b)^5 = a^5 + 5a^4b + 10a^3b^2 + 10a^2b^3 + 5ab^4 + b^5. \tag{7.65d}$$

Proof (less formal, more conceptual). We have

$$(a+b)^n = \underbrace{(a+b)(a+b)(a+b)\cdots(a+b)}_{n \text{ times}}. \tag{7.66}$$

The right-hand side is the sum of products

$$c_1 c_2 c_3 \cdots c_n, \tag{7.67}$$

where each c_i, $1 \leq i \leq n$, is either a or b. By the commutativity of multiplication, we may write each such term as $a^{n-i}b^i$, for some $0 \leq i \leq n$, since in view of $(n-i)+i = n$ such binomials are the products of n numbers, each of which is either an a or a b. Now, given $0 \leq i \leq n$, the number of terms of the form $a^{n-i}b^i$ is equal to the number of ways to choose $c_1 c_2 c_3 \cdots c_n$ to consist of i b's and $n-i$ a's. Thus, the number of terms of the form $a^{n-i}b^i$ is equal to $\binom{n}{i}$. The Binomial Theorem follows. □

b	$a^k b + \cdots +$	$\binom{k}{i-2} a^{k+2-i} b^{i-1} +$	$\binom{k}{i-1} a^{k+1-i} b^i +$	$\binom{k}{i} a^{k-i} b^{i+1}$	$+ \cdots + \binom{k}{1} a\, b^k +$	b^{k+1}
a	$a^{k+1} +$	$\binom{k}{1} a^k b + \cdots +$	$\binom{k}{i-1} a^{k+2-i} b^{i-1}$	$+ \binom{k}{i} a^{k+1-i} b^i +$	$\binom{k}{i+1} a^{k-i} b^{i+1}$	$+ \cdots + a\, b^k$

$$a^k + \cdots + \binom{k}{i-2} a^{k+2-i} b^{i-2} + \binom{k}{i-1} a^{k+1-i} b^{i-1} + \binom{k}{i} a^{k-i} b^i + \binom{k}{i+1} a^{k-1-i} b^{i+1} + \cdots + b^k$$

Figure 7.4.9. Visualizing the inductive step of the proof of the Binomial Theorem 7.26. Using Theorem 7.23, we can infer the formula to prove the inductive step $(a+b) \sum_{i=0}^{k} \binom{k}{i} a^{k-i} b^i = \sum_{i=0}^{k+1} \binom{k+1}{i} a^{k+1-i} b^i$ from this figure.

We now start thinking about a formal proof of the Binomial Theorem. Regarding the terms in Figure 7.4.9, by Theorem 7.23 we have

$$a^k b + \binom{k}{1} a^k b = \binom{k+1}{1} a^k b, \qquad \binom{k}{1} a\, b^k + a\, b^k = \binom{k+1}{1} a\, b^k, \tag{7.68}$$

$$\binom{k}{i-2} a^{k+2-i} b^{i-1} + \binom{k}{i-1} a^{k+2-i} b^{i-1} = \binom{k+1}{i-1} a^{k+2-i} b^{i-1}, \tag{7.69}$$

$$\binom{k}{i-1} a^{k+1-i} b^i + \binom{k}{i} a^{k+1-i} b^i = \binom{k+1}{i} a^{k+1-i} b^i, \tag{7.70}$$

etc.

Proof (by induction). If you like to think about proofs visually, you may refer to Figure 7.4.9 as you read this proof.

We leave it to the reader to check the base case $n = 1$. By induction assume that $k \geq 0$ is such that

$$(a+b)^k = \sum_{i=0}^{k} \binom{k}{i} a^{k-i} b^i.$$

We compute

$$\begin{aligned}
(a+b)^{k+1} &= (a+b)^k (a+b) \\
&= \sum_{i=0}^{k} \binom{k}{i} a^{k-i} b^i (a+b) \quad \text{(inductive hypothesis)} \\
&= \sum_{i=0}^{k} \binom{k}{i} a^{(k+1)-i} b^i + \sum_{i=0}^{k} \binom{k}{i} a^{k-i} b^{i+1} \quad \text{(distributive law).}
\end{aligned}$$

We want the exponents in $a^{k-i}b^{i+1}$ above to match up with the exponents in $a^{(k+1)-i}b^i$ for the $k+1$ case of the Binomial Theorem. So we define

$$i' = i + 1,$$

so that summing i from 0 to k is the same as summing i' from 1 to $k+1$. We get that the last line in the display above is

$$\begin{aligned}
&= \sum_{i=0}^{k} \binom{k}{i} a^{(k+1)-i} b^i + \sum_{i'=1}^{k+1} \binom{k}{i'-1} a^{(k+1)-i'} b^{i'} \text{ (writing in terms of } i' \text{ instead of } i) \\
&= \sum_{i=0}^{k} \binom{k}{i} a^{(k+1)-i} b^i + \sum_{i=1}^{k+1} \binom{k}{i-1} a^{(k+1)-i} b^i \quad \text{(renaming } i' \text{ to be } i) \\
&= \sum_{i=0}^{0} \binom{k}{i} a^{(k+1)-i} b^i + \sum_{i=1}^{k} \binom{k}{i} a^{(k+1)-i} b^i + \sum_{i=1}^{k} \binom{k}{i-1} a^{(k+1)-i} b^i \\
&\quad + \sum_{i=k+1}^{k+1} \binom{k}{i-1} a^{(k+1)-i} b^i \\
&= \binom{k}{0} a^{(k+1)-0} b^0 + \sum_{i=1}^{k} \left(\binom{k}{i} + \binom{k}{i-1} \right) a^{(k+1)-i} b^i \\
&\quad + \binom{k}{(k+1)-1} a^{(k+1)-(k+1)} b^{k+1} \\
&= \binom{k+1}{0} a^{(k+1)-0} b^0 + \sum_{i=1}^{k} \binom{k+1}{i} a^{(k+1)-i} b^i + \binom{k+1}{k+1} a^{(k+1)-(k+1)} b^{k+1} \\
&= \sum_{i=0}^{k+1} \binom{k+1}{i} a^{(k+1)-i} b^i,
\end{aligned}$$

where the penultimate equality is by Theorem 7.23 and because

$$\binom{k}{0} = 1 = \binom{k+1}{0} \quad \text{and} \quad \binom{k}{(k+1)-1} = \binom{k}{k} = 1 = \binom{k+1}{k+1}. \qquad \square$$

Analogous to Pascal's triangle in §7.4.4, the expansions in the Binomial Theorem can be visualized as Figure 7.0.1 at the beginning of this chapter (compare with (7.65)).

It is interesting to consider the Binomial Theorem in modular arithmetic. Especially useful is the case where the modulus is a prime.

Exercise 7.18 (Binomial Theorem modulo a prime). *Let p be a prime, and let $x, y \in \mathbb{Z}$. Prove that*

$$(x+y)^p \equiv x^p + y^p \bmod p. \tag{7.71}$$

Hint: Use Exercise 7.16. *Some "prime" rows of Pascal's triangle are*

$$\begin{array}{c} 1\ 3\ 3\ 1 \\ 1\ 5\ 10\ 10\ 5\ 1 \\ 1\ \ 7\ \ 21\ \ 35\ \ 35\ \ 21\ \ 7\ \ 1 \\ 1\ \ 11\ \ 55\ \ 165\ \ 330\ \ 462\ \ 462\ \ 330\ \ 165\ \ 55\ \ 11\ \ 1 \end{array}$$

If we count the 1 *at the top of Pascal's triangle as the* 0*-th row, then the display above pictures rows* 3, 5, 7, *and* 11.

Exercise 7.19 (Proof Fermat's Little Theorem using the Binomial Theorem). *We give another proof of Fermat's Little Theorem* 6.63. *Let p be a prime and let a be a positive integer.*

(1) *Prove that* $(a+1)^p \equiv a^p + 1 \bmod p$.
(2) *Prove by induction on a that* $a^p \equiv a \bmod p$ *for all positive integers a.*
(3) *Assuming in addition that a is not a multiple of p, prove that* $a^{p-1} \equiv 1 \bmod p$.

Exercise 7.20 (More binomial identities). *Let $k, \ell \in \mathbb{Z}^{\geq}$. We have the obvious identity*

$$(1+x)^{k+\ell} = (1+x)^k(1+x)^\ell.$$

By applying the Binomial Theorem to each of the binomials $(1+x)^{k+\ell}$, $(1+x)^k$, *and* $(1+x)^\ell$ *and by matching the resulting coefficients of* x^m, *where* $0 \leq m \leq k+\ell$, *for both sides of the identity, prove the following formula. For* $0 \leq m \leq k+\ell$,

$$\binom{k+\ell}{m} = \sum_{i=\max\{m-\ell,0\}}^{\min\{k,m\}} \binom{k}{i}\binom{\ell}{m-i}. \tag{7.72}$$

For example, if $k=3$, $\ell=4$, *and* $m=5$, *this says that*

$$\binom{7}{5} = \sum_{i=1}^{3} \binom{3}{i}\binom{4}{5-i}. \tag{7.73}$$

Indeed, this is true because

$$21 = 3 + 12 + 6.$$

Show, as a special case, that we have the formula

$$\binom{2n}{n} = \sum_{i=0}^{n} \binom{n}{i}^2.$$

7.4.9. Leibniz's rule for higher-order derivatives of products. We now give a calculus application. Using essentially the same proof as for the Binomial Theorem, we can prove a product formula for the higher derivatives of one-variable functions.

Theorem 7.27. *Let f and g be one-variable functions that have derivatives to all orders (such functions are called **smooth**). For all $n \in \mathbb{Z}^+$,*

$$(fg)^{(n)} = \sum_{k=0}^{n} \binom{n}{k} f^{(k)} g^{(n-k)}.$$

Proof. A proof of this formula is at the general Leibniz rule Wikipedia page. The proof we give here is essentially the same.

We prove the theorem by induction.

The base case, $n = 1$, is true because

$$(fg)^{(1)} = (fg)' = f\,g' + f'\,g = \binom{1}{0} f^{(0)} g^{(1-0)} + \binom{1}{1} f^{(1)} g^{(1-1)} = \sum_{k=0}^{1} \binom{1}{k} f^{(k)} g^{(1-k)}$$

by the usual Leibniz rule.

Inductive step. Suppose for some $m \in \mathbb{Z}^+$ that

$$(fg)^{(m)} = \sum_{k=0}^{m} \binom{m}{k} f^{(k)} g^{(m-k)}.$$

We then compute that

$$\begin{aligned}
(fg)^{(m+1)} &= ((fg)^{(m)})' \\
&= \left(\sum_{k=0}^{m} \binom{m}{k} f^{(k)} g^{(m-k)} \right)' && \text{by the inductive hypothesis} \\
&= \sum_{k=0}^{m} \binom{m}{k} \left(f^{(k)} g^{(m-k)} \right)' && \text{by the linearity of the derivative operator} \\
&= \sum_{k=0}^{m} \binom{m}{k} ((f^{(k)})' g^{(m-k)} + f^{(k)} (g^{(m-k)})') && \text{by the usual Leibniz rule} \\
&= \sum_{k=0}^{m} \binom{m}{k} (f^{(k+1)} g^{(m-k)} + f^{(k)} g^{(m-k+1)}) && \text{since } (f^{(k)})' = f^{(k+1)}, \text{ etc.} \\
&= \sum_{k=0}^{m} \binom{m}{k} f^{(k+1)} g^{(m-k)} + \sum_{k=0}^{m} \binom{m}{k} f^{(k)} g^{(m+1-k)}.
\end{aligned}$$

At this point, we need to decide how to proceed to obtain the formula in the desired form. We observe that the term with $f^{(k)} g^{(m+1-k)}$, which is the second term on the right side of the last displayed line, is of the desired form but has the wrong coefficient $\binom{m}{k}$. We rewrite the first term on the right side of the last displayed line to help us. Namely, by letting $k = \ell - 1$, we have

$$\sum_{k=0}^{m} \binom{m}{k} f^{(k+1)} g^{(m-k)} = \sum_{\ell=1}^{m+1} \binom{m}{\ell - 1} f^{(\ell)} g^{(m+1-\ell)} = \sum_{k=1}^{m+1} \binom{m}{k-1} f^{(k)} g^{(m+1-k)},$$

where the last equality is by relabelling ℓ as k. Thus

$$(fg)^{(m+1)} = \sum_{k=1}^{m+1} \binom{m}{k-1} f^{(k)} g^{(m+1-k)} + \sum_{k=0}^{m} \binom{m}{k} f^{(k)} g^{(m+1-k)}.$$

The formula we desire is a sum from $k = 0$ to $k = m + 1$. We have a mismatch of the summation limits in the two sums above. So we separate the special cases

$k = m+1$ and $k = 0$, which each only appear in one sum, to get

$$\begin{aligned}
(fg)^{(m+1)} &= \sum_{k=1}^{m} \binom{m}{k-1} f^{(k)} g^{(m+1-k)} + \binom{m}{m} f^{(m+1)} g^{(0)} + \binom{m}{0} f^{(0)} g^{(m+1)} \\
&\quad + \sum_{k=1}^{m} \binom{m}{k} f^{(k)} g^{(m+1-k)} \\
&= \binom{m}{0} f^{(0)} g^{(m+1)} + \sum_{k=1}^{m} \binom{m}{k-1} f^{(k)} g^{(m+1-k)} + \sum_{k=1}^{m} \binom{m}{k} f^{(k)} g^{(m+1-k)} \\
&\quad + \binom{m}{m} f^{(m+1)} g^{(0)} \\
&= \binom{m+1}{0} f^{(0)} g^{(m+1)} + \sum_{k=1}^{m} \left(\binom{m}{k-1} + \binom{m}{k} \right) f^{(k)} g^{(m+1-k)} \\
&\quad + \binom{m+1}{m+1} f^{(m+1)} g^{(0)} \\
&= \binom{m+1}{0} f^{(0)} g^{(m+1)} + \sum_{k=1}^{m} \binom{m+1}{k} f^{(k)} g^{(m+1-k)} + \binom{m+1}{m+1} f^{(m+1)} g^{(0)} \\
&= \sum_{k=0}^{m+1} \binom{m+1}{k} f^{(k)} g^{(m+1-k)},
\end{aligned}$$

where for the penultimate equality we used the recursive formula of Theorem 7.23. This proves the inductive step and hence, by induction, completes the proof. □

Remark 7.28. The interested readers (or should we say the bored readers?)[2] may read on their own about **multinomial coefficients** and the multinomial theorem.

7.4.10. More relations between binomial coefficients.

Solved Problem 7.29. *Let X and Y be disjoint finite sets with $|X| = 3$ and $|Y| = 4$. Can you formulate an approach to compute $\binom{7}{5}$ in terms of choosing pairs of subsets of X and Y whose unions have exactly five elements?*

Solution. Imagine a subset A of X of cardinality i and a subset B of Y of cardinality j. By the addition principle, the cardinality of $A \cup B$ equals 5 if and only if $i + j = 5$. In other words, if $i := |A|$, then $|B| = 5 - i$. Now, $|A| \le |X| \le 3$, and $|B| \le |Y| \le 4$. Thus, we have the restrictions $i \le 3$ and $5 - i \le 4$, which are equivalent to $1 \le i \le 3$.

Since X and Y are disjoint, by Exercise 7.12 there is a one-to-one correspondence between cardinality-5 subsets C of $X \cup Y$ and pairs (A, B), where $A \subset X$ and $B \subset Y$, with $|A| = i$, $|B| = 5 - i$, and $1 \le i \le 3$. That is, there exists a bijection between $\mathcal{P}_5(X \cup Y)$ and the disjoint union

$$(7.74) \qquad \big(\mathcal{P}_1(X) \times \mathcal{P}_4(Y)\big) \cup \big(\mathcal{P}_2(X) \times \mathcal{P}_3(Y)\big) \cup \big(\mathcal{P}_3(X) \times \mathcal{P}_2(Y)\big).$$

[2] Interested in something more; bored with something less. ☺

From this and the addition principle, we obtain the formula

$$(7.75)\qquad \begin{aligned}\binom{7}{5} &= |\mathcal{P}_5(X \cup Y)| \\ &= |(\mathcal{P}_1(X) \times \mathcal{P}_4(Y)) \cup (\mathcal{P}_2(X) \times \mathcal{P}_3(Y)) \cup (\mathcal{P}_3(X) \times \mathcal{P}_2(Y))| \\ &= |\mathcal{P}_1(X) \times \mathcal{P}_4(Y)| + |\mathcal{P}_2(X) \times \mathcal{P}_3(Y)| + |\mathcal{P}_3(X) \times \mathcal{P}_2(Y)| \\ &= |\mathcal{P}_1(X)| \cdot |\mathcal{P}_4(Y)| + |\mathcal{P}_2(X)| \cdot |\mathcal{P}_3(Y)| + |\mathcal{P}_3(X)| \cdot |\mathcal{P}_2(Y)| \\ &= \binom{3}{1}\binom{4}{4} + \binom{3}{2}\binom{4}{3} + \binom{3}{3}\binom{4}{2}.\end{aligned}$$

Observe that, indeed, both sides of the equation above equal 21.

We can imagine the formula in the following way. We can think of $X \cup Y$ as having 3 red rooms and 4 blue rooms, where red and blue correspond to X and Y, respectively. A cardinality-5 subset of people in $X \cup Y$ consists of i people in X and $5 - i$ people in Y, where $1 \leq i \leq 3$. See Figure 7.4.10.

Figure 7.4.10. An example of a cardinality-5 subset with 2 people in X and 3 people in Y is pictured.

Solved Problem 7.30. *Prove the binomial coefficient identity:*

$$(7.76)\qquad \binom{n+1}{k+1} = \sum_{i=k}^{n} \binom{i}{k} = \binom{k}{k} + \binom{k+1}{k} + \binom{k+2}{k} + \cdots + \binom{n}{k}.$$

Solution. Consider the power set $\mathcal{P}_{k+1}(\mathbb{N}_{n+1})$, whose cardinality is equal to $\binom{n+1}{k+1}$ by definition. We partition the set $\mathcal{P}_{k+1}(\mathbb{N}_{n+1})$ into disjoint subsets $\mathcal{S}_k, \mathcal{S}_{k+1}, \mathcal{S}_{k+2} \ldots, \mathcal{S}_n$, where

$$(7.77)\qquad \mathcal{S}_i := \{A \in \mathcal{P}_{k+1}(\mathbb{N}_{n+1}) : \max A = i + 1\}.$$

Indeed, since the maximum of a subset of $\mathbb{N}_{n+1}$ is unique and since a subset having cardinality $k + 1$ implies that its maximum is at least $k + 1$, we have that

$$(7.78)\qquad \mathcal{P}_{k+1}(\mathbb{N}_{n+1}) = \bigcup_{i=k}^{n} \mathcal{S}_i = \mathcal{S}_k \cup \mathcal{S}_{k+1} \cup \mathcal{S}_{k+2} \cup \cdots \cup \mathcal{S}_n$$

is a disjoint union. Thus, by the addition principle for sets,

$$(7.79)\qquad \binom{n+1}{k+1} = \sum_{i=k}^{n} |\mathcal{S}_i|.$$

For each $k \leq i \leq n$, we define the function

$$(7.80)\qquad F_i : \mathcal{S}_i \to \mathcal{P}_k(\mathbb{N}_i)$$

by

$$(7.81)\qquad F_i(A) = A - \{i + 1\}.$$

The function F_i is a bijection with inverse equal to the function $G_i : \mathcal{P}_k(\mathbb{N}_i) \to \mathcal{S}_i$ defined by $G_i(B) = B \cup \{i+1\}$. Thus,

$$|\mathcal{S}_i| = |\mathcal{P}_k(\mathbb{N}_i)| = \binom{i}{k}. \tag{7.82}$$

The result follows.

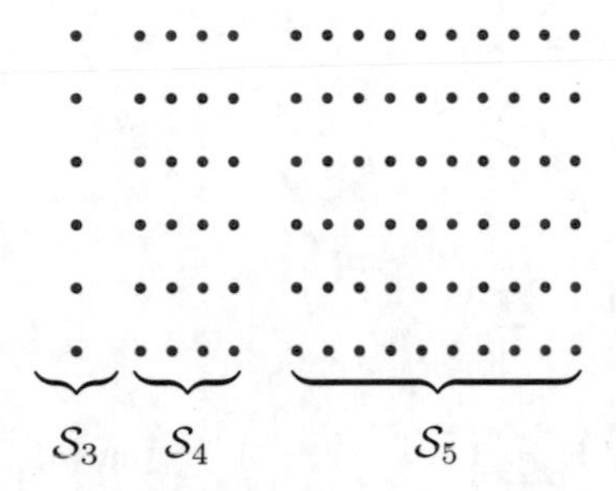

Figure 7.4.11. The power set $\mathcal{P}_4(\mathbb{N}_6)$ is partitioned into the disjoint union $\mathcal{S}_3 \cup \mathcal{S}_4 \cup \mathcal{S}_5$. We have $|\mathcal{S}_3| = \binom{3}{3} = 1$, $|\mathcal{S}_4| = \binom{4}{3} = 4$, and $|\mathcal{S}_5| = \binom{5}{3} = 10$.

Solved Problem 7.31. *Prove the identity*

$$\begin{aligned} f_n &= \sum_{i=0}^{\lfloor \frac{n-1}{2} \rfloor} \binom{n-i-1}{i} \\ &= \binom{n-1}{0} + \binom{n-2}{1} + \binom{n-3}{2} + \cdots + \binom{n-1-\lfloor \frac{n-1}{2} \rfloor}{\lfloor \frac{n-1}{2} \rfloor} \end{aligned} \tag{7.83}$$

for all $n \in \mathbb{Z}^+$, where f_n denotes the n-th Fibonacci number and $\lfloor \cdot \rfloor$ is the floor function.

Solution. Let $F(n)$ be the statement that formula (7.83) is true. Let $P(k)$ be the statement that $F(2k-1)$ and $F(2k)$ are both true. We prove the result by induction on k.

Base case: $k = 1$. We need to establish $P(1)$ and $P(2)$. Firstly, $P(1)$ is the statement that $f_1 = \sum_{i=0}^{0} \binom{-i}{i}$, which is true since both sides of the equation equal 1. Secondly, $P(2)$ is the statement that $f_2 = \sum_{i=0}^{0} \binom{1-i}{i}$, which is also true since again both sides of the equation equal 1.

Strong inductive step: Suppose that $k \in \mathbb{Z}^+$ is such that $P(k)$ is true; that is, $F(2k-1)$ and $F(2k)$ are both true. Now,

$$\left\lfloor \frac{2k-2}{2} \right\rfloor = \left\lfloor \frac{2k-1}{2} \right\rfloor = k-1. \tag{7.84}$$

Thus, our aforestated inductive hypothesis says that

(7.85a) $$f_{2k-1} = \sum_{i=0}^{k-1} \binom{2k-i-2}{i},$$

(7.85b) $$f_{2k} = \sum_{i=0}^{k-1} \binom{2k-i-1}{i}.$$

We compute that

$$\begin{aligned}
f_{2k+1} &= f_{2k-1} + f_{2k} \\
&= \sum_{i=0}^{k-1} \binom{2k-i-2}{i} + \sum_{i=0}^{k-1} \binom{2k-i-1}{i} \\
&= \sum_{i=1}^{k} \binom{2k-i-1}{i-1} + \sum_{i=0}^{k-1} \binom{2k-i-1}{i} \quad \text{(replaced } i \text{ by } i-1 \text{ in the first sum)} \\
&= 2 + \sum_{i=1}^{k-1} \left(\binom{2k-i-1}{i-1} + \binom{2k-i-1}{i} \right).
\end{aligned}$$

By applying Theorem 7.23, we may rewrite this as

(7.86) $$f_{2k+1} = 2 + \sum_{i=1}^{k-1} \binom{2k-i}{i} = \sum_{i=0}^{k} \binom{2k-i}{i}.$$

This proves $F(2k+1)$.

It remains to prove $F(2k+2)$. We compute that

$$\begin{aligned}
f_{2k+2} &= f_{2k} + f_{2k+1} \\
&= \sum_{i=0}^{k-1} \binom{2k-i-1}{i} + \sum_{i=0}^{k} \binom{2k-i}{i} \\
&= \sum_{i=1}^{k} \binom{2k-i}{i-1} + \sum_{i=0}^{k} \binom{2k-i}{i} \\
&= 1 + \sum_{i=1}^{k} \left(\binom{2k-i}{i-1} + \binom{2k-i}{i} \right) \\
&= 1 + \sum_{i=1}^{k} \binom{2k+1-i}{i} \\
&= \sum_{i=0}^{k} \binom{2(k+1)-i-1}{i}.
\end{aligned}$$

This proves $F(2k+2)$. Thus, the proof of $P(k+1)$ is complete. By induction, we are done.

Remark 7.32. Let

$$g_k := \sum_{i=0}^{\lfloor \frac{k-1}{2} \rfloor} \binom{k-i-1}{i}. \tag{7.87}$$

We can illustrate the inductive step by considering the calculations corresponding to the special case $k = 3$. Namely, we show that $g_7 = g_5 + g_6$ and $g_8 = g_6 + g_7$. That is, at this stage, the sequence $\{g_n\}$ satisfies the same recursion relations as the Fibonacci sequence $\{f_n\}$. By (7.87), we have

$$g_5 = \binom{4}{0} + \binom{3}{1} + \binom{2}{2},$$
$$g_6 = \binom{5}{0} + \binom{4}{1} + \binom{3}{2}.$$

Firstly, we compute using Theorem 7.23 that

$$\begin{aligned} g_5 + g_6 &= \binom{5}{0} + \left(\binom{4}{0} + \binom{4}{1}\right) + \left(\binom{3}{1} + \binom{3}{2}\right) + \binom{2}{2} \\ &= \binom{6}{0} + \binom{5}{1} + \binom{4}{2} + \binom{3}{3} \\ &= g_7, \end{aligned}$$

where the last equality is by (7.87). Similarly, we can show that $g_8 = g_6 + g_7$. We leave this as an exercise for the reader. Since $g_1 = g_2 = 1 = f_1 = f_2$ and since the sequences $\{g_n\}$ and $\{f_n\}$ satisfy the same recursion relations, we conclude that they are equal as sequences.

Solved Problem 7.33. *Prove that for every $n \in \mathbb{Z}^{\geq}$,*

$$\sum_{k=0}^{n} \binom{n}{k} 2^k = 3^n. \tag{7.88}$$

Solution. Define the set

$$X := \{(A, B) : A \subset B \subset \mathbb{N}_n\}. \tag{7.89}$$

Firstly, we compute the cardinality of X as follows. We partition the set X by the subsets defined by

$$X_k := \{(A, B) \in X : |B| = k\} = \{(A, B) \in X : B \in \mathcal{P}_k(\mathbb{N}_n)\} \tag{7.90}$$

for $0 \leq k \leq n$. Indeed, we have that

$$X = \bigcup_{k=0}^{n} X_k = X_0 \cup X_1 \cup X_2 \cup \cdots \cup X_n \tag{7.91}$$

is a disjoint union. For each $B \in \mathcal{P}_k(\mathbb{N}_n)$, let

$$X^B := \{(A, B) : A \subset B\}. \tag{7.92}$$

Then the function $F^B : X^B \to \mathcal{P}(B)$ defined by $F^B(A) := (A, B)$ is a bijection. Therefore,

$$|X^B| = |\mathcal{P}(B)| = 2^{|B|} = 2^k. \tag{7.93}$$

On the other hand, we have that

$$X_k = \bigcup_{B \in \mathcal{P}_k(\mathbb{N}_n)} X^B \tag{7.94}$$

is a disjoint union (exercise). Hence, by the addition principle for sets, we have

$$|X_k| = \sum_{B \in \mathcal{P}_k(\mathbb{N}_n)} |X^B| = \sum_{B \in \mathcal{P}_k(\mathbb{N}_n)} 2^k = |\mathcal{P}_k(\mathbb{N}_n)|\, 2^k = \binom{n}{k} 2^k. \tag{7.95}$$

Therefore,

$$|X| = \sum_{k=0}^{n} |X_k| = \sum_{k=0}^{n} \binom{n}{k} 2^k. \tag{7.96}$$

Secondly, we can compute the cardinality of X as follows. Each pair $(A, B) \in X$ is in one-to-one correspondence with a sequence of n choices, where the i-th choice is whether:

(1) $i \in A$ (and hence $i \in B$),
(2) $i \in B - A$,
(3) $i \notin B$ (and hence $i \notin A$).

Observe that these three choices are mutually exclusive since

$$\mathbb{N}_n = A \cup (B - A) \cup (\mathbb{N}_n - B)$$

is a disjoint union. Since for each $1 \leq i \leq n$ there are exactly 3 choices, there are a total of 3^n choices. Since there is a one-to-one correspondence between sequences of n such choices and elements $(A, B) \in X$, we conclude that $|X| = 3^n$.

The result follows since the two ways of counting the cardinality of X must yield the same number.

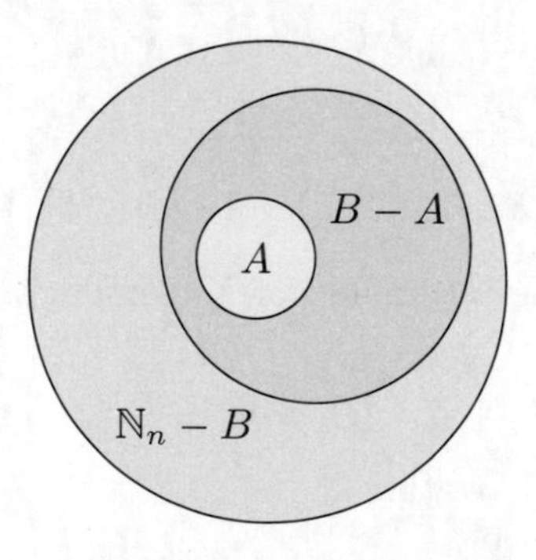

Figure 7.4.12. The trichotomy of choices R.B.G.

7.5. Counting functions

Instead of counting stars, we count functions. Given two non-empty sets K and Y, the set of all functions from K to Y is denoted by $\text{Fun}(K, Y)$. That is,

$$\text{Fun}(K, Y) := \{f \,|\, f : K \to Y \text{ is a function}\}. \tag{7.97}$$

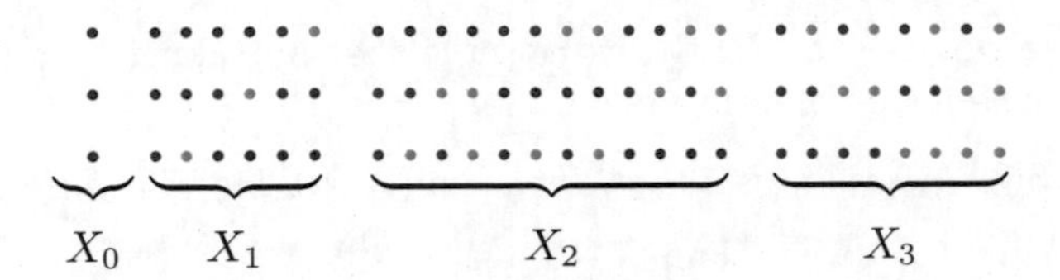

Figure 7.4.13. For $n = 3$ the partition of X into $X_0 \cup X_1 \cup X_2 \cup X_3$. We have $|X_0| = \binom{3}{0}2^0 = 1$, $|X_1| = \binom{3}{1}2^1 = 6$, $|X_2| = \binom{3}{2}2^2 = 12$, and $|X_3| = \binom{3}{3}2^3 = 8$.

7.5.1. Counting all functions.

The main result of this subsection is the following funky theorem (funky theorems come from Funkytown).

Theorem 7.34. *Let K and Y be non-empty finite sets. Then* $\text{Fun}(K, Y)$ *is a finite set and*

$$|\text{Fun}(K, Y)| = |Y|^{|K|}. \tag{7.98}$$

The reason why the theorem is true is that if the domain of a function has cardinality k and the codomain of the function has cardinality n, then a function is given by a sequence of k choices, where for each choice there are n possible choices. Since the choices are independent, the total number of possible sequences of choices is equal to

$$\underbrace{n \cdot n \cdots n}_{k \text{ times}} = n^k. \tag{7.99}$$

Here is the actual proof:

Proof. Since K is a non-empty finite set, we may express it as

$$K = \{x_1, x_2, \ldots, x_k\}, \tag{7.100}$$

where $k = |K|$. We define a bijection between $\text{Fun}(K, Y)$ and Y^k as follows. Define the function $B : \text{Fun}(K, Y) \to Y^k$ by

$$B(f) = (f(x_1), f(x_2), \ldots, f(x_k)). \tag{7.101}$$

And define

$$C : Y^k \to \text{Fun}(K, Y) \tag{7.102}$$

by

$$C(y_1, y_2, \ldots, y_k) = \text{ the function } g \text{ defined by } g(x_i) = y_i \text{ for } 1 \leq i \leq k. \tag{7.103}$$

We leave it to you to check that B and C are inverses of each other, and hence they are bijections.

The theorem is proven since by Corollary 7.7 we have that

$$|\text{Fun}(K, Y)| = |Y^k| = |Y|^k = |Y|^{|K|}. \tag{7.104}$$

□

Exercise 7.21. *Prove that the functions B and C in the proof of Theorem 7.34 are inverses of each other.*

7.5.2. The power set. Firstly, we observe that by taking $Y = \{0, 1\}$ in Theorem 7.34, we obtain the following.

Corollary 7.35. *If X is a finite set, then* $\operatorname{Fun}(X, \{0,1\})$ *is a finite set and*

$$|\operatorname{Fun}(X, \{0,1\})| = 2^{|X|}. \tag{7.105}$$

Let X be a set, and let A be a subset of X. The **characteristic function** of A is defined by

$$\chi_A(x) = \begin{cases} 1 & \text{if } x \in A, \\ 0 & \text{if } x \notin A. \end{cases} \tag{7.106}$$

Define the functions $\Phi : \operatorname{Fun}(X, \{0,1\}) \to \mathcal{P}(X)$ by

$$\Phi(f) := f^{-1}(1), \tag{7.107}$$

the pre-image of 1 under f. Here $\Psi : \mathcal{P}(X) \to \operatorname{Fun}(X, \{0,1\})$ by

$$\Psi(A) := \chi_A, \tag{7.108}$$

the characteristic function of A.

Lemma 7.36. *The functions Φ and Ψ are inverses of each other and hence are bijections. Since* $\operatorname{Fun}(X, \{0,1\})$ *is finite, this implies that $\mathcal{P}(X)$ is finite and*

$$|\mathcal{P}(X)| = |\operatorname{Fun}(X, \{0,1\})| = 2^{|X|}. \tag{7.109}$$

Exercise 7.22. *Prove Lemma* 7.36.

7.5.3. Counting injective functions. Let m, n be positive integers. Define the set of functions

$$F := \{f \mid f : \mathbb{N}_m \to \mathbb{N}_n \text{ is a function}\}. \tag{7.110}$$

Now let $m \leq n$ be positive integers. Define the subset of injective functions

$$I := \operatorname{Inj}(\mathbb{N}_m, \mathbb{N}_n) := \{f : \mathbb{N}_m \to \mathbb{N}_n \mid f \text{ is an injection}\}. \tag{7.111}$$

By thinking of each injective function from $\mathbb{N}_m$ to $\mathbb{N}_n$ as corresponding to a way of lining up m people out of a group of n people, it is easy to see the following.

Theorem 7.37. *The number of injective functions from $\mathbb{N}_m$ to $\mathbb{N}_n$, i.e., the cardinality of I, is equal to*

$$|I| = n(n-1)(n-2)\cdots(n-m+1). \tag{7.112}$$

Proof. We forgo giving a rigorous proof in lieu of a more intuitive argument. We sketch the rigorous proof in Exercise 7.23. We can count the number of injective functions as follows. Firstly, there are n choices for $f(1)$. Secondly, given any choice of $f(1)$, there are $n-1$ choices for $f(2)$ such that $f(2) \neq f(1)$, which is necessary for f to be an injection. Thirdly, given any such choices of $f(1), f(2)$, there are $n-2$ choices for $f(3)$ such that $f(3) \notin \{f(1), f(2)\}$, which is necessary for f to be an injection.[3] We continue in this fashion until we have made choices for $f(1), f(2), f(3), \ldots, f(m)$, where all of the values of f are distinct so that f is an injection. In particular, m-thly, there are $n-m+1$ choices for $f(m)$. The total

[3] This is *independent* of which choice you made for $f(1)$, and similarly for the subsequent choices.

number of different sets of m choices, with each set of choices corresponding to a different injection, is equal to the product of m integers:

$$n(n-1)(n-2)\cdots(n-m+1). \tag{7.113}$$

□

Table 7.5.1. Counting the number of injections from $\{1,2,3\}$ to $\{a,b,c,d\}$. There are 4 choices for $f(1)$, there are 3 choices for $f(2)$, and there are 2 choices for $f(3)$. In total, there are $24 = 4 \cdot 3 \cdot 2$ injections. The shaded set of choices represents the function f defined by $f(1) = c$, $f(2) = d$, and $f(3) = a$.

1	a						b						c						d					
2	b		c		d		a		c		d		a		b		d		a		b		c	
3	c	d	b	d	b	c	c	d	a	d	a	c	b	d	a	d	a	b	b	c	a	c	a	b

Exercise 7.23. *In this exercise, we give a rigorous proof of Theorem 7.37 by induction.*

(1) *(Base case) Prove Theorem 7.37 in the case where the cardinality of the domain is $m = 1$.*

(2) *(Inductive step) Suppose that $m \in \mathbb{Z}^+$ has the property that the number of injections from $\mathbb{N}_m$ to $\mathbb{N}_n$ is equal to $n(n-1)(n-2)\cdots(n-m+1)$ for all positive integers n. Prove that this formula is true for m replaced by $m+1$.*

Hint: Justify that we have the disjoint union

$$\mathrm{Inj}(\mathbb{N}_{m+1}, \mathbb{N}_n) = \bigcup_{j=1}^{n} I_j, \tag{7.114}$$

where

$$I_j := \{f \in \mathrm{Inj}(\mathbb{N}_{m+1}, \mathbb{N}_n) : f(m+1) = j\}. \tag{7.115}$$

Show that the cardinality of each set I_j is equal to the cardinality of the set $\mathrm{Inj}(\mathbb{N}_m, \mathbb{N}_{n-1})$.

See Figure 7.5.1 for a visualization of the injections from $\mathbb{N}_3$ to $\mathbb{N}_4$.

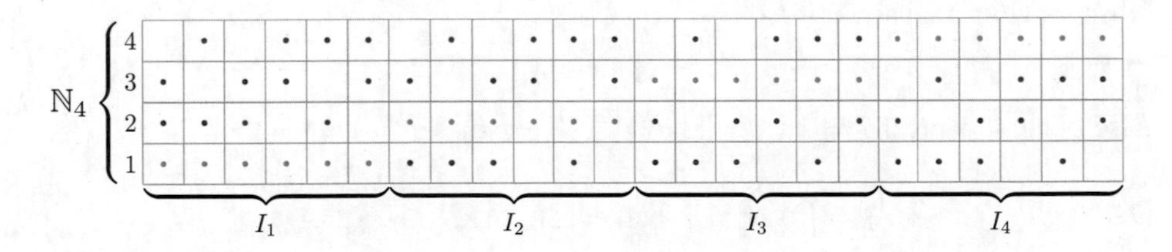

Figure 7.5.1. We visualize injections f from $\mathbb{N}_3$ to $\mathbb{N}_4$ by coloring $f(3)$ green, $f(1)$ red, and $f(2)$ blue.

If the cardinality of the domain and codomain are equal, then we immediately obtain the following consequence of Theorem 7.37.

Corollary 7.38. *The number of injective functions from $\mathbb{N}_n$ to $\mathbb{N}_n$ is equal to*

$$\text{Inj}(\mathbb{N}_n, \mathbb{N}_n) = n! \tag{7.116}$$

Since by Theorem 5.63 the set of injections equals the set of surjections, which in turn equals the set of bijections, this same count holds for the number of surjections and the number of bijections.

Recall that a bijection of a finite set is also called a permutation.

7.5.4. Counting surjective functions. Now let $m \geq n$ be positive integers. Define the subset of surjective functions

$$S := \{f : \mathbb{N}_m \to \mathbb{N}_n \mid f \text{ is a surjection}\}. \tag{7.117}$$

The set of non-surjections N is

$$N := F - S. \tag{7.118}$$

We can think of S as the set of ways one can give n monkeys a total of m numbered bananas, where each monkey gets at least one banana.

We can write N as a union of sets of functions as follows. Define

$$A_i := \{f : \mathbb{N}_m \to \mathbb{N}_n \mid i \notin \text{Im}(f)\} \tag{7.119}$$

for $1 \leq i \leq n$. Then

$$N = \bigcup_{i=1}^{n} A_i. \tag{7.120}$$

Indeed, if $f \in A_i$ for some i, then f is not a surjection, so $f \in N$. On the other hand, if $f \in N$, then f is not a surjection, so that there exists an i such that $i \notin \text{Im}(f)$; that is, $f \in A_i$.

Since all of the sets we considered above are subsets of the finite set $\text{Fun}(X, Y)$, these subsets are finite sets. Therefore we can use the inclusion-exclusion principle to compute the cardinality of N.

Given a non-empty subset I of $\mathbb{N}_n$, define

$$A_I := \{f : \mathbb{N}_m \to \mathbb{N}_n \mid I \cap \text{Im}(f) = \emptyset\}. \tag{7.121}$$

Observe that for each i,

$$A_{\{i\}} = A_i. \tag{7.122}$$

Also observe that if $I = \{i_i, i_2, \ldots, i_k\} \subset \mathbb{N}_n$, then

$$A_I = \bigcap_{i \in I} A_i = \bigcap_{j=1}^{k} A_{i_j} = A_{i_1} \cap A_{i_2} \cap \cdots \cap A_{i_k}. \tag{7.123}$$

Evermore liking to make observations, we further observe that

$$A_I = \text{Fun}(\mathbb{N}_m, \mathbb{N}_n - I). \tag{7.124}$$

Thus,

$$|A_I| = (n - |I|)^m. \tag{7.125}$$

Now we may apply the inclusion-exclusion principle (Theorem 7.14) to obtain that the cardinality of the set of non-surjections is

$$|N| = \left|\bigcup_{i=1}^{n} A_i\right| = \sum_{\emptyset \neq I \subset \mathbb{N}_n} (-1)^{|I|-1}|A_I|. \tag{7.126}$$

By applying (7.125) to this, we have

$$|N| = \sum_{\emptyset \neq I \subset \mathbb{N}_n} (-1)^{|I|-1}(n - |I|)^m. \tag{7.127}$$

Furthermore, given $0 \leq k \leq n$, there are $\binom{n}{k}$ subsets of order k in $\mathbb{N}_n$. That is, there are $\binom{n}{k}$ many I's with $|I| = k$. We conclude that

$$|N| = \sum_{k=1}^{n}(-1)^{k-1}\binom{n}{k}(n-k)^m = \sum_{k=1}^{n-1}(-1)^{k-1}\binom{n}{k}(n-k)^m. \tag{7.128}$$

Thus, the cardinality of the set of surjections is

$$|S| = |F| - |N| = n^m + \sum_{k=1}^{n-1}(-1)^k\binom{n}{k}(n-k)^m. \tag{7.129}$$

Recall that if $m = n$, then a surjection is a bijection. The number of bijections to $\mathbb{N}_n$ to itself is $n!$. Thus, as a special case, we obtain

$$n! = n^n + \sum_{k=1}^{n-1}(-1)^k\binom{n}{k}(n-k)^n. \tag{7.130}$$

Example 7.39. We list the surjections from $\mathbb{N}_3$ to $\mathbb{N}_2$ as follows. Given a surjection f, we list its images $f(1), f(2), f(3)$ in Table 7.5.2, which defines f.

Table 7.5.2. There are six surjections from $\mathbb{N}_3$ to $\mathbb{N}_2$.

$f(1)$	$f(2)$	$f(3)$
1	1	2
1	2	1
1	2	2
2	1	1
2	1	2
2	2	1

This agrees with (7.129), which says that the number of surjections is equal to

$$2^3 - \binom{2}{1}(2-1)^3 = 8 - 2 = 6. \tag{7.131}$$

Let $\{i, j\} = \{a, b\}$ as (unordered) sets. Then the sets defined analogously to (7.119) given by

$$A_i = \{f : \mathbb{N}_3 \to \{a, b\} : f(k) = j \text{ for all } k \in \mathbb{N}_3\} \tag{7.132}$$

Table 7.5.3. The surjections from $\mathbb{N}_3$ to $\{a, b\}$. The shaded cells represent the surjection f defined by $f(1) = b$, $f(2) = a$, and $f(3) = b$.

<table>
<tr><td>1</td><td colspan="3">a</td><td colspan="3">b</td></tr>
<tr><td>2</td><td>a</td><td colspan="2">b</td><td colspan="2">a</td><td>b</td></tr>
<tr><td>3</td><td>b</td><td>a</td><td>b</td><td>a</td><td>b</td><td>a</td></tr>
</table>

are each singleton sets consisting solely of the constant function with image $\{j\}$, for $i \in \{a, b\}$. We count by (7.126) that $|N| = |A_a| + |A_b| = 2$. So, again we see that the set of surjections from $\mathbb{N}_3$ to $\{a, b\}$ has cardinality

$$|S| = |\operatorname{Fun}(\mathbb{N}_3, \{a, b\})| - |N| = 8 - 2 = 6. \tag{7.133}$$

Exercise 7.24. *Similarly to Table* 7.5.2, *list the surjections from* $\mathbb{N}_4$ *to* $\mathbb{N}_2$ *and verify that the number of surjections agrees with formula* (7.129).

7.6. Counting problems

In this section we count the number of ways to sum a fixed number of integers to a given integer.

Example 7.40. Let n be a positive integer. How many integer solutions are there to the equation

$$x + y = n \tag{7.134}$$

with $x \geq 0$ and $y \geq 0$?

The answer to this question is easy in the sense that we can "list" the solutions:

$$(0, n),\ (1, n-1),\ (2, n-2),\ \ldots,\ (n, 0). \tag{7.135}$$

So we see that there are $n + 1$ solutions.

If you are not convinced that there are exactly $n + 1$ solutions, we can make the following more formal argument. Let S be the set of solutions, whose elements are listed above. Note that $S \subset \mathbb{Z}^+ \times \mathbb{Z}^+$. Define the function

$$\pi : S \to \mathbb{Z}^+ \tag{7.136}$$

by

$$\pi(x, y) = x. \tag{7.137}$$

(This function is the restriction of a projection function.) Then π maps S bijectively to the subset $\{0, 1, 2, \ldots, n\}$. The latter set has cardinality $n+1$, and hence so does S.

In terms of counting, we have $n + 1$ choices for x (zero through n, inclusive) and then y is determined by $y = n - x$. So the number of solutions is equal to

$$\binom{n+1}{1} = \binom{n+1}{n}. \tag{7.138}$$

We can also think of the question in the following tricky way. Imagine that we have n ping pong balls lined up in a row. To describe the solution $x = 3$ and $y = n - 3$ (assume $n \geq 3$), we put a ping pong paddle as a divider right after the third ping pong ball. So, in general, a solution (x, y) to $x + y = n$ with $x \geq 0$ and

$y \geq 0$ corresponds to choosing one spot out of $n+1$ to put the ping pong paddle, or equivalently, n spots out of $n+1$ to put the ping pong balls.

Example 7.41. How many triples of non-negative integers (x, y, z) that satisfy the equation

$$x + y + z = 7 \tag{7.139}$$

are there?

We use the same idea as in the previous example. In this case, we can model this question as follows. Suppose that we have 7 Pokémon. Suppose that each Pokémon is one of three types: (1) Xena fans, (2) Yoda fans, and (3) Zeus fans. How many totals (x, y, z) are there in having x Xena fans, y Yoda fans, and z Zeus fans? Each total corresponds to putting two dividers between the 7 Pokémon. For the dividers and Pokémon combined, we have 9. The number of ways of inserting two dividers is $\binom{9}{2}$.

For example, choosing the two dividers at the positions 5 and 7 corresponds to choosing $x = 4$ Xena fans $(1, 2, 3, 4)$ and $y = 1$ Yoda fans (6) and $z = 2$ Zeus fans $(8, 9)$. See Figure 7.6.1.

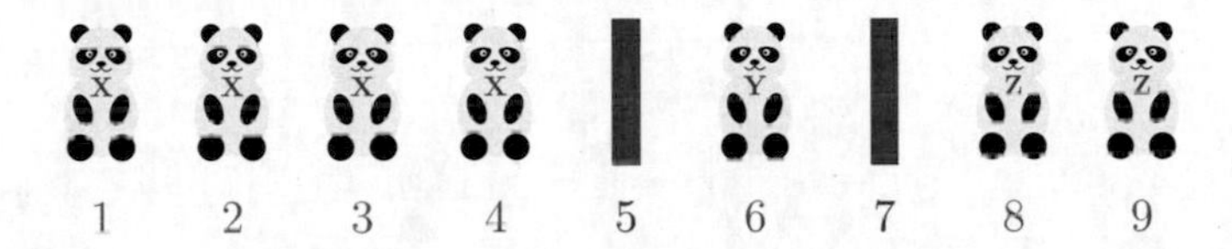

Figure 7.6.1. 4 Xena fans, 1 Yoda fans, and 2 Zeus fans, separated by dividers.

Lemma 7.42. *Let n and k be positive integers. The number of non-negative integer k-tuples $(x_1, x_2, \ldots, x_k)$ satisfying the equation*

$$\sum_{i=1}^{k} x_i = x_1 + x_2 + \cdots + x_k = n \tag{7.140}$$

is equal to

$$\binom{n+k-1}{k-1} = \binom{n+k-1}{n}. \tag{7.141}$$

Proof. The proof we give involves looking at the result in a clever way:

Imagine we have a row of n ones, which of course sum up to n. Add to this row of ones $k-1$ dividers, which we can denote by vertical lines. Then, in total, we have a row of $n+k-1$ ones and dividers altogether.

Claim. Sequences of n ones and $k-1$ dividers are in one-to-one correspondence with non-negative integer solutions $(x_1, x_2, \ldots, x_k)$ to (7.140).

To see the claim, we describe a bijection and its inverse between the set of sequences and the set of solutions.

Given a solution $(x_1, x_2, \ldots, x_k)$ as above, we put the first divider in the $x_1 + 1$ spot. For example, if $x_1 = 0$, then we put the first divider in the first spot.

We then put the second divider in the $x_1 + x_2 + 2$ spot. In this way we will have x_1 ones, the first divider, x_2 ones, the second divider, in that order.

Continuing this way, we see that we should put the $(k-1)$-st divider in the $x_1 + x_2 + \cdots + x_{k-1} + k - 1$ spot. We have defined a sequence of n ones and $k-1$ dividers.

Why $k-1$ dividers? Think of $k-1$ walls separating k people in a row. For example, 1 wall separates 2 people.

We now define a function which is the inverse of the function described above.

Suppose that we are given a sequence of n ones and $k-1$ dividers. Let the spots of the $k-1$ dividers be

$$y_1 < y_2 < \cdots < y_{k-1}. \tag{7.142}$$

We define

$$x_1 := y_1 - 1,$$
$$x_2 := y_2 - x_1 - 2,$$

and in general

$$x_i := y_i - x_1 - x_2 - \cdots - x_{i-1} - i \quad \text{for } 1 \leq i \leq k-1, \tag{7.143}$$

and

$$x_k = n - x_1 - x_2 - \cdots - x_{k-1}. \tag{7.144}$$

Then the function that this describes is the inverse of the previously described function. We leave this fact as an exercise. This completes the proof of the claim.

Now the claim implies that the set of sequences of ones and dividers and the set of non-negative integer solutions have the same cardinality. For the former set, each sequence corresponds to a choice of an n-element subset of $\mathbb{N}_{n+k-1}$ to put the ones. The number of such subsets is $\binom{n+k-1}{n}$. Equivalently (or should we say complementarily?), each sequence corresponds to a choice of a $(k-1)$-element subset of $\mathbb{N}_{n+k-1}$ to put the dividers. The number of such subsets is $\binom{n+k-1}{k-1}$. Of course, by Lemma 7.21 the two binomials we arrived at are equal. □

Example 7.43. For $k = 3$ and $n = 3$ the answer is $\binom{5}{2} = 10$. The list of 3-tuples summing to 3, in "increasing order", is

$$(0,0,3),\ (0,1,2),\ (0,2,1),\ (0,3,0),\ (1,0,2),$$
$$(1,1,1),\ (1,2,0),\ (2,0,1),\ (2,1,0),\ (3,0,0).$$

If we choose to think in terms of rows of ones and dividers, as in the proof of Lemma 7.42, then the list (in the same order) looks like this:

$$||111,\quad |1|11,\quad |11|1,\quad |111|,\quad 1||11,$$
$$1|1|1,\quad 1|11|,\quad 11||1,\quad 11|1|,\quad 111||.$$

7.7*. Using the idea of a bijection

We end this chapter with a fun application of using bijections of finite sets to find a strategy to win a game.

7.7.1. Who can sum to 15?

7.7.1.1. *The rules of the game.* Consider the following game you can play with your friend. The two of you take turns choosing from the integers 1 through 9, where once an integer is chosen, it is no longer available. The goal is to have three integers among all of the integers you have chosen to sum to the number 15 The game stops once one of you has three integers summing to 15 or when there are no longer any of the nine integers left.

1 2 3 4 5 6 7 8 9

Figure 7.7.1. The integers 1 through 9.

7.7.1.2. *A sample game.* For example, your friend starts by choosing the number 2. You then choose 5. Your friend chooses 8. You choose 6. The current situation looks like this:

Your friend's integers: 2, 8.
Your integers: 5, 6.
Integers remaining: 1, 3, 4, 7, 9.

It is your friend's turn. What should your friend do?

Your friend calculates $15 - 2 - 8 = 5$ and wishes they could choose the number 5 to win. But, alas, the 5 is safely in your possession.

Next, your friend looks at the situation from your point of view and calculates $15 - 5 - 6 = 4$, and they realize that you will win if you are able to choose the 4 next.

So your friend eagerly snatches up the 4 for their next choice. So now your friend has 2, 4, 8. They must like powers of 2!

Now it is your turn and you must decide what integer to choose among the remaining 1, 3, 7, 9. You know you cannot win on this choice as the 4 is gone. What do you do?

You decide that you are going to block your friend from summing to 15 on their next choice. Firstly, since your friend has 2 and 4, if they get 9, you are lost. Secondly, since your friend also has 4 and 8, if they get 3, you are also lost. Unfortunately, both 3 and 9 are available and you cannot choose both of them at once. So no matter what integer you choose, you will lose if your friend makes the correct subsequent choice!

7.7.1.3. *What are the ways to win?* This is easy to see and we list the ways below in lexicographical order. There are eight ways to choose single-digit integers to sum to 15:

1, 5, 9
1, 6, 8
2, 4, 9
2, 5, 8
2, 6, 7
3, 4, 8
3, 5, 7
4, 5, 6

7.7.1.4. *How do we win the game, or can we win the game?* One way of analyzing the game is to place the nine integers 1 through 9 on a tic-tac-toe board as in Figure 7.7.1. Observe that there are 3 rows, 3 columns, and 2 diagonals (totalling 8) and that if you sum the 3 integers in any row or in any column or in any diagonal, the result is 15. So winning at tic-tac-toe corresponds exactly to summing three integers to 15.

Now, with correct play, tic-tac-toe always ends in a draw. We will leave it to you to do a complete analysis of tic-tac-toe and verify this fact. For example, there are three types of first moves:

(1) The first player puts an "X" in the middle square. Then the second player has to put an "O" in a corner square, or they lose. Etc.
(2) The first player puts an "X" in a corner square. Then the second player has to put an "O" in the middle square, or they lose. Etc.
(3) The first player puts an "X" in a side square. Then the second player has to put an "O" in either the middle square, the opposite side square, an adjacent corner square, or they lose. Etc.

Table 7.7.1. The integers from 1 to 9 on a tic-tac-toe board (3×3 grid).

2	7	6
9	5	1
4	3	8

Exercise 7.25. *Prove that with correct play, tic-tac-toe always ends in a draw. With correct play, what is the outcome of the game of summing to 15?*

7.8. Hints and partial solutions for the exercises

Hint for Exercise 7.1. Suppose that X and Y are disjoint. Use the addition principle to obtain a contradiction.

Hint for Exercise 7.2. Firstly, show that the function f is well-defined; that is, if (x, y) is in the domain, then $f(x, y)$ is in the codomain.

It is rather easy to see that f is a bijection because it is essentially the identity function.

Apply the addition principle for n sets.

Hint for Exercise 7.3. We can compute that

$$\begin{aligned} X \cup Y &= ((X \cap Y) \cup (X \cap Y^c)) \cup ((Y \cap X) \cup (Y \cap X^c)) \\ &= (X \cap Y) \cup (X - Y) \cup (X \cap Y) \cup (Y - X) \\ &= (X - Y) \cup (Y - X) \cup (X \cap Y). \end{aligned}$$

Show that the pairwise intersections of the three sets on the right-hand side are all empty.

Hint for Exercise 7.4. Let A, B, C denote the sets of dinosaurs that eat apples, bananas, and cherries, respectively.

By (1), $a := |A| = 125$, $b := |B| = 131$, and $c := |C| = 136$.

By (2), $ab := |A \cap B| = 81$, $ac := |A \cap C| = 84$, and $bc := |A \cap C| = 88$.

By (3), $abc := |A \cap B \cap C| = 41$.

By the inclusion-exclusion principle, we have that the set of dinosaurs which eat at least one of the three fruits has cardinality

$$|A \cup B \cup C| = |A| + |B| + |C| - |A \cap B| - |A \cap C| - |B \cap C| + |A \cap B \cap C|.$$

Continue this calculation.

Hint for Exercise 7.5. The way we solved the previous exercise solves this exercise.

Hint for Exercise 7.6. The following diagram gives the counts:

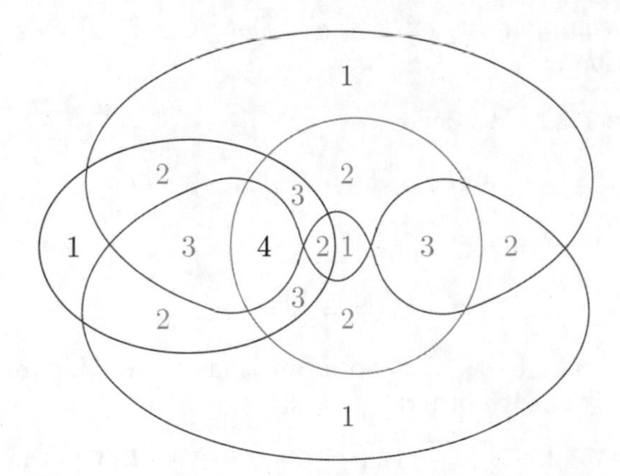

Hint for Exercise 7.7. We prove the $n = 3$ case. Let $Y_1 = X_1 \cap X_3$ and $Y_2 = X_2 \cap X_3$. By the inclusion-exclusion principle for 2 sets,

$$|X_1 \cup X_2 \cup X_3| = |X_1 \cup X_2| + |X_3| - |Y_1 \cup Y_2| \tag{7.145}$$

since $(X_1 \cup X_2) \cap X_3 = Y_1 \cup Y_2$. The result follows by applying the inclusion-exclusion principle for 2 sets to both $|X_1 \cup X_2|$ and $|Y_1 \cup Y_2|$, while observing that $Y_1 \cap Y_2 = X_1 \cap X_2 \cap X_3$.

Hint for Exercise 7.8. **(1)** The empty set $\emptyset$ is a subset of any set.

(2) Suppose $A \subset B$. Let $C \in \mathcal{P}(A)$. Show that $C \in \mathcal{P}(B)$.

(3) Use that $A \cap B \subset A$ and $A \cap B \subset B$.

Hint for Exercise 7.9. The answer is that we must have $(A, B) = (E, O)$. Use that $A \cup B = \mathbb{Z}$ and $B \subset O$ to prove that $E \subset A$. Similarly, prove that $O \subset B$. See Figure 7.8.1.

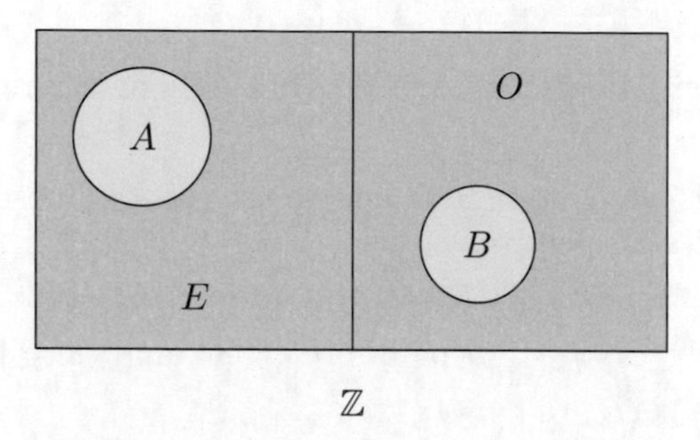

Figure 7.8.1. This Venn diagram shows that $\mathbb{Z}$ is the disjoint union of E and O. Using that $\mathbb{Z}$ is also the disjoint union of A and B, we deduce that $A = E$ and $B = O$; that is, the red shaded parts contain no elements.

Hint for Exercise 7.10. Define the "complement function" $c : X \to Y$ by

$$c(A) = A^c. \tag{7.146}$$

Use the addition principle to show that this function is well-defined.

Show that the complement function $c' : Y \to X$ defined by $c'(B) = B^c$ is the inverse of the function c.

Hint for Exercise 7.11. We define

$$H : S_{\text{Out}} \to \mathcal{P}_r(\mathbb{N}_{n-1})$$

by $H(C) := C$, and we define

$$K : \mathcal{P}_r(\mathbb{N}_{n-1}) \to S_{\text{Out}}$$

by $K(D) := D$. It is easy to see that both functions H and K are well-defined and that they are inverses of each other.

Hint for Exercise 7.12. (1) Show that $f(C) \in P(X) \times \mathcal{P}(Y)$. Denote $i := |C \cap X|$, so that $|C \cap X| \in \mathcal{P}_i(X)$. Show that $|C \cap Y| = k - i$, so that $C \in \mathcal{P}_{k-i}(Y)$. This implies that the function f is well-defined.

Let $(A, B) \in \bigcup_{i=0}^{k} \mathcal{P}_i(X) \times \mathcal{P}_{k-i}(Y)$. Then there exists $1 \leq i \leq k$ such that $A \in \mathcal{P}_i(X)$ and $B \in \mathcal{P}_{k-i}(Y)$. Show that $g(A, B) \subset X \cup Y$ and $|g(A, B)| = k$. This implies that the function g is well-defined.

(2) Show that $g(f(C)) = C$ and that $f(g(A, B)) = (A, B)$, using that X and Y are disjoint.

(3) Since $|X| = m$, $|Y| = n$, and $|X \cup Y| = m + n$, we may start a calculation by

$$\binom{m+n}{k} = |\mathcal{P}_k(X \cup Y)| = \left|\bigcup_{i=0}^{k} \mathcal{P}_i(X) \times \mathcal{P}_{k-i}(Y)\right|.$$

Continue.

Hint for Exercise 7.13. If $n < 0$, $r < 0$, or $r > n$, then each term in (7.45) is zero. We also check that (7.45) holds for $r = 0$.

Hint for Exercise 7.14. One starts the calculation by

$$\begin{aligned} f_{n+3} &= f_{n+1} + f_{n+2} \\ &= \sum_{j+k=n} \binom{k}{j} + \sum_{j+k=n+1} \binom{k}{j} \\ &= \sum_{j+k=n} \left(\binom{k}{j} + \binom{k}{j+1}\right), \end{aligned}$$

and one uses the previous exercise.

Hint for Exercise 7.15. Use Theorem 7.24.

Hint for Exercise 7.16. Use Theorem 7.24 and properties about division by primes.

Hint for Exercise 7.17. Use the previous exercise.

Hint for Exercise 7.18. Use the Binomial Theorem.

Hint for Exercise 7.19. Use part (1) for the inductive step in part (2).

Hint for Exercise 7.20. By the Binomial Theorem, we have that $\binom{k+\ell}{m}$ is the coefficient of x^m in the binomial expansion of $(1+x)^{k+\ell}$, for $0 \leq m \leq k + \ell$. On the other hand, by the Binomial Theorem, we have that

$$\begin{aligned} (1+x)^k \cdot (1+x)^\ell &= \sum_{i=0}^{k} \binom{k}{i} x^i \cdot \sum_{j=0}^{\ell} \binom{\ell}{j} x^j \\ &= \sum_{i=0}^{k} \sum_{j=0}^{\ell} \binom{k}{i} \binom{\ell}{j} x^{i+j}. \end{aligned}$$

Now, in more generality, suppose that we have a sum of the form

$$\sum_{i=0}^{k} \sum_{j=0}^{\ell} a(i,j). \tag{7.147}$$

We claim that we may rewrite this as the sum

$$\sum_{m=0}^{k+\ell}\sum_{i=0}^{m} a(i,m-i). \tag{7.148}$$

To see this, let $N_k := \{0,1,2,\dots,k\}$ for $k \in \mathbb{Z}^{\geq}$. Then

$$\sum_{i=0}^{k}\sum_{j=0}^{\ell} a(i,j) = \sum_{(i,j)\in N_k\times N_\ell} a(i,j). \tag{7.149}$$

Here, the sum over a finite set (such as $N_k \times N_\ell$) is well-defined since addition is commutative. We can partition the (index) set $N_k \times N_\ell$ as

$$N_k \times N_\ell = \bigcup_{m=0}^{k+\ell} I_m, \tag{7.150}$$

where

$$I_m := \{(i,j) \in N_k \times N_\ell : i+j=m\}. \tag{7.151}$$

Thus,

$$\begin{aligned}\sum_{(i,j)\in N_k\times N_\ell} a(i,j) &= \sum_{m=0}^{k+\ell}\sum_{i\in I_m} a(i,j)\\ &= \sum_{m=0}^{k+\ell}\left(\sum_{i=0}^{m} a(i,m-i)\right)\end{aligned}$$

since $(i,j) \in I_m$ means that $j = m-i$, where $0 \leq i \leq m$. This proves the claim.

Now, applying the claim yields

$$\begin{aligned}\sum_{i=0}^{k}\sum_{j=0}^{\ell}\binom{k}{i}\binom{\ell}{j}x^{i+j} &= \sum_{m=0}^{k+\ell}\sum_{i=0}^{m}\binom{k}{i}\binom{\ell}{m-i}x^{i+(m-i)}\\ &= \sum_{m=0}^{k+\ell}\sum_{i=0}^{m}\binom{k}{i}\binom{\ell}{m-i}x^{m}\\ &= \sum_{m=0}^{k+\ell}\sum_{i=\max\{m-\ell,0\}}^{\min\{k,m\}}\binom{k}{i}\binom{\ell}{m-i}x^{m},\end{aligned}$$

where for the last equality we used that

$$\binom{k}{i} = 0 \text{ for } i > k, \quad \binom{\ell}{m-i} = 0 \text{ for } i < m - \ell. \tag{7.152}$$

By equating the coefficients of x^m, this proves formula (7.72).

Hint for Exercise 7.21. (1) We compute that

$$B(C(y_1, y_2, \dots, y_k)) = B(g),$$

where the function g is defined by $g(x_i) = y_i$ for $1 \leq i \leq k$. Show that

$$B(C(y_1, y_2, \dots, y_k)) = (y_1, y_2, \dots, y_k).$$

(2) Show that

$$C(B(f)) = h,$$

where h is the function defined by

$$h(x_i) = f(x_i). \tag{7.153}$$

Hint for Exercise 7.22. We show that $\Phi \circ \Psi$ is the identity and leave it to you to show that $\Psi \circ \Phi$ is the identity. We calculate that

$$\Phi(\Psi(A)) = \Phi(\chi_A) = \chi_A^{-1}(1) = A. \tag{7.154}$$

Hint for Exercise 7.23.

Inductive step: Suppose, as our inductive hypothesis, that $m \in \mathbb{Z}^+$ has the property that the number of injections from $\mathbb{N}_m$ to $\mathbb{N}_n$ is equal to $n(n-1)(n-2)\cdots(n-m+1)$ for all positive integers n.

Firstly, show that

$$\operatorname{Inj}(\mathbb{N}_{m+1}, \mathbb{N}_n) = \bigcup_{j=1}^{n} I_j \tag{7.155}$$

is a disjoint union.

Next, show that the cardinality of each set I_j is equal to the cardinality of the set $\operatorname{Inj}(\mathbb{N}_m, \mathbb{N}_{n-1})$.

Let $1 \leq j \leq n$ and let $f \in I_j$. Since f is an injection, we have that

$$f|_{\mathbb{N}_m} : \mathbb{N}_m \to \mathbb{N}_n - \{j\} \cong \mathbb{N}_{n-1} \tag{7.156}$$

is a well-defined injection. Moreover, it is not difficult to see that the function defined by $f \mapsto f|_{\mathbb{N}_m}$ is a bijection from I_j to $\operatorname{Inj}(\mathbb{N}_m, \mathbb{N}_n - \{j\})$. Show that

$$|I_j| = (n-1)(n-2)(n-3)\cdots(n-m).$$

Finally, show that

$$\begin{aligned} |\operatorname{Inj}(\mathbb{N}_{m+1}, \mathbb{N}_n)| &= \sum_{j=1}^{n} |I_j| \\ &= n(n-1)(n-2)\cdots(n-(m+1)+1). \end{aligned}$$

Hint for Exercise 7.24. See Table 7.8.1.

Table 7.8.1. There are fourteen surjections from $\mathbb{N}_4$ to $\mathbb{N}_2$.

$f(1)$	$f(2)$	$f(3)$	$f(4)$
1	1	1	2
1	1	2	1
1	1	2	2
1	2	1	1
1	2	1	2
1	2	2	1
1	2	2	2
2	1	1	1
2	1	1	2
2	1	2	1
2	2	1	1
2	1	2	2
2	2	1	2
2	2	2	1

Hint for Exercise 7.25. With correct play, the game of summing to 15 always ends in a draw.

Chapter 8

Congruence Class Arithmetic, Groups, and Fields

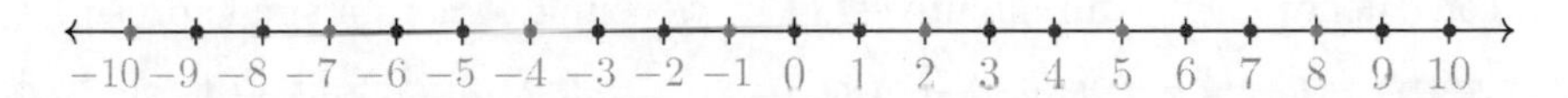

Figure 8.0.1. The congruence class of red numbers is $[0]_3$, the congruence class of blue numbers is $[1]_3$, and the congruence class of green numbers is $[2]_3$. In totality, we get the Notorious R.B.G.; i.e., the union is $R \cup B \cup G = \mathbb{Z}$.

Goals of this chapter: The arithmetic of congruence classes. Some abstract algebra: groups, rings, and fields. The Fundamental Theorem of Algebra. The law of quadratic reciprocity.

8.1. Congruence classes modulo m

The idea of congruence classes is to put in the same basket integers that are congruent. It's marvelous that we can do arithmetic with congruence classes.

8.1.1. The definition of congruence class. Consider looking at integers from the point of view of dividing by 3. We know by the Division Theorem that there are 3 possible remainders: 0, 1, and 2. Two integers are congruent modulo 3 if and only if they have the same remainder when divided by 3. What if we put all of the integers that are congruent to each other in the same basket? How many baskets would we have? For example, all of the integers that have remainder 0 are congruent to each other, and these are the integers in one of the baskets, which we can color red. A second basket consists of the integers with remainder 1, which we can color blue. The third, and final, basket consists of the integers with remainder 2, which we can color green. By the Division Theorem, any integer is in one of these three baskets, and for each basket the integers in it are all congruent to each other. Thus, we have 3 baskets, each of which is a subset of $\mathbb{Z}$. We denote the red, blue, green baskets by $[0]_3$, $[1]_3$, $[2]_3$, respectively. Each of these sets is a *congruence class*, which we now define for every modulus. See Figure 8.0.1.

Definition 8.1. Let m be a positive integer. Given $a \in \mathbb{Z}$, its **congruence class** modulo m, written as $[a]_m$, is the set of all integers congruent to a modulo m; that is,

$$[a]_m = \{x \in \mathbb{Z} \,|\, x \equiv a \mod m\}. \tag{8.1}$$

That is, we group together all of the integers congruent to a fixed integer to form a set, called the congruence class of that integer. For example, the congruence class of -5 (modulo 3) is

$$\begin{aligned}[-5]_3 &= \{x \in \mathbb{Z} : x \equiv -5 \mod 3\} \\ &= \{-5 + 3k : k \in \mathbb{Z}\} \\ &= \{\ldots, -14, -11, -8, -5, -2, 1, 4, 7, 10, 13, \ldots\}.\end{aligned}$$

Each congruence class is an infinite set. But we think of it as a single object!

Solved Problem 8.2. *Show that $[-5]_3 = [1]_3$. That is, the subset $[-5]_3$ of $\mathbb{Z}$ is equal to, as a set, the subset $[1]_3$ of $\mathbb{Z}$.*

Solution. Using $-5 \equiv$ mod3 and the transitivity of congruence (modulo 3), we compute that

$$\begin{aligned}[-5]_3 &= \{x \in \mathbb{Z} : x \equiv -5 \mod 3\} \\ &= \{x \in \mathbb{Z} : x \equiv 1 \mod 3\} \\ &= [1]_3.\end{aligned}$$

In Figure 8.0.1, each of the sets $[-5]_3$ and $[1]_3$ is the set of blue integers. □

We start off with some simple properties.

Theorem 8.3. *Let m be a positive integer.*

(1) *For any integer a, we have $a \in [a]_m$.*

(2) *$x \in [a]_m$ if and only if m divides $x - a$.*

(3) *The congruence class of a is equal to*

$$[a]_m = \{a + km : k \in \mathbb{Z}\}. \tag{8.2}$$

Proof. (1) By the reflexive property of congruence, we have that $a \equiv a \mod m$. Therefore $a \in [a]_m$.

(2) We have that

$$\begin{aligned} x \in [a]_m \quad &\Leftrightarrow \quad x \equiv a \mod m \\ &\Leftrightarrow \quad m \text{ divides } x - a.\end{aligned}$$

(3) Let $x \in [a]_m$. This is equivalent to $x \equiv a \mod m$. This, in turn, is equivalent to m divides $x - a$. This is also equivalent to the fact that there exists $k \in \mathbb{Z}$ such that $x - a = km$; that is, $x = a + km$. This says that $x \in \{a + km : k \in \mathbb{Z}\}$. In conclusion, we have shown that $x \in [a]_m$ is equivalent to $x \in \{a + km : k \in \mathbb{Z}\}$. □

Example 8.4 ($m = 2$). Since we are familiar with even and odd integers, we first consider the case where $m = 2$.

(1) $[0]_2$ *is the set of even integers.* Indeed, $x \equiv 0 \bmod 2$ if and only if $x = 2k$ for some integer k; that is, x is even. Of course, $0 \in [0]_2$.

(2) $[1]_2$ *is the set of odd integers.* Indeed, $x \equiv 1 \bmod 2$ if and only if $x - 1 = 2k$, i.e., $x = 2k + 1$, for some integer k. The result follows from the characterization (4.5b) of odd integers.

Observe the following:

(3) $[a]_2 = [0]_2$ if and only if a is an even integer.

(4) $[a]_2 = [1]_2$ if and only if a is an odd integer.

In words, the congruence class of an integer is equal to the congruence class of zero if and only if the integer is even. The congruence class of an integer is equal to the congruence class of one if and only if the integer is odd.

Proof of (3). ($\Rightarrow$) Suppose that $[a]_2 = [0]_2$. Since, by Theorem 8.3(1), $a \in [a]_2$, we thus have that $a \in [0]_2$, which implies that a is even.

($\Leftarrow$) Suppose that a is an even integer. Then there exists an integer k such that $a = 2k$. Let $x \in [a]_2$. This is equivalent to $x = a + 2p = 2(k + p)$ for some integer p, which is equivalent to x being an even integer; that is, $x \in [0]_2$. Here we used part (1). We have proved $[a]_2 = [0]_2$. □

Exercise 8.1. *Prove Example* 8.4(4); *that is,* $[a]_2 = [1]_2$ *if and only if* a *is an odd integer.*

We can visualize the example above as follows. Let us choose to make even numbers red and to make odd numbers blue. Then, for every even integer a, we have that $[a]_2$ is the set of red integers. Similarly, for every odd integer a, we have that $[a]_2$ is the set of blue integers.

$$\{\ldots, -4 -3, -2, -1, 0, 1, 2, 3, 4, \ldots\}. \tag{8.3}$$

For any positive modulus m, we can perform a similar visualization with m colors. For example, Figure 8.0.1 shows for $m = 3$ the following color coding:

$$\{\ldots, -5, -4 -3, -2, -1, 0, 1, 2, 3, 4, 5, \ldots\}. \tag{8.4}$$

8.1.2. Fundamental properties of congruence classes. Firstly, we understand congruence classes in terms of congruence. To get going, observe the following simple result.

Lemma 8.5. *The congruence class of an integer in a congruence class is equal to that congruence class. In fact, we have the characterization*

$$a \in [b]_m \quad \Leftrightarrow \quad [a]_m = [b]_m\,. \tag{8.5}$$

Proof. ($\Leftarrow$) follows from $a \in [a]_m$.

($\Rightarrow$) Suppose $a \in [b]_m$. Then there exists an integer j such that $a = b + jm$.

Thus

(8.6) $[a]_m = \{a+km : k \in \mathbb{Z}\} = \{b+(j+k)m : k \in \mathbb{Z}\} = \{b+\ell m : \ell \in \mathbb{Z}\} = [b]_m$.

□

For example,

$$a \in [-2]_3 \quad \Leftrightarrow \quad [a]_3 = [-2]_3 .$$

The equivalent conditions in the display are each equivalent to $a \equiv 1 \bmod 3$.

Recall from (4.11) that the set of remainders modulo m is

$$R_m = \{0, 1, 2, 3, \ldots, m-1\}, \tag{8.7}$$

and we have the remainder function $\mathbf{r} : \mathbb{Z} \to R_m$ defined by

$$\mathbf{r}(a) = r, \tag{8.8}$$

where r is the unique remainder given by the Division Theorem.

Theorem 8.6.

(1) *Two integers having equal congruence classes is equivalent to them being congruent:*

$$a \equiv b \bmod m \quad \Leftrightarrow \quad [a]_m = [b]_m \quad \Leftrightarrow \quad [a]_m \cap [b]_m \neq \emptyset. \tag{8.9}$$

Recall that two integers are congruent if and only if they have the same remainder.

(2) *Equivalently, two integers having unequal congruence classes is equivalent to them being incongruent:*

$$a \not\equiv b \bmod m \quad \Leftrightarrow \quad [a]_m \neq [b]_m \quad \Leftrightarrow \quad [a]_m \cap [b]_m = \emptyset.$$

Two integers are incongruent if and only if they have different remainders.

(3) *We have*

$$[a]_m = [\mathbf{r}(a)]_m \tag{8.10}$$

for $a \in \mathbb{Z}$, and the union

$$\mathbb{Z} = [0]_m \cup [1]_m \cup [2]_m \cup \cdots \cup [m-1]_m \tag{8.11}$$

is a disjoint union. That is, the collection of congruence classes partitions *the set of integers. See §8.4 below for the general definition of partition.*

Proof. (1) Let a and b be integers. Then

$$\begin{aligned} a \equiv b \bmod m \quad &\Leftrightarrow \quad m \text{ divides } a - b \\ &\Leftrightarrow \quad a \in [b]_m \\ &\Leftrightarrow \quad [a]_m \cap [b]_m \neq \emptyset. \end{aligned}$$

Moreover,

$$\begin{aligned}[a]_m \cap [b]_m \neq \emptyset &\Rightarrow \exists k, \ell \in \mathbb{Z},\ a + km = b + \ell m \\ &\Rightarrow \exists k, \ell \in \mathbb{Z},\ a = b + (\ell - k)m \\ &\Rightarrow a \in [b]_m \\ &\Leftrightarrow [a]_m = [b]_m\,,\end{aligned}$$

where the last $\Leftrightarrow$ follows from Lemma 8.5.

Part (2) is evidently equivalent to part (1) since any biconditional $P \Leftrightarrow Q$ is equivalent to the biconditional not $P \Leftrightarrow$ not Q.

(3) Let $a \in \mathbb{Z}$. Then $a \equiv \mathbf{r}(a) \bmod m$ by (6.13). Hence, part (1) implies that $a \in [a]_m = [\mathbf{r}(a)]_m$. Since $\mathbf{r}(a) \in \mathbb{Z}_m$, this proves

$$\mathbb{Z} = [0]_m \cup [1]_m \cup [2]_m \cup \cdots \cup [m-1]_m\,. \tag{8.12}$$

Now, if $b, c \in \mathbb{Z}_m$ satisfy $b \neq c$, then $b \not\equiv c \bmod m$ by Lemma 6.4, which implies that $[b]_m \cap [c]_m = \emptyset$ by part (2). This shows that the union in (8.12) is a disjoint union. □

Example 8.7 ($m = 3$). Revisiting the mod 3 discussion at the beginning of this chapter, we have by Theorem 8.6(1) that

$$[a]_3 = [b]_3 \Leftrightarrow a \equiv b \bmod 3 \Leftrightarrow a,\ b \text{ have the same remainder after dividing by 3.}$$

So, $R := [0]_3$ is the set of all multiples of 3, $B := [1]_3$ is the set of all multiples of 3 plus 1, and $G := [2]_3$ is the set of all multiples of 3 plus 2. In other words, the red integers have remainder 0, the blue integers have remainder 1, and the green integers have remainder 2. See Figure 8.1.1 for a visualization of these congruence classes. We have, for example,

$$16 \in [1]_3, \quad -7 \in [2]_3.$$

We also have that $\mathbb{Z} = R \cup B \cup G$ is a disjoint union (as in (8.11)).

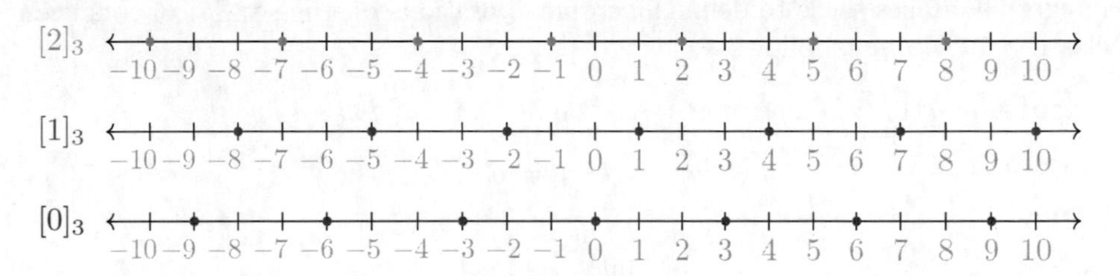

Figure 8.1.1. Visualizing Notorious R.B.G. as a partition of $\mathbb{Z}$. Compare with Figure 8.0.1.

Exercise 8.2. *Let S be the set of all integers n with the property that*

$$\mathbb{Z} = [n]_3 \cup [16]_3 \cup [-16]_3. \tag{8.13}$$

Find this set S.

Exercise 8.3. *Find a positive integer m with the property that*

$$[0]_3 \cap E = [0]_m, \tag{8.14}$$

where E is the set of even integers. Prove your answer.

8.1.3. The set of congruence classes.

By Theorem 8.6(3), the **set of congruence classes** is

$$\mathbb{Z}_m := \{[0]_m, [1]_m, [2]_m, \ldots, [m-1]_m\}. \tag{8.15}$$

This set has m elements. Each *element* of $\mathbb{Z}_m$ is in and of itself a *subset* of $\mathbb{Z}$.

Example 8.8. Let m be a positive integer. The set

$$[4]_m = \{\ldots, 4-2m, 4-m, 4, 4+m, 4+2m, \ldots\}$$

itself comprises just one *element* of $\mathbb{Z}_m$. However, *as a set*, $[4]_m \subset \mathbb{Z}$ has infinitely many elements.

Observe that there is a (natural) bijection between the set of congruence classes $\mathbb{Z}_m$ and the set of remainders R_m.

Lemma 8.9. *The function $f : R_m \to \mathbb{Z}_m$, defined by*

$$f(r) = [r]_m \quad \text{for } 0 \le r < m, \tag{8.16}$$

is a bijection.

Exercise 8.4. *Prove Lemma 8.9.*

Exercise 8.5. *Prove that for every integer b,*

$$\mathbb{Z}_3 = \{[b]_3, [b+1]_3, [b+2]_3\}. \tag{8.17}$$

Can you generalize this mod 3 *statement to a* mod m *statement for every positive integer m?*

8.1.4. Addition and multiplication of congruence classes.

Let m be a positive integer. It makes sense to define algebraic operations on the set $\mathbb{Z}_m$ of congruence classes modulo m as follows.

Theorem 8.10. *The operations $+$, $-$, and $\cdot$ on $\mathbb{Z}_m$ defined by*

$$[a]_m + [b]_m := [a+b]_m, \tag{8.18a}$$

$$[a]_m - [b]_m := [a-b]_m, \tag{8.18b}$$

$$[a]_m \cdot [b]_m := [a \cdot b]_m \tag{8.18c}$$

are well-defined. So we can add, subtract, and multiply congruence classes.

Proof. So what is the issue? A congruence class is a set, and there are an infinite number of ways to write a congruence class. Namely, consider the congruence class $[a]_m$ of an integer a. This congruence class is equal to the congruence class $[a+mk]_m$ for each integer k. Since there are an infinite number of integers, there are therefore an infinite number of ways to write the congruence class $[a]_m$. So, when we define congruence class addition by (8.18a), we have to make sure that the definition makes sense, that is, is independent of the way we write the congruence class.

Suppose that $[a]_m = [a']_m$ and $[b]_m = [b']_m$. We then have that $a' = a + mk$ and $b' = b + m\ell$ for some integers k and ℓ.

(a) Thus, firstly, we have

$$\begin{aligned}[a' + b']_m &= [(a + mk) + (b + m\ell)]_m \\ &= [a + b + m(k + \ell)]_m \\ &= [a + b]_m,\end{aligned}$$

where we used Theorem 8.6(1) to obtain the last equality. This proves that the addition of congruence classes is well-defined.

(b) Secondly, we leave the proof that the subtraction of congruence classes is well-defined as an exercise.

(c) Thirdly, we have

$$\begin{aligned}[a' \cdot b']_m &= [(a + mk) \cdot (b + m\ell)]_m \\ &= [a \cdot b + m(a\ell + bk + mk\ell)]_m \\ &= [a \cdot b]_m.\end{aligned}$$

This proves that the multiplication of congruence classes is well-defined. □

Figure 8.1.2. The modulo m congruence classes George and Martha.

To recap the idea of adding congruence classes, suppose that George and Martha are congruence classes[1]; i.e., George, Martha $\in \mathbb{Z}_m$. How do we add them to get a congruence class George + Martha $\in \mathbb{Z}_m$? Since George, Martha $\in \mathbb{Z}_m$, there exist integers G and M such that

$$\text{George} = [G]_m \quad \text{and} \quad \text{Martha} = [M]_m. \tag{8.19}$$

We define

$$\text{George} + \text{Martha} = [G + M]_m. \tag{8.20}$$

Here is the catch. Given George $\in \mathbb{Z}_m$, there is not a unique integer G such that George $= [G]_m$. For example, George $= [G + m]_m = [G + 2m]_m$ also. In fact, there are an infinite number of integers that represent George. The same goes for Martha. So, for addition of congruence classes to be well-defined, the question we need to answer is this:

If G and G' both represent George and if M and M' both represent Martha, do $G + M$ and $G' + M'$ both represent the same congruence class? If so, then we can define George + Martha to be this congruence class.

The affirmative answer is provided by Theorem 8.10.

[1] In the series of children's books by James Marshall, George and Martha are hippopotami.

Using the Notorious R.B.G. (red, blue, green) color coding, the addition and multiplication tables for $\mathbb{Z}_3$ are

$+_3$	R	B	G
R	R	B	G
B	B	G	R
G	G	R	B

and

$\cdot_3$	R	B	G
R	R	R	R
B	R	B	G
G	R	G	B

The addition and multiplication tables for $\mathbb{Z}_5$ are given by Figure ??. Observe that the multiplication tables for $(R_5, \times_5)$ in Example 6.10 and for $(\mathbb{Z}_5, \cdot)$ in Figure ?? are equivalent.

$+$	$[0]_5$	$[1]_5$	$[2]_5$	$[3]_5$	$[4]_5$
$[0]_5$	$[0]_5$	$[1]_5$	$[2]_5$	$[3]_5$	$[4]_5$
$[1]_5$	$[1]_5$	$[2]_5$	$[3]_5$	$[4]_5$	$[0]_5$
$[2]_5$	$[2]_5$	$[3]_5$	$[4]_5$	$[0]_5$	$[1]_5$
$[3]_5$	$[3]_5$	$[4]_5$	$[0]_5$	$[1]_5$	$[2]_5$
$[4]_5$	$[4]_5$	$[0]_5$	$[1]_5$	$[2]_5$	$[3]_5$

$\cdot$	$[0]_5$	$[1]_5$	$[2]_5$	$[3]_5$	$[4]_5$
$[0]_5$	$[0]_5$	$[0]_5$	$[0]_5$	$[0]_5$	$[0]_5$
$[1]_5$	$[0]_5$	$[1]_5$	$[2]_5$	$[3]_5$	$[4]_5$
$[2]_5$	$[0]_5$	$[2]_5$	$[4]_5$	$[1]_5$	$[3]_5$
$[3]_5$	$[0]_5$	$[3]_5$	$[1]_5$	$[4]_5$	$[2]_5$
$[4]_5$	$[0]_5$	$[4]_5$	$[3]_5$	$[2]_5$	$[1]_5$

Figure 8.1.3. Addition and multiplication tables for $\mathbb{Z}_5$.

Figure 8.1.4 visualizes a multiplication table. What is the modulus?

Exercise 8.6. *Prove: If $a, b \in \mathbb{Z}$ are such that $[a]_m = [b]_m$, then* $\gcd(a, m) = \gcd(b, m)$.

In particular, if $a, b \in \mathbb{Z}$ are such that $[a]_m = [b]_m$, then $\gcd(a, m) = 1$ *if and only if* $\gcd(b, m) = 1$.

Exercise 8.7. *Prove that $[a]_m \cdot [b]_m = [1]_m$ implies that* $\gcd(a, m) = \gcd(b, m) = 1$. *Hint: Apply Theorem* 6.16.

Exercise 8.8. *Find all pairs $([a]_{24}, [b]_{24})$ such that*

$$[a]_{24} \cdot [b]_{24} = [0]_{24} \quad \text{and} \quad 0 < a \le b < 24. \tag{8.21}$$

Hint: There are 27 pairs!

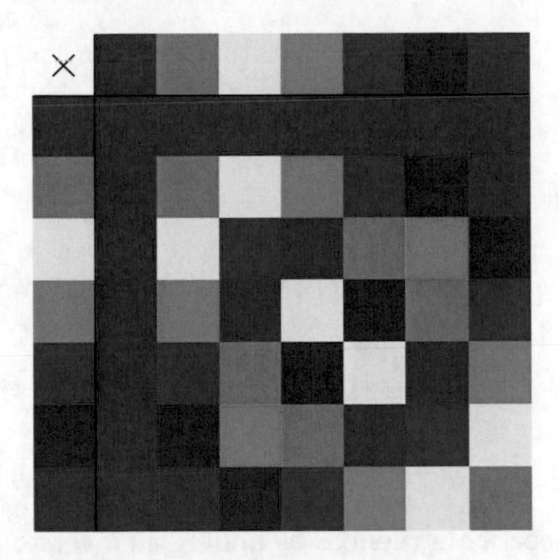

Figure 8.1.4. Explain why this colorful diagram represents a multiplication table rather than a chessboard. ☺

8.2. Inverses of congruence classes

Recall from Definition 0.15 that integers a and b are *inverses* of each other modulo m if and only if

$$ab \equiv 1 \bmod m. \tag{8.22}$$

In this case, by definition of the multiplication of congruence classes, we have

$$[a]_m \cdot [b]_m = [1]_m.$$

This motivates:

Definition 8.11. If $[a]_m \cdot [b]_m = [1]_m$, then we say that $[a]_m$ is **invertible** and that $[b]_m$ is the **inverse** of $[a]_m$.

Example 8.12. (1) $[1]_7$ is its own inverse.

(2) $[2]_7$ and $[4]_7$ are inverses of each other since $2 \cdot 4 = 8 = 7 + 1$.

(3) $[3]_7$ and $[5]_7$ are inverses of each other since $3 \cdot 5 = 15 = 2 \cdot 7 + 1$.

(4) $[6]_7$ is its own inverse since $6 \cdot 6 = 36 = 5 \cdot 7 + 1$. (It is easier to see this from $(-1)(-1) = 1$; why?)

Example 8.13. $[3]_{12}$ does not have an inverse. Neither does $[6]_{14}$.

Exercise 8.9. *Show that* $[a]_m$ *has an inverse if and only if* $\gcd(a, m) = 1$. *Hint: Apply Theorem* 6.16. *See also Exercise* 8.7.

Lemma 8.14. *If a congruence class* $[a]_m$ *has an inverse, then its inverse is unique.*

Proof. Suppose that $[b]_m$ and $[c]_m$ are both inverses of $[a]_m$. Then $ab \equiv 1 \bmod m$ and $ac \equiv 1 \bmod m$. This implies that $a(b - c) \equiv 0 \bmod m$; that is,

$$m \mid a(b - c).$$

On the other hand, $ab \equiv 1 \bmod m$ implies that $\gcd(a, m) = 1$ (exercise). Thus, by Theorem 4.23, we have that

$$m \mid (b - c).$$

This proves that $[b]_m = [c]_m$. □

Remark 8.15. Assuming that you have solved it already, Exercise 6.17 gives a quicker proof of the lemma. Conversely, the proof of the lemma gives a solution to Exercise 6.17.

Exercise 8.10. (1) *For each congruence class* $[a]_8$ (mod 8) *that has an inverse, find its inverse.*

(2) *For each congruence class* $[a]_9$ (mod 9) *that has an inverse, find its inverse.*

8.2.1. Solving linear congruences by finding an inverse. We find it convenient to denote the inverse of $[a]_m$ by $[a^{-1}]_m$, as long as we don't confuse a^{-1} with the usual multiplicative inverse (which is not an integer unless $a = \pm 1$).

Suppose that $[a]_m$ has an inverse $[a^{-1}]_m$. Then $aa^{-1} \equiv 1 \bmod m$. Hence,

$$x = a^{-1}b$$

is a solution to the linear congruence equation

$$ax \equiv b \bmod m. \tag{8.23}$$

Indeed, we have that

$$ax \equiv aa^{-1}b \equiv b \bmod m.$$

This solution is unique modulo m (see below). So the set of solutions to (8.23) is

$$[a^{-1}b]_m = \{a^{-1}b + km : k \in \mathbb{Z}\}. \tag{8.24}$$

Now we show that the solution to (8.23) is unique modulo m: If $ax \equiv b$ mod m and $ay \equiv b \bmod m$, then $a(x - y) \equiv 0 \bmod m$. Since $\gcd(a, m) = 1$, we conclude that $x - y \equiv 0 \bmod m$.

Remark 8.16. Note that what this means for congruence classes is: If a and m are coprime, then for every $b \in \mathbb{Z}$ there exists a unique solution $[x]_m \in \mathbb{Z}_m$ to

$$[a]_m \cdot [x]_m = [b]_m .$$

Solved Problem 8.17. *Prove that* $[67]_{131}$ *is invertible, and find its inverse. Use this to solve the linear congruence*

$$67x \equiv 3 \bmod 131.$$

Solution. By the Euclidean algorithm,

$$\begin{aligned} 131 &= 67 \cdot 1 + 64, \\ 67 &= 64 \cdot 1 + 3, \\ 64 &= 3 \cdot 21 + 1, \\ 3 &= 1 \cdot 3 + 0. \end{aligned}$$

Hence $\gcd(67, 131) = 1$. By reversing the Euclidean algorithm, we get

$$\begin{aligned} 1 &= 64 - 3 \cdot 21 \\ &= (-21)67 + (22)64 \\ &= (22)131 + (-43)67. \end{aligned}$$

Therefore, $[-43]_{131} = [88]_{131}$ is the unique inverse of $[67]_{131}$. This implies that $x = -43$ is a solution to $67x \equiv 1 \bmod 131$, so that $x = -129$ is a solution to $67x \equiv 3 \bmod 131$. Thus, the set of all solutions to $67x \equiv 3 \bmod 131$ is

$$[-129]_{131} = \{-129 + 131k : k \in \mathbb{Z}\} = \{2 + 131k : k \in \mathbb{Z}\} = [2]_{131}.$$

Indeed, we check the particular solution

$$67 \cdot 2 = 134 \equiv 3 \bmod 131. \tag{8.25}$$

Since $\gcd(67, 131) = 1$, we confirm that the general solution differs from 2 by a multiple of 131. □

Exercise 8.11. *For each $0 < a < 7$, use inverses to find the general solution to*

$$ax \equiv 3 \bmod 7. \tag{8.26}$$

If you prefer, you may state your answer by giving for each $0 < a < 7$ the set of all solutions to (8.26) as a congruence class.

Solved Problem 8.18. *Suppose that $[a]_m \cdot [b]_m = [c]_m$. Prove that $\gcd(a, m)$ divides c.*

Solution. By the hypothesis, there exists an integer q such that

$$ab - c = mq; \quad \text{that is,} \quad ab - mq = c. \tag{8.27}$$

This implies that $\gcd(a, m)$ divides c. Observe that we also have that $\gcd(b, m)$ divides c. □

8.2.2. The set of invertible congruence classes. We define the set of invertible congruence classes modulo m by

$$\mathbb{Z}_m^* := \{[x]_m \in \mathbb{Z}_m : [x]_m \text{ is invertible}\}. \tag{8.28}$$

By Lemma 8.14, the inverse of each element $\mathbb{Z}_m^*$ is unique. Given $[a]_m \in \mathbb{Z}_m^*$, we denote its unique inverse by $[a^{-1}]_m$. By Exercise 8.9, $[a]_m$ is invertible if and only if $\gcd(a, m) = 1$. Recall from (6.116) that $R_m^* := \{x \in R_n : \gcd(x, m) = 1\}$. We can now prove the following.

Lemma 8.19. *The function $\phi : R_m^* \to \mathbb{Z}_m^*$ defined by $\phi(x) = [x]_m$ is a bijection.*

Proof. Firstly, we observe that if $x \in R_m^*$, then $[x]_m \in \mathbb{Z}_m^*$ since $\gcd(x, m) = 1$ implies that $[x]_m$ is invertible. Hence, ϕ is a well-defined function.

Secondly, it is easy to see that ϕ is a bijection. For suppose $x, y \in R_m*$ satisfy $\phi(x) = \phi(y)$. Then $[x]_m = [y]_m$, which implies $x \equiv y \bmod m$. Since $x, y \in R_m* \subset R_m$, this implies that $x = y$. Hence, ϕ is an injection. To see that ϕ is a surjection, suppose that $[x]_m \in \mathbb{Z}_m^*$. Then $[x]_m = [\mathbf{r}(x)]_m$ and $\mathbf{r}(x) \in R_m$. Since $[x]_m \in \mathbb{Z}_m^*$, we have $\gcd(x, m) = 1$, so that $\gcd(\mathbf{r}(x), m) = 1$, which in turn implies that $\mathbf{r}(x) \in R_m^*$. We now conclude that $\phi(\mathbf{r}(x)) = [\mathbf{r}(x)]_m = [x]_m$. This proves that ϕ is a surjection. □

Example 8.20. If p is a prime, then

$$\mathbb{Z}_p^* = \{[1],[2]_p,\ldots,[p-1]_p\}. \tag{8.29}$$

Solved Problem 8.21. *Write out the multiplication table for* $(\mathbb{Z}_9^*,\cdot)$*, where multiplication* $\cdot$ *is defined by* (8.18c).

Solution. We have

$$\mathbb{Z}_9^* = \{[1]_9,[2]_9,[4]_9,[5]_9,[7]_9,[8]_9\}. \tag{8.30}$$

The multiplication table is

(8.31)

$(\mathbb{Z}_9^*,\cdot)$	$[1]_9$	$[2]_9$	$[4]_9$	$[5]_9$	$[7]_9$	$[8]_9$
$[1]_9$	$[1]_9$	$[2]_9$	$[4]_9$	$[5]_9$	$[7]_9$	$[8]_9$
$[2]_9$	$[2]_9$	$[4]_9$	$[8]_9$	$[1]_9$	$[5]_9$	$[7]_9$
$[4]_9$	$[4]_9$	$[8]_9$	$[7]_9$	$[2]_9$	$[1]_9$	$[5]_9$
$[5]_9$	$[5]_9$	$[1]_9$	$[2]_9$	$[7]_9$	$[8]_9$	$[4]_9$
$[7]_9$	$[7]_9$	$[5]_9$	$[1]_9$	$[8]_9$	$[4]_9$	$[2]_9$
$[8]_9$	$[8]_9$	$[7]_9$	$[5]_9$	$[4]_9$	$[2]_9$	$[1]_9$

As we have implicitly assumed above for $m = 9$, multiplication $\cdot$ on $\mathbb{Z}_m$ restricts to $\mathbb{Z}_m^*$:

Exercise 8.12. *In* (8.18c) *multiplication* $\cdot$ *is defined on* $\mathbb{Z}_m$. *Prove that if* $[a]_m, [b]_m \in \mathbb{Z}_m^*$*, then* $[a]_m \cdot [b]_m \in \mathbb{Z}_m^*$.

8.3. Reprise of the proof of Fermat's Little Theorem

We now reformulate the proof of Fermat's Little Theorem 6.63 in the language of congruence classes.

Let p be a prime and let $a \in \mathbb{Z}^+$ be such that a is not a multiple of p; i.e., p does not divide a.

Let us look at the set $\mathbb{Z}_p$ of congruence classes modulo p. There are exactly p congruence classes and

$$\mathbb{Z}_p = \{[0]_p, [1]_p, \ldots, [p-1]_p\}.$$

We define the function $f_a : \mathbb{Z}_p \to \mathbb{Z}_p$ by

$$f_a([x]_p) = [ax]_p = [a]_p \cdot [x]_p. \tag{8.32}$$

8.3.1. The idea by an example of the proof of Fermat's Little Theorem.

Example 8.22 ($a = 3$ and $p = 5$). Here, we have that

$$f_3 : \mathbb{Z}_5 \to \mathbb{Z}_5$$

is given by

$$\begin{aligned} f_3([0]_5) &= [0]_5 , \\ f_3([1]_5) &= [3]_5 , \\ f_3([2]_5) &= [6]_5 = [1]_5 , \\ f_3([3]_5) &= [9]_5 = [4]_5 , \\ f_3([4]_5) &= [12]_5 = [2]_5 . \end{aligned}$$

So, in order, the function f_3 maps the sequence $[0]_5 , [1]_5 , [2]_5 , [3]_5 , [4]_5$ to the sequence

$$[0]_5 , [3]_5 , [1]_5 , [4]_5 , [2]_5 .$$

Observe that f_3 is a bijection from $\mathbb{Z}_5$ to $\mathbb{Z}_5$ (itself). Further observe that we can read off this sequence from the fourth row of the multiplication table in Figure 8.1.3.

Recall that $3^{5-1} \equiv 3^4 \equiv 81 \equiv 1 \bmod 5$. A robust way of proving this is the following congruence class calculation:

$$\begin{aligned} [4!]_5 &= [1]_5 [2]_5 [3]_5 [4]_5 && \text{definition of multiplying congruence classes} \\ &= [3]_5 [1]_5 [4]_5 [2]_5 && \text{commutativity of multiplying congruence classes} \\ &= [3]_5 [6]_5 [9]_5 [12]_5 && \text{observed equality of congruence classes using } f_3 \\ &= [3 \cdot 6 \cdot 9 \cdot 12]_5 && \text{definition of multiplying congruence classes} \\ &= [3^4 \cdot 1 \cdot 2 \cdot 3 \cdot 4]_5 && \text{factoring out 3's} \\ &= [3^4 \cdot 4!]_5 && \text{definition of factorial.} \end{aligned}$$

So, in terms of congruences, we have

$$4! \equiv 4! \cdot 3^4 \bmod 5.$$

On the other hand, since 5 is prime, we have $\gcd(4!, 5) = 1$ by Corollary 2.18. Hence

$$1 \equiv 3^4 \bmod 5$$

by Corollary 6.21.

8.3.2. Congruence class proof of Fermat's Little Theorem. Now we show the derivation in the previous subsection that $3^4 \equiv 1 \bmod 5$ is robust:

Theorem 8.23 (Fermat's Little Theorem redux). *If p is a prime and if a is a positive integer which is not a multiple of p, then $a^{p-1} \equiv 1 \bmod p$.*

Proof. (1) We claim that

$$f_a : \mathbb{Z}_p \to \mathbb{Z}_p \text{ is a bijection.} \tag{8.33}$$

Since the domain $\mathbb{Z}_p$ has the same number of elements as the codomain $\mathbb{Z}_p$, it suffices to show that f_a is an injection by Theorem 5.63.

Proof that f_a is an injection: Suppose $f_a([x_1]_p) = f_a([x_2]_p)$. Then

$$ax_1 \equiv ax_2 \bmod p.$$

Since a is not a multiple of p, it follows from Proposition 2.17 that $\gcd(a, p) = 1$. Hence, by Corollary 6.21, we have

$$x_1 \equiv x_2 \mod p.$$

This implies $[x_1]_p = [x_2]_p$. We have proved that f_a is an injection.

(2) Since f_a is a bijection and since $f_a([0]_p) = [0]_p$, we have that the set $\{[1]_p, [2]_p, \ldots, [p-1]_p\}$ is equal to the set $\{[1a]_p, [2a]_p, \ldots, [(p-1)a]_p\}$. Therefore we obtain the second equality in the following display (the first equality is by the definition of multiplication of congruence classes):

$$\begin{aligned}[(p-1)!]_p &= [1]_p \cdot [2]_p \cdots [p-1]_p \\ &= [1a]_p \cdot [2a]_p \cdots [(p-1)a]_p .\end{aligned}$$

Now

$$[1a]_p \cdot [2a]_p \cdots [(p-1)a]_p = [(p-1)! \cdot a^{p-1}]_p,$$

so that

$$(p-1)! \equiv (p-1)! \cdot a^{p-1} \mod p.$$

On the other hand, since p is prime, we have $\gcd((p-1)!, p) = 1$ by Corollary 2.18. Hence

$$1 \equiv a^{p-1} \mod p.$$

This completes the congruence class proof of Fermat's Little Theorem. □

8.4. Equivalence relations, equivalence classes, and partitions

In this section we consider the general notion of an equivalence relation. This notion is important in many areas of mathematics.

8.4.1. Relations.

Definition 8.24. A **relation** on a set X can be defined as a subset R of the cartesian product set $X \times X$. We then define a to be related to b, written as

$$a \sim b, \tag{8.34}$$

if $(a, b) \in R$.

This is a super general definition as we have not imposed any conditions on R.

Example 8.25. Let $X = \mathbb{Z}$. Let m be a positive integer. Define the relation $R \subset \mathbb{Z} \times \mathbb{Z}$ to be the subset

$$R := \{(a, b) \in \mathbb{Z} \times \mathbb{Z} : a \equiv b \bmod m\}. \tag{8.35}$$

So $a \sim b$ if and only if $a \equiv b \bmod m$. In this sense, congruence (modulo m) is a relation. Furthermore, recall that congruence has the properties of being reflexive, symmetric, and transitive.

8.4.2. Equivalence relations. More generally, we have the following:

Definition 8.26. Let X be a set. We say that a relation $R \subset X \times X$ is:

(1) **Reflexive** if: $\forall x \in X$, $x \sim x$.
(2) **Symmetric** if: $\forall x, y \in X$, $x \sim y$, then $y \sim x$.
(3) **Transitive** if: $\forall x, y, z \in X$, $x \sim y$ and $y \sim z$, then $x \sim z$.

Definition 8.27. We say that a relation is an **equivalence relation** if it is reflexive, symmetric, and transitive.

So, of course:

Example 8.28 (Congruence is an equivalence relation)**.** Given a positive integer m, congruence modulo m is an equivalence relation on $\mathbb{Z}$.

On the other hand, here is a pitfall.

Non-Theorem 8.29. *For any relation $\sim$ on a set X, the symmetry property implies the reflexivity property.*

Non-Proof. Assume the symmetry property. Let $x \in X$. For any $y \in X$, $x \sim y$ implies $y \sim x$. In particular, for $y = x$ we obtain $x \sim x$. This proves the reflexivity property. ☺ □

The problem with the proof above is that by taking $y = x$ in the symmetry property, we obtain the implication

$$x \sim x \text{ implies } x \sim x\ ,$$

which is a tautology and which certain does not imply that $x \sim x$ is true! (More abstractly, the implication $P \Rightarrow P$ being true does not imply that the statement P is true!)

8.4.3. Equivalence classes. Suppose that R is an equivalence relation on a set X. An **equivalence class** is defined to be a set of the form

$$[x] := \{y \in X : y \sim x\}. \tag{8.36}$$

Example 8.30 (Congruence classes are equivalence classes)**.** The equivalence classes of the congruence equivalence relation on $\mathbb{Z}$ are called congruence classes.

Non-Example 8.31 (Keep your friends close and your enemies closer)**.** *Consider the set X of all people and define the relation $\sim$ on X by $x \sim y$ if and only if x and y are enemies. Then this relation is not transitive, for the enemy of your enemy is your friend!* ☺

Moreover, if x is their own (worst) enemy, then $x \sim x$. However, if there exists a person who is their own (best) friend, then the relation $\sim$ is not reflexive!

Exercise 8.13. *Define a relation $\sim$ on $\mathbb{R}^n$ by*

$$\mathbf{x} \sim \mathbf{y} \quad \textit{if and only if} \quad |\mathbf{x}| = |\mathbf{y}|. \tag{8.37}$$

Prove that $\sim$ is an equivalence relation on $\mathbb{R}^n$.

Describe the equivalence classes of $\sim$. Hint: Spheres and, in one case, a point.

Exercise 8.14. *Define a relation $\sim$ on the set of real $n \times n$ invertible matrices by*

$$A \sim P^{-1}AP, \tag{8.38}$$

where P is also a real $n \times n$ invertible matrix. Prove that $\sim$ is an equivalence relation on this set.

We say that A and $P^{-1}AP$ are similar matrices (a.k.a. ***conjugate matrices****, where being conjugate is a more general group theory notion).*

8.4.4. Partitions. Let X be a set. A collection $\{Y_\alpha\}_{\alpha \in A}$ of subsets of X is called a **partition** of X if X is the disjoint union of these sets. That is,

(1)

$$X = \bigcup_{\alpha \in A} Y_\alpha, \tag{8.39}$$

(2) $Y_\alpha \cap Y_\beta \neq \emptyset$ implies $\alpha = \beta$.

Example 8.32 (Pre-images of a function form a partition)**.** If $f : X \to Y$, then

$$X = \bigcup_{y \in Y} f^{-1}(y) \tag{8.40}$$

is a disjoint union. That is, the collection $\{f^{-1}(y)\}_{y \in Y}$ is a partition of X.

As a further example of this, using the remainder map $\mathbf{r} : \mathbb{Z} \to \mathbb{Z}_m$, we set that the collection of sets

$$\{\mathbf{r}^{-1}(a) = [a]_m\}_{a \in R_m} \tag{8.41}$$

is a partition of $\mathbb{Z}$. This set is the same as the collection $\mathbb{Z}_m$ of congruence classes modulo m.

More generally:

Lemma 8.33. *Given an equivalence relation $\sim$ on a set X, the set of equivalence classes*

$$\mathcal{E} := \{[x] : x \in X\} \tag{8.42}$$

is a partition of X.

Exercise 8.15. *Prove Lemma* 8.33.

Note that, for every $x \in X$, for each $y \in X$ such that $y \sim x$, $[y]$ is considered as the same object as $[x]$ in the collection $\mathcal{E}$. For example, if $X = \mathbb{Z}$ and $\sim$ is congruence modulo 17, then [3] is the same object as [20].

8.4.5. Quotients.

Definition 8.34. We call the set of equivalence classes $\mathcal{E}$ the **quotient set** of X under the equivalence relation $\sim$ and we also denote it by $X/\sim$.

Example 8.35 (Set of congruence classes)**.** Given a positive integer m, let $X = \mathbb{Z}$ and let $\sim$ be congruence modulo m. Then the quotient set $X/\sim$ is $\mathbb{Z}_m$, the set of congruence classes modulo m.

8.4.6. Circles and tori.

Example 8.36 (The circle as a quotient of an interval). Let $X = [0, 1]$, the closed unit interval in $\mathbb{R}$. Define the equivalence relation $\sim$ on X by

$$x \sim x \text{ for all } x \in [0,1], \qquad 0 \sim 1. \tag{8.43}$$

The quotient set is

$$X/\sim\, = \{\{x\}\}_{x\in(0,1)} \cup \{0,1\}. \tag{8.44}$$

That is, $X/\sim$ is the union of the singletons of points in the interior $(0, 1)$ of $[0, 1]$ together with the 2-element set of the endpoints 0 and 1. So, we can "obtain" $X/\sim$ from X by "gluing" together 0 and 1.

Figure 8.4.1. Left: The unit interval $X = [0, 1]$. Right: A circle of length 1 obtained by identifying together the endpoints of the unit interval $X/\sim$.

Example 8.37 (The circle as a quotient of the real line). Let $X = \mathbb{R}$ and define an equivalence relation on $\mathbb{R}$ by

$$x \sim y \quad \Leftrightarrow \quad x - y \in \mathbb{Z}. \tag{8.45}$$

It is customary to write the quotient set $\mathbb{R}/\sim$ as $\mathbb{R}/\mathbb{Z}$. Can you explain why $\mathbb{R}/\mathbb{Z}$ is essentially the same as $[0, 1]/\sim$ in Example 8.35?

Exercise 8.16 (A circle of circumference L). *Let L be a positive real number. Define a relation $\sim$ on $\mathbb{R}$ by*

$$x \sim y \quad \textit{if and only if} \quad x - y \in L\mathbb{Z}, \tag{8.46}$$

where $L\mathbb{Z} := \{kL : k \in \mathbb{Z}\}$.

Prove that $\sim$ is an equivalence relation on $\mathbb{R}$.

In what way is the set of equivalence classes of $\sim$ the same as a circle of circumference L?

Is any of this analogous to taking a string of length L and gluing its ends together?

Example 8.38 (A 2-dimensional torus). Let $X = [0, 1] \times [0, 1]$, the unit square in the plane $\mathbb{R}^2$. Define an equivalence relation $\sim$ on X by

$$\begin{aligned}(x, y) &\sim (x, y) \text{ for all } (x, y) \in X,\\ (x, 0) &\sim (x, 1) \text{ for all } x \in [0, 1],\\ (0, y) &\sim (1, y) \text{ for all } y \in [0, 1].\end{aligned}$$

The quotient set $T := [0, 1] \times [0, 1]/\sim$ is called a **torus**.

Equivalently, we may define an equivalence relation $\approx$ on $\mathbb{R}^2$ by (compare with Example 8.37)

$$(x, y) \approx (x + k, y + \ell) \quad \text{for } (x, y) \in \mathbb{R}^2 \quad \text{and } k, \ell \in \mathbb{Z}. \tag{8.47}$$

We leave it to the reader to define a bijection between T and $\mathbb{R}^2/\approx$.

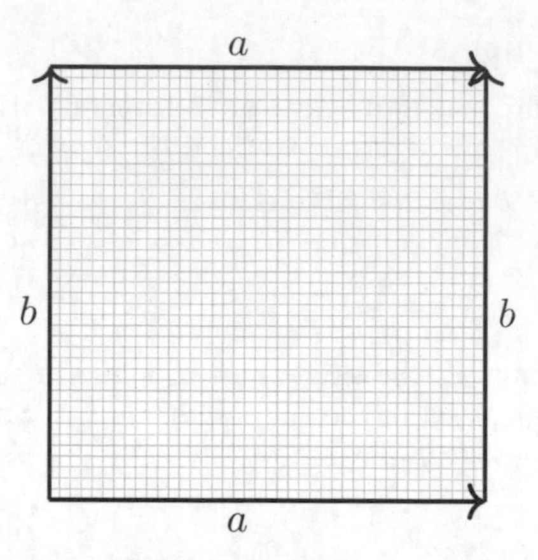

Figure 8.4.2. The quotient set T is obtained from the unit square by identifying the two sides labeled a together and the two sides labeled b together.

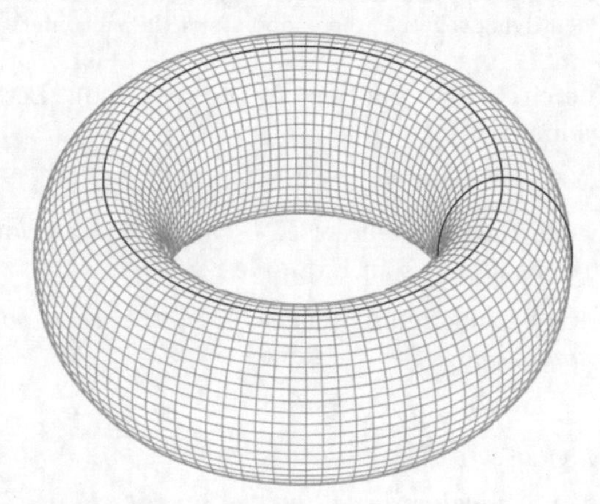

Figure 8.4.3. The torus rendered as a surface in 3-dimensional Euclidean space. Author: Krishnavedala. Available under the Creative Commons CC0 1.0 Universal Public Domain Dedication.

8.4.7. Klein bottles.

Re-imagining a Tom Waits quote:

> I'd rather have a Klein bottle in front of me than a fundamental trichotomy.

Let us consider Example 8.38 with a twist. Namely, let $X = [0,1] \times [0,1]$, and define a relation $\sim$ on X by

$$(x,y) \sim (x,y) \text{ for all } (x,y) \in X,$$
$$(x,0) \sim (1-x,1) \text{ and } (1-x,1) \sim (x,0) \text{ for all } x \in [0,1],$$
$$(0,y) \sim (1,y) \text{ and } (1,y) \sim (0,y) \text{ for all } y \in [0,1].$$

Let $K := [0,1] \times [0,1]/\sim$ be the quotient set. The set K is called the Klein bottle.

Exercise 8.17. *Prove that $\sim$ is an equivalence relation on X.*

Equivalently, we may describe the Klein bottle as a quotient of $\mathbb{R}^2$. Define the equivalence relation $\approx$ on $\mathbb{R}^2$ by

(8.48)
$$(x,y) \approx \left((-1)^{\ell}x + \frac{1-(-1)^{\ell}}{2} + k, y+\ell\right) \quad \text{for } (x,y) \in \mathbb{R}^2 \quad \text{and } k,\ell \in \mathbb{Z}.$$

For example, if ℓ is even, then $(x,y) \approx (x+k, y+\ell)$ for all $k \in \mathbb{Z}$. If ℓ is odd, then $(x,y) \approx (1-x+k, y+\ell)$ for all $k \in \mathbb{Z}$. We leave it to the reader to check that $\approx$ is an equivalence relation and to define a bijection between K and $\mathbb{R}^2/\approx$.

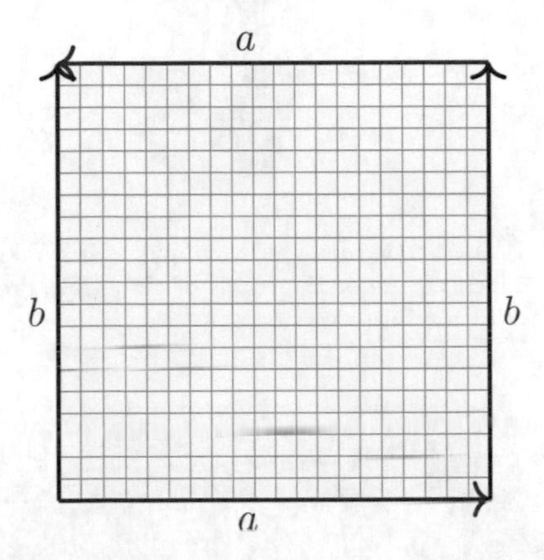

Figure 8.4.4. The quotient set K is obtained by identifying sides of a square according to the arrows. For example, the bottom side is flipped horizontally when identified with the top side.

Exercise 8.18. *Convince yourself that (in some sense!) the quotient set K can be rendered as the surface in Figure* 8.4.5. *Hint: The Klein bottle link above shows some steps.*

As a further example, we have the Möbius band, which is the quotient set

(8.49)
$$M = ([0,1] \times \mathbb{R})/\approx,$$

where $\approx$ is defined by (8.48) with $k=0$ and $\ell \in \mathbb{Z}$.

Equivalently, we may define Möbius band as the quotient of $[0,1] \times [0,1]$ by the equivalence relation $\simeq$ defined by

$$(x,y) \simeq (x,y) \text{ for all } (x,y) \in X,$$
$$(x,0) \simeq (1-x,1) \text{ and } (1-x,1) \simeq (x,0) \text{ for all } x \in [0,1].$$

This is how we make Möbius bands in real life; see Figure 8.4.6.

We leave it as an exercise for the reader to show that the projective plane $\mathbb{R}P^2$ (see §5.2.8) may be considered as the quotient of $[0,1] \times [0,1]$ by the equivalence relation $\sim$ defined by

$$(x,y) \sim (x,y) \text{ for all } (x,y) \in X,$$
$$(x,0) \sim (1-x,1) \text{ and } (1-x,1) \simeq (x,0) \text{ for all } x \in [0,1],$$
$$(0,y) \sim (1,1-y) \text{ and } (1,1-y) \simeq (0,y) \text{ for all } y \in [0,1].$$

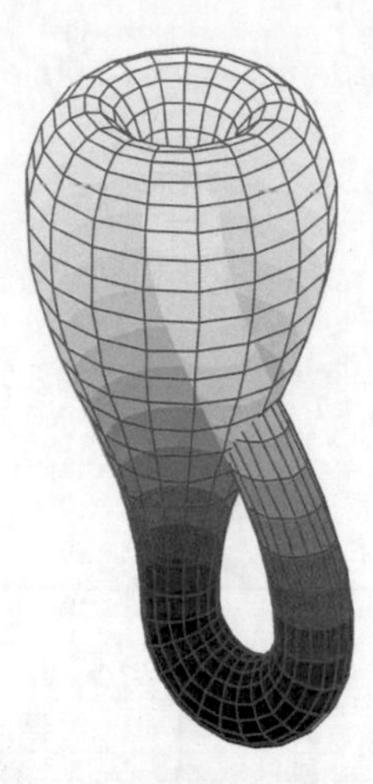

Figure 8.4.5. A Klein bottle made with gnuplot 4.0. Wikimedia Commons, author: Tttrung. Licensed under the terms of the GNU Free Documentation License, Version 1.2 or any later version.

Figure 8.4.6. A Möbius band. Credit: Wikimedia Commons. Photo by David Benbennick. Licensed under Creative Commons Attribution Share Alike 3.0 Unported (https://creativecommons.org/licenses/by-sa/3.0/deed.en) license.

8.5. Elementary abstract algebra

Throughout this book, we have been considering algebraic structures. In this section we will consider various basic examples and their abstractions. Roughly speaking, these abstractions are based on "how many" properties we decide to impose and of course the nature of these properties.

8.5.1. The integers and other examples of algebraic structures. The most basic algebraic structure we have been discussing is the integers. The integers $\mathbb{Z}$ is a set and we have a pair of **binary operations** on $\mathbb{Z}$ of **addition**, denoted by $+$, and **multiplication**, denoted by $\cdot$ or $\times$. That is, $+$ assigns to an ordered pair of integers (a, b) an integer $a + b$ which is their sum.

The integer 0, read as "zero", is special in that it is the **additive identity**: For any integer a, we have

$$0 + a = a + 0 = a. \tag{8.50}$$

Addition is **commutative**: For any integers a and b,

$$a + b = b + a. \tag{8.51}$$

Addition is **associative**: For any integers a, b, and c,

$$(a + b) + c = a + (b + c). \tag{8.52}$$

For each integer a there exists an **additive inverse** $-a$ satisfying

$$a + (-a) = (-a) + a = 0. \tag{8.53}$$

In general, a set with a binary operation which has an identity element, is commutative and associative, and for which each element has an inverse is called an **abelian group** (see §8.5.2, below the abstract definition). In particular, the integers $\mathbb{Z}$ with addition $+$ is an abelian group. We remark that "abelian" is a synonym for "commutative".

Here is another example of an abelian group:

Example 8.39. Let $X = \{e, o\}$, which is a set with two elements, one called "e" and one called "o". Define the binary operation $\oplus$ taking an ordered pair (a, b), where $a, b \in X$, to $a \oplus b$, by the **addition table** in Table 8.5.1.

Table 8.5.1. Addition table for the abelian group $(\{e, o\}, \oplus)$.

$\oplus$	e	o
e	e	o
o	o	e

That is, the addition table defines

$$e \oplus e = e, \quad e \oplus o = o, \quad o \oplus e = o, \quad o \oplus o = e. \tag{8.54}$$

One can see by inspection and a short calculation that e is the additive identity and $\oplus$ is commutative and associative. We also see that each of e and o is its own inverse. The color coding in (8.54) is such that purple means red and blue. Note also that (8.54) explains how we should read the addition table in Table 8.5.1.

Exercise 8.19. *Verify the statements above that the set X with the "addition" binary operation $\oplus$ given by the table in Example 8.39 satisfies the properties that e is the additive identity and $\oplus$ is both commutative and associative. And each of e and o is its own inverse. Thus, $(X, \oplus)$ is an abelian group, as defined in §8.5.2 below.*

Exercise 8.20. *Write out an addition table for parity, where the two elements being multiplied are "even" and "odd", using the theorems in §1.2.2. How does your addition table for parity compare to the addition table in Table 8.5.1? Try to be as specific as possible with your comparison.*

Exercise 8.21. *Write out a multiplication table for parity, where the two elements being multiplied are "even" and "odd", using the theorems in* §1.2.3.

Now we turn our attention to multiplication. For the integers, multiplication $\cdot$ assigns to an ordered pair of integers (a, b) an integer $a \cdot b$ which their product. For example, it assigns 15 to $(3, 5)$.

The integer 1 is special in that it is the **multiplicative identity**: For any integer a, we have

$$1 \cdot a = a \cdot 1 = a. \tag{8.55}$$

Multiplication is **commutative**: For any integers a and b,

$$a \cdot b = b \cdot a. \tag{8.56}$$

Multiplication is **associative**: For any integers a, b, and c,

$$(a \cdot b) \cdot c = a \cdot (b \cdot c). \tag{8.57}$$

We say that integers a and b are inverses of each other if

$$a \cdot b = b \cdot a = 1. \tag{8.58}$$

Since $1 \cdot 1 = 1$ and $(-1) \cdot (-1) = 1$, 1 has the multiplicative inverse 1, and -1 has the multiplicative inverse -1. But no other integers have multiplicative inverses! That is, if a is an integer that is neither 1 nor -1, then there does not exist an integer b satisfying (8.58). That is the bad news. But the good news is that addition and multiplication cooperate to give some nice properties. Namely, we have the **distributive property**: For any integers a, b, and c, we have

$$a \cdot (b + c) = a \cdot b + a \cdot c. \tag{8.59}$$

Since multiplication is commutative, we obtain "right" distributivity as well as the above "left" distributivity:

$$(b + c) \cdot a = b \cdot a + c \cdot a. \tag{8.60}$$

More generally, when a set with a pair of binary operations $+$ and $\cdot$ satisfies the properties above, we say that it is a commutative ring. In particular, $(\mathbb{Z}, +, \cdot)$ is a commutative ring, called the **ring of integers**. So, by a ring, we do not mean something you wear to signify that you are married, or the sound a doorbell makes. Rather it indicates the presence of two binary operations which we call addition and multiplication on a set satisfying the properties above.

Exercise 8.22. *Let $\otimes$ denote the multiplication you defined for parity in Exercise* 8.21. *Show that the set $X = \{e, o\}$ with the binary operations of "addition" $\oplus$ and "multiplication" $\otimes$ satisfies all of the same axioms stated above that the set of integers $\mathbb{Z}$ with addition $+$ and multiplication $\times$ satisfies. For this reason, we call $(X, \oplus, \otimes)$ a commutative ring.*

Solved Problem 8.40 (Multiplication on the set of remainders modulo 3). *Define the set $R_3 := \{0, 1, 2\}$. Define the binary operation $\odot$ on R_3 by*

$$a \odot b := \begin{cases} ab & \text{if } ab < 3, \\ ab - 3 & \text{if } ab \geq 3. \end{cases} \tag{8.61}$$

Write out the multiplication table for $(R_3, \odot)$.

Solution. It is easy to see that if $a \in R_3$, then

$$a \odot 0 = 0 \odot a = 0, \qquad a \odot 1 = 1 \odot a = a. \tag{8.62}$$

The only remaining product to compute is

$$2 \odot 2 = 2 \cdot 2 - 3 = 1. \tag{8.63}$$

So, piecing all of this together, we obtain that the multiplication table

$\odot$	0	1	2
0	$0 \odot 0$	$0 \odot 1$	$0 \odot 2$
1	$1 \odot 0$	$1 \odot 1$	$1 \odot 2$
2	$2 \odot 0$	$2 \odot 1$	$2 \odot 2$

is given by

$\odot$	0	1	2
0	0	0	0
1	0	1	2
2	0	2	1

We remark that this is the same as congruence class multiplication on $\mathbb{Z}_3$.

Exercise 8.23. *Let m be a positive integer. Define the set*

$$R_m := \{0, 1, 2, \ldots, m-1\}. \tag{8.64}$$

Another way to describe this set is

$$R_m = \{a \in \mathbb{Z} : 0 \le a < m\}. \tag{8.65}$$

Define the binary operation $\oplus$ on R_m by

$$a \oplus b := \begin{cases} a+b & \text{if } a+b < m, \\ a+b-m & \text{if } a+b \ge m. \end{cases} \tag{8.66}$$

Write out the addition tables for R_2 and R_3, both with the binary operation $\oplus$. See Table 8.5.1 *for an example of an addition table.*

Note that the binary operation $\oplus$ on R_m above is the same as congruence class addition on $\mathbb{Z}_m$.

Exercise 8.24. *Let m be a positive integer. Prove that if $a, b \in R_m$, then $a \oplus b \in R_m$. This just says that indeed $\oplus$ is a (well-defined) binary operation on R_m.*

Exercise 8.25. *Show that the binary operation $\oplus$ on R_m is commutative and associative. How does $(R_m, \oplus)$ generalize Example* 8.39*?*

In this subsection we have considered the *ring* of integers $(\mathbb{Z}, +, \cdot)$. If we only consider the additive structure $+$ on $\mathbb{Z}$ and forget about multiplication $\cdot$, we have what is known as a *group* structure on $\mathbb{Z}$.

8.5.2. What is a group anyway?

In general, we have the following.

Definition 8.41. A set G together with a binary operation $\odot : G \times G \to G$ is called a **group** if it satisfies the following axioms:

(1) (Associativity) For any $x, y, z \in G$,

$$(x \odot y) \odot z = x \odot (y \odot z). \tag{8.67}$$

(2) There exists an **identity element** e with the following property: For any $x \in G$,

$$e \odot x = x \odot e = x. \tag{8.68}$$

(3) Each $x \in G$ has an **inverse element**, which we denote by x^{-1}, satisfying

$$x \odot x^{-1} = x^{-1} \odot x = e. \tag{8.69}$$

8.5.3. Types of groups.

Definition 8.42. We say that a group $(G, \odot)$ is a **finite group** if G is a finite set.

We call its cardinality $|G|$ the **order** of the group.

If G is not finite, then we say that it is **infinite**.

Definition 8.43. We say that a group $(G, \odot)$ is **abelian** if $\odot$ is commutative; that is, for every $x, y \in G$,

$$x \odot y = y \odot x. \tag{8.70}$$

Definition 8.44. A group $(G, \odot)$ is a cyclic group if there exists an element $a \in G$ such that each element g of G is equal to at least one of the following forms:

(1) $g = e$.

(2) $g = a^k := \underbrace{a \odot a \odot \cdots \odot a}_{k \text{ times}}$, where $k \in \mathbb{Z}^+$.

(3) $g = (a^{-1})^\ell := \underbrace{a^{-1} \odot a^{-1} \odot \cdots \odot a^{-1}}_{\ell \text{ times}}$, where $\ell \in \mathbb{Z}^+$.

We call the element a a **generator** of G, and we write $G = \langle a \rangle$ (cf. (8.110) below).

We next consider examples of the definitions above.

8.5.4. Jostling examples and non-examples of groups.

Example 8.45 (The additive integers). Let $G = \mathbb{Z}$ and let $\odot = +$, the usual addition of integers. This is an infinite abelian cyclic group. The identity element is given by $e = 0$. The inverse is given by $x^{-1} = -x$. A choice of generator for $\mathbb{Z}$ is 1. The only other possible choice of generator for $\mathbb{Z}$ is -1.

Taking the generator to be 1, we have:

(1) If $g > 0$, then $g = 1^g := \underbrace{1 + 1 + \cdots + 1}_{g \text{ times}}$.

(2) If $g < 0$, then $g = (-1)^{(-g)} := \underbrace{-1 - 1 - \cdots - 1}_{-g \text{ times}}$.

Non-Example 8.46. *Let $G = \mathbb{Z}$ and let $\odot = \cdot$ be multiplication. Then $(G, \odot) = (\mathbb{Z}, \cdot)$ is not a group. This is because, although $e = 1$ is an identity element, the element 0 does not have an inverse. That is, there does not exist any element, call this mythical object 0^{-1}, such that*

$$0 \cdot 0^{-1} = 1. \tag{8.71}$$

Note that neither does 2 have an inverse since $\frac{1}{2}$ is not an integer. In fact, the only integers with multiplicative inverses are 1 and -1.

Example 8.47 (Set of congruence classes with addition)**.** Now let m be a positive integer, and let $\mathbb{Z}_m$ be the set of congruence classes modulo m as defined in (8.15). Let $+$ be the binary operation on $\mathbb{Z}_m$ defined by (8.18a). Then

$$(\mathbb{Z}_m, +)$$

is a finite abelian cyclic group of order m, where the inverse is given by $([a]_m)^{-1} = [-a]_m$.

Exercise 8.26. *Prove that, as claimed in Example 8.47, for every $m \in \mathbb{Z}^+$, $(\mathbb{Z}_m, +)$ is an abelian group.*

Non-Example 8.48. *Let $\cdot$ be the binary operation on $\mathbb{Z}_m$ defined by (8.18c), where $m \geq 2$. Then*

$$(\mathbb{Z}_m, \cdot)$$

is not a group.

Exercise 8.27. *Prove that, as claimed in Non-Example 8.48, for every $m \in \mathbb{Z}^+$ with $m \geq 2$, $(\mathbb{Z}_m, \cdot)$ is not a group. Hint: The element $[1]_m$ is the identity, but what element has no inverse?*

Example 8.49 (Non-zero congruence classes mod 5 with multiplication)**.** Let

$$\mathbb{Z}_5^* := \{[1]_5, [2]_5, [3]_5, [4]_5\} \tag{8.72}$$

be the set of non-zero congruence classes modulo 5. From the multiplication table in Figure 8.1.3, we see that $\mathbb{Z}_5^*$ with the binary operation $\cdot$ is an abelian group. For example, the identity element is $[1]_5$ and the list of inverses is

$$[1]_5^{-1} = [1]_5, \quad [2]_5^{-1} = [3]_5, \quad [3]_5^{-1} = [2]_5, \quad [4]_5^{-1} = [4]_5. \tag{8.73}$$

Non-Example 8.50. *Let*

$$\mathbb{Z}_4^\# := \{[1]_4, [2]_4, [3]_4\}. \tag{8.74}$$

We see that $\mathbb{Z}_4^\#$ with the binary operation $\cdot$ is not a group. In particular, while $[1]_4$ is still an identity element, the element $[2]_4$ does not have an inverse.

We can rectify this by defining instead

$$\mathbb{Z}_4^* := \{[1]_4, [3]_4\}. \tag{8.75}$$

In this case, $(\mathbb{Z}_4^*, \cdot)$ is an abelian group!

Exercise 8.28. *Prove that $(\mathbb{Z}_4^*, \cdot)$ is an abelian group.*

Exercise 8.29. *Define the dihedral group D_2 as follows. Its four elements are the functions from $\mathbb{R}^2$ to $\mathbb{R}^2$ defined by*

$$1 : (x, y) \mapsto (x, y), \tag{8.76}$$

$$a : (x, y) \mapsto (-x, y), \tag{8.77}$$

$$b : (x, y) \mapsto (x, -y), \tag{8.78}$$

$$c : (x, y) \mapsto (-x, -y). \tag{8.79}$$

That is, $1 = I_{\mathbb{R}^2}$ is the identity function, a is a reflection about the y-axis, b is a reflection about the x-axis, and c is rotation by $180°$ (a.k.a. the antipodal function). Define multiplication to be the composition of functions. For example, we have

$$a \cdot b := a \circ b : (x, y) \overset{b}{\mapsto} (x, -y) \overset{a}{\mapsto} (-x, -y), \tag{8.80}$$

so we see that $a \cdot b = c$.

(1) *Write out the multiplication table for $(D_2, \cdot)$.*
(2) *Prove that $(D_2, \cdot)$ is a group.*

Exercise 8.30. *Let m be a composite number, and define*

$$\mathbb{Z}_m^{\#} := \{[i]_m : 0 < i < m\}. \tag{8.81}$$

Prove that $(\mathbb{Z}_m^{\#}, \cdot)$ is not a group. Hint: Not only are there element(s) with no inverse(s), congruence class multiplication $\cdot$ on the set $\mathbb{Z}_m^{\#}$ is not well-defined.

Example 8.51 (Non-zero congruence classes mod a prime with multiplication). Let p be a prime. Define $\mathbb{Z}_p^*$ to be $\mathbb{Z}_p - \{[0]_p\}$, that is, the set of non-zero congruence classes modulo p. Then

$$(\mathbb{Z}_p^*, \cdot)$$

is an abelian group.

(1) The identity element is $[1]_p$.
(2) We claim that each element $[a]_p \in \mathbb{Z}_p^*$ has an inverse. To see this, recall that a and p are coprime and hence by Theorem 6.48 there exists $b \in \mathbb{Z}$ such that

$$ab \equiv 1 \bmod p. \tag{8.82}$$

This implies that

$$[a]_p \cdot [b]_p = [b]_p \cdot [a]_p = [1]_p. \tag{8.83}$$

(3) The rest of the group axioms are easy to verify.

Non-Example 8.52. *The set of complex numbers $\mathbb{C}$ with complex multiplication $\cdot$, as defined in §8.7.3.1, is not a group. For 0 does not have an inverse!*

Example 8.53 (Non-zero complex numbers with multiplication). Let $\mathbb{C}^* := \mathbb{C} - \{0\}$ be the set of non-zero complex numbers. Then $(\mathbb{C}^*, \cdot)$ is a group. The inverse of a non-zero complex number z is

$$z^{-1} = \frac{\overline{z}}{|z|^2}, \tag{8.84}$$

where $\overline{z}$ is the complex conjugate defined by (8.171).

Non-Example 8.54. *Can we see another example instead?* ☺

Example 8.55 (General linear groups of matrices)**.** Important examples of groups are matrix groups. A couple of examples are:

(1) The (real) general linear group $GL(n, \mathbb{R})$, which is the set of invertible real $n \times n$ matrices with multiplication given by matrix multiplication. Recall that a square matrix is invertible if and only if its determinant is non-zero.

(2) The complex general linear group $GL(n, \mathbb{C})$. This is the set of invertible complex $n \times n$ matrices with multiplication given by matrix multiplication.

We will not discuss these groups in detail in this book.

Exercise 8.31. *Prove that if a group $(G, \odot)$ is cyclic, then it is abelian.*

8.5.5. Basic properties of groups. Let $(G, \odot)$ be a **monoid**; that is, the binary operation $\odot$ is associative and there exists an identity element e. Let $a \in G$. We say that b is a **right inverse** of a if

$$a \odot b = e. \tag{8.85}$$

We say that c is a **left inverse** of a if

$$c \odot a = e. \tag{8.86}$$

Suppose that b is a right inverse of a and that c is a left inverse of a. Then we have

$$b = e \cdot b = (c \cdot a) \cdot b = c \cdot (a \cdot b) = c \cdot e = c. \tag{8.87}$$

That is, for every element in a monoid, if it has both a right inverse and a left inverse, then they are equal.

Exercise 8.32. *Let $(G, \cdot)$ and $(H, *)$ be groups. Their* ***product group*** *is the set $G \times H$ with multiplication $\circ$ defined by*

$$(g_1, h_1) \circ (g_2, h_2) := (g_1 \cdot g_2, h_1 * h_2). \tag{8.88}$$

Prove that this product group is indeed a group.

8.5.6. Subgroups (less is more). Let H be a subset of a group $(G, \odot)$. We say that $(H, \odot)$ is a subgroup of $(G, \odot)$ if $(H, \odot)$ itself is a group. By necessity, this means:

(1) H is closed under multiplication; that is, for every $x, y \in H$, $xy \in H$.

(2) $e \in H$, where e is the identity element of G.

(3) If $x \in H$, then $x^{-1} \in H$, where x^{-1} is the inverse of x in G.

Often we omit the multiplication symbol $\odot$ in the notation for a group.

If H is a subgroup of G, we denote this by

$$H \leq G.$$

Example 8.56. Let m be a positive integer. Then $(m\mathbb{Z}, +)$, where

$$m\mathbb{Z} := \{mx : x \in \mathbb{Z}\}, \tag{8.89}$$

is a subgroup of the additive group of integers $(\mathbb{Z}, +)$. Observe that, as sets,

$$m\mathbb{Z} = [0]_m, \tag{8.90}$$

the congruence class of 0 modulo m, which is defined in Definition 8.1.

For example, $2\mathbb{Z}$ is the additive group of even integers.

Observe that if a positive integer r is a multiple of m, then $(r\mathbb{Z}, +)$ is a subgroup of $(m\mathbb{Z}, +)$.

Exercise 8.33. *Let E and O denote the set of even and odd integers, respectively.*

(1) *Is $(E, +)$ a subgroup of $(\mathbb{Z}, +)$?*
(2) *Is $(O, +)$ a subgroup of $(\mathbb{Z}, +)$?*
(3) *Is $(\mathbb{Z}^{\geq}, +)$ a subgroup of $(\mathbb{Z}, +)$?*

Example 8.57. Let $m = ab$ be a composite positive integer, where $a, b \geq 2$. Then

$$\langle [a]_m \rangle := \{[ka]_m : 0 \leq k < b\}$$

is a subgroup of $(\mathbb{Z}_m, +)$ of order b (cf. (8.110)).

Exercise 8.34. *Prove the statements in Example* 8.57.

We say that a subgroup H of a group G is **non-trivial** if $H \neq \{e\}$, where e is the identity element.

Lemma 8.58 (Classification of subgroups of $\mathbb{Z}$). *If $(H, +)$ is a non-trivial subgroup of $(\mathbb{Z}, +)$, then there exists a positive integer m such that*

$$H = m\mathbb{Z}. \tag{8.91}$$

Proof. Since the subgroup H is non-trivial, there exists a non-zero integer $a \in H$. Since H is a group, $-a \in H$. Define $b = |a|$. Then $b \in H \cap \mathbb{Z}^+$. Since $H \cap \mathbb{Z}^+$ is a non-empty set of positive integers, by the well-ordering principle it has a minimum element. Let m denote the minimum element of the set $H \cap \mathbb{Z}^+$.

Claim. $H = m\mathbb{Z}$.

Firstly, since $m \in H$ and H is a group, we have $m\mathbb{Z} \subset H$.

Secondly, we prove that $H \subset m\mathbb{Z}$. Suppose for a contradiction that there exists an integer c with $c \in H$ and $c \notin m\mathbb{Z}$. By the Division Theorem, there exist integers q and r satisfying

$$c = mq + r \quad \text{and} \quad 0 < r < m. \tag{8.92}$$

The reason why $r > 0$ is that $c \notin m\mathbb{Z}$. Since $c, m \in H$, we have

$$r = c - mq \in H. \tag{8.93}$$

However, this contradicts m being the minimum element of $H \cap \mathbb{Z}^+$ since $0 < r < m$. Since we have a contradiction, we conclude that $H \subset m\mathbb{Z}$. □

Exercise 8.35. *Show that the unit complex numbers (the set of $z \in \mathbb{C}$ satisfying $|z| = 1$) with complex multiplication, form a subgroup of $(\mathbb{C}^*, \cdot)$.*

8.5.7. Finite groups.

Theorem 8.59 (Lagrange). *If G is a finite group and if $H \leq G$ is a subgroup of G, then their orders (see Definition* 8.42*) have the property that*

$$|H| \text{ divides } |G|.$$

Example 8.60. Let m be a positive integer and let a be a positive divisor of m. Then $\langle [a]_m \rangle$ is a subgroup of $\mathbb{Z}_m$ of order $\frac{m}{a}$, which is a positive integer dividing m. For example, taking $a = 3$ and $m = 15$, we see that

$$\langle [3]_{15} \rangle = \{[0]_{15}, [3]_{15}, [6]_{15}, [9]_{15}, [12]_{15}\}$$

is a subgroup of $\mathbb{Z}_{15}$ of order 5, where the group multiplication is given by congruence class addition.

We now discuss some concepts we will need for the proof of Lagrange's Theorem, which will be given at the end of this subsection. Using the subgroup H, we can define an equivalence relation on G by

$$y \sim x \text{ if and only if there exists } h \in H \text{ such that } y = xh.$$

We verify that $\sim$ is:

(1) *Reflexive*: $x \sim x$ since $x = xe$ for all $x \in G$ and the identity element e of G is in H.

(2) *Symmetric*: If $x \sim y$, then there exists $h \in H$ such that $x = yh$. This implies that $y = xh^{-1}$. Since H is a subgroup, $h^{-1} \in H$ and hence $y \sim x$.

(3) *Transitive*: If $x \sim y$ and $y \sim z$, then there exist $h, k \in H$ such that $x = yh$ and $y = zk$. Thus $x = zkh$. Since H is a subgroup, $kh \in H$. Thus $x \sim z$.

If $x \in G$, we denote the equivalence class of x by

$$xH = [x] = \{y \in G : y \sim x\} = \{xh : h \in H\}, \tag{8.94}$$

and we call it the (left) **coset** of x. Since $\sim$ is an equivalence relation, the collection of distinct cosets partition the set G.

Example 8.61. Although we are presently interested in finite groups, for simplicity we first consider the infinite group $(\mathbb{Z}, +)$. Let H be a non-trivial subgroup of $\mathbb{Z}$. By Lemma 8.58, we have $H = m\mathbb{Z}$ for some positive integer m.

Let $x \in \mathbb{Z}$. Then the left coset

$$xH = \{x + h : h \in m\mathbb{Z}\} = \{x + mk : k \in \mathbb{Z}\} = [x]_m \tag{8.95}$$

is the congruence class of x modulo m.

Example 8.62. Recall from Example 8.47 that $(\mathbb{Z}_m, +)$ is a finite abelian cyclic group of order m. Assume, more generally, that $(G, \odot)$ is a finite abelian cyclic group of order m. Suppose that $a \in G$ is a generator of G. Then

$$G = \{a^k : 0 \leq k < m\}, \tag{8.96}$$

where $a^0 := e$ is the identity element.

Let c be a positive divisor of m, and define

$$H := \langle a^c \rangle. \tag{8.97}$$

Let $b := m/c$, which is a positive integer. Then

$$H = \{a^{cj} : 0 \leq j < b\}. \tag{8.98}$$

The distinct cosets are

$$a^i H, \quad 0 \leq i < c. \tag{8.99}$$

Now let us return to the general case of a subgroup H of a finite group G. We leave it as an exercise that the function

$$f_x : H \to xH \tag{8.100}$$

defined by

$$f_x(h) = xh \quad \text{for } h \in H \tag{8.101}$$

is a bijection. From this and Theorem 5.62, we conclude that for each $x \in G$,

$$|xH| = |H| \tag{8.102}$$

(is independent of x).

Exercise 8.36. *Prove that $f_x : H \to xH$ defined by* (8.101) *is a bijection.*

The **index** of the subgroup H of the finite group G is the number of distinct cosets and is denoted by

$$|G : H|. \tag{8.103}$$

Proof of Theorem 8.59. Since there are $|G : H|$ cosets, since each coset has $|H|$ elements, and since the cosets partition G, we see that

$$|G| = |H| \cdot |G : H|; \quad \text{that is,} \quad |G : H| = \frac{|G|}{|H|}. \tag{8.104}$$

Since $|G : H|$ is a positive integer, we conclude that $|H|$ divides $|G|$. This completes the proof of Theorem 8.59. □

Exercise 8.37. *Find all of the subgroups of the dihedral group $(D_2, \cdot)$ defined in Exercise* 8.29.

8.5.8. Normal subgroups and quotient groups. We say that a subgroup H of a group G is **normal** if for all $g \in G$ we have

$$gH = Hg, \tag{8.105}$$

where $Hg := \{hg : h \in H\}$ is the **right coset** of g. Equivalently, $ghg^{-1} \in H$ for all $g \in G$. If G is abelian, then any subgroup of G is normal.

Given a normal subgroup H of G, we can define the quotient group G/H to be the set of left cosets:

$$G/H = \{gH : g \in G\}. \tag{8.106}$$

The group multiplication on this set is defined by

$$g_1 H \cdot g_2 H = (g_1 g_2)H. \tag{8.107}$$

As an exercise, you may check that this is a well-defined group by using the assumption that the subgroup H is normal.

Example 8.63. Let m be a positive integer. Since $m\mathbb{Z}$ is a normal subgroup of the additive group of integers $(\mathbb{Z}, +)$, we have the quotient group $(\mathbb{Z}_m, +) := \mathbb{Z}/m\mathbb{Z}$, where $m\mathbb{Z}$ is defined by (8.89) as $m\mathbb{Z} = \{mx : x \in \mathbb{Z}\}$. This quotient group is naturally isomorphic to the additive group of congruence classes, where a left coset $a(m\mathbb{Z})$ corresponds to the congruence class $[a]_m$ for $a \in \mathbb{Z}$.

For an example for the problems that arise when we try to define the quotient group using a subgroup that is not normal, see Non-Example 8.78 below.

8.5.9. Subgroup generated by an element. Let $(G, \odot)$ be a group. Given $a \in G$, we define the **powers** of a as follows. The positive powers are defined by

$$a^1 := a, \quad a^2 := a \odot a, \quad a^3 := a \odot a \odot a, \quad \ldots, \quad a^k := \underbrace{a \odot a \odot \cdots \odot a}_{k \text{ times}}. \tag{8.108}$$

We define the negative powers by

$$a^{-k} := \underbrace{a^{-1} \odot a^{-1} \odot \cdots \odot a^{-1}}_{k \text{ times}} \tag{8.109}$$

for $k \in \mathbb{Z}^+$. The 0-th power is $a^0 := e$.

Let $(G, \odot)$ be a group and let $a \in G$. The subgroup of G generated by a is

$$\langle a \rangle := \{a^k : k \in \mathbb{Z}\}. \tag{8.110}$$

Note that $\langle a \rangle$ is a cyclic group.

Example 8.64. If m is a positive integer in the additive group of integers $(\mathbb{Z}, +)$, then

$$\langle m \rangle = m\mathbb{Z}. \tag{8.111}$$

Lemma 8.65. *If $(G, \odot)$ is a finite group and if $a \in G$, then there exists a least positive integer n such that*

$$a^n = e. \tag{8.112}$$

Moreover,

$$n = |\langle a \rangle|. \tag{8.113}$$

The order n of the group $\langle a \rangle$ is also called the **order** *of the element a.*

Proof. To show that there exists a *least* positive integer n satisfying (8.112), by the well-ordering principle (see §4.1.4) it suffices to show there exists *some* n satisfying (8.112).

Suppose for a contradiction that there does not exist any $n \in \mathbb{Z}^+$ such that $a^n = e$. Then we claim that

$$a^i \neq a^j \quad \text{for } i, j \in \mathbb{Z}^+ \text{ satisfying } i \neq j. \tag{8.114}$$

To see this, suppose $i, j \in \mathbb{Z}^+$ satisfy $a^i = a^j$. Without loss of generality, we may assume that $i \leq j$. We compute that

$$e = a^i(a^{-1})^i = a^j(a^{-1})^i = a^{j-i}. \tag{8.115}$$

Since $j - i \geq 0$ and since there does not exist any $n \in \mathbb{Z}^+$ such that $a^n = e$, we obtain that $j - i = 0$; that is, $i = j$. We conclude that some $n \in \mathbb{Z}^+$ satisfies (8.112).

Let n be the least positive integer satisfying (8.112). Then

$$\langle a \rangle = \{e, a, a^2, \ldots, a^{n-1}\}, \tag{8.116}$$

where the elements on the right-hand side are all distinct. Therefore $|\langle a \rangle| = n$. □

Example 8.66. Let m be a positive integer, and let $(\mathbb{Z}_m, +)$ denote the additive group of congruence classes modulo m. Then the order of an element $[a]_m$ is equal to $\frac{m}{\gcd(a,m)}$. We leave it as an exercise for you to prove this.

Theorem 8.67. *Let $(G, \odot)$ be a finite group, and let $a \in G$. Then*

$$a^{|G|} = e. \tag{8.117}$$

Proof. We have that $\langle a \rangle$ is a subgroup of G. Hence

$$|\langle a \rangle| \text{ divides } |G|.$$

On the other hand, by Lemma 8.65,

$$a^{|\langle a \rangle|} = e. \tag{8.118}$$

Since $|\langle a \rangle|$ divides $|G|$, there exists a positive integer k such that

$$|\langle a \rangle|\, k = |G|. \tag{8.119}$$

Therefore

$$a^{|G|} = (a^{|\langle a \rangle|})^k = e^k = e. \tag{8.120}$$

□

8.5.10. The multiplicative group $\mathbb{Z}_m^*$. Recall from (8.28) that $\mathbb{Z}_m^*$ is the set of invertible congruence classes modulo m. Equivalently,

$$\mathbb{Z}_m^* := \{[a]_m \in \mathbb{Z}_m : \gcd(a, m) = 1\}, \tag{8.121}$$

and this set is well-defined since by Exercise 8.6 we have the following:

If $a, b \in \mathbb{Z}$ are such that $[a]_m = [b]_m$, then $\gcd(a, m) = 1$ if and only if $\gcd(b, m) = 1$.

When $m = p$ is a prime, the definition in (8.121) agrees with the definition in (8.51).

Lemma 8.68. *Let $m \in \mathbb{Z}^+$. We have that $(\mathbb{Z}_m^*, \cdot)$ is an abelian group, where multiplication $\cdot$ is given by congruence class multiplication.*

Exercise 8.38. *Prove Lemma 8.68. Hint: The main point is that any element of $\mathbb{Z}_m^*$ has a multiplicative inverse.*

Example 8.69. (1) We have

$$\mathbb{Z}_6^* = \{[1]_6, [5]_6\}, \tag{8.122}$$

where the multiplication table is

(8.123)

$(\mathbb{Z}_6^*, \cdot)$	$[1]_6$	$[5]_6$
$[1]_6$	$[1]_6$	$[5]_6$
$[5]_6$	$[5]_6$	$[1]_6$

(2) We also have

$$\mathbb{Z}_{10}^* = \{[1]_{10}, [3]_{10}, [7]_{10}, [9]_{10}\}, \tag{8.124}$$

where the multiplication table is

(8.125)

$(\mathbb{Z}_{10}^*, \cdot)$	$[1]_{10}$	$[3]_{10}$	$[7]_{10}$	$[9]_{10}$
$[1]_{10}$	$[1]_{10}$	$[3]_{10}$	$[7]_{10}$	$[9]_{10}$
$[3]_{10}$	$[3]_{10}$	$[9]_{10}$	$[1]_{10}$	$[7]_{10}$
$[7]_{10}$	$[7]_{10}$	$[1]_{10}$	$[9]_{10}$	$[3]_{10}$
$[9]_{10}$	$[9]_{10}$	$[7]_{10}$	$[3]_{10}$	$[1]_{10}$

Exercise 8.39. *As in Example* 8.69 *for* $\mathbb{Z}_6^*$ *and* $\mathbb{Z}_{10}^*$, *write out the multiplication tables for both* $\mathbb{Z}_{12}^*$ *and* $\mathbb{Z}_{15}^*$.

Exercise 8.40. *Let* $m \in \mathbb{Z}^+$, *and suppose that* R *is a reduced residue system modulo* m *as defined in* §6.10.1. *Prove that*

$$\{[a]_m : a \in R\} = \mathbb{Z}_m^*. \tag{8.126}$$

8.5.11. Another proof of Fermat's Little Theorem. The following gives another proof of Fermat's Little Theorem 6.63.

Exercise 8.41. *Let* p *be a prime number. Recall from Example* 8.51 *that* $(\mathbb{Z}_p^*, \cdot)$ *is a group.*

(1) *Let* $[a]_p \in \mathbb{Z}_p^*$. *Consider the subgroup* $\langle [a]_p \rangle$. *Explain why* $k := |\langle [a]_p \rangle|$ *divides* $p - 1$.
(2) *Deduce that* $[a^{p-1}]_p = ([a]_p)^{p-1} = [1]_p$.

8.5.12. Proof of Euler's Theorem. Let $n \geq 2$ be an integer. Recall that Euler's totient function $\varphi : \mathbb{Z}^+ \to \mathbb{Z}^+$ is given in Definition 6.71 by $\varphi(n)$ is the number of integers k satisfying $1 \leq k \leq n$ and $\gcd(k, n) = 1$. Recall also that

$$R_m^* := \{a \in \mathbb{N}_m : \gcd(a, m) = 1\}. \tag{8.127}$$

We have seen that the map from R_m^* to $\mathbb{Z}_m^*$ defined by $a \mapsto [a]_m$ is a bijection. In particular,

$$|\mathbb{Z}_m^*| = |R_m^*|. \tag{8.128}$$

By (8.128), we have

$$\varphi(m) = |R_m^*| = |\mathbb{Z}_m^*|. \tag{8.129}$$

Recall Euler's Theorem 6.79 (generalizing Fermat's Little Theorem), which we stated earlier and we can now prove using the group theory we have developed.

Theorem 8.70. *If positive integers* a *and* $m \geq 2$ *are coprime, then*

$$a^{\varphi(m)} \equiv 1 \bmod m. \tag{8.130}$$

In other words,

$$[a]_m^{\varphi(m)} = [1]_m. \tag{8.131}$$

Proof. By Lemma 8.68, $(\mathbb{Z}_m^*, \cdot)$ is a group. Since a and m are coprime, we have $[a]_m \in \mathbb{Z}_m^*$. Thus, by Theorem 8.67, we have

$$[a]_m^{\varphi(m)} = [a]_m^{|\mathbb{Z}_m^*|} = [1]_m \tag{8.132}$$

is the identity element of $\mathbb{Z}_m^*$. □

8.5.13. Group morphisms. Let $(G, \cdot)$ and $(H, *)$ be groups. A function $f : G \to H$ is a **homomorphism** if f "preserves" the multiplication structures; that is,

$$f(x \cdot y) = f(x) * f(y). \tag{8.133}$$

If a homomorphism f is an injection, then we say that f is a **monomorphism**.

If a homomorphism f is a surjection, then we say that f is an **epimorphism**.

If a homomorphism f is a bijection, then we say that f is an **isomorphism** and $(G, \cdot)$ and $(H, *)$ are **isomorphic** (as groups).

Example 8.71. The group $(m\mathbb{Z}, +)$ in Example 8.56 is isomorphic to $(\mathbb{Z}, +)$.

Exercise 8.42. *Prove that for each positive integer m, there are exactly two isomorphisms from $(\mathbb{Z}, +)$ to $(m\mathbb{Z}, +)$.*

Exercise 8.43. *Prove that $(\mathbb{Z}_4^*, \cdot)$, where $\mathbb{Z}_4^*$ is as in (8.75), is isomorphic as a group to $(\mathbb{Z}_2, +)$. Recall that group isomorphisms are defined in §8.5.13.*

Example 8.72. Let m be a positive integer, and let i denote the imaginary unit. Let

$$Z_m := \{e^{i\,2\pi k/m} \in \mathbb{C} : k \in \mathbb{Z}_m\}. \tag{8.134}$$

The set Z_m is closed under complex multiplication $\cdot$, and $(Z_m, \cdot)$ forms a finite group of order m, called the **group of m-th roots of unity**.

Define the function

$$f : \mathbb{Z}_m \to Z_m \tag{8.135}$$

by

$$f([k]_m) = e^{i\,2\pi k/m}. \tag{8.136}$$

Then (see Exercise 8.44 below)

$$f : (\mathbb{Z}_m, +) \to (Z_m, \cdot) \tag{8.137}$$

is not only a well-defined homomorphism, it is an isomorphism!

Observe that for each k,

$$(e^{i\,2\pi k/m})^m = 1. \tag{8.138}$$

For this reason, we call each $e^{i\,2\pi k/m}$ a **root of unity**.

For example, Z_3 consists of the 3-rd roots of unity:

$$e^{i\,2\pi\cdot 0/3} = 1, \quad e^{i\,2\pi\cdot 1/3} = \frac{1}{2} + \frac{\sqrt{3}}{2}i, \quad e^{i\,2\pi\cdot 2/3} = \frac{1}{2} - \frac{\sqrt{3}}{2}i. \tag{8.139}$$

Exercise 8.44. *Prove that, as claimed above, the function $f : (\mathbb{Z}_m, +) \to (Z_m, \cdot)$ defined by (8.137) is a well-defined group isomorphism.*

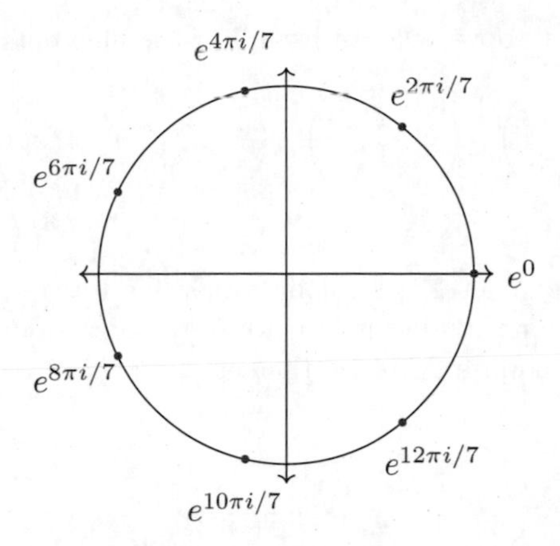

Figure 8.5.1. The 7-th roots of unity. The group $(Z_7, \cdot)$ is isomorphic to $(\mathbb{Z}_7, +)$.

8.5.14. The symmetric group. Let X be any set. A **permutation** σ of X is defined to simply be a bijection $\sigma : X \to X$.

Let $S(X)$ be the set of bijections from X to itself. Recall from Theorem 5.52(3) that the composition of two bijections is a bijection. Thus, composition $\circ$ is a binary operation on $S(X)$.

Lemma 8.73. *For any set X, $(S(X), \circ)$ is a group, called the symmetric group of X.*

Proof. We verify the group properties in Definition 8.41.

(1) By Theorem 5.39(1), composition is associative.
(2) Let 1_X denote the identity map. By Theorem 5.39(1), we have
$$1_X \circ \sigma = \sigma \circ 1_X = \sigma.$$
(3) Let $\sigma \in S(X)$. Since $\sigma : X \to X$ is a bijection, by Theorem 5.53, there exists an inverse $\sigma^{-1} : X \to X$. This is indeed the group inverse:

$$\sigma \circ \sigma^{-1} = \sigma^{-1} \circ \sigma = 1_X. \tag{8.140}$$ □

Example 8.74. Let $X = \mathbb{N}_3 = \{1, 2, 3\}$. Let us express a function $f : \mathbb{N}_3 \to \mathbb{N}_3$ by the 2×3 matrix

$$\begin{pmatrix} 1 & 2 & 3 \\ f(1) & f(2) & f(3) \end{pmatrix}. \tag{8.141}$$

For example, the matrix

$$\begin{pmatrix} 1 & 2 & 3 \\ 2 & 3 & 2 \end{pmatrix} \tag{8.142}$$

denotes the function g defined by $g(1) = 2$, $g(2) = 3$, $g(3) = 2$. Observe that g is not a bijection and hence is not a permutation. So $g \notin S(\mathbb{N}_3)$.

Using the notation above, we can list all of the elements of $S(\mathbb{N}_3)$ as follows:
(8.143)

$$\begin{pmatrix} 1 & 2 & 3 \\ 1 & 2 & 3 \end{pmatrix}, \begin{pmatrix} 1 & 2 & 3 \\ 1 & 3 & 2 \end{pmatrix}, \begin{pmatrix} 1 & 2 & 3 \\ 2 & 1 & 3 \end{pmatrix}, \begin{pmatrix} 1 & 2 & 3 \\ 2 & 3 & 1 \end{pmatrix}, \begin{pmatrix} 1 & 2 & 3 \\ 3 & 1 & 2 \end{pmatrix}, \begin{pmatrix} 1 & 2 & 3 \\ 3 & 2 & 1 \end{pmatrix}.$$

So $|S(\mathbb{N}_3)| = 6$.

Example 8.75. More generally, let n be a positive integer, and let X be a finite set with n elements. For this discussion, we only care about the cardinality of X, so for convenience we choose X to be the set $\mathbb{N}_n = \{1, 2, \ldots, n\}$ consisting of the first n positive integers.

We denote

$$S_n := S(\mathbb{N}_n), \tag{8.144}$$

which is called the symmetric group *on n elements* (or *over n symbols*).

By definition, a permutation σ of the set $\mathbb{N}_n$ is defined simply to be a bijection $\sigma : \mathbb{N}_n \to \mathbb{N}_n$. Observe that σ is a bijection if and only if the n elements $\sigma(1), \sigma(2), \ldots, \sigma(n)$ are distinct.

We can count the number of bijections as follows:

(1) Firstly, there are n choices for $\sigma(1)$: namely any one of $1, 2, \ldots, n$.

(2) Secondly, independent of the choice of $\sigma(1)$, there are $n-1$ choices for $\sigma(2)$ since σ being bijective implies that $\sigma(2) \neq \sigma(1)$. So $\sigma(2)$ can be any element in the set

$$\mathbb{N}_n - \{\sigma(1)\}.$$

(3) Thirdly, independent of the choices of $\sigma(1)$ and $\sigma(2)$, there are $n-2$ choices for $\sigma(3)$ since σ being bijective implies that $\sigma(3)$ is an element in the set

$$\mathbb{N}_n - \{\sigma(1), \sigma(2)\}.$$

(4) Continuing in this way (and no, we will not try to formalize this argument by induction!), we see that for the n successive choices of $\sigma(i)$, $1 \leq i \leq n$, since σ is a bijection, the numbers of possible choices are shown in Table 8.5.2.

Table 8.5.2. The top row entries are the elements of $\mathbb{N}_n$: i, where $1 \leq i \leq n$. The bottom row entries are the corresponding number of possible choices for $\sigma(i)$.

choice	1	2	3	$\cdots$	$n-1$	n
number of choices	n	$n-1$	$n-2$	$\cdots$	2	1

(5) By the multiplicative law, we see that the cardinality

$$|S(\mathbb{N}_n)| = n \cdot (n-1) \cdot (n-2) \cdots 2 \cdot 1 = n!. \tag{8.145}$$

Alternatively, (8.145) follows from Corollary 7.38.

This also agrees with Example 8.74, where we saw that $|S(\mathbb{N}_3)| = 6 = 3!$.

We say that a permutation τ of a set X is a **transposition** if there exist two distinct elements x_1 and x_2 of X such that

$$\tau(x) = x \text{ if } x \in X - \{x_1, x_2\}, \quad \tau(x_1) = x_2, \quad \tau(x_2) = x_1. \tag{8.146}$$

That is, τ switches two elements while keeping all other elements the same.

We end this subsection with two fundamental results, which we will not prove.

Fact 8.76. *If X is a finite set, then any permutation of X is the composition of transpositions.*

Fact 8.77 (Parity of permutations). *Let σ be a permutation of a finite set X with $|X| \geq 2$. If σ is the composition of k transpositions, then either*

(1) *k is always even, in which case we say that σ is an **even permutation**, or*

(2) *k is always odd, in which case we say that σ is an **odd permutation**.*

Non-Example 8.78. *Using the symmetric group S_3, we can give a simple example for why we need a subgroup to be normal in order to define the quotient group. Let $i, j, k \in \mathbb{N}_3$ be distinct. Let τ_{ij} denote the tranposition $\begin{pmatrix} i & j & k \\ j & i & k \end{pmatrix}$, let π_{ijk} denote the permutation $\begin{pmatrix} i & j & k \\ j & k & i \end{pmatrix}$, and let $e = 1_{\mathbb{N}_3}$ denote the identity element. Consider the subgroup $H = \{e, \tau_{12}\}$ of S_3 of order 2. Using that multiplication in the group S_3 is given by composition bijections of $\mathbb{N}_3$, we see that the left cosets of H in S_3 are*

$$\begin{aligned} eH &= \tau_{12}H = H, \\ \pi_{123}H &= \tau_{13}H = \{\pi_{123}, \tau_{13}\}, \\ \pi_{132}H &= \tau_{23}H = \{\pi_{132}, \tau_{23}\}. \end{aligned}$$

From this we can easily deduce that left coset multiplication is not well-defined. For example, computing multiplication using different representatives of the same cosets, we have

$$\pi_{123}\pi_{132}H = H, \quad \text{whereas} \quad \tau_{13}\tau_{23}H = \pi_{132}H \neq H.$$

8.5.15*. Applications of group theory. Group theory has been applied to study the mathematics of Rubik's cube and the mathematics of Sudoku. We invite you to click on the Wikipedia links to learn more.

8.6. Rings, principal ideal domains, and all that

Recall that we discussed the *ring* of integers in §8.5.1. In general, we have:

Definition 8.79. A **ring** is a set R with binary operations $+$ called **addition** and $\cdot$ called **multiplication** such that the following hold:

(1) $(R, +)$ is an abelian group (see Definition 8.43).

(2) $(R, \cdot)$ is a monoid.

(3) (Distributivity) For all $x, y, z \in R$,

$$x \cdot (y + z) = x \cdot y + x \cdot z, \qquad (y + z) \cdot x = y \cdot x + z \cdot x. \tag{8.147}$$

For example, $\mathbb{Z}$, $\mathbb{Q}$, and $\mathbb{R}$, with the usual addition $+$ and multiplication $\cdot$, are all rings.

A ring is an integral domain if $x \neq 0$ and $y \neq 0$ implies that $x \cdot y \neq 0$.

Lemma 8.80 (Cancellation property). *For any $x \neq 0$, if $x \cdot y = x \cdot z$, then $y = z$.*

Proof. Exercise! □

For example, $(\mathbb{Z}, +, \cdot)$ is an integral domain; in fact it is the prototypical example.

If p is a prime, then $(\mathbb{Z}_p, +, \cdot)$ is an integral domain.

Example 8.81. The ring $\mathbb{Z}[x]$ of polynomials in a single variable with integer coefficients is an integral domain with the usual addition and multiplication. We leave it to the reader to check the ring structure. Moreover, clearly the product of two non-zero polynomials with integer coefficients is a non-zero polynomial with integer coefficients.

On the other hand, if m is a positive integer, then $(\mathbb{Z}_m, +, \cdot)$ is a ring. However, unless m is a prime, it is not an integral domain. Indeed, if $m = ab$ where $a, b > 1$ is a composite number, then $[a]_m \neq [0]_m$ and $[b]_m \neq [0]_m$, but $[a]_m \cdot [b]_m = [0]_m$.

Definition 8.82. Let $(R, +, \cdot)$ be a ring and let S be a subset that is closed under both $+$ and $\cdot$; that is, the sums and products of elements in S remain in S. We say that $(S, +, \cdot)$ is a **subring** if it is a ring. By the uniqueness of the multiplicative identity, the multiplicative identity of S is equal to the multiplicative identity of R.

Example 8.83. The ring $(\mathbb{Z}, +, \cdot)$ of integers has only itself as a subring. Indeed, if S is a subring of $\mathbb{Z}$, then $1 \in S$. This implies that finite sums of 1 are in S. Since the negatives of integers in S are also in S, and hence $0 \in S$, we conclude that $S = \mathbb{Z}$.

Example 8.84. The ring $(\mathbb{Z}, +, \cdot)$ of integers is a subring of the ring of rational numbers $(\mathbb{Q}, +, \cdot)$, which in turn is a subring of the ring of real numbers $(\mathbb{R}, +, \cdot)$.

Non-Example 8.85. *Consider the ring $(\mathbb{Q}, +, \cdot)$, and define the equivalence relation $\sim$ on $\mathbb{Q}$ by $a \sim b$ if and only if $a - b \in \mathbb{Z}$. As a set, we can consider the quotient set $\mathbb{Q}/\sim$ as $\mathbb{Q} \cap [0, 1)$. However, addition and multiplication on $\mathbb{Q}$ do not descend to the quotient $\mathbb{Q}/\sim$. Note that $[1] = [0]$, so we would have $[0] = [1] \cdot [r] \neq [1 \cdot r] = [r]$ unless $r \in \mathbb{Z}$. This shows in general that one cannot take the quotient of a ring by a subring in a meaningful way.*

To be able to take quotients of a ring by a subset, we have the following notion.

Definition 8.86. A (two-sided) **ideal** I of a ring $(R, +, \cdot)$ is a subset satisfying the following:

(1) $(I, +)$ is a subgroup of $(R, +)$.

(2) For every $i \in I$ and $r \in R$, we have both $r \cdot i \in I$ and $i \cdot r \in I$.

Observe that if an ideal I of a ring R contains the multiplicative identity e, then $I = R$. For example, $\mathbb{Z}$ is not an ideal of the ring $(\mathbb{Q}, +, \cdot)$. In fact, the only ideals of $(\mathbb{Q}, +, \cdot)$ are $\{0\}$ and itself.

We leave it to the reader to imagine what the definitions of **left ideal** and **right ideal** are (you may check your guess at the "Ideal (ring theory)" Wikipedia link for example). Of course, if the ring is commutative, then left ideals and right ideals are each two-sided ideals.

Example 8.87. For the ring of integers $(\mathbb{Z}, +, \cdot)$, $m\mathbb{Z}$ is an ideal for each positive integer m.

We say that an integral domain is a principal ideal domain if each ideal is generated by a single element. That is, for each ideal I, there exists an element a such that $I = \langle a \rangle$.

Example 8.88. The ring of integers $(\mathbb{Z}, +, \cdot)$ is a principal ideal domain.

8.6.1. The quotient ring. Let $(R, +, \cdot)$ be a ring and let I be an ideal of R. The quotient set $R/I := R/\sim$ has as elements the equivalence classes

$$[r] = r + I = \{r + i | i \in I\}. \tag{8.148}$$

We leave it as an exercise to the reader to show that if we define addition and multiplication on R/I by

$$[r] + [s] := [r + s] \quad \text{and} \quad [r] \cdot [s] := [r \cdot s], \tag{8.149}$$

respectively, then $(R/I, +, \cdot)$ is a ring. This ring is called the **quotient ring**.

8.6.2. Ideals in the ring of integers. We show that Example 8.87 gives all of the ideals in the ring of integers. Let I be an ideal of the ring of integers $(\mathbb{Z}, +, \cdot)$, which is not trivial; i.e., $I \neq \{0\}$.

Exercise 8.45. *Prove that* $I \cap \mathbb{Z}^+ \neq \emptyset$.

Let $S := I \cap \mathbb{Z}^+$. By the well-ordering principle, $m := \min S$ exists.

Exercise 8.46. *Prove that* $I = m\mathbb{Z}$. *Hint: Firstly, show that* $m\mathbb{Z} \subset I$.

Definition 8.89. We say that an ideal I in a ring $(R, +, \cdot)$ is a **prime ideal** if $a, b \in R$ and $a \cdot b \in I$ implies that $a \in I$ or $b \in I$.

Exercise 8.47. *Prove that if an integer* $m \geq 2$ *is composite, then the ideal* $m\mathbb{Z}$ *is not a prime ideal.*

Exercise 8.48. *Let* p *be a prime number. Prove that the ideal* $p\mathbb{Z}$ *is a prime ideal.*

The quotient ring $\mathbb{Z}/m\mathbb{Z}$ is denoted by $(\mathbb{Z}_m, +, \cdot)$. Note that the ring $(\mathbb{Z}_6, +, \cdot)$ has zero divisors in the sense that $[2] \cdot [3] = [0]$. On the other hand, if p is prime, then $(\mathbb{Z}_p, +, \cdot)$ does not have zero divisors. Exercise: Prove this!

8.7. Fields

In algebra, fields do not mean (sadly!) Fields of Dreams or even force fields for that matter!

8.7.1. Familiar examples. In §8.5 we discussed *groups*, which have a single binary operation, with $(\mathbb{Z}, +)$ as an infinite prototype and $(\mathbb{Z}_m, +)$ as a finite prototype.

Earlier, in §8.5.1 we briefly discussed the notion of a commutative ring, which has two binary operations, one called addition and one called multiplication. Here, the ring of integers $(\mathbb{Z}, +, \cdot)$ is a prototype (there are more general rings of integers). Other prototypical rings are the ring of rational numbers $\mathbb{Q}$ and the ring of real numbers $\mathbb{R}$, both with the same addition and multiplication as for the integers $\mathbb{Z}$.[2]

For the ring of integers, one could say it has the following defect: Not all integers have multiplicative inverses inside the set of integers. In fact, the only integers with multiplicative inverses are 1 and -1. On the other hand, any non-zero rational number has a multiplicative inverse in $\mathbb{Q}$. Similarly, any non-zero real number has a multiplicative inverse in $\mathbb{R}$.

8.7.2. Definition of a field. If we axiomatize these properties, we obtain the notion of a field.

Definition 8.90. A **field** is a set K with binary operations $+$ and $\cdot$ on K satisfying the following properties:

(1) (Associativity) For any $x, y, z \in K$,

$$(x + y) + z = x + (y + z), \tag{8.150}$$

$$(x \cdot y) \cdot z = x \cdot (y \cdot z). \tag{8.151}$$

(2) (Commutativity) For any $x, y \in K$,

$$x + y = y + x, \tag{8.152}$$

$$x \cdot y = y \cdot x. \tag{8.153}$$

(3) (Identity element) There exist elements $0 \in K$ and $e \in K$, with $0 \neq e$, satisfying: For all $x \in K$,

$$x + 0 = 0 + x = x, \tag{8.154}$$

$$x \cdot e = e \cdot x = x. \tag{8.155}$$

(4) (Existence of an inverse) $0 + 0 = 0$, and for each $x \in K - \{0\}$, there exist elements $-x$ and x^{-1} satisfying

$$x + (-x) = (-x) + x = 0, \tag{8.156}$$

$$x \cdot x^{-1} = x^{-1} \cdot x = e. \tag{8.157}$$

(5) (Distributivity) For any $x, y, z \in K$,

$$x \cdot (y + z) = x \cdot y + x \cdot z. \tag{8.158}$$

[2] More precisely, the addition and multiplication operations *extend* from the case of the integers.

We note that since addition and multiplication are commutative, in the "identity element" and "existence of an inverse" axioms it was redundant to switch the order of the additions and multiplications.

In other words, a field is a commutative ring with the properties that $0 \neq e$ and each non-zero element has a multiplicative inverse.

8.7.3. The fields $\mathbb{Q}$, $\mathbb{R}$, and $\mathbb{C}$. As we have mentioned in the previous subsection, both $\mathbb{Q}$ and $\mathbb{R}$ are fields. We leave it to the reader to spend a moment to check this.

8.7.3.1. *The complex numbers* $\mathbb{C}$. One problem with the field of real numbers is that not all polynomial equations have solutions! Take for example the equation

$$x^2 + 1 = 0. \tag{8.159}$$

This quadratic polynomial equation has no solutions in the field of real numbers.

This can be rectified by defining the *complex* numbers.

We can define the field of complex numbers $(\mathbb{C}, +, \cdot)$ by introducing an indeterminate i, called the imaginary unit, and making a definition using elementary constructions in formal algebra. We sidestep this method by defining the field of complex numbers in the following way.

As a set, we let $\mathbb{C}$ be equal to the cartesian product $\mathbb{R}^2 = \mathbb{R} \times \mathbb{R}$ (see §5.2). We define **complex addition** $+$ to be the usual vector addition:

$$(x_1, y_1) + (x_2, y_2) = (x_1 + x_2, y_1 + y_2). \tag{8.160}$$

With the motivation of $(0, 1)$ corresponding to i satisfying $i^2 = -1$, we define **complex multiplication** $\cdot$ by

$$(x_1, y_1) \cdot (x_2, y_2) = (x_1 x_2 - y_1 y_2,\ x_1 y_2 + y_1 x_2). \tag{8.161}$$

As a special case, we see that

$$(0, 1) \cdot (0, 1) = (-1, 0). \tag{8.162}$$

We leave it as an exercise for the reader to show that $(\mathbb{C}, +, \cdot)$ is a field. This is an easy exercise in the sense that each field axiom in Definition 8.90 is straightforward to check, but you do have to check a handful of axioms.

A natural way to think of the set of real numbers as a subset of the set of complex numbers is to define the injection

$$\phi : \mathbb{R} \to \mathbb{C}$$

by

$$\phi(x) = (x, 0). \tag{8.163}$$

Via this injection ϕ, we can think of a real number as a special case of a complex number. Namely, given $x \in \mathbb{R}$, we identify it with $\phi(x) = (x, 0) \in \mathbb{C}$ and still call the image of x under ϕ to be x. So $x \in \mathbb{C}$ with this notation. In this sense,

we consider $\mathbb{R}$ as a subset of $\mathbb{C}$

even though it is really $\phi(\mathbb{R}) \subset \mathbb{C}$.

We also use the notation

$$i := (0, 1). \tag{8.164}$$

With this notation, we write $x + yi$ for (x, y). Complex multiplication becomes

$$(x_1 + y_1 i) \cdot (x_2 + y_2 i) = (x_1 x_2 - y_1 y_2) + (x_1 y_2 + y_1 x_2)\, i. \tag{8.165}$$

We abbreviate for $x, y \in \mathbb{R}$,

$$x + 0i =: x, \qquad 0 + yi =: yi. \tag{8.166}$$

We also abbreviate $0 + 1i = 1i =: i$. The **origin** is the element $0 = 0 + 0i$.

Let $z := x + yi \in \mathbb{C}$. We define

$$z^2 := z \cdot z = (x + yi) \cdot (x + yi) = (x^2 - y^2) + 2xyi. \tag{8.167}$$

In particular,

$$i^2 = -1; \tag{8.168}$$

that is, $z = i$ is a solution to

$$z^2 + 1 = 0. \tag{8.169}$$

We leave it to the reader to check that $z = -i$ is another solution to (8.169).

We define $z^0 := 1$ and for $k \in \mathbb{Z}^+$,

$$z^k := \underbrace{z \cdot z \cdots z}_{k \text{ times}}, \tag{8.170}$$

where we have taken advantage of the fact that multiplication is associative to omit the brackets. To be more rigorous, we should define z^k by induction:

(1) $z^0 := 1$.

(2) $z^k := z \cdot z^{k-1}$ for $k \in \mathbb{Z}^+$.

The **complex conjugate** of a complex number $z = x + yi$ is defined by

$$\overline{z} := x - yi. \tag{8.171}$$

The **modulus** (or **absolute value**) of a complex number $z = x + yi$ is defined by

$$|z| := \sqrt{x^2 + y^2}. \tag{8.172}$$

This is equal to the distance from (x, y) to $(0, 0)$ in the Euclidean plane $\mathbb{R}^2$. Observe that

$$|z|^2 = \overline{z}z = z\overline{z}. \tag{8.173}$$

The **principal value of the argument**

$$\operatorname{Arg}(z)$$

of a non-zero complex number $z = x + yi$ is equal to the unique angle (in radians) from the positive x-axis to the vector (x, y) satisfying

$$\operatorname{Arg}(z) \in (-\pi, \pi].$$

Example 8.91. (1) $\operatorname{Arg}(-4) = \pi$.

(2) $\operatorname{Arg}\left(1, -\sqrt{3}\right) = -\frac{\pi}{3}$.

(3) $\operatorname{Arg}\left(-\frac{3}{\sqrt{2}}, \frac{3}{\sqrt{2}}\right) = \frac{3\pi}{4}$.

The **argument**

$$\arg(z)$$

of a non-zero complex number $z = x + yi$ is the set of all angles from the positive x-axis to the vector (x, y). So $\arg(z)$ is a multi-valued function.

Example 8.92. (1) $\arg(-4) = \{\pi + 2\pi n : n \in \mathbb{Z}\}$.

(2) $\arg\left(1, -\sqrt{3}\right) = \{-\frac{\pi}{3} + 2\pi n : n \in \mathbb{Z}\}$.

(3) $\arg\left(-\frac{3}{\sqrt{2}}, \frac{3}{\sqrt{2}}\right) = \{\frac{3\pi}{4} + 2\pi n : n \in \mathbb{Z}\}$.

For any complex number z, we have

$$z = |z| \left(\cos(\arg(z)) + \sin(\arg(z))i\right). \tag{8.174}$$

The modulus and argument have the following beautiful properties of Lemma 8.93 (we leave it to you to work out or look up the proofs).

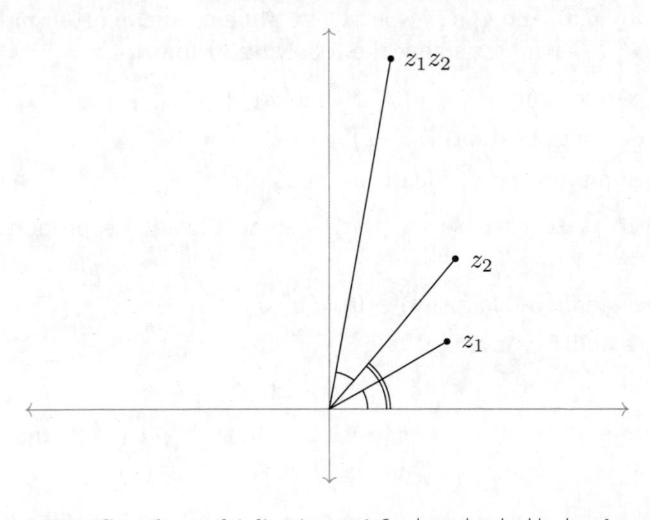

Figure 8.7.1. Complex multiplication satisfies $|z_1 z_2| = |z_1|\,|z_2|$ and $\arg(z_1 z_2) = \arg(z_1) + \arg(z_2)$.

Lemma 8.93. *Let z_1 and z_2 be complex numbers. Then*

(1) $|z_1 z_2| = |z_1|\,|z_2|$.

(2) $\arg(z_1 z_2) = \arg(z_1) + \arg(z_2)$.

(3) (*Triangle inequality*) $|z_1 + z_2| \leq |z_1| + |z_2|$.

The triangle inequality is actually a special case of the triangle inequality for Euclidean n-space ($n = 2$); see Corollary 5.25.

Example 8.94. Let $z_1 = 1 + \sqrt{3}i$ and $z_2 = -3 + 3i$. We calculate that

$$z_1 z_2 = (1 + \sqrt{3}i)(-3 + 3i) = -3(1 + \sqrt{3}) + 3(1 - \sqrt{3})i.$$

Observe that

$$|z_1| = 2, \quad |z_2| = 3, \quad |z_1 z_2| = 6,$$

which agrees with Lemma 8.93(1). Further observe that

$$\operatorname{Arg}(z_1) = \frac{\pi}{3}, \qquad \operatorname{Arg}(z_2) = \frac{3\pi}{4}.$$

Thus, by Lemma 8.93(2), we have

$$\operatorname{Arg}(z_1 z_2) = \frac{\pi}{3} + \frac{3\pi}{4} - 2\pi = -\frac{11}{12}\pi$$

where we needed to subtract the 2π to obtain an answer in the interval $(-\pi, \pi]$. Thus, by (8.174), we have

$$\cos\left(-\frac{11}{12}\pi\right) = -\frac{1+\sqrt{3}}{2}, \qquad \sin\left(-\frac{11}{12}\pi\right) = \frac{1-\sqrt{3}}{2}.$$

8.7.3.2. *Field homomorphisms.* Roughly speaking, a homomorphism from one algebraic structure to another of the same type (whether it be groups, rings, fields, etc.) is a function which preserves the algebraic structures.

Let $(K, +, \cdot)$ and $(K', +', \cdot')$ be fields and denote their additive and multiplicative identities by 0, $0'$ and e, e', respectively. A field homomorphism from K to K' is a function $\phi : K \to K'$ satisfying the following properties:

(1) (Preserves addition) $\phi(x + y) = \phi(x) +' \phi(y)$.

(2) (Preserves multiplication) $\phi(x \cdot y) = \phi(x) \cdot' \phi(y)$.

(3) (Preserves multiplicative identities) $\phi(e) = e'$.

We leave it as an exercise for you to show that these properties imply the following:

(1) (Preserves additive identities) $\phi(0) = 0'$.

(2) (Preserves additive inverses)

$$\phi(-x) = -\phi(x). \tag{8.175}$$

(3) (Preserves multiplicative inverses) If $x \neq 0$, then $\phi(x) \neq 0'$ and

$$\phi(x^{-1}) = (\phi(x))^{-1}. \tag{8.176}$$

Here, $-$ and $^{-1}$ denote the additive and multiplicative inverses, respectively, in both fields K and K'.

Example 8.95. The injective function $\phi : \mathbb{R} \to \mathbb{C}$ defined by (8.163) is a field homomorphism.

If a field homomorphism $\phi : K \to K'$ is a bijection, then we say that ϕ is a **field isomorphism**. In this case we say that the fields K and K' are **isomorphic**.

It turns out that for a field isomorphism ϕ, its inverse ϕ^{-1} is also a homomorphism and hence an isomorphism.

8.7.3.3. *Subfields.* Let L be a non-empty subset of a field $(K, +, \cdot)$. We say that $(L, +, \cdot)$ is a **subfield** of $(K, +, \cdot)$ if L is closed under both addition and multiplication and if $(L, +, \cdot)$ is itself a field. That is (we leave it as an exercise to show that the following list of conditions is equivalent):

(1) (Closed under addition and multiplication) If $x, y \in L$, then $x + y \in L$ and $x \cdot y \in L$.
(2) (Additive inverse exists) If $x \in L$, then $-x \in L$. Since L is non-empty, this implies $0 \in L$.
(3) $e \in L$.
(4) (Multiplicative inverse exists) If $x \in L - \{0\}$, then $x^{-1} \in L$.

Example 8.96. The subset $\phi(\mathbb{R})$ of $\mathbb{C}$, where ϕ is defined by (8.163), is a subfield. Since we like to identify $\mathbb{R}$ with $\phi(\mathbb{R})$,

we can consider $\mathbb{R}$ as a subfield of $\mathbb{C}$.

8.7.3.4. *Polynomials.* The following result is so important that we may call it the Big Fundamental.

8.7.3.5. *The Fundamental Theorem of Algebra.* The following is the Fundamental Theorem of Algebra.

Theorem 8.97. *Let $a_0, a_1, \ldots, a_n$ be complex numbers, where $n \in \mathbb{Z}^+$ and $a_n \neq 0$. Then the complex polynomial defined by*

$$f(z) := \sum_{k=0}^{n} a_k z^k = a_n z^n + a_{n-1} z^{n-1} + \cdots + a_1 z + a_0 \quad \text{for } z \in \mathbb{C} \tag{8.177}$$

has a zero; i.e., there exists $z \in \mathbb{C}$ such that $f(z) = 0$.

Proof. Using that f is a polynomial, we will show:

Step 1: *The function*

$$z \mapsto |f(z)|$$

achieves its minimum at some point $z_0 \in \mathbb{C}$. That is, there is a $z_0 \in \mathbb{C}$ where $f(z_0)$ is closest to the origin. Of course, later we want to show that $f(z_0)$ *is* the origin.

The infimum of $z \mapsto |f(z)|$ exists since it is a non-negative function and hence bounded from below by 0. The point is to show that the infimum is attained at some point; i.e., the minimum exists (the minimum of a function might not exist in general).

Slightly simplifying assumption: Note that since $a_n \neq 0$, we have

$$\left| \frac{f(z)}{a_n} \right| = \frac{|f(z)|}{|a_n|} \quad \text{and} \quad \inf_{z \in \mathbb{C}} \frac{|f(z)|}{|a_n|} = \frac{1}{|a_n|} \inf_{z \in \mathbb{C}} |f(z)|. \tag{8.178}$$

Thus, by dividing $f(z)$ by a_n, we may assume without loss of generality that

$$a_n = 1.$$

Recall, from Lemma 8.93(1), that $|zw| = |z|\,|w|$. In particular, $|-z| = |z|$, and $|z^n| = |z|^n$ for $n \in \mathbb{Z}^+$ (the fastidious reader may prove this last fact by induction). By the triangle inequality (Lemma 8.93(3) and, for us fastidious ones, induction), we have

$$|z_1 + \cdots + z_n| \leq |z_1| + \cdots + |z_n|. \tag{8.179}$$

Exercise 8.49. *Prove that for every $z_1, z_2, \ldots, z_n \in \mathbb{C}$,*

$$|z_1 + \cdots + z_n| \geq |z_n| - |z_1| - \cdots - |z_{n-1}|. \tag{8.180}$$

Hint:

$$(z_1 + \cdots + z_n) + (-z_1) + \cdots + (-z_{n-1}) = z_n. \tag{8.181}$$

Denote $r := |z|$, which is equal to the distance from z to the origin 0. The idea is that size-wise the leading order term of f, i.e., z^n, should dominate the rest of the terms when r is large. Since

$$f(z) = z^n + \sum_{k=0}^{n-1} a_k z^k,$$

using the triangle inequality (see Exercise 8.49) we may estimate

$$\begin{aligned} |f(z)| &\geq |z^n| - \sum_{k=0}^{n-1} |a_k z^k| \\ &= r^n - \sum_{k=0}^{n-1} |a_k|\, r^k \\ &= r^n \left(1 - \sum_{k=0}^{n-1} |a_k|\, r^{k-n}\right). \end{aligned}$$

Since each power $k - n$ of r in the sum on the right-hand side of the last line is negative, we have

$$\lim_{r \to \infty} \sum_{k=0}^{n-1} |a_k|\, r^{k-n} = 0.$$

This implies the key fact that

$$\lim_{z \to \infty} |f(z)| = \infty. \tag{8.182}$$

Recall the following extreme-value theorem.

Theorem 8.98. *Let $f : X \to \mathbb{R}$ be a continuous function on a closed and bounded subset X of $\mathbb{R}^n$. The f attains both its minimum and its maximum.*

With this, we can proceed to show that the minimum of $|f|$ is attained. Without loss of generality, we may assume that $f(0) \neq 0$ for otherwise we are done.

Define the constant $M := |f(0)| > 0$. Since $\lim_{z\to\infty} |f(z)| = \infty$, there exists a positive real number R such that:

(8.183) $$\text{If } |z| \geq R, \text{ then } |f(z)| \geq M = |f(0)|.$$

Consider the closed ball

$$\bar{B} := \{z \in \mathbb{C} : |z| \leq R\}.$$

Since the set $\bar{B}$ is closed and bounded, by Theorem 8.98, the restriction $f|_{\bar{B}}$ of f to $\bar{B}$ achieves its minimum, since f is a continuous function. That is, there exists $z_0 \in \mathbb{C}$ such that

(8.184) $$|f(z_0)| = \inf_{z\in\bar{B}} |f(z)|.$$

We now show that z_0 is a minimum point of $|f|$ on all of $\mathbb{C}$. Using (8.183), we see that

(8.185) $$\inf_{z\in\mathbb{C}-\bar{B}} |f(z)| \geq M = |f(0)| \geq \inf_{z\in\bar{B}} |f(z)| = |f(z_0)|,$$

with the second inequality because $0 \in \bar{B}$. Since $\mathbb{C} = \bar{B} \cup (\mathbb{C} - \bar{B})$, by (8.184) and (8.185) we conclude that

$$|f(z_0)| = \inf_{z\in\mathbb{C}} |f(z)|.$$

This completes Step 1.

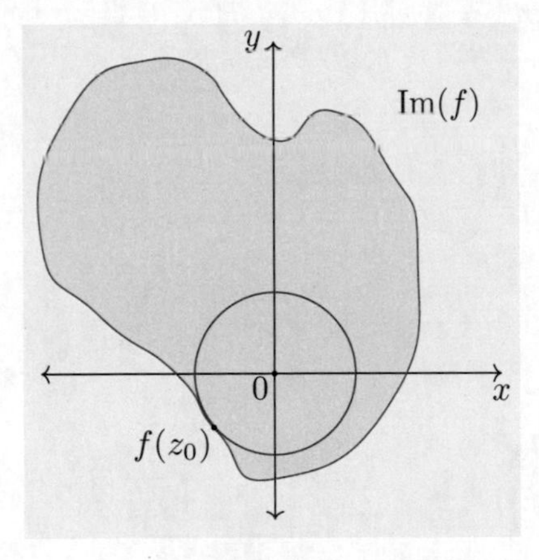

Figure 8.7.2. The land-dwelling function f sends inputs to the grassy area and avoids the lake in the middle. Here, $f(z_0)$, who has just learned to swim, is the most adventurous output (the origin 0 is the deepest part of the lake).

The theorem follows immediately from:

Step 2: *If there does not exist a zero of $f(z)$, then we have a contradiction.*

Suppose, for a contradiction, that there does not exist a zero of $f(z)$. Then $|f(z_0)| > 0$. So we may adjust the polynomial $f(z)$ by defining

$$g(z) := \frac{f(z + z_0)}{f(z_0)}.$$

We have $g(0) = 1$ and

(8.186) $$|g(z)| \geq 1 = |g(0)| \quad \text{for all } z \in \mathbb{C};$$

this is true because $|f(z + z_0)| \geq |f(z_0)|$ for all z.

Since g is a degree-n polynomial, we have

$$g(z) = \sum_{k=0}^{n} b_k z^k = b_n z^n + b_{n-1} z^{n-1} + \cdots + b_1 z + b_0,$$

where $b_n \neq 0$. Note that $b_0 = g(0) = 1$.

Let j be the smallest positive integer such that $b_j \neq 0$. Then $b_1 = \cdots = b_{j-1} = 0$, and we have

$$g(z) = b_n z^n + \cdots + b_j z^j + 1.$$

Since $\frac{-b_j}{|b_j|}$ is a unit vector, we may write it as

$$\frac{-b_j}{|b_j|} =: e^{-i(j\theta)},$$

where i is the imaginary unit complex number and θ is a real number. Why we do this will be clear in a moment. That is,

(8.187) $$b_j e^{i(j\theta)} = -|b_j|.$$

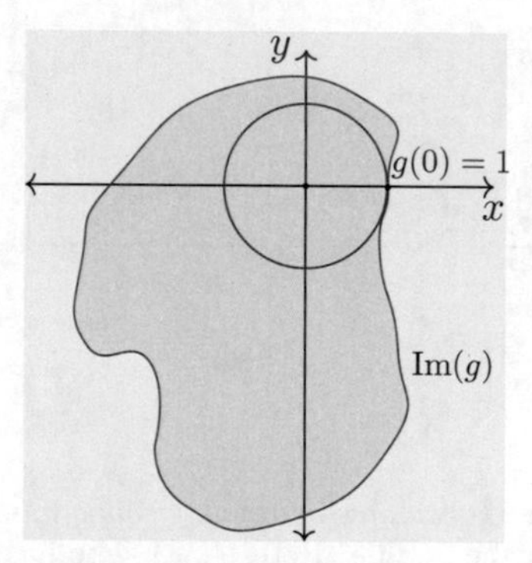

Figure 8.7.3. By a rotation and stretch, we can change the polynomial f to the polynomial g, where $g(z) = 1 + b_j z^j + \cdots b_n z^n$ and $b_j \neq 0$. When z is small, $g(z) \approx 1 + b_j z^j$. For suitable $z \approx 0$, $g(z)$ has jumped into the lake (see (8.188)), a contradiction.

Keep in mind that we are aiming for a contradiction. Consider points of the form $z = re^{i\theta}$, where $r > 0$. We have

$$g\left(re^{i\theta}\right) = b_n r^n e^{i(n\theta)} + \cdots + b_j r^j e^{i(j\theta)} + 1.$$

So, assuming that $r > 0$ is small enough so that $r < |b_j|^{-1/j}$, i.e., $-|b_j|r^j + 1 > 0$, we have by (8.187) that

$$\begin{aligned}\left|g\left(re^{i\theta}\right)\right| &\leq \sum_{k=j+1}^{n}\left|b_k r^k e^{i(k\theta)}\right| + \left|b_j r^j e^{i(j\theta)} + 1\right| \\ &= \sum_{k=j+1}^{n}|b_k|\, r^k - |b_j|\, r^j + 1 \\ &= \left(\sum_{k=j+1}^{n}|b_k|\, r^{k-j} - |b_j|\right) r^j + 1,\end{aligned}$$

where $|b_j| > 0$. Since

$$\lim_{r\to 0}\left(\sum_{k=j+1}^{n}|b_k|\, r^{k-j} - |b_j|\right) = -|b_j| < 0$$

and since the function $r \mapsto \sum_{k=j+1}^{n}|b_k|\, r^{k-j} - |b_j|$ is continuous, we have for $r > 0$ sufficiently small and θ chosen as above that

$$\left|g\left(re^{i\theta}\right)\right| < 1. \tag{8.188}$$

This contradicts $|g(z)| \geq 1$ for all $z \in \mathbb{C}$ from (8.186). The Fundamental Theorem of Algebra is proven! $\square$

8.7.4. Finite fields. An interesting topic is that of factoring polynomials over finite fields. We do not discuss this here.

8.7.5. Algebraic structures that are not quite fields. Of special mention are the quaternions. They form a non-commutative associative division algebra. Also interesting are the octonions, which form a non-associative non-commutative division algebra.

8.7.5.1. *The quaternions.*

Example 8.99. Although we will not go into detail about the quaternions $\mathbb{H}$, their elements can be represented by expressions of the form

$$x + yi + zj + wk, \tag{8.189}$$

where $x, y, z, w \in \mathbb{R}$ and i, j, k satisfy

$$i^2 = j^2 = k^2 = -1 \tag{8.190}$$

and

$$i \cdot j = -j \cdot i = k, \quad j \cdot k = -k \cdot j = i, \quad k \cdot i = -i \cdot k = j. \tag{8.191}$$

(There are more efficient ways to describe quaternion multiplication.) In any case, addition $+$ corresponds to the usual vector addition of (x, y, z, w) and multiplication $\cdot$ uses the rules above.

However, the quaternions $(\mathbb{H}, +, \cdot)$ are not a field since multiplication is not commutative.

8.8*. Quadratic residues and the law of quadratic reciprocity

In this section we work over the (finite) field $\mathbb{Z}_p$, where p is a prime. In this case we say that we are working over **characteristic** p. When we work over the field of complex numbers $\mathbb{C}$, we say that we are working over **characteristic** 0. More generally, we have the following.

Definition 8.100. Given a field K with additive and multiplicative identities 0 and e, respectively, we say that it has **characteristic** $p \in \mathbb{Z}^+$ if p is the smallest positive integer such that

$$\underbrace{e + e + \cdots + e}_{p \text{ summands}} = 0. \tag{8.192}$$

If no such p exists, then we say that the field K has characteristic 0.

Example 8.101. For all primes p, the field $\mathbb{Z}_p$ has characteristic p. The fields $\mathbb{Q}$, $\mathbb{R}$, and $\mathbb{C}$ all have characteristic 0.

We are interested in understanding quadratic equations in the field $\mathbb{Z}_p$, for example, zeroes of quadratic polynomial equations.

8.8.1. Lagrange's Theorem on zeroes of polynomials over $\mathbb{Z}_p$.

Lagrange's Theorem in number theory says the following. (We can think of this as saying that a weaker form of part of the Fundamental Theorem of Algebra (Theorem 8.97) holds in $\mathbb{Z}_p$.)

Theorem 8.102. *Let*

$$A(x) := a_n x^n + a_{n-1} x^{n-1} + \cdots + a_2 x^2 + a_1 x + a_0 \tag{8.193}$$

be a polynomial with integer coefficients $a_0, a_1, \ldots, a_n$, where $n \in \mathbb{Z}^+$. Let p be a prime. Then either

(1) *p divides a_i for all $1 \le i \le n$,*[3] *or*

(2) *the congruence equation*

$$A(x) \equiv 0 \bmod p \tag{8.194}$$

has at most n incongruent solutions.

In other words, given a degree-n polynomial over $\mathbb{Z}_p$, where p is a prime, either all the coefficients are zero modulo p (in this case we say that the polynomial is *trivial*) or there are at most n distinct roots to the polynomial modulo p.

Non-Example 8.103. *For $m = 9$, the quadratic equation $x^2 \equiv 0 \bmod m$ has the three incongruent solutions $0, 3, 6$. More generally, the quadratic equation $x^2 \equiv 0 \bmod p^2$, where p is a prime, has the p incongruent solutions $0, p, 2p, \ldots, (p-1)p$.*

Proof of Theorem 8.102. Without loss of generality, we may assume that $a_n \not\equiv 0 \bmod p$ in the statement of the theorem. For, otherwise, if $a_n \equiv 0 \bmod p$, then the term $a_n x^n \equiv 0 \bmod p$ for every $x \in \mathbb{Z}$.

We prove the theorem by induction on $n \in \mathbb{Z}^+$.

[3] In this case, $A(x) \equiv 0 \bmod p$ for all $x \in \mathbb{Z}$.

Base case: $n = 1$. Let

$$A(x) = a_1 x + a_0, \quad \text{where } a_1 \not\equiv 0 \bmod p.$$

Suppose that x_1 and x_2 solve the congruence equation (8.194). Then

$$a_1 x_1 + a_0 \equiv 0 \equiv a_1 x_2 + a_0 \bmod p. \tag{8.195}$$

This implies that

$$a_1 x_1 \equiv a_1 x_2 \bmod p. \tag{8.196}$$

Since $a_1 \not\equiv 0 \bmod p$ and p is a prime, we have $\gcd(a_1, p) = 1$, and so by Corollary 6.21, we obtain

$$x_1 \equiv x_2 \bmod p. \tag{8.197}$$

Thus A has a unique root modulo p (by Lemma 6.38, there exists a root).

Inductive step: Suppose the statement of the theorem is true for $n = k$, where $k \in \mathbb{Z}^+$.

Let

$$A(x) = a_{k+1}x^{k+1} + a_k x^k + \cdots + a_2 x^2 + a_1 x + a_0, \quad \text{where } a_{k+1} \not\equiv 0 \bmod p. \tag{8.198}$$

Without loss of generality, we may assume that A has a root x_0 modulo p. We may now use the **Polynomial Division Theorem** to obtain that there exists a degree-k polynomial $Q(x)$ with integer coefficients and a degree-0 polynomial $R(x) = c \in \mathbb{Z}$ (a constant) such that

$$A(x) = (x - x_0)Q(x) + c. \tag{8.199}$$

Since $A(x_0) \equiv 0 \bmod p$, we have $c \equiv 0 \bmod p$, and hence

$$A(x) \equiv (x - x_0)Q(x) \bmod p. \tag{8.200}$$

Since p is a prime, by (8.200) any root modulo p of $A(x)$ is equal to x_0 or a root modulo p of $Q(x)$. By the inductive hypothesis, there are at most k incongruent roots modulo p of $Q(x)$. Thus, $A(x)$ has at most $k + 1$ incongruent roots modulo p. This proves the inductive step, and hence by induction we are done. □

Figure 8.8.1. Carl Friedrich Gauss 1777–1855; portrait by Christian Albrecht Jensen (1840). Wikimedia Commons, Public Domain.

8.8.2. Simplifying quadratic equations in $\mathbb{Z}_p$. Let p be an odd prime. Consider the general quadratic congruence equation in characteristic p:

$$ax^2 + bx + c \equiv 0 \bmod p, \tag{8.201}$$

where $a, b, c \in \mathbb{Z}$ and $a \not\equiv 0 \bmod p$. Since $\gcd(a, p) = 1$, there exist integers b' and c' such that $b \equiv ab' \bmod p$ and $c \equiv ac' \bmod p$. Using $\gcd(a, p) = 1$ again, we see that (8.201) is equivalent to

$$x^2 + b'x + c' \equiv 0 \bmod p. \tag{8.202}$$

Since p is an odd prime, we have $\gcd(2, p) = 1$, so that there exists an integer b'' such that $b' \equiv 2b'' \bmod p$. So we may rewrite the congruence equation (8.201) as

$$(x + b'')^2 - (b'')^2 + c' \equiv 0 \bmod p. \tag{8.203}$$

Thus, if we let $x' := x + b''$ and $q := (b'')^2 - c'$, then we obtain the equivalent and more simpler looking equation

$$(x')^2 \equiv q \bmod p. \tag{8.204}$$

For this reason, we only look at quadratic equations of this last form.

Example 8.104. Consider the quadratic congruence equation

$$x^2 \equiv a \bmod 17 \tag{8.205}$$

for all possible choices of a. We list all solutions modulo 17 for $0 < a < 17$:

$$\begin{array}{llll} 1^2 \equiv 16^2 \equiv 1, & 2^2 \equiv 15^2 \equiv 4, & 3^2 \equiv 14^2 \equiv 9, & 4^2 \equiv 13^2 \equiv 16, \\ 5^2 \equiv 12^2 \equiv 8, & 6^2 \equiv 11^2 \equiv 2, & 7^2 \equiv 10^2 \equiv 15, & 8^2 \equiv 9^2 \equiv 13. \end{array}$$

On the other hand, for a equal to 3, 5, 6, 7, 10, 11, 12, 14, there are no solutions. Observe that for each choice of $0 < a < 17$, there are either 0 or 2 solutions to $x^2 \equiv a \bmod 17$ with $0 < x < 17$.

8.8.3. Quadratic residues. Let m be a positive integer. We say that an integer q is a quadratic residue modulo m if there exists an integer x such that

$$x^2 \equiv q \bmod m. \tag{8.206}$$

In other words, an integer q is a quadratic residue modulo m if in the "$\mathbb{Z}_m$ world" q is a perfect square!

Let us call q a **quadratic residue remainder** modulo m if q is a quadratic residue modulo m and $0 \le q < m$. Since $0^2 \equiv 0 \bmod m$ is a rather trivial case, we will assume that $q > 0$. Note also that $1^2 \equiv 1 \bmod m$, so that 1 is a quadratic residue remainder modulo m for all $m \ge 2$. So we have the following:

0 and 1 are quadratic residue remainders for every modulus m.

So, a quadratic residue remainder modulo m is a perfect square in the remainder modulo m world R_m.

We say that q is a **non-trivial quadratic residue** modulo m if m does not divide q (i.e., $q \not\equiv 0 \bmod m$).

By the example at the end of the preceding subsection, the quadratic residues modulo 17 that are strictly between 0 and 17 are 1, 2, 4, 8, 9, 13, 15, and 16.

If you like to think ahead, here is a question for you:

Let p be a prime. When is -1 a quadratic residue?

After all, -1 is the "next simplest" integer after 0 and 1.

Figure 8.8.2 is a table of the non-trivial quadratic residue remainders modulo m for the first nine integers $m \geq 2$.

(8.207)

m	quadratic residues
2	1
3	1
4	1
5	$1, 4$
6	$1, 3, 4$
7	$1, 2, 4$
8	$1, 4$
9	$1, 4, 7$
10	$1, 4, 5, 6, 9$

Figure 8.8.2. The non-trivial quadratic residue remainders modulo m for $2 \leq m \leq 10$.

In particular, the quadratic residues modulo 10 are the integers $1, 4, 5, 6, 9$. This means that the last digit of any perfect square is always one of these five integers.

Example 8.105. We can use quadratic residues to rule out integers from being perfect squares. For example:

(1) The integers $10^7 + 1$ and $10^7 + 3$ are not perfect squares. More generally, any integer of the form $3k + 2$ is not a perfect square.

(2) The integers $10^7 + 2$ and $10^7 + 3$ are not perfect squares. More generally, any integer of the form $4k + 2$ or $4k + 3$ is not a perfect square.

(3) The integers $10^7 + 7$ and $10^7 + 8$ are not perfect squares. More generally, any integer of the form $5k + 2$ or $5k + 3$ is not a perfect square.

(4) The integers $10^7 + 10$ and $10^7 + 13$ are not perfect squares. More generally, any integer of the form $6k + 2$ or $6k + 5$ is not a perfect square.

(5) On the other hand, $10^7 + 4569 = 3163^2$ is a perfect square. We see that it is equal to $1+4+5+6+9 \equiv 25 \equiv 1 \bmod 3$, $1 \bmod 4$, $4 \bmod 5$, $4567 \equiv 1 \bmod 6$. With the help of a calculator, we see that $10^7 + 4569 = 3163^2$ is equal to $7 \cdot 1429224 + 1$, so that it is $1 \bmod 7$. It is equal to $8 \cdot 1250571 + 1$, so that it is $1 \bmod 8$. It is equal to $9 \cdot 1111618 + 7$, so that it is $7 \bmod 9$. Clearly it is $9 \bmod 10$. Of course, all of these congruence formulas are consistent with $10^7 + 4569$ being a perfect square.

8.8.4. The Legendre symbol. We can keep track of what are and what are not quadratic residues as follows. The Legendre symbol is defined by the following: For a positive integer m and an integer q,

$$\left(\frac{q}{m}\right) := \begin{cases} 1 & \text{if } q \text{ is a non-trivial quadratic residue modulo } m, \\ -1 & \text{if } q \text{ is not a quadratic residue modulo } m, \\ 0 & \text{if } q \text{ is a multiple of } m. \end{cases} \tag{8.208}$$

Example 8.106. (1) For all $m \geq 2$, since $1^2 = 1 < m$, we have

$$1 = \left(\frac{1}{m}\right). \tag{8.209}$$

(2) Since $2^2 = 4$, $3^2 - 6 = 3$, and $3^2 - 7 = 2$, we have

$$1 = \left(\frac{4}{5}\right) = \left(\frac{3}{6}\right) = \left(\frac{4}{6}\right) = \left(\frac{2}{7}\right) = \left(\frac{4}{7}\right). \tag{8.210}$$

(3) Since the non-zero quadratic residue remainder modulo 3 is 1 and modulo 4 is 1 and since the non-zero quadratic residue remainders modulo 6 are $1, 3, 4$ and modulo 7 are $1, 2, 4$, we have

$$-1 = \left(\frac{2}{3}\right) = \left(\frac{2}{4}\right) = \left(\frac{3}{4}\right) = \left(\frac{2}{6}\right) = \left(\frac{5}{6}\right) = \left(\frac{3}{7}\right) = \left(\frac{5}{7}\right) = \left(\frac{6}{7}\right). \tag{8.211}$$

(4) Since the non-zero quadratic residue remainders modulo 10 are $1, 4, 5, 6, 9$, we have

$$\left(\frac{q}{10}\right) = \begin{cases} 1 & \text{if } q = 1, 4, 5, 6, 9, \\ -1 & \text{if } q = 2, 3, 7, 8. \end{cases} \tag{8.212}$$

Recall that Fermat's Little Theorem 6.63 says that if p is a prime and if a is a positive integer which is not a multiple of p, then

$$a^{p-1} \equiv 1 \bmod p. \tag{8.213}$$

Now assume in addition that $p \neq 2$, so that p is odd. Then $\frac{p-1}{2}$ is an integer and we may rewrite this equation as the "quadratic" equation

$$\left(a^{\frac{p-1}{2}} - 1\right)\left(a^{\frac{p-1}{2}} + 1\right) \equiv \left(a^{\frac{p-1}{2}}\right)^2 - 1 \equiv 0 \bmod p. \tag{8.214}$$

Since p is a prime, we thus have

$$a^{\frac{p-1}{2}} - 1 \equiv 0 \bmod p \quad \text{or} \quad a^{\frac{p-1}{2}} + 1 \equiv 0; \tag{8.215}$$

that is, if $a \in \mathbb{Z}^+$ is not a multiple of an odd prime p, then

$$a^{\frac{p-1}{2}} \equiv \pm 1 \bmod p. \tag{8.216}$$

For example, if $a \in \mathbb{Z}^+$ is not a multiple of the odd prime 7, then $a^3 \equiv \pm 1 \bmod 7$. Table 8.8.1 shows the actual values ($+1$ or -1).

Table 8.8.1. Euler's criterion (see Theorem 8.109 below) for $p = 7$: $a^3 \equiv 1 \bmod 7$ ($-1 \bmod 7$) if and only if a is (is not) a quadratic residue.

a	a^3	$a^3 \bmod 7$	$0 < b < 7$, where $b^2 \equiv a$
1	1	1	1
2	8	1	3
3	27	-1	none
4	64	1	2
5	125	-1	none
6	216	-1	none

8.8.5. Number of solutions to a quadratic equation modulo a prime.

We are particularly interested in the case of prime moduli.

Lemma 8.107. *Let p be an odd prime, and let a be an integer that is not a multiple of p. Then there are either 0 or 2 solutions to the quadratic congruence equation*

$$x^2 \equiv a \bmod p. \tag{8.217}$$

In other words, by definition (8.208) there are

$$\left(\frac{a}{p}\right) + 1 \tag{8.218}$$

solutions to (8.217).

Proof. Suppose that x_1 and x_2 are both solutions to (8.217). Then we compute that

$$(x_1 - x_2)(x_1 + x_2) \equiv x_1^2 - x_2^2 \equiv 0 \bmod p. \tag{8.219}$$

Thus, since p is a prime, either

$$x_1 \equiv x_2 \bmod p \quad \text{or} \quad x_1 \equiv -x_2 \bmod p. \tag{8.220}$$

Now, if there are no solutions to (8.217), then we are done. So, we suppose that there exists a solution x_1 to (8.217). By (8.220), there are at most two solutions. On the other hand, $p - x_1$ is also a solution, and since p is odd, this solution is distinct from x_1. Therefore, there are exactly two distinct solutions modulo p. □

Non-Example 8.108. *On the other hand, for the modulus $m = 6$ (which is not a prime), the quadratic congruence equation $x^2 \equiv 3 \bmod 6$ has the unique solution $x \equiv 3 \bmod 6$.*

8.8.6. Euler's criterion.

Euler's criterion is a criterion for determining whether it is $+1$ or -1 in formula (8.216): $a^{\frac{p-1}{2}} \equiv \pm 1 \bmod p$. Table 8.8.1 is the special case $p = 7$ of the following.

Theorem 8.109. *Let $p > 2$ be a prime, and let a be an integer coprime to p. Then*

$$a^{\frac{p-1}{2}} \equiv 1 \bmod p \quad \text{if } a \text{ is a quadratic residue,} \tag{8.221a}$$

$$a^{\frac{p-1}{2}} \equiv -1 \bmod p \quad \text{if } a \text{ is not a quadratic residue.} \tag{8.221b}$$

Proof. (*Slow version, where we intersperse examples*) Let a be an integer coprime to p, and let

(8.222) $$S_a := \{c \in \mathbb{N}_{p-1} : c^2 \not\equiv a \bmod p\},$$

where $\mathbb{N}_{p-1} = R_p - \{0\} = 1, 2, 3, \ldots, p-1$. Observe that if a is not a quadratic residue, then $S_a = \mathbb{N}_{p-1}$.

Step 1 = Claim 1. *The cardinality of* S_a *is even.*

For example, by Lemma 6.44 we have that

(8.223) $$S_1 = \mathbb{N}_{p-1} - \{1, p-1\} = \{2, 3, \ldots, p-2\},$$

which has cardinality $p-3$, which is even since p is an odd prime.

Another example is S_a for $p = 7$. In this case, the sets in (8.222) are

$$S_1 = \{2,4,3,5\}, \quad S_2 = \{1,2,5,6\}, \quad S_3 = \{1,3,2,5,4,6\},$$
$$S_4 = \{1,4,3,6\}, \quad S_5 = \{1,5,2,6,3,4\}, \quad S_6 = \{1,6,2,3,4,5\},$$

which all have even cardinality.

Key idea to prove Step 1. *If* $c \in S_a$ *satisfies* $c^2 \not\equiv a \bmod p$*, then there exists a unique* $x \in S_a$ *such that*

(8.224) $$cx \equiv a \bmod p \quad \text{and} \quad x \neq c.$$

Proof of the key idea. Since $\gcd(c, p) = 1$, by Lemma 6.38 there exists an integer x such that $cx \equiv a \bmod p$. By Theorem 6.41, the solution x is unique modulo p. Since $c^2 \not\equiv a \bmod p$, we have $x \neq c$.

Now suppose for a contradiction that $x^2 \equiv a \bmod p$. Then $x^2 \equiv cx \bmod p$. Since $\gcd(c, p) = 1$, this implies that $x \equiv c \bmod p$ (by Corollary 6.21), so that $x = c$ since both integers are in $\mathbb{N}_{p-1}$, a contradiction. Thus $x^2 \not\equiv a \bmod p$, which implies that $x \in S_a$. We have proved the key idea.

Now, the key idea implies that the elements of S_a are partitioned into (unordered) pairs $\{c, d\}$ where $cd \equiv a \bmod p$ and $c \neq d$. (The reason why we have a partition is that if $cd_1 \equiv a \equiv cd_2 \bmod p$, then $d_1 \equiv d_1 \bmod p$.) This proves Step 1 that $|S_a|$ is even.

Step 2. *Writing* $(p-1)!$ *modulo* p *in a clever way.*

Clearly, we have the disjoint union $\mathbb{N}_{p-1} = S_a \cup (\mathbb{N}_{p-1} - S_a)$. Thus, $p-1$ factorial may be written as

(8.225) $$(p-1)! = \prod_{i=1}^{p-1} i = \prod_{j \in S_a} j \cdot \prod_{k \in \mathbb{N}_{p-1} - S_a} k.$$

Consider the first product on the right-hand side. We have that $\prod_{j \in S_a} j$ is equal to the product of $\frac{|S_a|}{2}$ pairs whose products are congruent to a modulo p, and hence

(8.226) $$\prod_{j \in S_a} j \equiv a^{\frac{|S_a|}{2}} \bmod p.$$

Step 3. *Deducing the theorem.*

Case 1: *a is not a quadratic residue.* In this case, $S_a = \mathbb{N}_{p-1}$, so that $|S_a| = p-1$. Hence

$$(p-1)! = \prod_{j \in S_a} j \equiv a^{\frac{|S_a|}{2}} \equiv a^{\frac{p-1}{2}} \bmod p. \tag{8.227}$$

Case 2: *a is a quadratic residue.* In this case, there exists $r \in \mathbb{N}_{p-1}$ such that $r^2 \equiv a \bmod p$. This also implies that $(p-r)^2 \equiv a \bmod p$. Since p is odd, we have that $p - r \not\equiv r \bmod p$.

We claim that r and $p-r$ are the only elements of $\mathbb{N}_{p-1}$ whose squares are congruent to a modulo p. Therefore $|S_a| = p-3$.

Indeed, suppose that $s \in \mathbb{N}_{p-1}$ satisfies $s^2 \equiv a \bmod p$. Then

$$(s+r)(s-r) \equiv s^2 - r^2 \equiv a - a \equiv 0 \bmod p. \tag{8.228}$$

Since p is a prime, this implies that

$$s \pm r \equiv 0 \bmod p. \tag{8.229}$$

Since $s \subset \mathbb{N}_{p-1}$, we conclude that $s = r$ or $s = p - r$.

Therefore, when a is a quadratic residue, we have that

$$\prod_{j \in S_a} j \equiv a^{\frac{|S_a|}{2}} \equiv a^{\frac{p-3}{2}} \bmod p. \tag{8.230}$$

We also have that

$$\prod_{k \in \mathbb{N}_{p-1} - S_a} k \equiv r(-r) \equiv -r^2 \equiv -a \bmod p. \tag{8.231}$$

Therefore,

$$(p-1)! \equiv \prod_{j \in S_a} j \cdot \prod_{k \in \mathbb{N}_{p-1} - S_a} k \equiv a^{\frac{p-3}{2}}(-a) \equiv -a^{\frac{p-1}{2}} \bmod p. \tag{8.232}$$

In particular, since 1 is a quadratic residue, we obtain

$$(p-1)! \equiv -1 \bmod p. \tag{8.233}$$

By the way, this reproves[4] the $\Leftarrow$ direction of Wilson's Theorem 6.69.

With this knowledge, we revisit (8.227) and (8.232) to conclude that:

(1) If a is not a quadratic residue, then (8.227) implies that

$$-1 \equiv a^{\frac{p-1}{2}} \bmod p. \tag{8.234}$$

(2) If a is a quadratic residue, then (8.232) implies that

$$1 \equiv a^{\frac{p-1}{2}} \bmod p. \tag{8.235}$$

We have completed the proof of the theorem. □

[4]By "reprove" we mean to prove again, and not to reprimand or censure. ☺

By the definition of the Legendre symbol (8.208), an immediate consequence of the theorem is:

Corollary 8.110. *If p is an odd prime and if a is an integer, then*

$$\left(\frac{a}{p}\right) \equiv a^{\frac{p-1}{2}} \bmod p. \tag{8.236}$$

8.8.7. Statement of the law of quadratic reciprocity. An important fact about quadratic residues is the following.

Theorem 8.111 (Law of quadratic reciprocity). *If p and q are distinct odd primes, then the Legendre symbol satisfies the identity*

$$\left(\frac{p}{q}\right)\left(\frac{q}{p}\right) = (-1)^{\frac{(p-1)(q-1)}{4}} = (-1)^{\frac{p-1}{2}\cdot\frac{q-1}{2}}. \tag{8.237}$$

We prove this fundamental theorem in the next subsection.

Example 8.112. Taking $p = 5$ and $q = 7$, we have

$$\left(\frac{5}{7}\right)\left(\frac{7}{5}\right) = (-1)\cdot(-1) = 1 = (-1)^{\frac{(5-1)(7-1)}{4}}, \tag{8.238}$$

where we used $\left(\frac{7}{5}\right) = \left(\frac{2}{5}\right)$.

Figure 8.8.3 is a table of the prime factorizations of integers of the form $n^2 - 5$ for $3 \leq n \leq 10$.

(8.239)

n	$n^2 - 5$	prime factorization
3	4	2^2
4	11	2^2
5	20	$2^2 \cdot 5$
6	31	31
7	44	$2^2 \cdot 11$
8	59	59
9	76	$2^2 \cdot 19$
10	95	$5 \cdot 19$

Figure 8.8.3. The prime factorizations of perfect square minus 5.

Remarkably, the prime factors are always 2, 5, or a prime that ends in 1 or 9. Thus, primes ending in 3 or 7 cannot occur. In other words:

If p is a prime dividing $n^2 - 5$, then $p \not\equiv 3, 7 \bmod 10$.

We leave it to the reader to conduct an easy search of the literature for a proof of this well-known fact.

8.8.8. A proof of the law of quadratic reciprocity. Recall from (8.121) that for all positive integers m,

$$\mathbb{Z}_m^* := \{[k]_m \in \mathbb{Z}_m : \gcd(k,m) = 1\}. \tag{8.240}$$

By Lemma 8.68, $(\mathbb{Z}_m^*, \cdot)$ is an abelian group, where $\cdot$ denotes congruence class multiplication. By Definition 6.71, Euler's totient function $\varphi(m)$ is equal to its order $|\mathbb{Z}_m^*|$.

Let p and q be distinct odd primes. Then $\mathbb{Z}_p^* := \mathbb{Z}_p - \{[0]_p\}$ is an abelian group (cf. Example 8.51). Note that $|\mathbb{Z}_p^*| = p-1$. Let

$$P := \frac{p-1}{2} \quad \text{and} \quad Q := \frac{q-1}{2},$$

so that $2P = p-1$ and $2Q = q-1$. Since p and q are odd, we have that P and Q are integers. Recall that the law of quadratic reciprocity (8.237), which we will now prove, states that

$$\left(\frac{p}{q}\right)\left(\frac{q}{p}\right) = (-1)^{PQ}. \tag{8.241}$$

As defined in Exercise 8.32, we have the (abelian) product group $\mathbb{Z}_p^* \times \mathbb{Z}_q^*$. By definition,

$$\mathbb{Z}_p^* \times \mathbb{Z}_q^* = \{([i]_p, [j]_q) : 1 \le i \le 2P, 1 \le j \le 2Q\}. \tag{8.242}$$

Per definition, multiplication in $\mathbb{Z}_p^* \times \mathbb{Z}_q^*$ is defined by

$$([i']_p, [j']_q) \cdot ([i]_p, [j]_q) = ([i'i]_p, [j'j]_q).$$

Since p and q are distinct primes, by Theorem 6.76 we have

$$|\mathbb{Z}_{pq}^*| = (p-1)(q-1) = |\mathbb{Z}_p^*|\,|\mathbb{Z}_q^*|. \tag{8.243}$$

In fact, recalling the proof of Theorem 6.76, we observe that by the Chinese Remainder Theorem 6.53, for any integers a, b, the system of linear congruence equations

$$k \equiv a \bmod p, \tag{8.244a}$$
$$k \equiv b \bmod q \tag{8.244b}$$

has a solution k_0. Moreover, an integer k is a solution to (8.244) if and only if $k \equiv k_0 \bmod pq$. Thus, we can define a function $\Phi : \mathbb{Z}_p^* \times \mathbb{Z}_q^* \to \mathbb{Z}_{pq}^*$ by

$$\Phi([a]_p, [b]_q) = [k]_{pq}, \tag{8.245}$$

where k is a solution to (8.244). Indeed, given $[a]_p \in \mathbb{Z}_p^*$ and $[b]_q \in \mathbb{Z}_q^*$, the Chinese Remainder Theorem implies that $[k]_{pq} \in \mathbb{Z}_{pq}$ is well-defined. Moreover, suppose for a contradiction that $\gcd(k, pq) > 1$. Since p and q are primes, this implies that $p|k$ or $q|k$, either of which contradicts (8.244), $p \nmid a$, and $q \nmid b$. Thus $\gcd(k, pq) = 1$, so that $[k]_{pq} \in \mathbb{Z}_{pq}^*$. We have proved that Φ is a well-defined function.[5]

[5] To wit, the unique solvability of an equation yields to a well-defined function.

We leave it as an exercise to show that Φ is a group homomorphism (as defined in §8.5.13). We then observe that its inverse function is the homomorphism $\Psi = \Phi^{-1} : \mathbb{Z}_{pq}^* \to \mathbb{Z}_p^* \times \mathbb{Z}_q^*$ defined by

$$\Psi([k]_{pq}) = \big([k]_p, [k]_q\big). \tag{8.246}$$

Therefore, $\Phi : \mathbb{Z}_p^* \times \mathbb{Z}_q^* \to \mathbb{Z}_{pq}^*$ is a group isomorphism.

For convenience, we denote with a bit of "abuse of notation"

$$(i,j) := \big([i]_p, [j]_q\big) \in \mathbb{Z}_p^* \times \mathbb{Z}_q^*.$$

Consider the normal subgroup $U := \{(1,1), (-1,-1)\}$ of $\mathbb{Z}_p^* \times \mathbb{Z}_q^*$ and the corresponding quotient group (see §8.5.8)

$$G = (\mathbb{Z}_p^* \times \mathbb{Z}_q^*)/U, \tag{8.247}$$

which is an abelian group. Note that $|U| = 2$ and $|G| = 2PQ$. We have

$$G = \big\{(i,j)U \,:\, 1 \le i \le 2P,\, 1 \le j \le Q\big\}, \tag{8.248}$$

expressed as distinct elements, where we used that the coset $(i,j)U$ is equal to the coset $(p-i, q-j)U$ (recall that $[-i]_p = [p-i]_p$).[6] In G, consider the product of its $2PQ$ elements, which is the coset

$$g := \left(\prod_{i=1}^{2P}\prod_{j=1}^{Q}(i,j)\right)U. \tag{8.249}$$

Now, summing over j first and over i second,

$$\begin{aligned}\prod_{i=1}^{2P}\prod_{j=1}^{Q}(i,j) &= \prod_{i=1}^{2P}\left(i^Q, \prod_{j=1}^{Q} j\right)\\ &= \prod_{i=1}^{2P}\left(i^Q, Q!\right)\\ &= \left((2P)!^Q, Q!^{2P}\right).\end{aligned} \tag{8.250}$$

On the other hand, using $i \equiv -(q-i) \bmod q$ for $1 \le i \le Q$, we see that

$$Q!^2 \equiv \prod_{i=1}^{Q} i \prod_{i=1}^{Q}\big(-(q-i)\big) \equiv (-1)^Q(2Q)! \bmod q,$$

since $\prod_{i=1}^{Q}(q-i) = 2Q\cdot(2Q-1)\cdots(Q+1)$. From all of this we obtain that the product of the $2PQ$ elements of G is equal to the coset

$$g = \big((2P)!^Q, (2Q)!^P(-1)^{PQ}\big)U \in G. \tag{8.251}$$

Claim.

$$G = \big\{\big([k]_p, [k]_q\big)U \,:\, 1 \le k \le (pq-1)/2,\, \gcd(k,pq) = 1\big\}. \tag{8.252}$$

By the claim, we have that

$$g = \prod_k (k,k)U = \left(\prod_k k, \prod_k k\right)U, \tag{8.253}$$

[6]This allows us to sum j from 1 to Q instead of $2Q$.

where the products in the display are over all integers k satisfying $1 \leq k \leq (pq-1)/2$ and $\gcd(k, pq) = 1$. Now, each product above may be expressed as (dividing by a congruence class means multiplying by its inverse; see also Remark 8.113 below)

(8.254)

$$\prod_k k = \frac{\prod_{i=1}^{2P} i \cdot \prod_{i=1}^{2P}(1p+i) \cdot \prod_{i=1}^{2P}(2p+i) \cdots \prod_{i=1}^{2P}((Q-1)p+i)\prod_{i=1}^{P}(Qp+i)}{1q \cdot 2q \cdots Pq}$$

$$\equiv \frac{(2P)!^Q}{q^P} \bmod p,$$

by cancelling the factors $P!$ in the numerator and denominator $\left(\prod_{i=1}^{P}(Qp+i) \equiv P! \bmod P\right)$ and since for each $0 \leq j \leq Q-1$, we have

$$\prod_{i=1}^{2P}(jp+i) \equiv (2P)! \bmod p.$$

Similarly to (8.254), we have

$$\prod_k k \equiv \frac{(2Q)!^P}{p^Q} \bmod q. \tag{8.255}$$

Remark 8.113. For the first equality in (8.254), we used the fact that we have the set equality

$$\{k \in \mathbb{Z} : 1 \leq k \leq (pq-1)/2, \gcd(k, pq) = 1\}$$

$$= \left(\bigcup_{0 \leq j < Q} \bigcup_{1 \leq i \leq 2P} \{jp+i\} \cup \bigcup_{1 \leq i \leq P} \{Qp+i\} \right) - \{q, 2q, \ldots, Pq\}.$$

This uses the fact that the integers coprime to pq are those that are not multiples of p or q.

Recall that Euler's criterion (Theorem 8.109) says that

$$q^P \equiv \left(\frac{q}{p}\right) \bmod p \quad \text{and} \quad p^Q \equiv \left(\frac{p}{q}\right) \bmod q. \tag{8.256}$$

From (8.253), (8.254), (8.255), and this we obtain that (note that 1 and -1 are each their own inverses modulo p or q)

$$g = \left((2P)!^Q \left(\frac{q}{p}\right), (2Q)!^P \left(\frac{p}{q}\right)\right) U. \tag{8.257}$$

Finally, by comparing (8.251) and (8.257), we obtain

$$\left(\frac{q}{p}\right)\left(\frac{p}{q}\right) = (-1)^{PQ}, \tag{8.258}$$

which is the law of quadratic reciprocity! The proof of Theorem 8.111 is complete. ☺

8.8.9. Supplements of the law of quadratic reciprocity. Fundamental facts are often true for a multitude of reasons. In this subsection we give alternative proofs of some special cases of the law of quadratic reciprocity, called *supplements*. In particular, we consider the solvability of the quadratic congruence equations $x^2 \equiv -1 \bmod p$ and $x^2 \equiv 2 \bmod p$.

Table 8.8.2. Examples of the affirmative case of the *first supplement*: Solutions to $x^2 \equiv -1 \bmod p$ for the first several primes that are congruent to 1 mod 4. If x is a solution, then so is $p - x \equiv -x$.

p	Solutions to $x^2 \equiv -1 \bmod p$	$x^2 + 1$ for the solution
5	2, 3	5, 10
13	5, 8	26, 65
17	4, 13	17, 170
29	12, 17	145, 290
37	6, 31	37, 962
41	9, 32	82, 1025
53	23, 30	530, 901
61	11, 50	122, 2501

Proposition 8.114 (First supplement to quadratic reciprocity). Let p be an odd prime. The quadratic congruence equation

$$x^2 \equiv -1 \bmod p \tag{8.259}$$

has a solution if and only if

$$p \equiv 1 \bmod 4. \tag{8.260}$$

In this case, by Lemma 8.107 there are exactly two solutions modulo p.

In other words,

$$\left(\frac{-1}{p}\right) = 1 \quad \text{if and only if} \quad p \equiv 1 \bmod 4. \tag{8.261}$$

Proof. By Euler's criterion (in the form of Corollary 8.110) with $a = -1$,

$$\left(\frac{-1}{p}\right) \equiv (-1)^{\frac{p-1}{2}} \bmod p. \tag{8.262}$$

Since the possible values for both sides of this congruence equation are in the set $\{-1, 1\}$ and since $p > 2$, we thus have

$$\left(\frac{-1}{p}\right) = (-1)^{\frac{p-1}{2}}. \tag{8.263}$$

The lemma follows since the right-hand side is equal to 1 if $p \equiv 1 \bmod 4$, whereas it is equal to -1 if $p \equiv 3 \bmod 4$. □

Table 8.8.3. Examples of the affirmative case of the *second supplement*: Solutions to $x^2 \equiv 2 \bmod p$ for the first several primes that are congruent to $\pm 1 \bmod 8$.

p	Solutions to $x^2 \equiv 2 \bmod p$	$x^2 - 2$ for the solution
7	3, 4	7, 14
17	6, 11	34, 119
23	5, 18	23, 322
31	8 23	62, 527
41	17, 24	287, 574
47	7, 40	47, 1598

Proposition 8.115 (Second supplement to quadratic reciprocity)**.** Let p be an odd prime. The quadratic congruence equation

$$x^2 \equiv 2 \bmod p \tag{8.264}$$

has a solution if and only if

$$p \equiv \pm 1 \bmod 8. \tag{8.265}$$

Exercise 8.50. *Prove that* $p-(-1)^{\frac{p-1}{2}}$ *is divisible by* 8 *if and only if* $p \equiv \pm 1 \bmod 8$.

In other words, $\left(\frac{2}{p}\right) = 1$ if and only if $p \equiv \pm 1 \bmod 8$. An elegant way of saying this is the formula

$$\left(\frac{2}{p}\right) = (-1)^{\frac{p^2-1}{8}}. \tag{8.266}$$

Exercise 8.51. *Prove that if* m, n *are odd integers congruent modulo* 8, *then*

$$(-1)^{\frac{m^2-1}{8}} = (-1)^{\frac{n^2-1}{8}}. \tag{8.267}$$

To prove the proposition, we will use the following result.

Lemma 8.116. *Let* p *be an odd prime. The quadratic congruence equation*

$$x^2 \equiv 2 \bmod p \tag{8.268}$$

has a solution if and only if the number of solutions to the quadratic congruence equation

$$x^2 + y^2 \equiv 2 \bmod p \tag{8.269}$$

is divisible by 8.

Proof. The key idea is an observation about the set of solutions to (8.269). Suppose that (x, y) is a solution to (8.269) of the "first type", which we define by the conditions

$$x \not\equiv 0 \bmod p, \quad y \not\equiv 0 \bmod p, \quad \text{and} \quad x \not\equiv \pm y \bmod p. \tag{8.270}$$

Then (x, y) is one of the 8 distinct solutions

$$(x, y),\ (x, -y),\ (-x, y),\ (-x, -y),\ (y, x),\ (y, -x),\ (-y, x),\ (-y, -x). \tag{8.271}$$

For example, for $p = 7$ we have no solutions of the type above. We only have the 4 solutions

$$(\pm 1, \pm 1) \tag{8.272}$$

and the 4 solutions

$$(\pm 3, 0), \quad (0, \pm 3). \tag{8.273}$$

(Note that $-3 \equiv 4 \bmod 7$.) So we have a total of 8 solutions.

By (8.270), a solution to (8.269) is not of the first type if and only if $x \equiv \pm y \bmod p$ or exactly one of x and y is congruent to 0 modulo p.

So next, we say that a solution to (8.269) is of the "second type" if it is of the form

$$(x, x), \ (x, -x), \tag{8.274}$$

where $x^2 \equiv 1 \bmod p$. This implies

$$x = \pm 1. \tag{8.275}$$

Since the prime p is greater than 2, there are 4 of this second type of solution.

If a solution to (8.269) is not of the first two types, then it is of the "third type", which we define to be of the form

$$(x, 0), \quad (-x, 0), \quad (0, x), \quad (0, -x), \tag{8.276}$$

where

$$x^2 \equiv 2 \bmod p; \tag{8.277}$$

that is, x satisfies (8.268).

So, altogether, we have a multiple of 8 numbers of solutions of the first type, we have 4 solutions of the second type, and we have 4 solutions of the third type if (8.264) has a solution, whereas there are 0 solutions of the third type if (8.264) does not have a solution. We have proved the lemma:

> Equation (8.268) has a solution if and only if the number of solutions to (8.269) is divisible by 8. □

We now give a proof of the second supplement to quadratic reciprocity.

Proof of Proposition 8.115. By Lemma 8.116, (8.268) has a solution if and only if the number of solutions to (8.269) is divisible by 8.

So we consider the quadratic congruence equation (8.269) for (x, y). Make the change of variables

$$x \equiv a + 1, \quad y \equiv at + 1 \bmod p. \tag{8.278}$$

Then equation (8.269) is equivalent to the equation for $(a, t) \in \mathbb{Z}^2 = \mathbb{Z} \times \mathbb{Z}$ that is modulo p,

$$2 \equiv (a+1)^2 + (at+1)^2 \equiv a^2(t^2+1) + a(2t+2) + 2. \tag{8.279}$$

Assume that $a \not\equiv 0$. Then this equation is equivalent to the easier to solve (in hindsight) congruence equation

$$a(t^2+1) \equiv -2(t+1) \bmod p. \tag{8.280}$$

Suppose that t is such that $t^2 \equiv -1 \bmod p$. Then (8.280) implies that $t \equiv -1 \bmod p$, which is a contradiction since $p > 2$. Therefore

$$t^2 \not\equiv -1 \bmod p. \tag{8.281}$$

Since p is a prime, for every integer t such that $t^2 \not\equiv -1 \bmod p$, there exists a unique integer a modulo p such that (a, t) solves (8.280).

Because, by (8.218), there are $\left(\frac{-1}{p}\right)+1$ solutions to $t \equiv -1 \bmod p$, we conclude that there are

$$p - 1 - \left(\left(\frac{-1}{p}\right) + 1\right) = p - \left(\frac{-1}{p}\right) = p - (-1)^{\frac{p-1}{2}} \tag{8.282}$$

solutions to (8.280), where the last equality is by (8.263).

So, from Lemma 8.116, there are solutions to (8.268) if and only if $p-(-1)^{\frac{p-1}{2}}$ is divisible by 8. Finally, we leave it to the reader to check that this condition is equivalent to $p \equiv \pm 1 \bmod 8$. □

8.9. Hints and partial solutions for the exercises

Hint for Exercise 8.1. Let $a \in \mathbb{Z}$. Then $[a]_2 = [1]_2$ if and only if there exists an integer k such that

$$a - 1 = 2k. \tag{8.283}$$

Continue.

Hint for Exercise 8.2. The condition on n is equivalent to

$$\mathbb{Z} = [n]_3 \cup [1]_3 \cup [2]_3. \tag{8.284}$$

Hint for Exercise 8.3. $m = 6$. Explain why.

Hint for Exercise 8.4. The inverse of the function f is the function $g : \mathbb{Z}_m \to R_m$ defined by $g([r]_m) = r$ for $0 \le r < m$. Show that f and g are inverses of each other.

Hint for Exercise 8.5. One way to prove this is to observe that $|\mathbb{Z}_3| = 3$ and to show that the three congruence classes on the right-hand side of (8.17) are distinct.

This generalizes to the following: For any positive integer m and any integer b,

$$\mathbb{Z}_m = \{[b]_m, [b+1]_m, \ldots, [b+m-1]_m\}. \tag{8.285}$$

Hint for Exercise 8.6. Suppose that $a, b \in \mathbb{Z}$ are such that $[a]_m = [b]_m$. By definition, this means that

$$a - b = qm \quad \text{for some } q \in \mathbb{Z}. \tag{8.286}$$

Apply Theorem 4.16.

Hint for Exercise 8.7. Suppose that $[a]_m[b]_m = [1]_m$. This is equivalent to $ab \equiv 1 \bmod m$. Continue and then apply Theorem 4.20.

Hint for Exercise 8.8. Such pairs are in one-to-one correspondence with pairs of integers (a, b), where $0 < a, b < 24$, satisfying

$$24 \text{ divides } ab. \tag{8.287}$$

Such pairs (a, b) are

$$
\begin{aligned}
&(2,12),\\
&(3,8),\ (3,16),\\
&(4,6),\ (4,12),\ (4,18),\\
&(6,8),\ (6,12),\ (6,16),\ (6,20),\\
&(8,9),\ (8,12),\ (8,15),\ (8,18),\ (8,21),\\
&(9,16),\\
&(10,12),\\
&(12,12),\ (12,14),\ (12,16),\ (12,18),\ (12,20),\ (12,22),\\
&(15,16),\\
&(16,18),\ (16,21),\\
&(18,20).
\end{aligned}
$$

Hint for Exercise 8.9. $[a]_m$ has an inverse if and only if there exists an integer b such that $[a]_m[b]_m = [1]_m$. Continue and use Theorem 4.20.

Hint for Exercise 8.10. (1) $[1^{-1}]_8 = [1]_8$, $[3^{-1}]_8 = [3]_8$, $[5^{-1}]_8 = [5]_8$, $[7^{-1}]_8 = [7]_8$.

Hint for Exercise 8.11. For example, for $a = 2$, $[2^{-1}]_7 = [4]_7$. Thus $x \equiv 4 \cdot 3 \bmod 7$, which is equivalent to $x \equiv 5 \bmod 7$.

Hint for Exercise 8.12. If $\gcd(a, m) = \gcd(b, m) = 1$, then $\gcd(ab, m) = 1$.

Hint for Exercise 8.13. The properties of reflexivity, symmetry, and transitivity are easy to check.

Observe that $|\mathbf{x}| \geq 0$ for all $\mathbf{x} \in \mathbb{R}^n$. So the equivalence classes of $\sim$ are the sets

$$\{\mathbf{x} \in \mathbb{R}^n : |\mathbf{x}| = r\}, \tag{8.288}$$

where r is a non-negative real number. When $r > 0$, this is the ($(n-1)$-dimensional) sphere of radius r. When $r = 0$, this is a single point, the origin.

Hint for Exercise 8.14. Reflexivity: Let $A \in GL(n, \mathbb{R})$. Taking $P = I$ the identity matrix, we have

$$A \sim I^{-1}AI = A. \tag{8.289}$$

Symmetry: Suppose that $A, B \in GL(n, \mathbb{R})$ are such that $A \sim B$. Then there exists an invertible real $n \times n$ matrix P such that $B = P^{-1}AP$. This implies

$$PBP^{-1} = P(P^{-1}AP)P^{-1} = (PP^{-1})A(PP^{-1}) = IAI = A. \tag{8.290}$$

Since $P = (P^{-1})^{-1}$, this says that

$$A = Q^{-1}BQ, \quad \text{where } Q = P^{-1}. \tag{8.291}$$

Since Q is an invertible real $n \times n$ matrix, we conclude that $B \sim A$.

Transitivity: Suppose that $A, B, C \in GL(n, \mathbb{R})$ are such that $A \sim B$ and $B \sim C$. Then there exists an invertible real $n \times n$ matrices P and Q such that $B = P^{-1}AP$ and $C = Q^{-1}BQ$. This implies that

$$C = Q^{-1}BQ = C = Q^{-1}(P^{-1}AP)Q = (PQ)^{-1}A(PQ). \quad (8.292)$$

Since PQ is an invertible real $n \times n$ matrix, we obtain that $A \sim C$.

We conclude that $\sim$ is an equivalence relation.

Hint for Exercise 8.15. Let $\sim$ be an equivalence relation on a set X. Firstly, clearly

$$\bigcup_{x \in X} [x] \quad (8.293)$$

since $x \in [x]$ for all $x \in X$.

Secondly, suppose that $x, y \in X$ are such that $[x] \cap [y] \neq \emptyset$. Then there exists $z \in X$ such that $z \in [x]$ and $z \in [y]$; that is, $x \sim z$ and $z \sim y$. By the transitivity of $\sim$, we obtain that $x \sim y$.

Now, if $w \in [x]$, then $w \sim x$. Since $x \sim y$, we have $w \sim y$, so that $w \in [y]$. Therefore, $[x] \subset [y]$. Similarly, $[y] \subset [x]$. Therefore $[x] = [y]$. This proves that $\mathcal{E} = \{[x] : x \in X\}$ is a partition of X.

Hint for Exercise 8.16. We leave the proof of reflexivity and symmetry to you. Transitivity is true because

$$\begin{aligned} x \sim y \text{ and } y \sim z &\Rightarrow \exists k, \ell \in \mathbb{Z},\ x - y = kL,\ y - z = \ell L \\ &\Rightarrow \exists k, \ell \in \mathbb{Z},\ x - z = (k + \ell)L \\ &\Rightarrow x \sim z \end{aligned}$$

and because the following is true: if $k, \ell \in \mathbb{Z}$, then $k + \ell \in \mathbb{Z}$. This proves that $\sim$ is an equivalence relation.

Let

$$\mathbb{R}/L\mathbb{Z} = \mathbb{R}/\sim \quad (8.294)$$

denote the set of equivalence classes of $\sim$. We can define a bijection between this set and the circle $\mathcal{C}$ of circumference L in the complex plane $\mathbb{C}$. Let

$$\mathcal{C} = \left\{ z \in \mathbb{C} : |z| = \frac{L}{2\pi} \right\}. \quad (8.295)$$

Define

$$b : \mathbb{R}/L\mathbb{Z} \to \mathcal{C} \quad (8.296)$$

by

$$b([x]) = e^{ix} \quad \text{for } x \in \mathbb{R}. \quad (8.297)$$

One checks that the function b is well-defined and in fact a bijection.

Yes, this is analogous to taking a string of length L and gluing its ends together. We have the quotient function

$$\pi : \mathbb{R} \to \mathbb{R}/L\mathbb{Z} \quad (8.298)$$

defined by

$$\pi(x) = [x]. \tag{8.299}$$

Restricting this function to

$$\pi : [0, L] \to \mathbb{R}/L\mathbb{Z}, \tag{8.300}$$

we obtain a surjection. One checks that this function is an injection (and hence a bijection) except for the fact that

$$\pi(0) = \pi(L). \tag{8.301}$$

So, we can consider $\mathbb{R}/L\mathbb{Z}$ as being obtained from the interval $[0, L]$ by "gluing" (or "identifying") the two points 0 and L together.

Hint for Exercise 8.17. Relexivity is obvious, and symmetry is easy. For transitivity, a fact that contributes to this is: The implication

$$(x, 0) \sim (-x, 1) \text{ and } (-x, 1) \sim (x, 0) \Rightarrow (x, 0) \sim (x, 0)$$

is true.

Hint for Exercise 8.18. Figure 8.9.1 shows a sequence of pictures from the Klein bottle Wikipedia page.

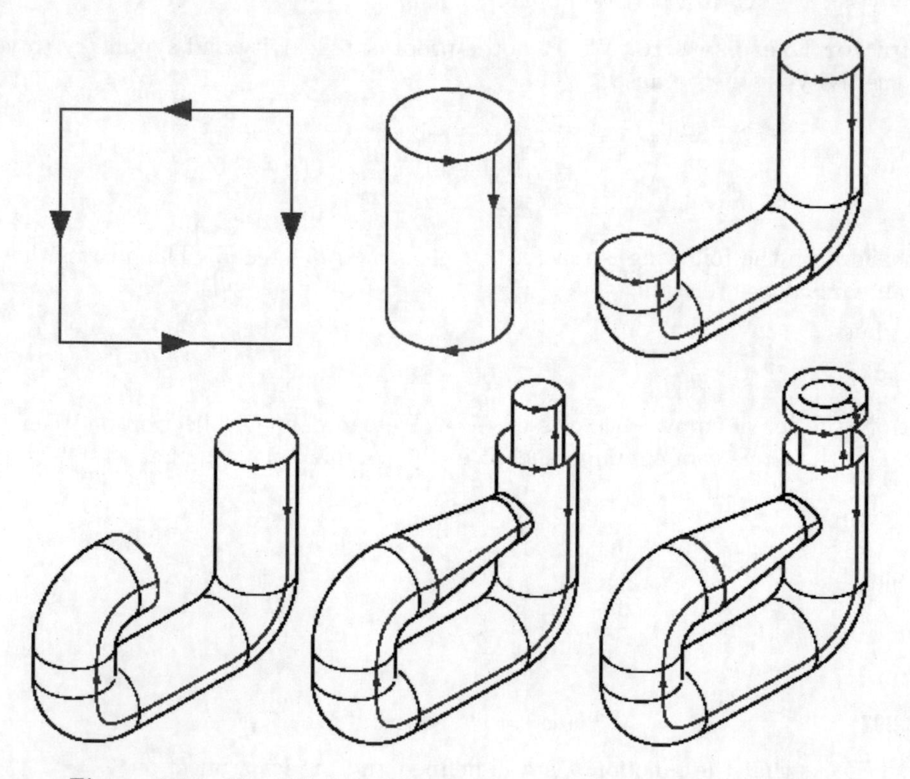

Figure 8.9.1. Visualizing the Klein bottle as a quotient. Wikimedia Commons. Author: Inductiveload. Work released by author into the public domain.

Hint for Exercise 8.19. This verification is relatively routine.

Hint for Exercise 8.20. The addition table for parity is given by Table 8.9.1. The tables are basically the same.

Table 8.9.1. Addition table for parity.

$\oplus$	even	odd
even	even	odd
odd	odd	even

Hint for Exercise 8.21. This is similar to the previous exercise, where here we replace addition by multiplication.

Hint for Exercise 8.22. This verification is also relatively straightforward.

Hint for Exercise 8.23. The addition table for R_3 is given by Table 8.9.2.

Table 8.9.2. Addition table for R_3.

$\oplus$	0	1	2
0	0	1	2
1	1	2	0
2	2	0	1

Hint for Exercise 8.24. The main point is that if $a, b \in R_m$ satisfy $a + b \geq m$, then $0 \leq a + b - m < m$.

Hint for Exercise 8.25. For $m = 2$, $(R_m, \oplus)$ is essentially the same as Example 8.39.

Hint for Exercise 8.26. One easily checks that $[0]_m$ is the unit for $+$. Associativity and commutativity of $+$ for congruence classes follows from associativity and commutativity of $+$ for integers, respectively. One also sees that the inverse of $[a]_m$ is $[-a]_m$ since $-m$ is the inverse of m for addition of integers and since $[0]_m$ is the unit for $+$. We conclude that $(\mathbb{Z}_m, +)$ is an abelian group.

Hint for Exercise 8.27. We see that $[1]_m$ is the unit for $\cdot$. However, $[0]_m$ does not have an inverse. Therefore, $(\mathbb{Z}_m, \cdot)$ is not a group.

Hint for Exercise 8.28. The point is that $(\mathbb{Z}_4^*, \cdot)$ is defined so that multiplicative inverses exist.

Hint for Exercise 8.29. The multiplication table for $(D_2, \cdot)$ is given by Table 8.9.3.

Hint for Exercise 8.30. We have $m = ab$, where $1 < a, b < m$. In particular, neither $[a]_m$ nor $[b]_m$ has a multiplicative inverse.

Hint for Exercise 8.31. The point is that two elements, where both elements are powers of the generator a and where the powers are any integers, must commute.

Table 8.9.3. Multiplication table for $(D_2, \cdot)$.

$\cdot$	1	a	b	c
1	1	a	b	c
a	a	1	c	b
b	b	c	1	a
c	c	b	a	1

Hint for Exercise 8.32. Let e_G and e_H be the identity elements of G and H, respectively. Then one easily checks that (e_G, e_H) is the identity element of $G \times H$.

Given an element (g, h), one also easily checks that (g^{-1}, h^{-1}) is its inverse. Finally, associativity is straightforward:

$$\begin{aligned}((g_1, h_1) \circ (g_2, h_2)) \circ (g_3, h_3) &= (g_1 \cdot g_2, h_1 * h_2) \circ (g_3, h_3) \\ &= ((g_1 \cdot g_2) \cdot g_3, (h_1 * h_2) * h_3) \\ &= (g_1 \cdot (g_2 \cdot g_3), h_1 * (h_2 * h_3)) \\ &= (g_1, h_1) \circ ((g_2, h_2) \circ (g_3, h_3)).\end{aligned}$$

Hint for Exercise 8.33. (1) Yes. (2) No. (3) No.

Hint for Exercise 8.34. This is intuitively easy, but perhaps clumsy to write a rigorous proof of.

Hint for Exercise 8.35. Let S^1 denote the unit complex numbers. Firstly, since $|1| = 1$, we have $1 \in S^1$, and since 1 is the unit of $\mathbb{C}^*$, it is the unit of S^1.

Secondly, suppose that $z, w \in S^1$. Then $|z| = |w| = 1$. We compute that

$$|zw| = |z||w| = 1 \cdot 1 = 1. \tag{8.302}$$

So $zw \in S^1$, which shows that S^1 is closed under complex multiplication.

Thirdly, given $z \in S^1$, since $z \neq 0$, its inverse $z^{-1} \in \mathbb{C}^*$ exists. We check that

$$|z^{-1}| = \left|\frac{\bar{z}}{|z|^2}\right| = \frac{|\bar{z}|}{|z|^2} = \frac{|z|}{|z|^2} = 1 \tag{8.303}$$

since $|z| = 1$. Thus $z^{-1} \in S^1$. This proves that $(S^1, \cdot)$ is a subgroup of $(\mathbb{C}^*, \cdot)$.

Hint for Exercise 8.36. By the definition of xH, it is easy to see that f_x is a surjection. We now prove that f_x is an injection. Suppose that $f_x(h_1) = f_x(h_2)$. Then $xh_1 = xh_2$. Multiplying this by x^{-1} yields $h_1 = h_2$.

Hint for Exercise 8.37. The subgroups of D_2 are

$$\{1\},\ \{1, a\},\ \{1, b\},\ \{1, c\},\ D_2. \tag{8.304}$$

Hint for Exercise 8.38. Again, $[1]_m$ is the unit for $\cdot$. Let $[a]_m \in \mathbb{Z}_m^*$. By the definition of $\mathbb{Z}_m^*$, there exists a unique element $[b]_m \in \mathbb{Z}_m^*$ such that

$$[b]_m \cdot [a]_m = [a]_m \cdot [b]_m = [1]_m. \tag{8.305}$$

Thus, every element has an inverse. Associativity and commutativity of $\cdot$ for congruence classes follows from associativity and commutativity of $\cdot$ for integers. Therefore, $(\mathbb{Z}_m^*, \cdot)$ is an abelian group.

Hint for Exercise 8.39. The multiplication table for $\mathbb{Z}_{12}^*$ is

(8.306)

$(\mathbb{Z}_{12}^*, \odot)$	$[1]_{12}$	$[5]_{12}$	$[7]_{12}$	$[11]_{12}$
$[1]_{12}$	$[1]_{12}$	$[5]_{12}$	$[7]_{12}$	$[11]_{12}$
$[5]_{12}$	$[5]_{12}$	$[1]_{12}$	$[11]_{12}$	$[7]_{12}$
$[7]_{12}$	$[7]_{12}$	$[11]_{12}$	$[5]_{12}$	$[7]_{12}$
$[11]_{12}$	$[7]_{12}$	$[5]_{12}$	$[1]_{12}$	$[7]_{12}$

The multiplication table for $\mathbb{Z}_{15}^*$ is

(8.307)

$(\mathbb{Z}_{15}^*, \odot)$	$[1]_{15}$	$[2]_{15}$	$[4]_{15}$	$[7]_{15}$	$[8]_{15}$	$[11]_{15}$	$[13]_{15}$	$[14]_{15}$
$[1]_{15}$	$[1]_{15}$	$[2]_{15}$	$[4]_{15}$	$[7]_{15}$	$[8]_{15}$	$[11]_{15}$	$[13]_{15}$	$[14]_{15}$
$[2]_{15}$	$[2]_{15}$	$[4]_{15}$	$[8]_{15}$	$[14]_{15}$	$[1]_{15}$	$[7]_{15}$	$[11]_{15}$	$[13]_{15}$
$[4]_{15}$	$[4]_{15}$	$[8]_{15}$	$[1]_{15}$	$[13]_{15}$	$[2]_{15}$	$[14]_{15}$	$[7]_{15}$	$[11]_{15}$
$[7]_{15}$	$[7]_{15}$	$[14]_{15}$	$[13]_{15}$	$[4]_{15}$	$[11]_{15}$	$[2]_{15}$	$[1]_{15}$	$[8]_{15}$
$[8]_{15}$	$[8]_{15}$	$[1]_{15}$	$[2]_{15}$	$[11]_{15}$	$[4]_{15}$	$[13]_{15}$	$[14]_{15}$	$[7]_{15}$
$[11]_{15}$	$[11]_{15}$	$[7]_{15}$	$[14]_{15}$	$[2]_{15}$	$[13]_{15}$	$[1]_{15}$	$[8]_{15}$	$[4]_{15}$
$[13]_{15}$	$[13]_{15}$	$[11]_{15}$	$[7]_{15}$	$[1]_{15}$	$[14]_{15}$	$[8]_{15}$	$[14]_{15}$	$[2]_{15}$
$[14]_{15}$	$[4]_{15}$	$[13]_{15}$	$[11]_{15}$	$[8]_{15}$	$[7]_{15}$	$[4]_{15}$	$[2]_{15}$	$[1]_{15}$

Hint for Exercise 8.40. Firstly, recall that the cardinality

$$|\mathbb{Z}_m^*| = \varphi(m). \tag{8.308}$$

Define the function $f : R \to \mathbb{Z}_m^*$ by

$$f(a) = [a]_m. \tag{8.309}$$

By definition, $|R| = \varphi(m)$. We claim that f is an injection, from which it follows that f is a bijection since the cardinalities of the domain and codomain are equal. We then are able to conclude that

$$\{[a]_m : a \in R\} = \mathbb{Z}_m^*. \tag{8.310}$$

To finish the exercise, we now prove that f is an injection. Suppose that $a, b \in R$ are such that

$$f(a) = f(b). \tag{8.311}$$

Then $[a]_m = [b]_m$. By definition, distinct elements of R are incongruent modulo m. So, if a and b are distinct, then they are incongruent modulo m. In this case, $[a]_m \neq [b]_m$, which is a contradiction. We thus conclude that $a = b$. This proves that f is an injection.

Hint for Exercise 8.41. Theorem 8.59.

Hint for Exercise 8.42. One isomorphism takes 1 to m, and the other isomorphism takes 1 to $-m$.

Hint for Exercise 8.43. Recall from (8.75) that

$$\mathbb{Z}_4^* := \{[1]_4, [3]_4\}. \tag{8.312}$$

One (easily) checks that the bijective function

$$f : \mathbb{Z}_2 \to \mathbb{Z}_4^* \tag{8.313}$$

defined by

$$f([0]_2) = [1]_4, \qquad f([1]_2) = [3]_4 \tag{8.314}$$

is an isomorphism; that is,

$$f([a]_2 + [b]_2) = f([a]_2) \cdot f([b]_2). \tag{8.315}$$

For example,

$$f([1]_2 + [1]_2) = f([0]_2) = [1]_4 = [3]_4 \cdot [3]_4. \tag{8.316}$$

Hint for Exercise 8.44. Recall that

$$Z_m = \{e^{i\,2\pi k/m} \in \mathbb{C} : k \in \mathbb{Z}_m\} \tag{8.317}$$

and $f : \mathbb{Z}_m \to Z_m$ is defined by

$$f([k]_m) = e^{i\,2\pi k/m}. \tag{8.318}$$

Firstly, we show that f is well-defined. The reason why this is an issue is because the input (on the left-side of the formula defining f) is $[k]_m$, whereas the right-hand side *a priori* depends on k instead of $[k]_m$.

Suppose that $[k]_m = [\ell]_m$. We need to show that $f([k]_m) = f([\ell]_m)$. This is equivalent to

$$e^{i\,2\pi k/m} = e^{i\,2\pi \ell/m}; \quad \text{that is,} \quad e^{i\,2\pi(k-\ell)/m} = 1. \tag{8.319}$$

This is true if and only if $(k-\ell)/m$ is an integer; i.e., $k-\ell$ is divisible by m. This is true by our hypothesis that $[k]_m = [\ell]_m$. We conclude that f is well-defined. In fact, since we didn't just prove an implication, but rather the biconditional

$$f([k]_m) = f([\ell]_m) \quad \Leftrightarrow \quad [k]_m = [\ell]_m, \tag{8.320}$$

we actually proved that f is an injection. Since $|\mathbb{Z}_m| = |Z_m| = m$, this implies that f is a bijection.

Secondly, given $[a]_m, [b]_m \in \mathbb{Z}_m$, we have

$$\begin{aligned} f([a]_m + [b]_m) &= f([a+b]_m) \\ &= e^{i\,2\pi(a+b)/m} \\ &= e^{i\,2\pi a/m} e^{i\,2\pi b/m} \\ &= f([a]_m) \cdot f([b]_m). \end{aligned}$$

Thus, f is a group homomorphism.

Since f is also a bijection, we conclude that f is an isomorphism (it is easy to see that the inverse function is a group homomorphism).

Hint for Exercise 8.45. Since $I \neq \{0\}$, there exists a non-zero integer a in I. Then $|a| \in I$.

Hint for Exercise 8.46. Suppose that $a \in I - m\mathbb{Z}$. Apply the Division Theorem to divide a by m to get a remainder r. What can you say about r to obtain a contradiction?

Hint for Exercise 8.47. Since m is composite, there exist integers $1 < a, b < m$ such that $ab = m$. Explain why $a \notin m\mathbb{Z}$ and $b \notin m\mathbb{Z}$.

Hint for Exercise 8.48. Suppose that integers a, b satisfy $ab = pk$ for some integer k. Then p divides ab. Continue.

Hint for Exercise 8.49. Using the hint and the (extended) triangle inequality, we compute that

$$\begin{aligned} |z_n| &= |(z_1 + \cdots + z_n) + (-z_1) + \cdots + (-z_{n-1})| \\ &\leq |z_1 + \cdots + z_n| + |-z_1| + \cdots + |-z_{n-1}| \\ &= |z_1 + \cdots + z_n| + |z_1| + \cdots + |z_{n-1}|. \end{aligned}$$

Therefore,

$$|z_1 + \cdots + z_n| \geq |z_n| - |z_1| - \cdots - |z_{n-1}|. \tag{8.321}$$

Hint for Exercise 8.50. ($\Leftarrow$) (i) Suppose that $p \equiv 1 \bmod 8$. Then $(-1)^{\frac{p-1}{2}} = 1$, so that

$$p - (-1)^{\frac{p-1}{2}} = p - 1 \tag{8.322}$$

is divisible by 8.

(ii) Suppose that $p \equiv -1 \bmod 8$. Then $(-1)^{\frac{p-1}{2}} = -1$, so that

$$p - (-1)^{\frac{p-1}{2}} = p + 1 \tag{8.323}$$

is divisible by 8.

($\Rightarrow$) Suppose that $p - (-1)^{\frac{p-1}{2}}$ is divisible by 8. Firstly, p must be odd. So we just need to rule out that $p \equiv \pm 3 \bmod 8$.

(i) Suppose that $p \equiv 3 \bmod 8$. Then $(-1)^{\frac{p-1}{2}} = -1$, so that

$$p - (-1)^{\frac{p-1}{2}} = p + 1 \tag{8.324}$$

is not divisible by 8, a contradiction.

(ii) Suppose that $p \equiv -3 \bmod 8$. Then $(-1)^{\frac{p-1}{2}} = 1$, so that

$$p - (-1)^{\frac{p-1}{2}} = p - 1 \tag{8.325}$$

is not divisible by 8, a contradiction.

Hint for Exercise 8.51. Note that if k is an odd integer, then $k = 2\ell + 1$ for some integer ℓ, so that

$$\frac{k^2 - 1}{8} = \frac{\ell(\ell + 1)}{2}, \tag{8.326}$$

which is an integer (since ℓ or $\ell + 1$ is even).

We compute that

$$\frac{(-1)^{\frac{m^2-1}{8}}}{(-1)^{\frac{n^2-1}{8}}} = (-1)^{\frac{m^2-n^2}{8}}$$
$$= (-1)^{(m+n)\frac{m-n}{8}}.$$

Since m and n are congruent modulo 8, we have that $\frac{m-n}{8}$ is an integer and that $m+n$ is even. Therefore the ratio in the display above is equal to 1.

Bibliography

[Ecc97] Peter J. Eccles, *An introduction to mathematical reasoning: Numbers, sets and functions*, Cambridge University Press, Cambridge, 1997, DOI 10.1017/CBO9780511801136. MR1607207

[FP96] Peter Fletcher and C. Wayne Patty, *Foundations of higher mathematics*, PWS Publishing Company, 1996.

[Pol14] G. Polya, *How to solve it: A new aspect of mathematical method*, with a foreword by John H. Conway, reprint of the second (2004) edition [MR2183670], Princeton Science Library, Princeton University Press, Princeton, NJ, 2014. MR3289212

Index

Selected Published Titles in This Series

61 **Bennett Chow,** Introduction to Proof Through Number Theory, 2023
59 **Oscar Gonzalez,** Topics in Applied Mathematics and Modeling, 2023
58 **Sebastian M. Cioabă and Werner Linde,** A Bridge to Advanced Mathematics, 2023
57 **Meighan I. Dillon,** Linear Algebra, 2023
55 **Joseph H. Silverman,** Abstract Algebra, 2022
54 **Rustum Choksi,** Partial Differential Equations, 2022
53 **Louis-Pierre Arguin,** A First Course in Stochastic Calculus, 2022
52 **Michael E. Taylor,** Introduction to Differential Equations, Second Edition, 2022
51 **James R. King,** Geometry Transformed, 2021
50 **James P. Keener,** Biology in Time and Space, 2021
49 **Carl G. Wagner,** A First Course in Enumerative Combinatorics, 2020
48 **Róbert Freud and Edit Gyarmati,** Number Theory, 2020
47 **Michael E. Taylor,** Introduction to Analysis in One Variable, 2020
46 **Michael E. Taylor,** Introduction to Analysis in Several Variables, 2020
45 **Michael E. Taylor,** Linear Algebra, 2020
44 **Alejandro Uribe A. and Daniel A. Visscher,** Explorations in Analysis, Topology, and Dynamics, 2020
43 **Allan Bickle,** Fundamentals of Graph Theory, 2020
42 **Steven H. Weintraub,** Linear Algebra for the Young Mathematician, 2019
41 **William J. Terrell,** A Passage to Modern Analysis, 2019
40 **Heiko Knospe,** A Course in Cryptography, 2019
39 **Andrew D. Hwang,** Sets, Groups, and Mappings, 2019
38 **Mark Bridger,** Real Analysis, 2019
37 **Mike Mesterton-Gibbons,** An Introduction to Game-Theoretic Modelling, Third Edition, 2019
36 **Cesar E. Silva,** Invitation to Real Analysis, 2019
35 **Álvaro Lozano-Robledo,** Number Theory and Geometry, 2019
34 **C. Herbert Clemens,** Two-Dimensional Geometries, 2019
33 **Brad G. Osgood,** Lectures on the Fourier Transform and Its Applications, 2019
32 **John M. Erdman,** A Problems Based Course in Advanced Calculus, 2018
31 **Benjamin Hutz,** An Experimental Introduction to Number Theory, 2018
30 **Steven J. Miller,** Mathematics of Optimization: How to do Things Faster, 2017
29 **Tom L. Lindstrøm,** Spaces, 2017
28 **Randall Pruim,** Foundations and Applications of Statistics: An Introduction Using R, Second Edition, 2018
27 **Shahriar Shahriari,** Algebra in Action, 2017
26 **Tamara J. Lakins,** The Tools of Mathematical Reasoning, 2016
25 **Hossein Hosseini Giv,** Mathematical Analysis and Its Inherent Nature, 2016
24 **Helene Shapiro,** Linear Algebra and Matrices, 2015
23 **Sergei Ovchinnikov,** Number Systems, 2015
22 **Hugh L. Montgomery,** Early Fourier Analysis, 2014
21 **John M. Lee,** Axiomatic Geometry, 2013
20 **Paul J. Sally, Jr.,** Fundamentals of Mathematical Analysis, 2013
19 **R. Clark Robinson,** An Introduction to Dynamical Systems: Continuous and Discrete, Second Edition, 2012

For a complete list of titles in this series, visit the
AMS Bookstore at **www.ams.org/bookstore/amstextseries/**.